NEUROMETHODS

For further volumes:
http://www.springer.com/series/7657

Neuromethods publishes cutting-edge methods and protocols in all areas of neuroscience as well as translational neurological and mental research. Each volume in the series offers tested laboratory protocols, step-by-step methods for reproducible lab experiments and addresses methodological controversies and pitfalls in order to aid neuroscientists in experimentation. *Neuromethods* focuses on traditional and emerging topics with wide-ranging implications to brain function, such as electrophysiology, neuroimaging, behavioral analysis, genomics, neurodegeneration, translational research, and clinical trials. *Neuromethods* provides investigators and trainees with highly useful compendiums of key strategies and approaches for successful research in animal and human brain function, including translational "bench to bedside" approaches to mental and neurological diseases.

Translational Research Methods for Neuropathic Pain

Edited by

Isaura Tavares

Institute of Histology and Embryology, University of Porto, Porto, Portugal

Hugo Leite-Almeid

Life and Health Sciences Research Institute (ICVS), School of Medicine, University of Minho, Campus Gualtar, Braga, Portugal; ICVS/3B's - PT Government Associate Laboratory, Braga/Guimarães, Portugal, Clinical Academic Center, Braga, Portugal

Humana Press

Editors
Isaura Tavares
Institute of Histology and Embryology
University of Porto
Porto, Portugal

Hugo Leite-Almeida
Life and Health Sciences Research Institute
(ICVS), School of Medicine
University of Minho, Campus Gualtar
Braga, Portugal

ICVS/3B's - PT Government Associate
Laboratory
Braga/Guimarães, Portugal

Clinical Academic Center
Braga, Portugal

ISSN 0893-2336 ISSN 1940-6045 (electronic)
Neuromethods
ISBN 978-1-0716-5311-1 ISBN 978-1-0716-5312-8 (eBook)
https://doi.org/10.1007/978-1-0716-5312-8

This Humana imprint is published by the registered company Springer Science+Business Media, LLC, part of Springer Nature.
The registered company address is: 1 New York Plaza, New York, NY 10004, U.S.A.

Preface to the Series

Experimental life sciences have two basic foundations: concepts and tools. The *Neuromethods* series focuses on the tools and techniques unique to the investigation of the nervous system and excitable cells. It will not, however, shortchange the concept side of things, as care has been taken to integrate these tools within the context of the concepts and questions under investigation. In this way, the series is unique in that it not only collects protocols but also includes theoretical background information and critiques that led to the methods and their development. Thus, it gives the reader a better understanding of the origin of the techniques and their potential future development. The *Neuromethods* publishing program strikes a balance between recent and exciting developments like those concerning new animal models of disease, imaging, in vivo methods, and more established techniques, including, for example, immunocytochemistry and electrophysiological technologies. New trainees in neurosciences still need a sound footing in these older methods in order to apply a critical approach to their results.

Under the guidance of its founders, Alan Boulton and Glen Baker, the *Neuromethods* series has been a success since its first volume was published through Humana Press in 1985. The series continues to flourish through many changes over the years. It is now published under the umbrella of Springer Protocols. While methods involving brain research have changed a lot since the series started, the publishing environment and technology have changed even more radically. Neuromethods has the distinct layout and style of the Springer Protocols program, designed specifically for readability and ease of reference in a laboratory setting.

The careful application of methods is potentially the most important step in the process of scientific inquiry. In the past, new methodologies led the way in developing new disciplines in the biological and medical sciences. For example, physiology emerged out of anatomy in the nineteenth century by harnessing new methods based on the newly discovered phenomenon of electricity. Nowadays, the relationships between disciplines and methods are more complex. Methods are now widely shared between disciplines and research areas. New developments in electronic publishing make it possible for scientists who encounter new methods to quickly find sources of information electronically. The design of individual volumes and chapters in this series takes this new access technology into account. Springer Protocols makes it possible to download single protocols separately. In addition, Springer makes its print-on-demand technology available globally. A print copy can therefore be acquired quickly and for a competitive price anywhere in the world.

Saskatoon, Canada *Wolfgang Walz*

Preface

It is our greatest pleasure to have served as editors for Springer Nature's *Neuromethods* series. It is particularly inspiring to do so in a challenging subject such as neuropathic pain. Neuropathic pain arises as a consequence of a malfunction in the somatosensory system caused by either a lesion or a disease. In other words, the pathophysiology of the condition involves the core processes of pain, including nociception, central transmission, ascending/descending modulation, and central processing. This has implications for the quality of the experienced pain, particularly its high aggressiveness, which significantly impacts patients' well-being. Additionally, it affects the effectiveness of analgesic pharmacological therapies, which currently is relatively limited. Pain research has long sought to address these issues. In the laboratory, experimental modeling of neuropathies, especially of traumatic nature, is simple. On the contrary, pain assessment in these conditions has proven to be challenging. The *Anaesthesia Dolorosa* [1, 2], for instance, was among the first experimental approaches developed with the intention of modeling neuropathic pain. It involved the complete denervation of the sciatic and saphenous nerves, which obviously hindered any possible evoked measures. Several neuropathy models emerged in the following decades, particularly of traumatic nature such as the Chronic Constriction Injury in 1988 [3], the Partial Sciatic Nerve Ligation in 1990 [4], the Spinal Nerve Ligation in 1992 [5], and the Spared Nerve Injury in 2000 [6]; there were also models of different etiologies, including toxic, viral, or diabetic neuropathies—see [7] for review. The use of evoked measurements as pain proxies became standard practice, particularly for assessing allodynia, which added translational value to the models—see for instance [8] on the somatosensory phenotype of patients with neuropathic pain by the application of standardized quantitative sensory tests. However, it soon became apparent that, while import, evoked measures represented only a part of the manifestations associated with neuropathies, and additional readouts were needed. In the late 1990s, several laboratories—see for instance [9]—introduced various behavioral paradigms to study anxiety- and depression-like behaviors in the context of chronic pain. In the decades that followed, strategies for behavioral characterization diversified, with significant emphasis placed on spontaneous behavior and the use of more ethological settings [7]. Many researchers believe that both models and pain-related outcomes have contributed to translational challenges in the field. For instance, a quick search in the literature reveals that traumatic neuropathy models significantly outnumber other types of models, which is in stark contrast to the clinical scenario where pain arising in the context of diabetic neuropathies or viral infections is much more common. It is not our intention to discuss the reasons for this incongruity or its impact on translation. However, our goal was to ensure that the book provides a comprehensive representation of pain models with proven neuropathic involvement and of strategies pertaining to the assessment of pain and related readouts. In this vein, as editors, it was our intention from the initial moments of planning and preparation of the book to collect a meaningful bulk of work on these two themes—models and pain readouts. We also wanted it to be well more than a mere collection of chapters. We planned the section and respective chapters to be complementary and informative, particularly to those readers interested in setting experiments in the field.

Models included in the first part have a more or less evident neuropathic component. In Chap. 1, the spared nerve injury (SNI) model is presented as the paradigmatic model of traumatic peripheral neuropathies. An overview of other neuropathic pain models is also given by the authors; readers may additionally consult [10]. Juliana Fiúza-Fernandes and Hugo Leite-Almeida provide a step-by-step guide for the SNI surgery. Note that the model originally developed/described by Isabelle Decosterd and Clifford J. Woolf [6], although the latest among the classical models (see above), was rapidly adopted by many laboratories due to its technical and phenotypical characteristics—simple, reproducible, and sustained over time.

In the next chapter, Sabrina L. McIlwrath and colleagues present the FRICT-ION (Foramen Rotundum Inflammatory Constriction Trigeminal InfraOrbital Nerve) model. While the SNI and FRICT-ION models share pathophysiological etiologies, the second is more akin to trigeminal neuralgia. The combined use of the two models can be of great interest for distinguishing factors related to the anatomical location of the lesion from those that are more systemic in nature.

Chapter 3 by Ana Ferreira provides an overlook of central traumatic neuropathy models, at the central nervous system (spinal cord), as well as the methodological approaches to each model. The authors also discuss pain assessment methods, which can be particularly challenging in these types of models.

Chapters 4 and 5 deal with etiologies of a systemic nature, particularly diabetic and toxicity-related neuropathies. The clinical counterparts of these models account for a significant proportion of neuropathic pain manifestations in the human population [11, 12]. Stéphanie A. Eid and Eva L. Feldman provide an overview of Type 1 and Type 2 diabetes animal models, encompassing different etiologies: chemical, genetic, and diet-induced. On the other hand, toxicity-related neuropathies are represented in Chap. 5 by paclitaxel, a chemotherapy medication used to treat different types of cancer by disrupting cell division; José Costa-Pereira and collaborators, in addition to the model and outcome measure (allodynia), describe magnetic resonance imaging (MRI) techniques, manganese-enhanced MRI, and ex vivo spectroscopy used for functional and metabolic evaluation of the model impact in the central nervous system.

The final group of chronic pain models comprises those without a direct nervous tissue lesion and where sensitization is secondary to events of other nature. Specifically, this includes skin incision (Chap. 6; by Kate Gates and Cholawat Pacharinsak), complex regional pain syndrome (Chap. 7; by Patrícia Rodrigues and colleagues), and arthritis (Chap. 8; by Sara Martínez-Ramos and Samuel García) models.

The second part of the book deals with pain-related manifestations, particularly those pertaining to (endo)phenotypical manifestations. Chronic pain can potentially affect all aspects of behavior, from those of a more spontaneous nature to those elicited in dedicated apparatus [7]. The former is represented in this book by ultrasonic-vocalizations (Chap. 11; by de Souza and colleagues) and burrowing behaviors (Chap. 13; by Kuo and colleagues), while the latter is represented by anxiety-/depressive-like behaviors (Chap. 14; by Srinivasan and colleagues) and executive function (Chap. 15; by Leite-Almeida). In all chapters, the authors presented comprehensive protocols complemented with an introductory theoretical frame. On a similar vein, Liptáková and colleagues describe the use of enriched environments in the context of experimental chronic pain (Chap. 17) and its impact on pain and related outcomes. In all cases above, populational heterogeneity is very common. While for many researchers such diversity is perceived as a nuisance that hampers conclusions, for a growing number of other researchers, heterogeneity has proven to be a valuable source of information. In some cases, the source of such heterogeneity can be easily traced to biological factors (e.g., sex; see

Chap. 9 by Emma Mannan and Loren J. Martin) and/or to pathophysiological characteristics (e.g., pain duration; see Chap. 14 by Srinivasan and colleagues). In other cases, the underlying factors are not so obvious. For instance, nearly 20 years ago, Keay's group reported alterations in social behavior and in sleep cycles in nearly half (55%) of a rodent population that underwent a neuropathic pain model; such alterations would persist in 30% of the animals [13]. Remarkably, this distribution remained stable in the following studies, suggesting the occurrence of a population-wide conserved characteristic. In Chap. 10, together with James Kang, Kevin Keay revisits this story and describes a modified version of the resident-intruder test that has provided consistent information on subpopulations of rats differently impacted by chronic pain. In a companion paper (Chap. 12), Damien Boorman and Kevin A. Keay tackle heterogeneity in the context of placebo/nocebo responders. In addition to behavioral readouts, there has been a long quest to obtain good endophenotypical readouts of pain, particularly if in parallel with behavioral manifestations. The chapter on corticostriatal connectivity by Cardoso-Cruz and colleagues (Chap. 16) is instructive on this matter. It describes a protocol not only for the obtention of data neurofunctional readouts in freely moving subjects but also for modulating the brain activity in an anatomical and time-specific manner.

We would like to start our final words with a paraphrase of Ronald Melzack and Patrick Wall's classical work title. The challenge of pain lies in its personal nature. However, such is often forgotten in preclinical studies. Chronic pain triggers a set of alterations that goes beyond pain itself, manifesting in each individual in a particular manner, conditioned by biological and environmental variables. As labor-intensive as it might be, a systematic study of such manifestations opens a window to central maladaptation with obvious advantages from both pathophysiological and therapeutic approaches. It is our utmost wish that readers find guidance in this book to pursue their studies on pain. Finally, we would like to thank all authors for their enthusiastic and rigorous contributions.

Braga, Portugal *Hugo Leite-Almeida*
Porto, Portugal *Isaura Tavares*

References

1. Wall PD et al (1979) Autotomy following peripheral nerve lesions: experimental anaesthesia dolorosa. Pain 7:103–111
2. Wall PD, Scadding JW, Tomkiewicz MM (1979) The production and prevention of experimental anesthesia dolorosa. Pain 6:175–182
3. Bennett GJ, Xie YK (1988) A peripheral mononeuropathy in rat that produces disorders of pain sensation like those seen in man. Pain 33:87–107
4. Seltzer Z, Dubner R, Shir Y (1990) A novel behavioral model of neuropathic pain disorders produced in rats by partial sciatic nerve injury. Pain 43:205–218
5. Kim SH, Chung JM (1992) An experimental model for peripheral neuropathy produced by segmental spinal nerve ligation in the rat. Pain 50:355–363
6. Decosterd I, Woolf CJ (2000) Spared nerve injury: an animal model of persistent peripheral neuropathic pain. Pain 87:149–158
7. Cunha AM, Pereira-Mendes J, Almeida A, Guimaraes MR, Leite- lmeida H (2020) Chronic pain impact on rodents' behavioral repertoire. Neurosci Biobehav Rev 119:101–127
8. Rolke R et al (2006) Quantitative sensory testing in the German research network on neuropathic pain (DFNS): standardized protocol and reference values. Pain 123:231–243
9. Kontinen VK, Kauppila T, Paananen S, Pertovaara A, Kalso E (1999) Behavioural measures of depression and anxiety in rats with spinal nerve ligation-induced neuropathy. Pain 80:341–346

10. Jaggi AS, Jain V, Singh N (2011) Animal models of neuropathic pain. Fundam Clin Pharmacol 25:1–28
11. Finnerup NB, Kuner R, Jensen TS (2021) Neuropathic pain: from mechanisms to treatment. Physiol Rev 101:259–301
12. Baskozos G et al (2023) Epidemiology of neuropathic pain: an analysis of prevalence and associated factors in UK Biobank. Pain Rep 8:e1066
13. Monassi CR, Bandler R, Keay KA (2003) A subpopulation of rats show social and sleep-waking changes typical of chronic neuropathic pain following peripheral nerve injury. Eur J Neurosci 17:1907–1920

Contents

Contributors

ROBERTO ANDREATINI • *Department of Pharmacology, Biological Sciences Sector, Federal University of Parana, Curitiba, Parana, Brazil*

NURIA ARIAS-RAMOS • *Instituto de Investigaciones Biomédicas "Alberto Sols", CSIC-UAM, Madrid, Spain; Conexión Nanomedicina, CSIC, Madrid, Spain*

SARA BERNARDO • *Ci&DEI, Instituto Politécnico de Viseu, Viseu, Portugal; Life and Health Sciences Research Institute (ICVS), School of Medicine, University of Minho, Campus Gualtar, Braga, Portugal*

DAMIEN C. BOORMAN • *School of Medical Sciences (Neuroscience), The University of Sydney, Sydney, NSW, Australia*

HELDER CARDOSO-CRUZ • *Pain Neurobiology Research Group, (i3S) Instituto de Investigação e Inovação em Saúde Universidade do Porto, Porto, Portugal; Departamento de Biomedicina – Unidade de Biologia Experimental, Faculdade de Medicina (FMUP), Universidade do Porto, Porto, Portugal*

MAGDA J. CASTELHANO-CARLOS • *Life and Health Sciences Research Institute (ICVS), School of Medicine, University of Minho, Braga, Portugal; ICVS/3B's - PT Government Associate Laboratory, Braga/Guimarães, Portugal*

SÍLVIA CHAMBEL • *Department of Biomedicine, Experimental Biology Unit, Faculty of Medicine of Porto, University of Porto, Porto, Portugal; Translational Neurourology, Instituto de Investigação e Inovação em Saúde-i3S and IBMC, Universidade do Porto, Porto, Portugal*

JULIANA GEREMIAS CHICHORRO • *Department of Pharmacology, Biological Sciences Sector, Federal University of Parana, Curitiba, Parana, Brazil*

JOSÉ TIAGO DA COSTA-PEREIRA • *Department of Biomedicine, Unit of Experimental Biology, Faculty of Medicine, University of Porto, Porto, Portugal; IBMC-Institute of Molecular and Cell Biology, University of Porto, Porto, Portugal; I3S- Institute of Investigation and Innovation in Health, University of Porto, Porto, Portugal; Faculty of Nutrition and Food Sciences, University of Porto, Porto, Portugal*

CÉLIA DUARTE CRUZ • *Department of Biomedicine, Experimental Biology Unit, Faculty of Medicine of Porto, University of Porto, Porto, Portugal; Translational Neurourology, Instituto de Investigação e Inovação em Saúde-i3S and IBMC, Universidade do Porto, Porto, Portugal*

ANA MARGARIDA CUNHA • *Ci&DEI, Instituto Politécnico de Viseu, Viseu, Portugal; Life and Health Sciences Research Institute (ICVS), School of Medicine, University of Minho, Campus Gualtar, Braga, Portugal; ICVS/3B's - PT Government Associate Laboratory, Braga/Guimarães, Portugal*

STÉPHANIE A. EID • *Department of Neurology, School of Medicine, University of Michigan, Ann Arbor, MI, USA*

MADALENA ESTEVES • *Ci&DEI, Instituto Politécnico de Viseu, Viseu, Portugal; Life and Health Sciences Research Institute (ICVS), School of Medicine, University of Minho, Campus Gualtar, Braga, Portugal*

EVA L. FELDMAN • *Department of Neurology, School of Medicine, University of Michigan, Ann Arbor, MI, USA*
SARA FERNANDES • *Life and Health Sciences Research Institute (ICVS), School of Medicine, University of Minho, Braga, Portugal; ICVS/3B's Associate Laboratory, PT Government Associated Laboratory, Braga/Guimarães, Portugal*
ANA FERREIRA • *Department of Biomedicine, Experimental Biology Unit, Faculty of Medicine of Porto, University of Porto, Porto, Portugal; Translational Neurourology, Instituto de Investigação e Inovação em Saúde-i3S and IBMC, Universidade do Porto, Porto, Portugal*
JULIANA FIÚZA-FERNANDES • *Life and Health Sciences Research Institute (ICVS), School of Medicine, University of Minho, Campus Gualtar, Braga, Portugal; ICVS/3B's—PT Government Associate Laboratory, Braga/Guimarães, Portugal*
JULIA FRARE • *Graduated Program in Pharmacology, Federal University of Santa Maria (UFSM), Santa Maria, RS, Brazil*
VASCO GALHARDO • *Pain Neurobiology Research Group, (i3S) Instituto de Investigação e Inovação em Saúde Universidade do Porto, Porto, Portugal; Departamento de Biomedicina – Unidade de Biologia Experimental, Faculdade de Medicina (FMUP), Universidade do Porto, Porto, Portugal*
SAMUEL GARCÍA • *Rheumatology & Immuno-Mediated Diseases Research Group (IRIDIS), Galicia Sur Health Research Institute (IIS Galicia Sur), SERGASUVIGO, Vigo, Spain; Rheumatology Department, University Hospital Complex of Vigo, Vigo, Spain*
KATHERINE GATES • *Department of Comparative Medicine, School of Medicine, Stanford University, Stanford, CA, USA*
MOHAMMAD ZAFAR IMAM • *Centre for Integrated Preclinical Drug Development (CIPDD), School of Biomedical Sciences, Faculty of Health, Medicine and Behavioural Sciences, The University of Queensland, St. Lucia, QLD, Australia*
JAMES W. M. KANG • *School of Medical Sciences (Neuroscience), The University of Sydney, Sydney, NSW, Australia*
KEVIN A. KEAY • *School of Medical Sciences (Neuroscience), The University of Sydney, Sydney, NSW, Australia*
EMILY J. KOUBEK • *Department of Neurology, School of Medicine, University of Michigan, Ann Arbor, MI, USA*
ANDY KUO • *Centre for Integrated Preclinical Drug Development (CIPDD), School of Biomedical Sciences, Faculty of Health, Medicine and Behavioural Sciences, The University of Queensland, St. Lucia, QLD, Australia*
HUGO LEITE-ALMEIDA • *Life and Health Sciences Research Institute (ICVS), School of Medicine, University of Minho, Campus Gualtar, Braga, Portugal; ICVS/3B's - PT Government Associate Laboratory, Braga/Guimarães, Portugal; Clinical Academic Center, Braga, Portugal*
VANESSA B. P. LEJEUNE • *Department of Pharmacology, Biological Sciences Sector, Federal University of Parana, Curitiba, Parana, Brazil*
SUSANA LIMA • *Institut des Neurosciences Cellulaires et Intégratives, UPR3212, Centre National de la Recherche Scientifique, Strasbourg, France; Unité de neurobiologie cellulaire, Centre de Recherche Université Laval Robert-Giffard, Québec, QC, Canada*
ANNAMÁRIA LIPTÁKOVÁ • *Life and Health Sciences Research Institute (ICVS), School of Medicine, University of Minho, Braga, Portugal; ICVS/3B's - PT Government Associate Laboratory, Braga/Guimarães, Portugal; Pharmacology and Therapeutics, Physiology, School of Medicine, Galway Neuroscience Centre, Centre for Pain Research, University of Galway, Galway, Ireland*

PILAR LÓPEZ-LARRUBIA • *Instituto de Investigaciones Biomédicas "Alberto Sols", CSIC-UAM, Madrid, Spain; Conexión Nanomedicina, CSIC, Madrid, Spain*
LOREN J. MARTIN • *Department of Psychological and Brain Sciences, University of Toronto Mississauga, Mississauga, ON, Canada*
SARA MARTÍNEZ-RAMOS • *Rheumatology & Immuno-Mediated Diseases Research Group (IRIDIS), Galicia Sur Health Research Institute (IIS Galicia Sur), SERGASUVIGO, Vigo, Spain; Rheumatology Department, University Hospital Complex of Vigo, Vigo, Spain*
VICTOR MATHIS • *Institut des Neurosciences Cellulaires et Intégratives, UPR3212, Centre National de la Recherche Scientifique, Strasbourg, France*
SABRINA L. MCILWRATH • *New Mexico VA Medical Center, Albuquerque, NM, USA*
CLARA MONTEIRO • *Pain Neurobiology Research Group, (i3S) Instituto de Investigação e Inovação em Saúde Universidade do Porto, Porto, Portugal; Departamento de Biomedicina – Unidade de Biologia Experimental, Faculdade de Medicina (FMUP), Universidade do Porto, Porto, Portugal*
MARENA MONTERA • *Department of Anesthesiology & Critical Care Medicine, University of New Mexico Health Sciences Center, Albuquerque, NM, USA*
ARJUN MURALIDHARAN • *Charles Perkins Centre, University of Sydney, Camperdown, NSW, Australia*
CHOLAWAT PACHARINSAK • *Department of Comparative Medicine, School of Medicine, Stanford University, Stanford, CA, USA*
PATRÍCIA RODRIGUES • *Graduated Program in Pharmacology, Federal University of Santa Maria (UFSM), Santa Maria, RS, Brazil*
NÁTHALY ANDRIGHETTO RUVIARO • *Graduated Program in Biological Sciences: Toxicological Biochemistry, Federal University of Santa Maria (UFSM), Santa Maria, RS, Brazil*
RAINER K. W. SCHWARTING • *Department of Psychology, Behavioral Neuroscience, Experimental and Biological Psychology, and Center for Mind, Brain and Behavior, Philipps-University of Marburg, Marburg, Germany*
NUNO SILVA • *Life and Health Sciences Research Institute (ICVS), School of Medicine, University of Minho, Braga, Portugal; ICVS/3B's Associate Laboratory, PT Government Associated Laboratory, Braga/Guimarães, Portugal*
MAREE T. SMITH • *Centre for Integrated Preclinical Drug Development (CIPDD), School of Biomedical Sciences, Faculty of Health, Medicine and Behavioural Sciences, The University of Queensland, St. Lucia, QLD, Australia*
NUNO SOUSA • *Life and Health Sciences Research Institute (ICVS), School of Medicine, University of Minho, Braga, Portugal; ICVS/3B's - PT Government Associate Laboratory, Braga/Guimarães, Portugal*
CAMILA PASQUINI DE SOUZA • *Department of Pharmacology, Biological Sciences Sector, Federal University of Parana, Curitiba, Parana, Brazil*
VANDANA SHREE VEDARTHAM SRINIVASAN • *Institut des Neurosciences Cellulaires et Intégratives, UPR3212, Centre National de la Recherche Scientifique, Strasbourg, France; Unité de neurobiologie cellulaire, Centre de Recherche Université Laval Robert-Giffard, Québec, QC, Canada*
ISAURA TAVARES • *Department of Biomedicine, Unit of Experimental Biology, Faculty of Medicine, University of Porto, Porto, Portugal; IBMC-Institute of Molecular and Cell Biology, University of Porto, Porto, Portugal; I3S- Institute of Investigation and Innovation in Health, University of Porto, Porto, Portugal*

GABRIELA TREVISAN • *Graduated Program in Pharmacology, Federal University of Santa Maria (UFSM), Santa Maria, RS, Brazil; Graduated Program in Biological Sciences: Toxicological Biochemistry, Federal University of Santa Maria (UFSM), Santa Maria, RS, Brazil*

KARIN N. WESTLUND • *Department of Anesthesiology & Critical Care Medicine, University of New Mexico Health Sciences Center, Albuquerque, NM, USA; New Mexico VA Medical Center, Albuquerque, NM, USA*

IPEK YALCIN • *Institut des Neurosciences Cellulaires et Intégratives, UPR3212, Centre National de la Recherche Scientifique, Strasbourg, France; Department of Psychiatry and Neuroscience, Université Laval, Québec, QC, Canada*

Chapter 1

The Spared Nerve Injury in the Context of Neuropathic Pain Models: A Step-By-Step Guide

Juliana Fiúza-Fernandes and Hugo Leite-Almeida

Abstract

Neuropathic pain preclinical models are very heterogeneous, which results from an attempt to mirror the pathophysiological diversity of related human diseases. Models of peripheral nerves trauma are among the most used. In fact, models such as the chronic constriction injury, spinal nerve ligation, or spared nerve injury became widespread in the preclinical neuropathic pain literature despite other etiologies (e.g., diabetic neuropathy, viral infections) being more relevant from a clinical stand of view. Reasons for such preference are not entirely clear but might be linked with the technical simplicity, reproducibility, and phenotypical stability over time of these models. In this chapter, we will present an overview of the nerve injury models and describe a step-by-step protocol using the spared nerve injury as reference.

Keywords Neuropathic pain, Peripheral nervous system, Lesions, Models, Allodynia, Hyperalgesia

1 Introduction: Chronic Neuropathic Pain—Pathophysiology and Models

Neuropathic pain is defined as a pain caused by a lesion or disease of the somatosensory nervous system—consult the International Association for the Study of Pain (IASP) website https://www.iasp-pain.org (resources/terminology); see also [1] for a commentary on the 2011 definition. It encompasses a large and diverse group of conditions both in terms of ethology, pathophysiology, and clinical manifestations—e.g., diabetes, viral infections (e.g., postherpetic neuralgia, HIV), stroke, peripheral nerve injury [2]—whose classification was recently revised in the 11th revision of the International Statistical Classification of Diseases and Related Health Problems (ICD) of the World Health Organization (WHO) [3]. It is estimated that neuropathic pain affects 7%–10% of the general population but prevalence increases in women and in individuals over 50 years old [2]. Despite the evident progresses in the understanding of its pathophysiology, neuropathic pain treatment is still considered challenging [4, 5]. Classical pharmacological therapies—antidepressants

Isaura Tavares and Hugo Leite-Almeida (eds.), *Translational Research Methods for Neuropathic Pain*, Neuromethods, vol. 235, https://doi.org/10.1007/978-1-0716-5312-8_1,

and anticonvulsants—are limited in their effects and therapeutical alternatives such as spinal cord stimulation [6], transcranial magnetic stimulation [7], or cognitive-behavioral therapy are under consideration with promising results—see also [5, 8].

1.1 Models of Traumatic Neuropathies

Animal models of neuropathic pain proved to be valuable research tools. These have been central not only for the characterization of the main pathophysiological mechanisms [9] but also to study both pharmacological therapies and alternative/complementary approaches with therapeutic potential—see for instance [10]. As stated above, clinical conditions leading to neuropathic pain are diverse (Table 1) and so are the underlying the pathophysiological mechanisms. Such diversity is reflected in the numerous animal models of chronic neuropathic pain developed in the past decades (see Fig. 1 and Table 2; [11, 12–24]). The first model of peripheral neuropathy consisted in the complete denervation of the distal limb simulating a condition

Table 1
A summary of the most common types of neuropathic pain from central and peripheral origins; adapted from [57]

Peripheral neuropathic pain	Central neuropathic pain
Acute and chronic inflammatory demyelinating polyradiculoneuropathy	Compressive myelopathy from spinal stenosis
Alcoholic polyneuropathy	HIV myelopathy
Complex regional pain syndrome	Multiple sclerosis-related pain
Entrapment neuropathies (e.g., carpal tunnel syndrome)	Postischemic myelopathy
HIV sensory neuropathy	Postradiation myelopathy
Iatrogenic neuralgias (e.g., post mastectomy or post thoracotomy)	Poststroke pain
Idiopathic sensory neuropathy	Posttraumatic spinal cord injury pain
Nerve compression or infiltration by tumor	Syringomyelia
Nutritional deficiency-related neuropathies	
Painful diabetic neuropathy	
Phantom limb pain	
Postherpetic neuralgia	
Posttraumatic neuralgias	
Postradiation plexopathy	
Radiculopathy (cervical, thoracic, or lumbosacral)	
Toxic exposure-related neuropathies	
Trigeminal neuralgia	

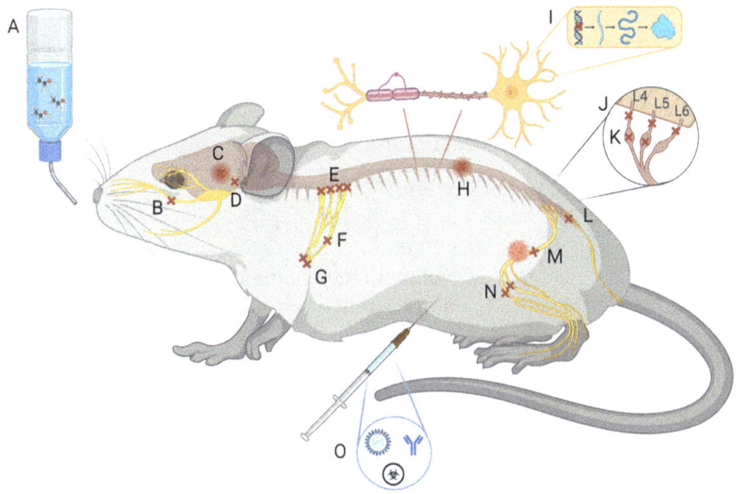

Fig. 1 Anatomical representation of neuropathy models commonly used in pain research. A—alcoholic neuropathy; B—trigeminal Neuralgia; C—thalamic syndrome; D—Parkinson's disease associated pain; E—brachial plexus avulsion; F—brachial plexus injury (crushing/ligation of the lower trunk); G—partial injury of the medial and ulnar nerves; H—spinal cord injury, dorsal root avulsion (T13 and L1 level); I—hereditary polyneuropathies; J—spinal nerve ligation, experimental lumbar radiculopathy, spinal nerve root transection, spinal nerve axotomy, lumbar spinal stenosis; K—dorsal root ganglia compression; L—inferior caudal trunk injury; M—complete nerve transection, chronic nerve compression, chronic constriction injury, sciatic nerve cryoneurolysis/cuffing/pinch/photochemically-induced ischemia, partial sciatic nerve ligation/transection, malignant sciatic neuropathy; N—spared nerve injury; O—autoimmune neuropathies, chemotherapy-induced neuropathies, diabetic sensory neuropathy models, pyridoxine intoxication neuropathy, VZV/HIV associated neuropathy. Abbreviations: *HIV* human immunodeficiency virus, *VZV* varicella Zoster virus. (Image produced in BioRender)

of anaesthesia *dolorosa* [25, 26]. Self-mutilation of the deafferented areas (autotomy) was used as an index of pain; evoked responses could not obviously be obtained [25, 27]. Subsequent models were also mostly based on traumatic injuries not only of peripheral nerves but also of nerve roots or spinal cord [21, 28]; such could involve sectioning, cuffing, ligation, freezing, inflammation, and/or ischemia (some of the most common used models of chronic neuropathic pain are presented in Table 2; references above). Additionally, drug toxicity and systemic metabolic disorders have also been used to trigger neuropathies. Importantly, in these models, abnormal evoked responses to both innocuous and noxious stimulation—allodynia and hyperalgesia, respectively—could be easily obtained and registered, which permitted to establish parallels with well-described clinical manifestations of neuropathies. In addition, these models also exhibit pain-related behavioral alterations/manifestations following injury, which have been used to characterize pain effect, as well as recovery [29, 30].

In this chapter, we are going to use the spared nerve injury (SNI; 31) model to demonstrate the procedures to induce the neuropathic lesion and to assess its result. The SNI model consists in the ligation

Table 2

Neuropathy models commonly used in chronic pain research. A brief overview of common neuropathy models is provided organized by etiology and anatomical region

Models of neuropathy	Description	Ref
Traumatic (compression, section, ligation, etc.)		
Central neuropathies		
Spinal cord injuries	Several models, including transection, compression, or photochemical lesions	[14]
Peripheral neuropathies (hindlimbs)		
Complete sciatic nerve transection	Sectioning of the sciatic and saphenous nerves	[26]
Chronic nerve (sciatic) compression	The sciatic nerve is compressed by 1 cm tubes of varying Ø; monkey variant using multifascicular median nerve [58];	[59]
Chronic constriction injury (CCI)	Four loosely constrictive ligatures around the common sciatic nerve spaced 1 mm; see also [60] for a CCI variant with different amount of sutures	[61]
Partial sciatic nerve ligation (PSNL)	Ligation of approximately 1/2 of the sciatic nerve; see [62] for partial saphenous nerve injury in mice	[63]
Spinal nerve ligation (SNL)	Ligation of L5/L6 spinal nerves; see [64] for adaptation in primates L7	[49]
Sciatic cryoneurolysis	Freezing the proximal sciatic nerve using a cryoprobe cooled to −60°C	[65]
Experimental lumbar radiculopathy	Two loose ligatures of 4/0 silk are placed around L4, L5, and L6 nerve roots (Silk ligature model). Four pieces of 4/0 chromic gut are laid adjacent to the nerve roots and secured by two loose ligatures of 4/0 chromic gut suture (chromic gut ligature model)	[66]
Spared nerve injury (SNI)	Ligation and distal section of the common peroneal and tibial nerves; see also [48] and [67] for alternative lesions of sciatic nerve branches	[45]
Partial sciatic nerve transection	A prolene 7/0 ligature is placed through the midpoint of the nerve just cranially to the branch running to the musculus biceps femoris. Half of the diameter of the nerve is transected in a ventrocranial direction up to the ligature and this is then removed	[68]
Common peroneal nerve ligation	Ligation of the common peroneal nerve with a chromic gut suture 5-0	[69]
Saphenous partial ligation	A tight ligature made with an 8/0 nylon suture is used to trap 30%–50% of the saphenous nerve	[70]

(continued)

Table 2
(continued)

Models of neuropathy	Description	Ref
Spinal nerve root transection	The lamina and transverse processes from L1 to S1 are exposed and the nerve roots are dissected proximally to the bony boundary, preserving the laminae. L4 (or L5, or L6) nerve root is then transected	[71]
Ventral root transection	Unilateral lumbar five ventral root transection	[72]
Sciatic nerve cuffing	Placement of a 2 mm long split section of polyethylene tubing around the common branch of the sciatic nerve	[73]
Photochemically-induced ischemia	A low power laser is used to make an ischemic lesion of the rat sciatic nerve	[74]
Dorsal root ganglia compression	L4/L5 intervertebral foramina are exposed and a stainless-steel rod is implanted in each vertebra to chronically compress the lumbar dorsal root ganglion (DRG)	[75]
Lumbar spinal stenosis	Cauda equina compression using a silicone block within the epidural spaces of L5–L6 vertebrae	[76]
Spinal nerve axotomy	L5 spinal nerve ligation (6-0 silk suture) and distal axotomy	[77]
Sciatic nerve pinch	Sciatic nerve is pinched for 3 s with a microsurgery clamp at the point where the nerve comes across the adductor brevis muscle	[78]
Tibial translocation neuroma	Ligation and transection of the posterior tibial nerve in the foot just proximal to the plantar bifurcation and translocation of the end of the ligated nerve to the area above the lateral malleolus	[79]
Dorsal root avulsion	The roots of T13 and L1 nerves, as they enter the spinal cord, are firmly grasped with fine-tipped curved forceps at the dorsal root entry zone and rapidly torn away from the dorsal horn. This model causes both central and peripheral neuropathic pain	[80]
Peripheral neuropathies (tail)		
Inferior caudal trunk injury (causalgia)	Resection of the left inferior caudal trunk between the S3 and S4 spinal nerves	[81]
Peripheral neuropathies (forelimbs)		
Brachial plexus injury	Brachial plexus avulsion: the lower trunk of the brachial plexus is grasped with forceps and extracted from the spinal cord by traction. This article also employs other models such as crushing and ligation of the lower trunk of the brachial plexus	[82]

(continued)

Table 2
(continued)

Models of neuropathy	Description	Ref
Median and ulnar partial injury	The halves of the ulnar and median nerves are transected by microscissors at about 5 mm proximal from the elbow	[83]
Orofacial		
Auditory neuropathy	Auditory nerve compression	[84]
Trigeminal neuralgia	**Infraorbital** nerve CCI: two chromic gut 5/0 ligatures are loosely tied around the infraorbital nerve; for review of trigeminal neuralgia models, see [85]	[86]
Biologically induced		
Varicella zoster-associated neuropathy	Subcutaneous injection of varicella zoster virus into the glabrous footpad of the left hind limb of the rats. See also [87]	[88]
HIV-related neuropathy	Application of human immunodeficiency virus type 1 glycoprotein 120 to the sciatic nerve	[89]
Malignant sciatic neuropathy	Inoculation of Meth A sarcoma cells to the immediate proximity of the sciatic nerve	[90]
Chronic postischemia pain	Prolonged hindpaw ischemia and reperfusion (model of complex regional pain syndrome-Type I). A tourniquet (tight-fitting O-ring) is placed on one hindlimb of anesthetized rats proximal to the ankle joint for 3 h, and then removed to allow reperfusion before anesthesia reversion	[91]
Autoimmune (antibody-mediated) neuropathy models	Use of antibodies against peripheral nerve glycolipids and myelin proteins to cause peripheral neuropathies. Note that not all of these might cause pain-related manifestations	[92]
Thalamic syndrome/poststroke pain syndrome	Hemorrhagic stroke lesion in the ventral posterolateral nucleus of the thalamus. For review, see [93]	[94]
Models of inherited neuropathies	Mutations in several myelin genes. Note that not all of these might cause pain-related manifestations	[95, 96]
Experimental autoimmune encephalomyelitis	Several models (see for review [97]) based on the immunization of susceptible animal strains with myelin components and encephalitogenic peptides, or by adaptive transfer of encephalitogenic T cells	[97]
Trigeminal neuralgia	Injection of cobra venom into the infraorbital nerve trunk	[98]

(continued)

Table 2
(continued)

Models of neuropathy	Description	Ref
Chemically-induced models		
Chemotherapy exposure-related	Repeated intravenous administration of vincristine; Cisplatin adaptation (intraperitoneal injections)—see [99]. See also for review [20]	[100]
Painful Alcoholic neuropathy	Rats are chronically fed with a Lieber–DeCarli liquid diet with ethanol (6.5% ethanol)	[101]
Sciatic nerve neuritis	An hemostatic oxidized cellulose (Oxycel) saturated with carrageenan or complete Freund's adjuvant is wrapped around the sciatic nerve	[102]
Excitotoxic spinal cord injury	Intraspinal or intrathecal injection of some quisqualate	[103]
Diabetic sensory neuropathy models	Intraperitoneal injection of Streptozotocin followed by insulin treatment to maintain animals at different degrees of glycemia. For a review, see also [104–106]	[107]
Sciatic inflammatory neuritis	Injection of zymosan (yeast cell walls) around the sciatic nerve	[108]
Pyridoxine (B6) intoxication	Repeated intraperitoneal injections of pyridoxine	[109]
Parkinson's disease associated pain	Injection of 6-hydroxy-dopamine into the substantia nigra, making a lesion in the nigrostriatal dopaminergic pathway; consult for other models [110]	[111]

and distal sectioning of two sciatic nerve branches—common peroneal and tibial—leaving a third branch intact—sural nerve. The procedure is technically very simple, and the behavioral outcomes are long lasting that are certainly among the reasons for its adoption.

2 Materials and Methods—Spared Nerve Injury Model

2.1 Material

Animals

The protocol has been widely used in several strains of both rats and mice, without major adaptations. The protocol described in the present text uses adult Wistar Han rats; attention should be paid to potential adaptions due to anatomical variations, drug dosages or other aspects when using other species/strains. Animals are pair-housed in standard plastic cages with food and water available ad libitum, in a room with controlled temperature (22 ± 1°C) and humidity (55%–60%), and in a 12-h light/dark cycle (lights on at 8 a.m.) as described in previous publication of the group (see for instance [31, 32]).

Drugs and Pharmaceutical Products

- Alcohol 70%
- Atipamezole hydrocloride (Antisedan®, Esteve Farma; 0.35 mg; i.p.)
- Chlorhexidine antiseptic solution
- Ketamine (Imalgene 1000® 100 mg mL^{-1}; Merial Portuguesa)
- Medetomidine (Dorbene® 1 mg mL^{-1}; Zoetis, Portugal)
- Ophtalmic ointment

Surgical Material

- Steel scalpel blades
- Micro Curette—1.0 mm Tip 13 cm (FST; 10081-10)
- Silk Suture Thread—4/0 (FST; 18020-40)
- Student Fine Scissors—Straight 11.5 cm (FST; 91460-11)
- Student Vannas Spring Scissors—Straight 9 cm (FST;91500-09)
- Student Halsted-Mosquito Hemostats—Serrated Curved 12.5 cm (FST, 91309-12)
- Standard Pattern Forceps—Straight 12 cm (FST; 11000-12)
- Iris Forceps—Serrated Straight 7 cm (FST; 11064-07)
- Coated vicryl (polyglactin 910)—4/0 Suture (Ethicon Inc, W9443)

Stainless material is autoclaved and maintained in hermetic containers prior to the surgery. Other sterilization and disinfection techniques can also be employed.

Other Materials/Equipment

- Shaver
- Nitrile Gloves
- Light (optional)
- Magnification glass (optional)

2.2 Animal Preparation

Surgical procedures are performed under deep anesthesia obtained with an intraperitoneal injection (i.p.) of a 1.5/1.0 mixture of ketamine/medetomidine at a dose of 1 mg kg^{-1}. In the absence of voluntary movements and postural reflexes (3–5 min postinjection), animals are challenged with a paw pinch stimulation; in the absence of a withdrawn response, surgical procedures are initiated [33]. It should be noted that other anesthetic regimens can be considered, but procedures should be adjusted in accordance. The implications for the outcomes should also be carefully pondered, and if necessary preliminary tests should be performed (see discussion).

Though the overall duration of the surgery is relatively short (10–20 min) and, therefore, the risk of corneal lesions due to dehydration is relatively minor—see for some studies on the theme [34, 35]—it is still advisable to protect the surface of the animals' eyes with ophthalmic ointment or artificial tears [36]. Next, the lateral area of the thigh is shaved in order to clean the surface. The animal is placed in a lateral position with the shaved leg facing upwards (Fig. 3A), and incision site is then disinfected (e.g., 70% alcohol or chlorhexidine solution). To avoid temperature loss during anesthesia, the animal should be placed on top of a heating pad, with the temperature set to 37°C. Respiratory pattern and other vital signs and anesthetic deepness should be monitored throughout the surgery [36, 37].

2.3 Surgery Protocol

- Using a scalpel blade/handle a longitudinally oriented incision is performed approximately in the thigh midline, in the upper half (Fig. 3B).
- The skin in gently lifted using a standard pattern forceps separated from the underneath muscle (biceps femoris) using a scissor (Fig. 3C). An overall area of approximately 3 cm^2 should be sufficient for an experimented individual to carry the next steps.
- A blunt dissection of the biceps femoris is performed (Fig. 3D–F); For this, the closed tips of the scissor are pressed against the muscle and open (do not cut the muscle). After a variable number of repetitions pressure/open, the sciatic nerve branches should be visible, particularly the peroneal nerve (running in medial to lateral in a frontal location to the incision; Fig. 3G, black arrow), and the sural nerve in a caudal location to the incision. Regarding the latter, great care should be taken to avoid damage (spared nerve). The tibial nerve is running medial and more deep relatively to the incision and might not be visible at this stage (Fig. 3G, white arrow; see also Fig. 3Q). Note that the tibial is the sciatic branch with the larger diameter, making it easier to recognize, even if the sciatic nerve and its branching site is not fully visible.
- The micro curette is used to isolate the peroneal nerve. First, the micro curette should be placed between the visible aspects of the nerve oriented ventrally; Next, it should be inserted and used to pull the nerve to the surface (Fig. 2H). Taking advantage of this position, a second micro curette can now be inserted under the nerve (Fig. 2I).
- A small portion of 4/0 silk suture (nonabsorbable) is placed around the tip of one of the micro curettes using the iris forceps; carefully move the micro curette and, together with it, the suture so that it crosses underneath the nerve. Tight ligate the

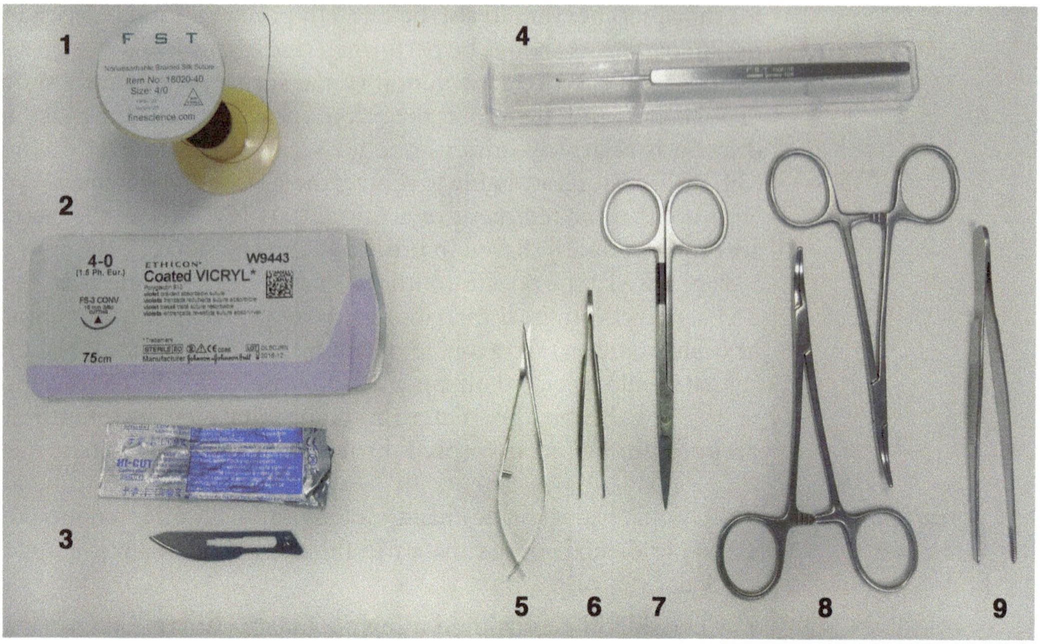

Fig. 2 Surgical material. 1. Silk Suture Thread—4/0 (FST; 18020-40); 2. Coated Vicryl (polyglactin 910) Suture (Ethicon Inc, W9443); 3. Steel scalpel blade; 4. Micro Curette—1.0 mm Tip 13 cm (FST; 10081-10); 5. Student Vannas Spring Scissors—Straight 9 cm (FST; 91500-09); 6. Iris Forceps—Serrated Straight 7 cm (FST; 11064-07); 7. Student Fine Scissors—Straight 11.5(FST; 91460-11); 8. Student Halsted-Mosquito Hemostats—Serrated Curved 12.5 cm (FST, 91309-12); 9. Standard Pattern Forceps—Straight 12 cm (FST; 11000-12)

nerve and cut it with a student Iris scissors, distal to the ligation (Fig. 2I–K) and remove the second micro curette.

- The procedure is now repeated for the tibial nerve (Fig. 2L–N).
- For sham surgery, tibial and peroneal nerves are exposed but not ligated or sectioned!
- Muscle and skin are now sutured in separated layers with an absorbable suture (e.g., coated vicryl) (Fig. 2O, P).
- Disinfect again the area again (chlorhexidine).

During the entire process, but particularly during the blunt dissection and nerve manipulation, great care should be taken to avoid direct damage or stretch of the sural nerve, which should remain intact (spared nerve).

2.4 Animal Recovery and Post-Surgical Care

- Atipamezole hydrocloride (100 μL) is injected subcutaneously and animals are placed at the respective home cages (or at an alternative cage if necessary). After the injection, animals should recover voluntary movements in 1–5 min.
- The home (recovery) cage should be maintained in a warm environment (e.g., under indirect red light). While the overall anesthesia period is relatively small, temperature control is important for postoperative pain, especially for small animals [38].

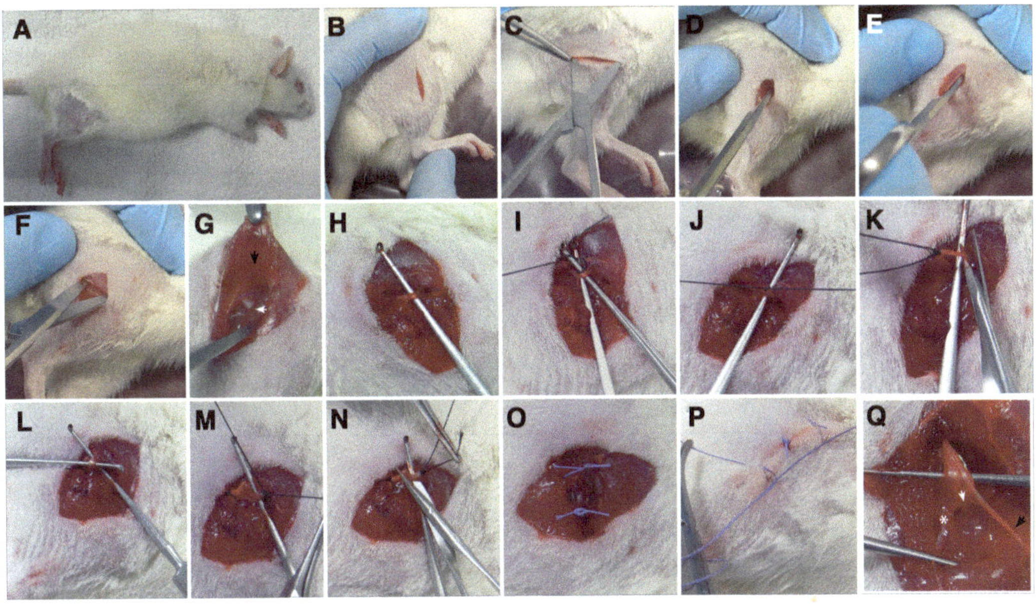

Fig. 3 A step-by-step visual guide for the spared nerve injury model execution in rats. In a deeply anaesthetized rat, (**A**) the lateral area thigh is shaved; (**B**) a longitudinally oriented incision is made; (**C**) separation of the skin from the muscle fascia; (**D–F**) blunt dissection of the bicep femoris muscle—alternate pressure (**D**) and open (**E, F**); (**G**) the peroneal (black arrow) and tibial (white arrow) nerves are visible; (**H**) peroneal nerve isolation, using a micro curette; (**I, J**) ligation of the peroneal nerve; (**K**) sectioning of the peroneal nerve, distal to the ligature; (**L–N**) tibial nerve isolation, ligation, and sectioning; (**O**) suturing of the muscle. Typically, one stitch is enough, however, for purposes of photographic illustration a bigger area was needed; (**P**) skin suturing; (**Q**) sciatic nerve isolation at its trifurcation. This procedure was performed in another animal, only for visualization of the relative spatial orientation of the nerves. Peroneal nerve (black arrow) runs more superficially. Tibial nerve (white arrow) running more deeply between the muscle layers. Sural nerve (white asterisk)—Great care must be taken to avoid damage of this nerve

- Provide food and water ad libitum.
- Inspect the sutures daily for the first week—pay attention to signs of infection and dehiscence.
- Inspect the paw for signs of autotomy and infection.
- See [39] for postoperative analgesic treatment.

2.5 Pain Assessment

Mechanical allodynia is the most commonly assessed pain-related manifestation to validate this model. This is frequently assessed using Von Frey up-and-down method [40]. Von Frey, similarly to the more commonly used pain assessment methods, relies on stimulus-evoked pain responses. There is a battery of pain-evoked test that differ essentially in the type of stimulus (e.g., heat, cold, chemicals, pressure—for review, see [29, 41]. Since most traumatic neuropathic pain models develop allodynia and hyperalgesia, these tests became widely used in this context. Although these greatly contributed to the unraveling of pain mechanisms and drug testing, they are unable to assess pain in other dimensions that are also clinically relevant (e.g., affective). In fact, studies using a reversible model of neuropathic pain showed that anxious- and depressive-like behaviors persisted even after the resolution of allodynia [42], emphasizing the need to assess pain-related comorbidities. In line with this, research from our lab using the SNI model also confirmed that these behavioral manifestations were also present in a subset of animals that do not manifest allodynia [43].

In addition to the above mentioned, other tests have been validated in the past decades in attempt to assess ongoing spontaneous pain, such as the conditioned place preference and the grimace scale [29, 44].

3 Conclusions

Traumatic peripheral neuropathies are the most commonly used rodent models of neuropathic pain. The chronic constriction injury (CCI; 46), partial sciatic nerve ligation (PSNL; 47), and spinal nerve ligation (SNL; 48) models are among the earliest models to be adopted by a vast community of researchers. The SNI model [45] was introduced later but was rapidly adopted in many laboratories. The reasons for such popularity are most probably related with its technical simplicity and reduced postoperative complications. Moreover, the SNI is highly reproducible, and the behavioral manifestations long-lasting [45] that is greatly advantageous in the context of chronic pain research. However, it should be noticed that, in some animals, allodynia is not marked if present at all—e.g., [31, 43, 46]; similarly, pain-related effect can vary dramatically across individuals—see [47] and subsequent reports from the group. There is good evidence that such heterogeneity is intrinsic to the animals/group and not related with procedural differences. Indeed, the affected/not-affect relation is stable across experiments/experimenters and, in the particular case of allodynia, a strain-dependent association has been demonstrated which suggests a contribution of the genetic background (references above).

We have, therefore, described the procedures to install and assess allodynia in the SNI [45] as a prototypic traumatic peripheral neuropathy models. While the described procedures can be easily adopted to other peripheral neuropathy models—e.g., [48]—it should be noted a number of alterations might have to be incorporated in the protocol particularly regarding the incision region and in the nerve isolation/lesion procedures. One of the disadvantages of these models—with the exception of the SNL model [49]—is that there is little control over the type of affected nerve (sensory/motor/autonomic). Indeed, nerves selected for lesion can dramatically affect the outcome [48]. Tibial and sural nerve lesion presented more robust mechanical/cold allodynia, as well as spontaneous pain that were not relieved by pharmacological sympathectomy. Specifically regarding SNI, the area for probing sensation (with Von Frey filaments for instance), the sural dermatome, is relatively small and sometimes difficult to access.

As a final word, the utility of chronic pain models is unquestionable. Nevertheless, some aspects need to be considered. Peripheral neuropathic pain traumatic models are by far the most commonly used models, which might have contributed to a translational problem, especially when considering that neuropathies due to trauma are clinically much less common than metabolic, viral, and chemical etiologies. In addition to the model—e.g., lesioned nerves [48]—and biological—e.g., duration [50], site [51], or age [52]—characteristics, readers should also be aware that other procedural parameters that were not discussed here can also affect the outcome. For instance, the selection of the anesthetics and postoperative care is not trivial and might have consequences in the animal behavior, particularly in pain-related manifestations. Regarding the first, there is evidence that isoflurane interferes with pain assessment using the rat grimace scale, even without a surgical event [53]. Regarding analgesic regimens, Wu and colleagues demonstrated that under 2%–4% isoflurane anesthesia, intrathecal bupivacaine administration 30 min prior to CCI significantly increased thresholds/latency to mechanical/thermal stimulation [54]. Additionally, opioid treatment was shown to induce long-term hyperalgesia [55, 56]. Nevertheless, a comparative study of postoperative analgesia regimens showed that for most systemic analgesic drugs, there is no compromise of the development of allodynia, thus recommending the use of these to prevent acute pain in the postoperative period [39].

Acknowledgments

This work has been funded by National funds, through the Foundation for Science and Technology (FCT)—project UIDB/50026/2020 and UIDP/50026/2020.

Conflict of Interest The authors have no conflicts of interest to declare that are relevant to the content of this chapter.

References

1. Jensen TS et al. (2011) A new definition of neuropathic pain. Pain 152(10):2204–2205
2. Colloca L et al. (2017) Neuropathic pain. Nat Rev Dis Primers 3:17002
3. Scholz J et al. (2019) The IASP classification of chronic pain for ICD-11: chronic neuropathic pain. Pain 160(1):53–59
4. Costigan M, Scholz J, Woolf CJ (2009) Neuropathic pain: a maladaptive response of the nervous system to damage. Annu Rev Neurosci 32:1–32
5. Attal N, Bouhassira D (2021) Advances in the treatment of neuropathic pain. Curr Opin Neurol 34(5):631–637
6. Kannan P, Bello UM, Winser SJ (2022) Physiotherapy interventions for pain relief in individuals with peripheral neuropathic pain: a systematic review and meta-analyses of randomized controlled trials. Contemp Clin Trials 125:107055
7. Jiang X et al. (2022) Effects of repetitive transcranial magnetic stimulation on neuropathic

pain: a systematic review and meta-analysis. Neurosci Biobehav Rev 132:130–141
8. Szok D, Tajti J, Nyari A, Vecsei L (2019) Therapeutic approaches for peripheral and central neuropathic pain. Behav Neurol 2019:8685954
9. Diaz MM et al. (2022) Toward composite pain biomarkers of neuropathic pain-focus on peripheral neuropathic pain. Front Pain Res (Lausanne) 3:869215
10. Matesanz-Garcia L et al. (2022) Effect of physiotherapeutic interventions on biomarkers of neuropathic pain: a systematic review of preclinical literature. J Pain 23(11):1833–1855
11. Bennett GJ, Chung JM, Honore M, Seltzer Z (2003) Models of Neuropathic Pain in the Rat. Curr Protoc Neurosci 22:9.14.1–9.14.16. https://doi.org/10.1002/0471142301.ns0914s22
12. Wang LX, Wang ZJ (2003) Animal and cellular models of chronic pain. Adv Drug Deliv Rev 55(8):949–965
13. Kim KJ, Yoon YW, Chung JM (1997) Comparison of three rodent neuropathic pain models. Exp Brain Res 113(2):200–206
14. Nakae A et al. (2011) The animal model of spinal cord injury as an experimental pain model. J Biomed Biotechnol 2011:939023
15. Bennett GJ (1993) An animal model of neuropathic pain: a review. Muscle Nerve 16(10):1040–1048
16. Dowdall T, Robinson I, Meert TF (2005) Comparison of five different rat models of peripheral nerve injury. Pharmacol Biochem Behav 80(1):93–108
17. Colleoni M, Sacerdote P (2010) Murine models of human neuropathic pain. Biochim Biophys Acta 1802(10):924–933
18. Jarvis MF, Boyce-Rustay JM (2009) Neuropathic pain: models and mechanisms. Curr Pharm Des 15(15):1711–1716
19. Bennett GJ (2010) Pathophysiology and animal models of cancer-related painful peripheral neuropathy. Oncologist 15(Suppl 2):9–12
20. Authier N et al. (2009) Animal models of chemotherapy-evoked painful peripheral neuropathies. Neurotherapeutics 6(4):620–629
21. Decosterd I, Berta T (2009) Animal models and neuropathic pain. In: Basbaum AI, Bushnell MC (Eds.)Eds. Science of pain. Academic Press, Oxford, 857–864
22. Klusakova I, Dubovy P (2009) Experimental models of peripheral neuropathic pain based on traumatic nerve injuries—an anatomical perspective. Ann Anat 191(3):248–259
23. Jaggi AS, Jain V, Singh N (2011) Animal models of neuropathic pain. Fundam Clin Pharmacol 25(1):1–28
24. Khan A, Hargreaves KM (2010) Animal models of orofacial pain. Methods Mol Biol 617:93–104
25. Wall PD et al. (1979) Autotomy following peripheral nerve lesions: experimental anaesthesia dolorosa. Pain 7(2):103–111
26. Wall PD, Scadding JW, Tomkiewicz MM (1979) The production and prevention of experimental anesthesia dolorosa. Pain 6(2):175–182
27. Kauppila T (1998) Correlation between autotomy-behavior and current theories of neuropathic pain. Neurosci Biobehav Rev 23(1):111–129
28. Mogil JS (2009) Animal models of pain: progress and challenges. Nat Rev Neurosci 10(4):283–294
29. Cunha AM, Pereira-Mendes J, Almeida A, Guimaraes MR, Leite-Almeida H (2020) Chronic pain impact on rodents' behavioral repertoire. Neurosci Biobehav Rev 119:101–127
30. Leite-Almeida H, Pinto-Ribeiro F, Almeida A (2015) Animal models for the study of comorbid pain and psychiatric disorders. Mod Trends Pharmacopsychiatry 30:1–21
31. Guimaraes MR et al. (2021) Chronic pain susceptibility is associated with anhedonic behavior and alterations in the accumbal ubiquitin-proteasome system. Pain 162(6):1722–1731
32. Cunha AM et al. (2020) Mesocorticolimbic monoamines in a rodent model of chronic neuropathic pain. Neurosci Lett 737:135309
33. Esteves M et al. (2019) MORPhA scale: behavioral and electroencephalographic validation of a rodent anesthesia scale. J Neurosci Methods 324:108304
34. Kamio K, Morita J, Nakanishi Y, Sasaki M, Wakamatsu M (2021) Corneal lesions related to an anesthetic mixture of medetomidine, midazolam, and butorphanol treatment in rats. J Toxicol Sci 46(12):561–568
35. Turner PV, Albassam MA (2005) Susceptibility of rats to corneal lesions after injectable anesthesia. Comp Med 55(2):175–182
36. Hellebrekers LJ, Booij LHDJ, Flecknell PA (2001) Anaesthesia, analgesia and euthanasia. In: van Zutphen LFM, Baumans V, Beynen AC (Eds.)Eds. Principles of Laboratory Animal Science. 2nd edn. Elsevier, Amsterdam, The Netherlands, 277–311
37. Hedenqvist P (2021) Laboratory animal analgesia, anesthesia, and euthanasia. In: Hau J, Schapiro SJ (Eds.)Eds. Handbook of laboratory animal science: essential principles and practices. 4th edn. CRC Press, Boca Raton, FL, USA, 343–378
38. Klune CB, Robbins HN, Leung VS, Pang DS (2020) Hypothermia during general anesthesia interferes with pain assessment in laboratory rats (*Rattus norvegicus*). J Am Assoc Lab Anim Sci JAALAS 59(6):719–725
39. Hestehave S et al. (2017) Is there a reasonable excuse for not providing post-operative

analgesia when using animal models of peripheral neuropathic pain for research purposes?. PLOS One 12(11):e0188113
40. Chaplan SR, Bach FW, Pogrel JW, Chung JM, Yaksh TL (1994) Quantitative assessment of tactile allodynia in the rat paw. J Neurosci Methods 53(1):55–63
41. Barrot M (2012) Tests and models of nociception and pain in rodents. Neuroscience 211:39–50
42. Dimitrov EL, Tsuda MC, Cameron HA, Usdin TB (2014) Anxiety- and depression-like behavior and impaired neurogenesis evoked by peripheral neuropathy persist following resolution of prolonged tactile hypersensitivity. J Neurosci 34(37):12304–12312
43. Guimaraes MR et al. (2019) Evidence for lack of direct causality between pain and affective disturbances in a rat peripheral neuropathy model. Genes Brain Behav 18(6):e12542
44. Tappe-Theodor A, King T, Morgan MM (2019) Pros and cons of clinically relevant methods to assess pain in rodents. Neurosci Biobehav Rev 100:335–343
45. Decosterd I, Woolf CJ (2000) Spared nerve injury: an animal model of persistent peripheral neuropathic pain. Pain 87(2):149–158
46. De Felice M et al. (2011) Engagement of descending inhibition from the rostral ventromedial medulla protects against chronic neuropathic pain. Pain 152(12):2701–2709
47. Monassi CR, Bandler R, Keay KA (2003) A subpopulation of rats show social and sleep-waking changes typical of chronic neuropathic pain following peripheral nerve injury. Eur J Neurosci 17(9):1907–1920
48. Lee BH, Won R, Baik EJ, Lee SH, Moon CH (2000) An animal model of neuropathic pain employing injury to the sciatic nerve branches. Neuroreport 11(4):657–661
49. Kim SH, Chung JM (1992) An experimental model for peripheral neuropathy produced by segmental spinal nerve ligation in the rat. Pain 50(3):355–363
50. Yalcin I et al. (2011) A time-dependent history of mood disorders in a murine model of neuropathic pain. Biol Psychiatry 70(10):946–953
51. Leite-Almeida H, et al. (2012) Differential effects of left/right neuropathy on rats' anxiety and cognitive behavior. Pain 153:2218–2225
52. Leite-Almeida H et al. (2009) The impact of age on emotional and cognitive behaviours triggered by experimental neuropathy in rats. Pain 144(1–2):57–65
53. Miller AL, Golledge HD, Leach MC (2016) The influence of isoflurane anaesthesia on the rat grimace scale. PLOS One 11(11):e0166652
54. Wu CC, et al. (2022) Preventive Intrathecal Injection of Bupivacaine Alleviated Microglia Activation and Neuropathic Pain in a Rat Model of Chronic Constriction Injury. Int J Mol Sci 23:7197.
55. Wilson NM, Ripsch MS, White FA (2016) Impact of opioid and nonopioid drugs on postsurgical pain management in the rat. Pain Res Treat 2016:8364762
56. Celerier E et al. (2000) Long-lasting hyperalgesia induced by fentanyl in rats: preventive effect of ketamine. Anesthesiology 92(2):465–472
57. Dworkin RH et al. (2003) Advances in neuropathic pain: diagnosis, mechanisms, and treatment recommendations. Arch Neurol 60(11):1524–1534
58. Mackinnon SE, Dellon AL, Hudson AR, Hunter DA (1985) A primate model for chronic nerve compression. J Reconstr Microsurg 1(3):185–195
59. Mackinnon SE, Dellon AL, Hudson AR, Hunter DA (1984) Chronic nerve compression—an experimental model in the rat. Ann Plast Surg 13(2):112–120
60. Grace PM, Hutchinson MR, Manavis J, Somogyi AA, Rolan PE (2010) A novel animal model of graded neuropathic pain: utility to investigate mechanisms of population heterogeneity. J Neurosci Methods 193(1):47–53
61. Bennett GJ, Xie YK (1988) A peripheral mononeuropathy in rat that produces disorders of pain sensation like those seen in man. Pain 33(1):87–107
62. Hulse R, Wynick D, Donaldson LF (2008) Characterization of a novel neuropathic pain model in mice. Neuroreport 19(8):825–829
63. Seltzer Z, Dubner R, Shir Y (1990) A novel behavioral model of neuropathic pain disorders produced in rats by partial sciatic nerve injury. Pain 43(2):205–218
64. Carlton SM, Lekan HA, Kim SH, Chung JM (1994) Behavioral manifestations of an experimental model for peripheral neuropathy produced by spinal nerve ligation in the primate. Pain 56(2):155–166
65. DeLeo JA et al. (1994) Characterization of a neuropathic pain model: sciatic cryoneurolysis in the rat. Pain 56(1):9–16
66. Kawakami M et al. (1994) Experimental lumbar radiculopathy. Behavioral and histologic changes in a model of radicular pain after spinal nerve root irritation with chromic gut ligatures in the rat. Spine 19(16):1795–1802
67. He L, et al. (2022). Modified Spared Nerve Injury Surgery Model of Neuropathic Pain in Mice. J Vis Exp, (179)
68. Lindenlaub T, Sommer C (2000) Partial sciatic nerve transection as a model of neuropathic

pain: a qualitative and quantitative neuropathological study. Pain 89(1):97–106
69. Vadakkan KI, Jia YH, Zhuo M (2005) A behavioral model of neuropathic pain induced by ligation of the common peroneal nerve in mice. J Pain 6(11):747–756
70. Walczak JS, Pichette V, Leblond F, Desbiens K, Beaulieu P (2005) Behavioral, pharmacological and molecular characterization of the saphenous nerve partial ligation: a new model of neuropathic pain. Neuroscience 132(4):1093–1102
71. Jou IM (2004) The effects from lumbar nerve root transection in rats on spinal somatosensory and motor-evoked potentials. Spine (Phila Pa 1976) 29(2):147–155
72. Li L, Xian CJ, Zhong JH, Zhou XF (2002) Effect of lumbar 5 ventral root transection on pain behaviors: a novel rat model for neuropathic pain without axotomy of primary sensory neurons. Exp Neurol 175(1):23–34
73. Benbouzid M et al. (2008) Sciatic nerve cuffing in mice: a model of sustained neuropathic pain. Eur J Pain 12(5):591–599
74. Gazelius B, Cui JG, Svensson M, Meyerson B, Linderoth B (1996) Photochemically induced ischaemic lesion of the rat sciatic nerve. A novel method providing high incidence of mononeuropathy. Neuroreport 7(15-17):2619–2623
75. Lin XY, Yang J, Li HM, Hu SJ, Xing JL (2012) Dorsal root ganglion compression as an animal model of sciatica and low back pain. Neurosci Bull 28(5):618–630
76. Takenobu Y, Katsube N, Marsala M, Kondo K (2001) Model of neuropathic intermittent claudication in the rat: methodology and application. J Neurosci Methods 104(2):191–198
77. Staunton CA, Barrett-Jolley R, Djouhri L, Thippeswamy T (2018) Inducible nitric oxide synthase inhibition by 1400W limits pain hypersensitivity in a neuropathic pain rat model. Exp Physiol 103(4):535–544
78. Xu X et al. (2012) Effects of exogenous galanin on neuropathic pain state and change of galanin and its receptors in DRG and SDH after sciatic nerve-pinch injury in rat. PLoS One 7(5):e37621
79. Dorsi MJ et al. (2008) The tibial neuroma transposition (TNT) model of neuroma pain and hyperalgesia. Pain 134(3):320–334
80. Wieseler J, Ellis A, Maier SF, Watkins LR, Falci S (2012) Unilateral T13 and L1 dorsal root avulsion: methods for a novel model of central neuropathic pain. Methods Mol Biol 851:171–183
81. Na HS, Han JS, Ko KH, Hong SK (1994) A behavioral model for peripheral neuropathy produced in rat's tail by inferior caudal trunk injury. Neurosci Lett 177(1-2):50–52
82. Rodrigues-Filho R, Santos AR, Bertelli JA, Calixto JB (2003) Avulsion injury of the rat brachial plexus triggers hyperalgesia and allodynia in the hindpaws: a new model for the study of neuropathic pain. Brain Res 982(2):186–194
83. Yi H, Kim MA, Back SK, Eun JS, Na HS (2011) A novel rat forelimb model of neuropathic pain produced by partial injury of the median and ulnar nerves. Eur J Pain 15(5):459–466
84. Matsumoto M, Sekiya T, Kojima K, Ito J (2008) An animal experimental model of auditory neuropathy induced in rats by auditory nerve compression. Exp Neurol 210(1):248–256
85. Gupta S, Bansal RN, Singh Sodhi SP, Brar GK (2022) Animal models—mimicking the pain of trigeminal neuralgia. Indian J Pharmacol 54(2):138–145
86. Vos BP, Strassman AM, Maciewicz RJ (1994) Behavioral evidence of trigeminal neuropathic pain following chronic constriction injury to the rat's infraorbital nerve. J Neurosci 14(5 Pt 1):2708–2723
87. Warner BE, Goins WF, Kramer PR, Kinchington PR (2023) A guide to preclinical models of zoster-associated pain and postherpetic neuralgia. Curr Top Microbiol Immunol 438:189–221
88. Fleetwood-Walker SM et al. (1999) Behavioural changes in the rat following infection with varicella-zoster virus. J Gen Virol 80(Pt 9):2433–2436
89. Wallace VC et al. (2007) Pharmacological, behavioural and mechanistic analysis of HIV-1 gp120 induced painful neuropathy. Pain 133(1–3):47–63
90. Shimoyama M, Tanaka K, Hasue F, Shimoyama N (2002) A mouse model of neuropathic cancer pain. Pain 99(1–2):167–174
91. Coderre TJ, Xanthos DN, Francis L, Bennett GJ (2004) Chronic post-ischemia pain (CPIP): a novel animal model of complex regional pain syndrome-type I (CRPS-I; reflex sympathetic dystrophy) produced by prolonged hindpaw ischemia and reperfusion in the rat. Pain 112(1–2):94–105
92. Willison HJ (2011) Biomarkers in experimental models of antibody-mediated neuropathies. J Peripher Nerv Syst 16(Suppl 1):60–62
93. De Vloo P, Morlion B, van Loon J, Nuttin B (2017) Animal models for central poststroke pain: a critical comprehensive review. Pain 158(1):17–29
94. Wasserman JK, Koeberle PD (2009) Development and characterization of a hemorrhagic rat model of central post-stroke pain. Neuroscience 161(1):173–183

95. Martini R (1997) Animal models for inherited peripheral neuropathies. J Anat 191(Pt 3):321–336
96. Meyer Zu Horste G, Nave KA (2006) Animal models of inherited neuropathies. Curr Opin Neurol 19(5):464–473
97. Mirabelli E, Elkabes S (2021) Neuropathic pain in multiple sclerosis and its animal models: focus on mechanisms, knowledge gaps and future directions. Front Neurol 12:793745
98. An JX et al. (2011) A new animal model of trigeminal neuralgia produced by administration of cobra venom to the infraorbital nerve in the rat. Anesth Analg 113(3):652–656
99. Authier N, Gillet JP, Fialip J, Eschalier A, Coudore F (2003) An animal model of nociceptive peripheral neuropathy following repeated cisplatin injections. Exp Neurol 182(1):12–20
100. Aley KO, Reichling DB, Levine JD (1996) Vincristine hyperalgesia in the rat: a model of painful vincristine neuropathy in humans. Neuroscience 73(1):259–265
101. Dina OA et al. (2000) Key role for the epsilon isoform of protein kinase C in painful alcoholic neuropathy in the rat. J Neurosci 20(22):8614–8619
102. Eliav E, Herzberg U, Ruda MA, Bennett GJ (1999) Neuropathic pain from an experimental neuritis of the rat sciatic nerve. Pain 83(2):169–182
103. Yezierski RP, Liu S, Ruenes GL, Kajander KJ, Brewer KL (1998) Excitotoxic spinal cord injury: behavioral and morphological characteristics of a central pain model. Pain 75(1):141–155
104. Islam MS (2013) Animal models of diabetic neuropathy: progress since 1960s. J Diabetes Res 2013:149452
105. Calcutt NA (2004) Experimental models of painful diabetic neuropathy. J Neurol Sci 220(1–2):137–139
106. Sullivan KA et al. (2007) Mouse models of diabetic neuropathy. Neurobiol Dis 28(3):276–285
107. Calcutt NA (2004) Modeling diabetic sensory neuropathy in rats. Methods Mol Med 99:55–65
108. Chacur M et al. (2001) A new model of sciatic inflammatory neuritis (SIN): induction of unilateral and bilateral mechanical allodynia following acute unilateral peri-sciatic immune activation in rats. Pain 94(3):231–244
109. Perry TA, Weerasuriya A, Mouton PR, Holloway HW, Greig NH (2004) Pyridoxine-induced toxicity in rats: a stereological quantification of the sensory neuropathy. Exp Neurol 190(1):133–144
110. Chia SJ, Tan EK, Chao YX (2020) Historical Perspective: Models of Parkinson's Disease. Int J Mol Sci 21:2464
111. Dieb W, Ouachikh O, Durif F, Hafidi A (2014) Lesion of the dopaminergic nigrostriatal pathway induces trigeminal dynamic mechanical allodynia. Brain Behav 4(3):368–380

Check for updates

Chapter 2

FRICT-ION: Simple and Reliable Trigeminal Nerve Model of Chronic Craniofacial Neuropathic Pain

Sabrina L. McIlwrath, Marena Montera and Karin N. Westlund

Abstract

An easily induced, reliable trigeminal neuropathic pain model with durable hypersensitivity is described here for the study of chronic orofacial pain. The model acronym is **FRICT-ION** (**F**oramen **R**otundum **I**nflammatory **C**onstriction **T**rigeminal **I**nfra**O**rbital **N**erve). The acronym also hints at the likely cause of the model's persistence: irritation of the nerve due to movement of the suture during normal chewing in the rat. The surgery is performed orally through a small surgical slit in the rat's cheek crease, sliding a piece of chromic gut suture along the trigeminal infraorbital nerve as it passes into the skull. Due to the surgery's simplicity, the chances of complications are low. There are no outward physical signs, making the FRICT-ION model ideal for study blinding in either rats or mice. The model allows testing of mechanical hypersensitivity with von Frey filaments through at least 10 weeks, with anxiety- and depression-like behaviors measurable 8 weeks after model induction. A step-by-step methodology and examination of behaviors is provided for reference.

Keywords Chronic pain, Orofacial pain, Neuropathic pain, Chronic cranial facial pain, Chronic gut, Infraorbital nerve, Anxiety, Depression

1 Introduction: Chronic Neuropathic Pain

1.1 Chronic Trigeminal Neuropathic Pain

Neuropathic pain is caused by disease, irritation, or lesion of peripheral sensory nerves [1]. If the pain persists over long periods of time, it will transition into chronic pain. Changes in the signaling molecules in the peripheral nerve cells eventually cause central sensitization and changes in the brain's pain and affective circuitry. One major obstacle to a better understanding of the mechanisms of chronic neuropathic pain, as well as the development of effective treatments, is the lack of available experimental animal models that mimic established chronic pain. While many nerve-injury models have been developed that allow study of nociceptive mechanisms and pain-related behaviors in the short term, they typically reverse within 3–4 weeks due to natural healing. The method described below is one of only a few models that allow long-term studies suitable for better understanding of clinical pain states and for testing experimental therapies.

Isaura Tavares and Hugo Leite-Almeida (eds.), *Translational Research Methods for Neuropathic Pain*, Neuromethods, vol. 235,
https://doi.org/10.1007/978-1-0716-5312-8_2,

Several innate differences in the trigeminal nerve are likely responsible for the ease in inducing the chronic persistence of the neuropathic pain model described here. The great importance of the trigeminal nerve to survival is postulated to be one characteristic responsible for the ease in producing persistence of trigeminal nerve injury pain and its effects on psychological well-being [2]. Anatomical proximity to the brain and the unique circuitry of the trigeminal system, with one less synapse forming more direct connectivity with the limbic circuitry, are also likely responsible for the intensity and persistence of trigeminal neuropathic pain [3].

1.2 Chronic Constriction of the Trigeminal Infraorbital Nerve

There have been previous successful chronic trigeminal pain models induced by loose chromic gut suture tie (chronic constriction injury) of the infraorbital nerve (ION) in rats (CCI-ION) beneath the eye [4, 5] or on the snout in rats (dIoN-CCI) and mice (DIONI) [6, 7]. However, the surgical approach to the ION in all of these models is technically challenging, and the vasculature is prone to severe bleeding. Likewise, blinding of the study is not possible with the evidence of the surgery visible.

Alternatively, the intraoral approach has also been used to induce the CCI-ION model in rats [8, 9]. Another previously introduced model induced trigeminal nerve sensitivity by sliding the irritative chromic gut suture into the ION fissure of the skull beneath the eye in mice, creating a trigeminal inflammatory compression (TIC) [10–13]. After induction of all of these trigeminal nerve injury models, mechanical hypersensitivity develops *within a week and persists indefinitely* in the ION's whisker pad receptive field. After trigeminal nerve injury, cognitive deficits are reported in rodents within 3 weeks. Anxiety- and depression-related symptoms emerge in subsequent weeks after injury (8 weeks), indicating brain origin and circuitry neuroplasticity are responsible for these behaviors accompanying chronic pain [5, 12–15]. Estimates are that in week 6–8 post-surgery, rats have experienced pain equivalent to 6–8 human years (each rat month equates to approximately 3 human years) and persistence of models in this time frame can be considered chronic [16, 17]. The interplay between psychological and physical functioning is a particular consideration in patients with orofacial pain [2]. The clinical condition is often referred to as "Suicide Syndrome" [18].

1.3 FRICT-ION Model in Rats

An improved surgical approach is the FRICT-ION (Foramen Rotundum Inflammatory Constriction of Trigeminal InfraOrbital Nerve) nerve injury model, first developed in mice [19]. This surgical method exposes the animals to only minutes of inhalant isoflurane anesthesia and produces bilateral orofacial hypersensitivity that we have shown to persist *at least 10 weeks in mice* without any apparent side effects or outward physical indications. This allows for group blinding in the studies. More importantly, the testing of potential analgesic compounds can be done at clinically relevant time points when pain-like behaviors are chronic and secondary behavioral changes (anxiety- and depression-like behaviors) can be measured. The FRICT-ION model can also be induced in rats and mastered as quickly and easily as in mice following the instructions below.

2 Materials

2.1 Animals

Animal experiments described were approved by the New Mexico Veteran's Administration Health Care System Animal Component of Research Protocol (ACORP #20-A322) and performed in accordance with the National Institute of Health Guide for the Care and Use of Laboratory Animals (NIH Publications No. 80-23) revised in 1996. Adult (~5–6 weeks old) male Sprague-Dawley rats (125–150 g, Envigo, Indianapolis, IN, USA) were housed on 12/12 h reverse light cycle and fed *ad libitum* with a soy protein-free diet (Teklad #2920x, Envigo). Our animals are acclimated to a reverse light cycle for 2 weeks, and all experimental procedures take place under red lighting during the animal's active dark cycle. This accommodates all investigator testing times during daylight hours. It is reported that rodent behavioral group differences decrease in late afternoon as their sleep cycle approaches [20].

2.2 Animals and Consumables

- Male Sprague-Dawley rats (125–150 g, 5–6 weeks, Envigo, Indianapolis, IN, USA).
- 4-0 chromic gut suture cut into 4 mm pieces (3 mm for mice) (Ethicon, Cat. # 635H).
- Isoflurane inhalant anesthetic (shopmedvet.com, Cat. # RXISO-250).
- 70% ethanol (local source).
- Nitrile exam gloves (Fishersci, Cat. #NGPF7002).
- Cotton-tipped applicators (Fishersci, Cat. #19-062-616).
- Cotton balls (Fishersci, Cat. #19-090-702).
- Oxygen tank.
- Labeling tape (Fishersci, Cat. # 15-901-10R).
- Surgical drape, Steri-Drape™ Utility Sheet, Absorbent Prevention, cut into 15 × 15 cm pieces (VWR, Cat. #76246-788).
- Tygon tubing with 3 mm inner diameter (Grainger, Cat. # 22XH87).
- Superglue (local source).
- Paperclips (local source).

2.3 Equipment

- OPTIKA SZN trinocular stereo zoom microscope on hinged overhanding stand (New York Microscope Company, Cat. # OPSZN-10).

- OPTIKA CL31 double arm LED illuminator (New York Microscope Company, Cat. # OPCL-31).
- Table top animal research portable anesthesia workstation "PAM" (Patterson Scientific, Cat. # AS-01-0007).
- Anesthesia induction chamber (Summit Medical Equipment Company, Cat. #AS-01-0530-LG).
- Universal rodent nose cone non-rebreathing system, 17 mm diameter (Summit Medical Equipment Company, Cat. #AS-01-0525).
- Tupper small hand pediatric fixation plate, 249 × 140 mm (Synovis Micro Companies Alliance, Cat. #THP-101.
- Glass bead sterilizer Germinator 500 (VWR, Cat. #102095-946).
- Autoclave.
- Small animal heating pad (Valley Vet Supply, Cat. #47375).
- Von Frey Filaments (Ugo Basile, Cat. #37450-275).
- Surgical tools:
    - #11 carbon steel scalpel blades (VWR, Cat. #21909-612).
    - Scalpel Handle No. 3 (VWR, Cat. #25607-947).
    - Biology Dumont #5 forceps (Fine Science Tools, Cat. #11252-30).
    - Student Vannas spring scissors, straight blade (Fine Science Tools, Cat. #91500-09).
    - Integra Miltex cotton & dressing pliers (Safco Dental Supply, Cat. # 66-317).
    - mandible retractor (construction is described under Surgical Preparations).

3 Methods

3.1 Surgical Preparations

- A retractor for the mandible is constructed using 20 cm Tygon tubing (3 mm inner diameter) (Fig. 1). A paperclip is cut into two pieces that are then shaped to form hooks. A piece of shaped wire is glued into each tubing end with superglue and allowed to harden. This allows the tubing to be hooked into the ribbed edge of the Tupper small hand fixation table.
- All surgical tools, cotton swabs, and surgical drape pieces are autoclaved. The tools can be autoclaved in one autoclave-safe container. Cotton swabs can be packaged five per commercial autoclave bag, and surgical drape pieces are bagged individually for autoclaving.

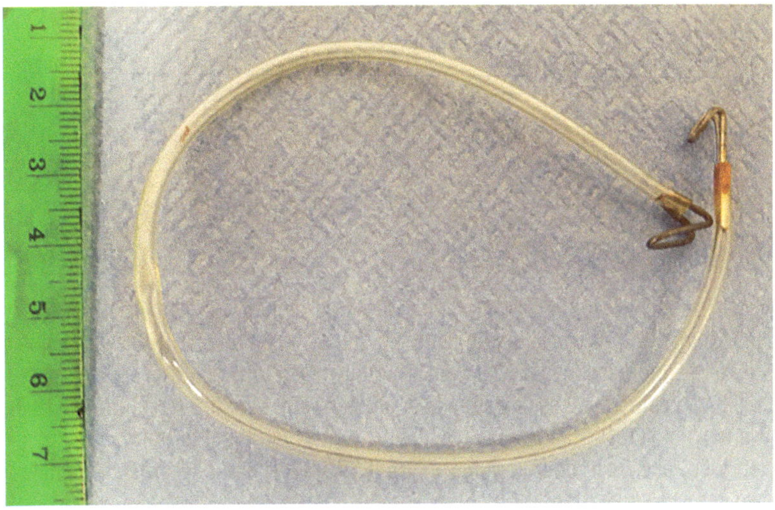

Fig. 1 Mandible retractor made from Tygon tubing and a paperclip

- When performing surgery on multiple animals on the same day, all tools are cleaned with 70% ethanol after each surgery, and the tips are sterilized for 5 s in a glass bead sterilizer. The tools are then placed on a new autoclaved piece of surgical drape to prevent contamination.
- An anesthesia machine is set up with a tube splitter. This allows for gas flow to be directed at either the induction chamber or the surgical area. The anesthesia induction chamber is placed in the fume hood along with the anesthesia gas scavenger tubes. The other tube goes to the surgical area and is fitted with the universal rodent nose cone non-rebreathing system. The nose cone directs gas flow to the animal and scavenges the excess isoflurane not inhaled. The outflow tube is placed under a fume hood or under an evacuation hood (Fig. 2).
- The surgery is performed on a rolling table directly in front of a fume hood. The surgery is performed underneath a dissection microscope on a sterilized surface that facilitates holding the mouth open. Here, the Tupper small hand fixation table with ribbed edge is used, which allows the mandible retractor to be hooked into the ribbed edge.
- The surgical recovery station is close by. A clean cage is set half on/half off a heating pad to warm the animal during recovery, allowing the rat to leave the heated area once recovered from anesthesia (Fig. 3). The bedding is moved away from the half of the cage that is on the heating pad where the rat is initially placed while anesthetized. This facilitates heat transduction and prevents the bedding material from impairing the animal's breathing during recovery.

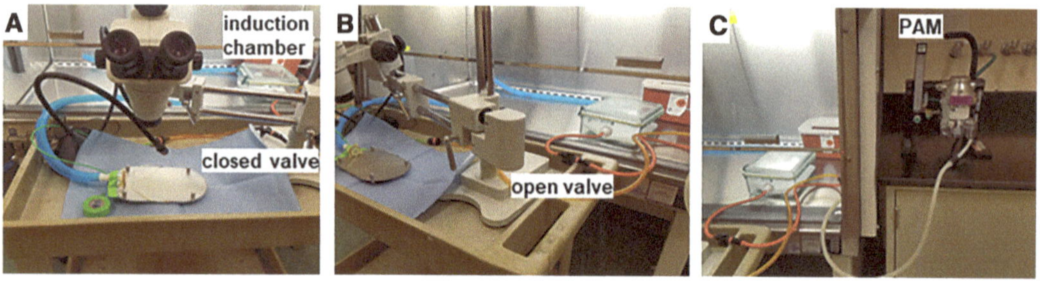

Fig. 2 Surgical setup. (**A**) Frontal view of the cart located directly in front of the fume hood. The nose cone is taped to the Tupper small hand pediatric fixation plate. The LED illuminator is on the left, and the boom stand of the stereo microscope is on the right. The thin green tube is the anesthesia gas mix flow to the nose cone. The blue tube is the anesthesia gas exhaust and terminates under the fume hood. (**B**) Side view of the surgical cart. The two anesthesia gas lines are visible. The valves in them determine the gas flow: to the induction chamber (red tube) and nose cone (orange tube). (**C**) The portable anesthesia machine (PAM) receives oxygen in this case from the wall oxygen supply

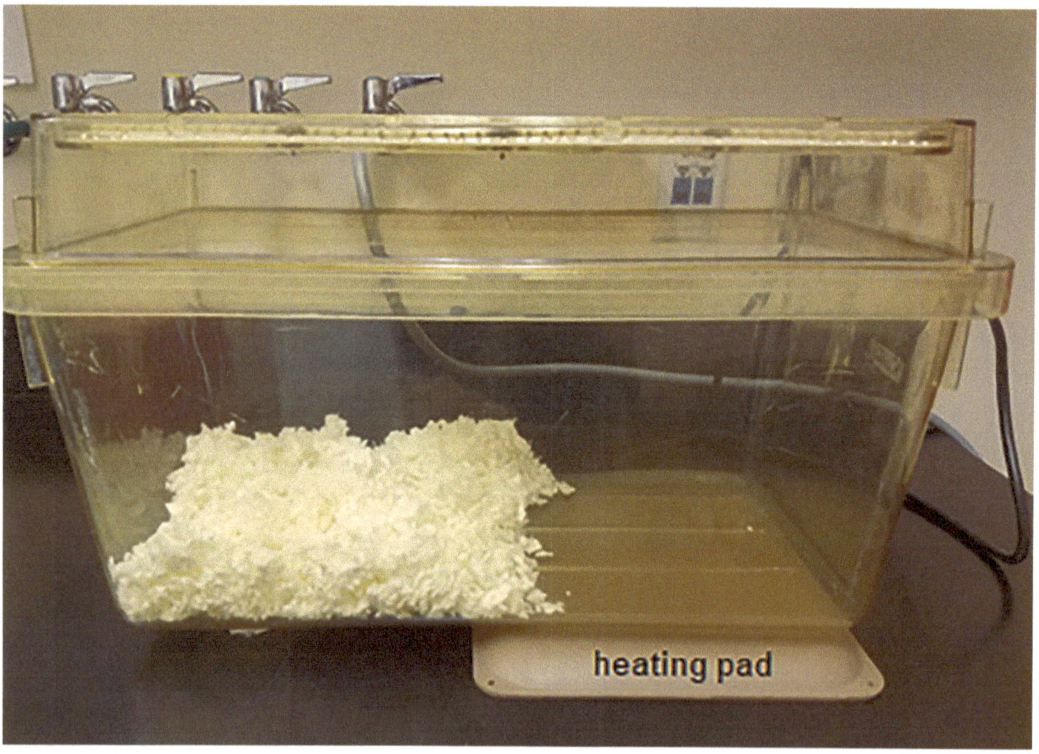

Fig. 3 Surgical recovery station. Adjacent to the surgical area, a clean cage placed half on and half off a heating pad serves as the recovery station. The bedding is moved to the side of the heating pad to improve heat transfer and allow unimpeded breathing. Immediately after surgery, the still anesthetized rat is placed on the cage floor over the heating pad

3.2 Surgery

- The anesthesia machine is turned on, and the gas flow is directed solely to the induction chamber. Oxygen flow valves are opened and turned to 0.8 L/min. This will later carry the anesthesia inhalant.

- The rat is placed in the induction chamber, and the lid is securely closed. Then the isoflurane level is turned on to 5% (maximal) for a fast anesthesia while the animal is closely observed.
- Once the animal has stopped moving and the breathing has slowed and become deep, the chamber can be gently rotated to see if the animal starts moving again and requires more time in the induction chamber. If it remains still, the gas flow to the induction chamber is turned off and directed solely to the surgical area tubing. The isoflurane content is decreased to 3.5%–4.5% depending on the animal.
- The rat is quickly moved to the surgical area on top of the Tupper small hand fixation table, and its nose is placed in the universal rodent nose cone. A toe is pinched to determine if the animal is areflexic or requires a higher concentration of inhalant isoflurane (Fig. 4).

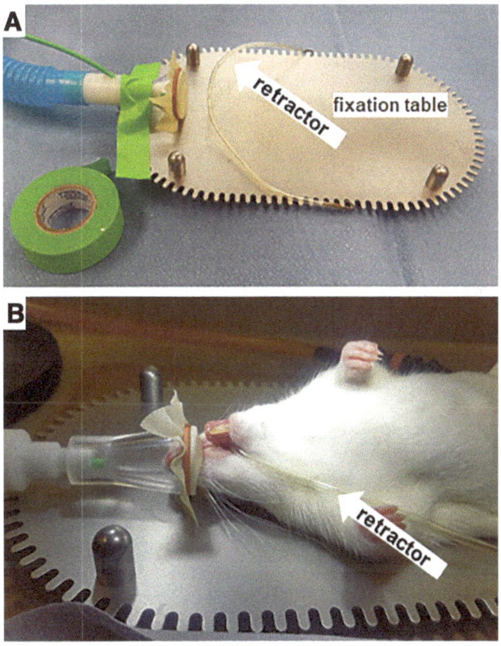

Fig. 4 Tupper small hand fixation table with universal rodent anesthesia cone and mandible retractor. (**A**) The nose cone is taped to the fixation table. Its diaphragm, constructed from glove material, is utilized to hold the rat's upper incisors. The mandible retractor is used to fix the rodent's mouth in an open position for the oral surgery. (**B**) The nose of the anesthetized rat is placed inside the nose cone so that the upper incisors are within the cone's diaphragm. The tubing of the mandible retractor tubing is threaded through the mouth and hooked to the ribbed edge of the fixation table to pull back the mandible and tongue

- The animal is placed in the supine position. The anesthesia cone is slipped over the nose so that the diaphragm covers the incisors. The mandible retractor is gently placed over the tongue and lower incisors as well as hooked to the ribbed edge to prop open the mouth for surgery (Fig. 4B).
- A small incision is made with the #11 scalpel blade anterior to the sphenomandibular ligament into the rat's inner buccal margin crease lateral to the hard palate, i.e., the roof of the mouth. The incision is widened by blunt dissection. To do this, the closed tips of a #5 forceps are inserted into the incision and slowly opened. This blunt dissection is then repeated. Vannas' student's scissors can be similarly used for blunt dissection. Minimal to no bleeding will occur. This should result in a 2–3 mm long opening that is deep enough to see the nerve bundles of the infraorbital nerve (ION) as well as the foramen rotundum where the ION exits the skull (Fig. 5).
- The foramen rotundum is exposed with blunt dissection. A small rolled up piece of cotton ball can be used to help. A 4 mm piece of dry 4-0 chromic gut suture is picked up with the #5 forceps and pushed into the foramen rotundum. Take care that the metal does not touch the nerve or any of its branches. Should the suture get wet by wicking up fluids in the surgical cavity, it will get soft and cannot be pushed into the foramen. In this case, take a new piece of dry suture and try again. Be careful that the suture stays in the rotundum foramen when releasing it from forceps.
- The surgery is complete. The surgical incision is left open as it was shown in mice that it heals within days [19]. Unhook the

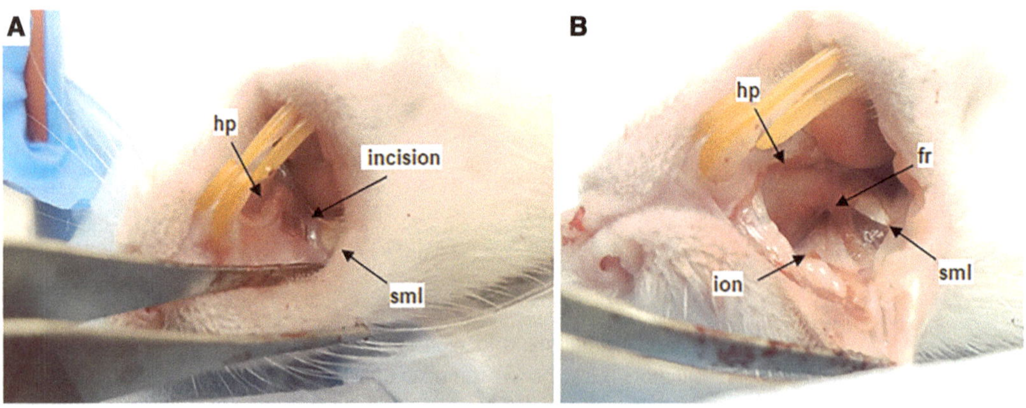

Fig. 5 Surgical guide to get to the foramen rotundum using a euthanized rat for easy visualization. (**A**) The buccal cheek is moved to the side to expose the sphenomandibular ligament. A small incision is made with the anterior to the ligament in the inner buccal margin crease next to the hard palate. (**B**) The initial incision is opened with blunt dissection to expose the nerve bundles of the maxillary branch of the ION and the foramen rotundum. Hp hard palate, sml sphenomandibular ligament, fr foramen rotundum, ion infraorbital nerve

mandible retractor and carefully take the nose out of the anesthesia cone, taking care that the upper incisors do not tear the diaphragm of the nose cone.

- Take the animal to the recovery station and place it prone in the cage overlaying the heating pad. Once it is fully awake and moving around, it can be returned to the animal room.
- Since this is a surgical model for the induction of chronic pain, no postsurgical analgesics are given, as this prevents the development of chronic pain.
- Clean the surgical tools with 70% ethanol, sterilize the tips by holding them in the glass bead sterilizer for 5 s, and place them on a new autoclaved piece of surgical drape. Before starting the next surgery, quickly flush the induction chamber with oxygen. Rats will not go into the induction chamber easily if they can smell isoflurane.
- The procedure for sham surgical rats differs from the FRICT-ION model only in that the nerve and foramen rotundum are exposed but no suture is inserted into the foramen rotundum. The procedure is otherwise identical.
- Behavioral and physiological well-being are closely monitored and recorded in all surgical animals for the next 5 days. Should more than 20% weight loss, unresolved porphyria (production of red tear staining just under the eyes), and/or lethargy occur, the veterinary staff should be consulted, and at their discretion, the animal may be euthanized. We have never had an occasion to find a rodent in this condition. Typically, following the recovery from surgery, animals do not demonstrate evidence of tissue injury or evidence of discomfort from pain, such as loss of appetite, decreased activity, porphyrin secretion, or piloerection. Study group blinding with this method is clearly an advantage.

3.3 Reflexive Behavioral Assay

Mechanical Response Threshold Testing

In order to determine the mechanical withdrawal threshold, the minimal force needed to elicit a response 50% of the time, a series of 8 von Frey filaments graded in size (with about 0.2 g increases) provides log step changes (Table 1).

Beginning at least 1 week prior to the surgery, animals are slowly acclimated to handling, restraint, and mechanical stimulation with von Frey fibers applied to the whisker pad, the innervation territory of the infraorbital nerve. Initially, each animal is restrained by gently wrapping it in a surgical towel and then held for several minutes. The second day each animal is restrained and after 5–10 min carefully probed on the whisker pad using the 4.31 von Frey filament which when bent elicits 2 g force. Naïve acclimated animals do not respond

Table 1
Von Frey filaments used to quantify mechanical sensitivity

Size	3.61	3.84	4.08	4.31	4.56	4.74	4.93	5.18
Force (g)	0.4	0.6	1.0	2.0	4.0	6.0	8.0	15.0

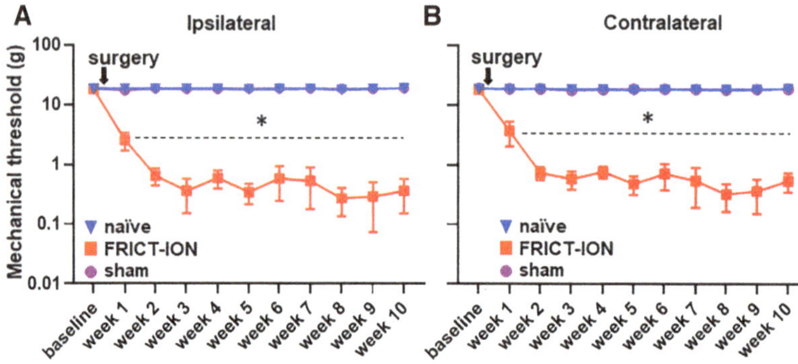

Fig. 6 Only rats with the FRICT-ION model develop bilateral mechanical hypersensitivity. (**A**) and (**B**) demonstrate that significantly less force is required to elicit a head withdrawal in animals with the FRICT-ION model. The mechanical withdrawal threshold is unchanged in surgical naïve rats and after sham surgery. $n = 7$/group; * $p < 0.05$ shown as a dotted line for all time points after week 1

to this stimulation, indicating that it is innocuous. On the third day, each animal is again restrained and after 10 min carefully probed on the whisker pad using the 4.31 von Frey filament, followed by the 4.93 von Frey filament, neither of which elicits responses in naïve, acclimated animals. On the fourth and fifth day, the mechanical threshold for head withdrawal is determined using the up-down method [9, 21]. After 10 min of restraint in the surgical towel to minimize stress-induced effects [22], testing always starts using the 4.56 filament that elicits 4.0 g force. Each filament is applied five times at several second intervals perpendicular to the whisker pad, being careful not to touch the vibrissae. If a positive response occurs, a head withdrawal to three of five stimuli, then the next weaker filament in the series is applied until the animal doesn't respond to the mechanical stimulation. At that point, the next higher filament is used. If it elicits a response, the lower filament is used again until four fibers have been applied after the change of response to mechanical stimulation of the whisker pad. The resulting response pattern is used to calculate the mechanical withdrawal threshold, the minimal amount of force needed to elicit a response in 50% of the time, using a curve fitting algorithm [23]. A decrease in the force required to elicit a head withdrawal response indicates increased sensitivity of the animal. Baseline values are measured before the surgery. After surgery, the mechanical threshold is determined once a week during the chronic experiment (Fig. 6). This reflexive test can be repeated several

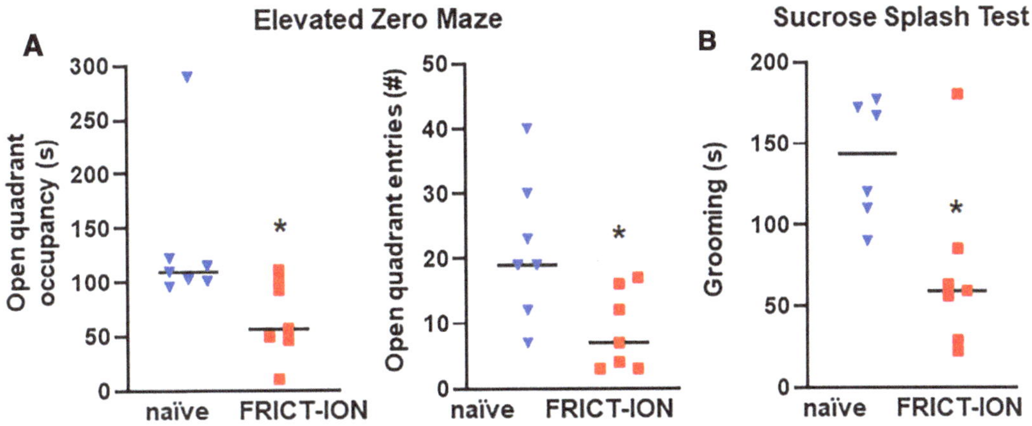

Fig. 7 Rats with the FRICT-ION model demonstrate anxiety- and depression-like behavior after week 8. (**A**) In the elevated zero maze, rats with FRICT-ION cumulative occupancy and number of entries of the open quadrants are significantly reduced compared to naïve control rats. (**B**) Rats with FRICT-ION model spent significantly less time grooming in the sucrose splash test. $n = 7$/group; * $p < 0.05$

times in a single day when the efficacy of a short-lasting compound is being determined. Within 1 week, the withdrawal threshold was significantly decreased. This hypersensitivity is maximal by week 3 after FRICT-ION surgery and persists until euthanasia, in post-surgical week 10 illustrated. This time course allows the evaluation of compounds to attenuate hypersensitivity at chronic time points, equivalent to years of pain experience in clinical patients.

Cognition and Emotional Dependent Behavioral Assay

Long-lasting hypersensitivity in animals also produces cognition-dependent impairments measurable with the novel object test, as well as anxiety- and depression-like behaviors. These are typically measured only once starting in week 6 after pain model induction to avoid practice effects [5, 14]. Data collected using two different standard tests are presented as examples. The elevated zero maze measures anxiety-like behaviors. Raised 1 m off the ground, the 10 cm wide walkway is divided into four equal-sized quadrants, two walled "safe" ones and two open "unsafe" ones. Rats with FRICT-ION model enter and spend less time in the open quadrants than naïve control animals (Fig. 7A), displaying anxiety-like behavior. The sucrose splash test is used to determine depression-like behavior. Time spent grooming after a 10% sucrose solution was spritzed on the base of the tail is significantly reduced in the following 10 min in rats with FRICT-ION compared to naïve controls (Fig. 7B).

4 Conclusions

The FRICT-ION model in rats is simple and reliable, producing craniofacial hypersensitivity that persists indefinitely. We believe the normal chewing is the cause of the long-term nerve irritation, enabled by daily meal chewing while normal weight gain is ongoing. Associated cognitive and affective behavioral changes develop that can be assessed after 4–6 weeks of an ongoing pain model. Yet the animals display no overt indication of distress or outward surgical evidence, allowing study of blinded groups. The durable model enables the researcher to determine the efficacy of pharmacological treatments to reduce chronic pain, anxiety-, and depression-like behaviors in time frames relevant to neuroplastic events that are well established, i.e., in clinically relevant time frames. Advantages of the model include the hypersensitivty that develops almost immediately and persists over 10 weeks. A limitation of the method in a rat is the possibility that the surgical exposure encounters more muscle, and if not deep enough, the surgeon might mistakenly assume the white ligament is the infraorbital nerve. In contrast, in the mouse, an open space surrounds the nerve and provides good visualization. Standard of care pharmacologics carbamazepine and gabapentin have been shown to reverse the hypersensitivity for 2–4 h in trigeminal nerve injury models, but hypersensitivity then returns. The reversal by carbamazepine indicates that the FRICT-ION model is a suitable model of trigeminal neuralgia.

Acknowledgments

This research is supported by NIH, NIDCR R21 DE028096 and HEAL UG3 NS123958; DoD CPMRP CP190116; and VA Merit grant BX002695.

Conflict of Interest The authors have no conflicts of interest to declare that are relevant to the content of this chapter.

References

1. Terminology | International Association for the Study of Pain. In: International Association for the Study of Pain (IASP). https://www.iasp-pain.org/resources/terminology/. Accessed 2 Mar 2022
2. Carlson CR (2007) Psychological factors associated with orofacial pains. Dent Clin North Am 51:145–160. https://doi.org/10.1016/j.cden.2006.09.001. vii
3. Rodriguez E, Sakurai K, Xu J et al. (2017) A craniofacial-specific monosynaptic circuit enables heightened affective pain. Nat Neurosci 20:1734–1743. https://doi.org/10.1038/s41593-017-0012-1
4. Vos BP, Strassman AM, Maciewicz RJ (1994) Behavioral evidence of trigeminal neuropathic pain following chronic constriction injury to the rat's infraorbital nerve. J Neurosci 14:2708–2723
5. Kniffin TC, Danaher RJ, Westlund KN et al. (2015) Persistent neuropathic pain influences persistence behavior in rats. J Oral Facial Pain Headache 29:183–192. https://doi.org/10.11607/ofph.1300
6. Ding W, You Z, Shen S et al. (2017) An improved rodent model of trigeminal neuropathic pain by unilateral chronic constriction injury of distal infraorbital nerve. J Pain 18:899–907. https://doi.org/10.1016/j.jpain.2017.02.427
7. Hardt S, Fischer C, Vogel A et al. (2019) Distal infraorbital nerve injury: a model for persistent facial pain in mice. Pain 160:1431–1447. https://doi.org/10.1097/j.pain.0000000000001518

8. Imamura Y, Kawamoto H, Nakanishi O (1997) Characterization of heat-hyperalgesia in an experimental trigeminal neuropathy in rats. Exp Brain Res 116:97–103. https://doi.org/10.1007/pl00005748
9. McIlwrath S, Montera M, Gott K et al. (2020) Manganese-enhanced MRI reveals changes within brain anxiety and aversion circuitry in rats with chronic neuropathic pain- and anxiety-like behaviors. NeuroImage 223:117343. https://doi.org/10.1016/j.neuroimage.2020.117343
10. Ma F, Zhang L, Oz HS et al. (2015) Dysregulated TNFα promotes cytokine proteome profile increases and bilateral orofacial hypersensitivity. Neuroscience 300:493–507. https://doi.org/10.1016/j.neuroscience.2015.05.046
11. Ma J, Kavelaars A, Dougherty PM, Heijnen CJ (2018) Beyond symptomatic relief for chemotherapy-induced peripheral neuropathy: targeting the source. Cancer 124:2289–2298. https://doi.org/10.1002/cncr.31248
12. Lyons DN, Kniffin TC, Zhang LP et al. (2015) Trigeminal inflammatory compression (TIC) injury induces chronic facial pain and susceptibility to anxiety-related behaviors. Neuroscience 295:126–138. https://doi.org/10.1016/j.neuroscience.2015.03.051
13. Lyons DN, Zhang L, Pandya JD et al. (2018) Combination drug therapy of pioglitazone and D-cycloserine attenuates chronic orofacial neuropathic pain and anxiety by improving mitochondrial function following trigeminal nerve injury. Clin J Pain 34:168–177. https://doi.org/10.1097/AJP.0000000000000515
14. Yalcin I, Bohren Y, Waltisperger E et al. (2011) A time-dependent history of mood disorders in a murine model of neuropathic pain. Biol Psychiatry 70:946–953. https://doi.org/10.1016/j.biopsych.2011.07.017
15. Yalcin I, Barthas F, Barrot M (2014) Emotional consequences of neuropathic pain: insight from preclinical studies. Neurosci Biobehav Rev 47:154–164. https://doi.org/10.1016/j.neubiorev.2014.08.002
16. Dutta S, Sengupta P (2016) Men and mice: relating their ages. Life Sci 152:244–248. https://doi.org/10.1016/j.lfs.2015.10.025
17. Hannaman MR, Fitts DA, Doss RM et al. (2016) The refined biomimetic NeuroDigm GEL™ model of neuropathic pain in a mature rat. F1000Res 5:2516. https://doi.org/10.12688/f1000research.9544.2
18. Adams H, Pendleton C, Latimer K et al. (2011) Harvey Cushing's case series of trigeminal neuralgia at the Johns Hopkins hospital: a surgeon's quest to advance the treatment of the "suicide disease". Acta Neurochir (Wien) 153:1043–1050. https://doi.org/10.1007/s00701-011-0975-8
19. Montera MA, Westlund KN (2020) Minimally invasive oral surgery induction of the FRICT-ION chronic neuropathic pain model. Bio Protoc 10:e3591. https://doi.org/10.21769/BioProtoc.3591
20. Ticca M (1976) [Effects of reversing the feeding cycle and the light period on the spontaneous activity of the rat (author's transl)]. S TA NU 6:153–160
21. Ma F, Zhang L, Lyons D, Westlund KN (2012) Orofacial neuropathic pain mouse model induced by trigeminal inflammatory compression (TIC) of the infraorbital nerve. Mol Brain 5: 44. https://doi.org/10.1186/1756-6606-5-44
22. Aloisi AM, Albonetti ME, Carli G (1994) Sex differences in the behavioural response to persistent pain in rats. Neurosci Lett 179:79–82. https://doi.org/10.1016/0304-3940(94)90939-3
23. Chaplan SR, Bach FW, Pogrel JW et al. (1994) Quantitative assessment of tactile allodynia in the rat paw. J Neurosci Methods 53:55–63. https://doi.org/10.1016/0165-0270(94)90144-9

Chapter 3

Traumatic Spinal Cord Injury as a Model for Central Neuropathic Pain

Ana Ferreira, Sara Fernandes, Sílvia Chambel, Nuno Silva and Célia Duarte Cruz

Abstract

Neuropathic pain represents an enormous challenge. It is caused by diseases or lesion to the central nervous system. The chronic nature of neuropathic pain and its resistance to conventional pain management approaches further exacerbate the challenges faced by patients. It is highly distressing for patients and produces a massive negative impact in quality of life. Neuropathic pain is a common complication of spinal cord injury, an event that dramatically changes the course of affected individuals. The mechanisms underlying neuropathic pain are still yet to be fully resolved and animal models of disease represent invaluable tools to advance our knowledge. While models of neurodegenerative disorders that also course with neuropathic pain, such as multiple sclerosis, are challenging due to the immune dysfunction associated, models of traumatic spinal cord injury serve as valuable tools for understanding neuropathic pain due to its well-defined etiology. They present an opportunity for elucidation of pathophysiological mechanisms and for development of new and better targeted therapeutic interventions. Here, we describe the most commonly used models to produce traumatic spinal cord injury and methods to assess peripheral sensitivity as a measure of neuropathic pain arising after spinal cord injury.

Keywords Spinal cord injury, Central neuropathic pain, Spinal cord transection, Spinal cord contusion, Adapted behavioral tests

1 Introduction

The International Association for the Study of Pain (IASP) defines pain as "a sensory and emotional experience associated with actual or potential tissue damage, described in terms of such damage, or similar to previous experience of actual or potential injury" [1]. Several factors contribute to different pain perception among individuals, including biological, spiritual, psychological, and social factors. For example, pain tolerance has been proven to rise with high testosterone levels, while for females, changes in estrogen have been shown to enhance pain intensity and perception [2]. Acute pain has an important protective role, but when it becomes a chronic condition

Isaura Tavares and Hugo Leite-Almeida (eds.), *Translational Research Methods for Neuropathic Pain*, Neuromethods, vol. 235, https://doi.org/10.1007/978-1-0716-5312-8_3,

and a pathology by itself, it is a distressing, maladaptive, and harmful experience for the individual, significantly impairing their quality of life [3]. Pain is typically classified as nociceptive, nociplastic, and neuropathic. The latter is a specific type of pain that arises after a lesion or disease of the somatosensory nervous system [4, 5]. Neuropathic pain is long-lasting, persisting, or recurring from more than 3 months and affecting more women and individuals above 50 years old [5]. Patients have poor quality of life, with multiple complaints associated to persistent pain, sleep disturbances, psychiatric disorders, among other complaints, requiring extensive drug prescription and healthcare support [6, 7]. Treatment is often ineffective and CNP is a significant burden to the affected individual and the society as a whole [6, 7].

Neuropathic pain can be divided into peripheral and central neuropathic pain (CNP), depending on whether it results from insults to the peripheral nervous system or if it arises within the central nervous system. CNP is particularly challenging because sometimes pain manifests much later than the original insult to the central nervous system and may involve simultaneous aberrant activity of peripheral neurons, which will further complicate the identification of operating pathophysiological mechanisms and the choice of the most effective pain-alleviating treatment for that patient. Specific causes of CNP include, among others, demyelinating diseases, neoplastic processes, vascular incidents, and spinal cord injury (SCI) [8]. Among these, CNP associated to traumatic spinal insult is one of the most well-studied, since animal models are widely available, with standardized protocols and techniques for inducing injury and assessing pain-related outcomes, which facilitates experimental reproducibility. This chapter will focus on traumatic SCI methods to study CNP.

1.1 Pain Arising after Spinal Cord Injury

The worldwide incidence of SCI has been estimated in 930 thousand new cases each year (13/100,000 inhabitants) and a prevalence of 27.04 million (368/100 K inhabitants) [9, 10]. Traumatic SCI is more frequent in males than females and the age profile of SCI patients follows a bimodal distribution, with one peak between 15 and 29 years of age, and a second peak, starting from 50 years onward [11–13]. For those age groups, traffic accidents and falls are, respectively, the most common SCI causes [12, 13].

Most SCI patients report nociceptive, neuropathic pain or a combination of both, and the combined prevalence of all types of pain is reported to be as high as 80% [14]. In what concerns specifically CNP, a recent systematic review with meta-analysis reported a prevalence of 53% [15]. CNP arising after SCI is spontaneous and patients describe burning sensations, shooting pain, and hyperalgesia [14]. It is typically divided in at-level, above-level, and below-level neuropathic pain. At-level pain usually emerges at early time points after SCI. It refers to pain felt in the dermatomes at the level of injury and includes central and peripheral components. Above-level neuropathic pain is now described as "other pain" [16], relating to injury management, such as wheelchair pulling or pain following surgery. Finally, below-level pain is typically of central origin, felt diffusely below the level of injury and appearing when chronicity has set [16, 17]. Below-level neuropathic pain is the most common form of pain in SCI patients [18]. Gabapentin, opioids, and pregabalin remain gold standard for SCI-associated pain treatment, but are often ineffective and do not prevent pain worsening [19].

1.2 The Role of the Animal Model

Animal models play a crucial role in understanding the mechanisms underlying CNP and in the development of new therapeutic approaches [20]. Because pain is subjective, encompassing several emotional components, experimental research must take into consideration that perception of noxious stimulation in animal models is not fully equivalent to the complex human experience [20]. There is, however, a growing body of studies addressing how emotions can affect nociception in animal models of disease [21–23], but translating the effects of cognitive and emotional function in experimental research remains a challenge. Still, despite limitations, animal models have allowed notable achievements, including the identification of molecular changes in the peripheral and central nervous systems, the involvement of glial cells and alterations in their interaction with neurons [24–30].

Commonly used models of SCI-induced CNP resort to the use of rodents (rat and mice), while other animals, including cats, dogs, swine, and nonhuman primates, are less frequently used [30, 31]. Indeed, rats are used in the majority of experimental studies [30]. Their use comes with several advantages, including affordable maintenance costs, ease of care, and a well-studied neuroanatomy of ascending and descending pain circuits [32]. Its bigger size, when compared to smaller rodents, allows for more complex SCI induction protocols, and the majority of behavioral tests assessing perception of potentially noxious stimuli, particularly when including conditioning, are more easily adjusted for rats [33]. Nonetheless, mice are becoming more popular and increasingly implemented in pain studies [30], as they offer the possibility of producing genetically modified models. These are helpful in studies evaluating the under- or overexpression of genes of interest, which might be relevant for SCI pathophysiology and ensuing outcomes, including CNP development [20]. However, the small size of mice might pose some technical problems for spinal injury procedures. Additionally, it has been reported that motor recovery in mice is delayed, when in comparison to rats, which should also be considered [34, 35]. The use of rodents raises the question of whether observations made in rats can be directly extrapolated to mice. A study focusing on thermal pain sensitivity, using three strains of rats and mice, demonstrated that thermal sensitivity seems to depend more on the strain and the species effect is less evident [36]. Despite the wealth of knowledge obtained in animal studies, the full translation of preclinical research to the clinical setting is not always possible and/or straightforward, mostly due to different pain pathophysiological mechanisms operating in humans and rodents [37]. Anatomical differences between rodents and humans have also been described, as those occurring in the medial *vs* lateral location of the corticospinal tracts [30].

Another aspect that needs to be considered in animal models of SCI-related CNP is the age and sex of the animals used. The majority of SCI patients are young males [11, 12]. Animal models should match the SCI clinical demographics, and most studies focusing on SCI-induced pain use male young rats [30]. However, women also suffer from SCI, and gender dysmorphism in pain processing and pathophysiology should be accounted for. For example, some studies have reported a higher functional recovery rate in female than in male rodents [38, 39]. This is also reflected on pain behaviors, as at-level allodynia was less evident in female mice than in mice after thoracic [40]. Similar findings have also been seen in rats [41] but in Humans, this matter has been less investigated. An early study suggested there were no gender differences in pain perception [42], but further studies are clearly needed to clarify this matter. The explanation for this sexual dimorphism is not evident, but it could reside in fine molecular mechanisms concerning the immune activation in response to injuries [43–45]. These differences may also reflect sex-related differences in emotional pain processing [46], which remains challenging to evaluate in animal models.

1.3 Methods to Induce Traumatic SCI Pain in the Animal Model

There is a significant debate to identify the most appropriate animal model to study SCI-induced neuropathic pain since each has distinctive characteristics that mimic the human condition [30]. In most published experimental studies, SCI is caused by trauma caused by transection (complete or incomplete), contusion, compression, and spinal cord displacement. Here, we will focus on spinal sectioning, contusion, and spinal compression as these are the most frequent SCI produced in experimental animals. In experimental studies, SCI is inflicted at low thoracic SCI segments [30]. This is mainly due to animal-welfare reasons, as cervical and high thoracic injuries can result in respiratory compromise, the mortality rate is higher, and there is a higher risk of SCI-induced immune suppression [47–50]. This is in contrast with human pathology as cervical or high thoracic segments are the most frequently affected in SCI patients [51], who report more intense neuropathic pain than SCI patients that had been affected at more caudal spinal segments [17].

1.3.1. Spinal Cord Transection

Complete transection (CT) models (Fig. 1A) are the easiest, and the most affordable reproducible approaches to induce experimental SCI

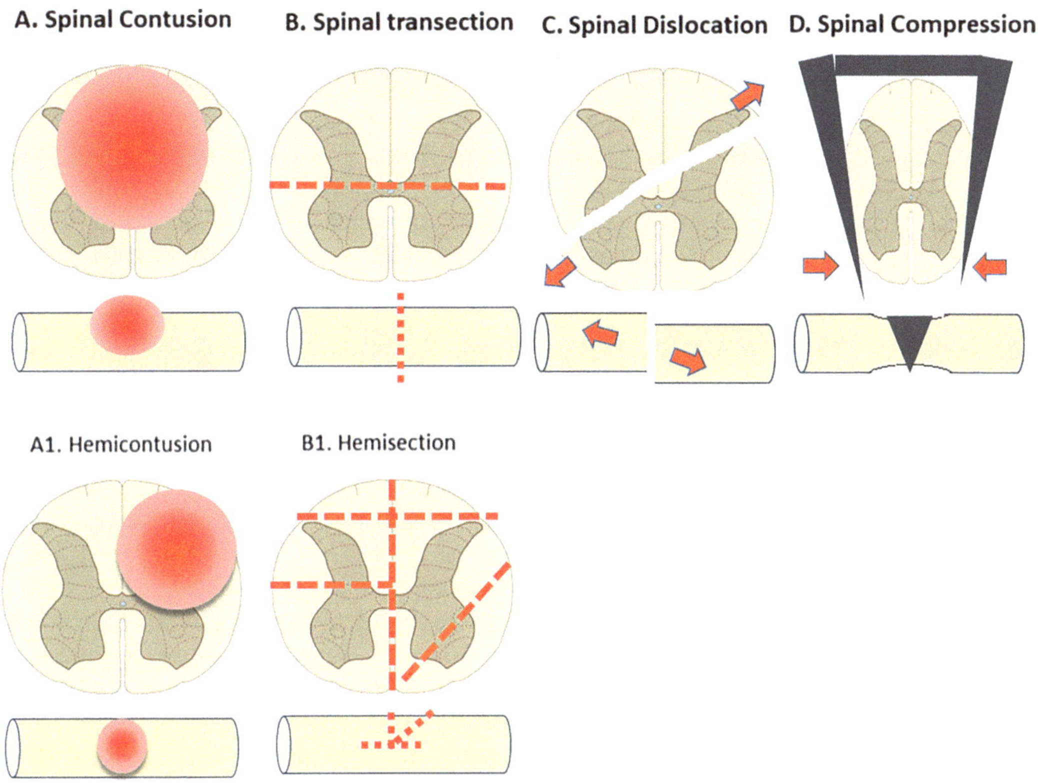

Fig. 1 Models to induce SCI-induced pain. (**A**) Spinal contusion consists in the application of a force-controlled impact onto the spinal cord, causing a blunt trauma. The injury spreads to several rostral and caudal spinal segments away from the injury site. Partial contusion models, also designated by hemicontusion (A1), allow for a more precise injury and a selective lesion of a particular area of interest and is produced using smaller probes. (**B**) In spinal Transection models, the spinal cord is sectioned with a scalpel or fine scissors, which can be complete or incomplete (hemisection; B1). In the case of hemisection, the direction of cut may vary but allows for the specific disruption of an area of interest. (**C**) Spinal displacement is characterized by a dislocation of the spinal tissue with two beams attached to the adjacent vertebrae via sliding clamp. The duration, strength, and direction of spinal displacement are previously defined. (**D**) Spinal compression makes use of vascular clips or forceps that embraces the desired spinal cord segment for the desired amount of time, causing spinal ischemia

[52]. These come as advantages when compared with other methods that require expensive tools, and a significantly higher number of animals to achieve reproducibility. CT models are particularly useful for the assessment of axonal regeneration, since there is a complete interruption between the caudal and rostral segments of the injured spinal cord. Yet, in clinical setting, the complete sectioning of the spinal cord is rare [53], making CTs less appropriate for translational investigation of complex pathophysiological mechanisms of human SCI, including those involved in the emergence and maintenance of CNP [48, 52, 54–56]. To increase the translational fitness of transection models, partial transection (hemisections, unilateral transections,

or partial transections) (Fig. 1A1) has been implemented and is becoming more popular. In the case of cervical injuries, these methods also ensure that more animals survive, given the high severity of a complete transection. Partial transections allow the lesioning of specific tracts and a direct comparison between the injured and the noninjured side [57–59]. Importantly, in partial transections motor dysfunction is typically observed in the ipsilateral side, whereas mechanical allodynia and thermal hyperalgesia are bilateral present [60–62].

1.3.2. Spinal Cord Contusion

Spinal cord contusion (Fig. 1B) is another frequently used method to induce SCI in rodents [20]. In addition to motor dysfunction, spinal contusion induces sensory dysfunction, such as tactile allodynia [63], thermal hyperalgesia [64], and CNP [41, 65] in 80% of the tested animals [66]. Contusion and compression models are the most translational methods, since human SCIs mostly occur due to blunt trauma (i.e., motor vehicle crash or sport injuries), where the spinal cord is damaged by an object or displaced bone or tissue [48, 52, 67, 68]. It should be considered, though, that, like in human pathology, spinal contusion carry a large variability in lesion extent and pathophysiological consequences, which frequently results in a need to increase the number of experimental animals [56]. This is particularly evident in hemicontusion models (Fig. 1B1), when it is possible to compromise a specific area of the spinal cord to study the consequences of such an injury.

The first experimental protocol to induce spinal cord contusion was described by Allen in 1911, and consisted in dropping a metal piece with a specific weight from a predetermined height onto the rat exposed the spinal cord [69]. More recently, sophisticated automated impactor devices have become available [70], such is the case of the MASCIS device [71, 72] or the horizon spinal impactor [34, 73]. These devices have less intertrial variability, produce a force- and dwell-timed controlled impact and allow for immediate detection of errors during the experiment [48, 56]. Nevertheless, automatic devices are costly and involve high maintenance costs, which precludes their widespread use. Thus, classical weigh drop devices derived from Allen protocol are still widely used nowadays, allowing easy manipulation of its components to induce lesions of different severities with low associated costs [31, 56].

1.3.3. Other Models

Other methods to study traumatic SCI-induced pain include spinal cord displacement and spinal compression model. In spinal cord displacement (Fig. 1C), two beams are attached to adjacent vertebrae and dislocated during a specific time with a specific velocity. The variation of these two parameters generates SCI with different degrees of severity [74], which course with mechanical and tactile allodynia [75].

For spinal cord compression, balloons, forceps, or vascular clips may be used, the latter being preferred in mice models [76]. They are placed dorsoventrally, involving all the perimeter of the cord (Fig. 1C). The devices are activated for the desired amount of time (e.g., 1 min), producing a spinal tightening, which results in spinal ischemia [77]. Spinal compression injury is proved to induce tactile allodynia and hyperalgesia [78] (See Table 1).

1.4 Methods to Access Peripheral Sensitization in Animal Models of SCI Pain

How to Assess Peripheral Sensitization in the SCI Animal Model

The use of animal models to investigate how CNP emerges after SCI involves behavioral testing to determine changes in pain perception. Even though that pain perception encompasses more than sensory mechanisms, pain is evaluated essentially by the assessment of sensory alterations in animal models of disease. Sensory tests are based evoked responses to peripheral stimuli (thermal or mechanical), that is presented to the paws, tail, or, less commonly, to the trunk [30]. Stimuli are typically presented in a graded fashion, which allows the detection of altered peripheral sensitivity as allodynia and hyperalgesia. While responses to peripheral stimulation are considered to be representative of evoked pain, most behavioral tests are not able to fully evaluate the emotional/affective components of pain. The behaviors recorded as positive responses to stimulation include licking or paw shaking, paw observation, jumping, among others [79]. Mechanical sensitivity is usually assessed with the Von Frey test and adaptations from it [78, 80–82]. Thermal sensitivity is determined by presenting hot or cold stimuli and positive responses determine the presence of thermal hyperalgesia or cold allodynia [83]. The hot plate, Hargreaves, and tail flick test are the most popular tests to study sensitivity to heat stimuli, whereas the cold stimuli are better studied with the cold plate and the acetone evaporating test [31, 84].

Challenges of Measuring Pain Responses in the SCI Animal Model

In both experimental and clinical conditions, SCI is accompanied by disruption of neuronal connections with various degrees of severity, which reflects on impaired ascending and descending neurotransmission and exacerbation of spinal reflexes, characterized by increments in the amplitude and duration of the nociceptive withdrawal response [85]. Importantly, as motor function may be impaired at early stages of disease progression, behavioral testing to evaluate the presence and intensity of pain states may not be possible until some motor function is recovered. One should not forget that behavioral testing to assess nociception, such as the Von Frey or Hargreaves tests, rely

Table 1

Summary of the characteristics of the most commonly used methods to induce SCI and produce animal models of central neuropathic pain of traumatic origin

SCI model	Description of the method	Advantages	Disadvantages	Reference protocol
Complete transection	Total section of the spinal cord using a scalpel or fine scissors after laminectomy	- Easy to perform - Affordable - High reproducibility	-Low translational value	[98–100,143]
Hemisection	Removal of 2–3 mm of the neuronal tissue of the chosen side of the hemicord	-Clinically relevant -Provides conservation of the contralateral side for comparison studies	-Low reproducibility -Technically challenging	[107–110]
Contusion by weight drop method	Dropping a metal weight from distinct heights from the exposed spinal cord after laminectomy	- Clinically relevant - Affordable - Allow lesions of different severities	-Large variability -Technically challenging	[113, 114]
Contusion by modified weight drop method	Dropping a metal weight from distinct heights from the exposed spinal cord after laminectomy	- Clinically relevant - Affordable - Allow lesions of different severities	-Technically challenging	[115, 116]
Contusion by automatic impacter	Automatic computer-controlled contusion after laminectomy	- Clinically relevant -Easy to use -Allow lesions of different severities and dwell times -Provides the exact measure of impact force and dwell time -High reproducibility	-Expensive -High maintenance costs	[118]

(continued)

Table 1 (continued)

SCI model	Description of the method	Advantages	Disadvantages	Reference protocol
Spinal compression	Spinal cord compression by a clip or forceps during a defined period after the dissection of the extradural plane between the dura mater and the adjacent vertebrae	- Easy to perform -Clinically relevant - Allow lesions of different severities	-Large variability	[120–122]

on complete plantar placement usually of the hind paws on the test surface and ability to move the hindlimbs. As SCI is accompanied by motor impairments, hind paw testing may not be possible at all time-points, the behavioral responses may be more subtle (like toe spreading) and forepaw stimulation might become an option to properly assess peripheral sensitivity [31, 84, 86]. Importantly, responses to peripheral stimulation to assess the presence and intensity of CNP may also be exacerbated by increased sensitivity to noise and vibration and autonomic dysfunction, the latter being driven by sympathetic hyperactivity [87]. In this case, animals may display anxiety-like and escape behaviors that can be wrongly interpreted as pain responses [30]. This is less evident in models of partial spinal injury such as contusion or hemisection [30]. Thus, assessment of SCI-induced CNP in animal models should consider these aspects to improve clinical translation.

Another critical consideration regards the choice of the experimental animal and the right battery of behavioral tests. In pain research, rats are often preferred due to their larger size that facilitates surgery and handling and, therefore, behavioral tests are mostly adapted for rats [84]. Most rat strains are easy to handle, which facilitates the acclimatization to behavioral apparatuses that often requires restraining. Rats tend to be cooperant and easier to habituate to the experimenter and the behavioral paradigm. Wistar and Sprague Dawley are commonly used rat strains, since they present a docile behavior and tend to move moderately inside the enclosures. However, due to the widespread use of genetic engineering and the production of transgenic animals, the use of mice has increased in the last decades

[88]. While manipulation of gene expression is a major advantage, mice are more territorial and tend to be more aggressive than rats, but they can still be used in pain research, provided adaptation of habituation protocols. As with rats, different strains exist and the most widely used include C57BL/6, BALB/c and CD1 strains [84].

2 Materials

2.1 Animals and Husbandry

Adult individuals should be preferred: rats with more than 6 months (weighting 200–300 g) and mice superior to 8 weeks (weighting 18–22 g). Smaller younger animals pose more difficulties as they risk higher weight losses after spinal lesion, which must do not exceed 20% of total body weight, what may compromise the protocol. Although SCI is more prevalent in man [89–91], the choice of the sex of the experimental is critical. If considering complete lesions, the experimenters need to ponder that during spinal shock the bladder presents little or no bladder reflex activity [92–95]. Urine needs to be removed by daily abdominal compression [96]. Due to anatomical reasons, this is more easily accomplished in female than male rodents. This is not so critical in partial lesions but should also be accounted for. Animals should be kept with ad libitum access to food and water in a temperature and humidity-controlled room, and all efforts must be made to reduce animal stress and suffering.

2.2 Anesthesia

SCI is induced under deep anesthesia. For that, a mixture of medetomidine (0.5 mg/kg for rats and 1 mg/kg for mice) and ketamine (75 mg/kg for both rodents), is administered via intraperitoneal (IP) injection. The corneal and pedal reflex should be carefully assessed, and the procedure should only be initiated when these reflexes are absent. To protect the eyes, a specific ointment should be applied to the corneas. Once the surgical procedures are completed, anesthesia is reverted with an intramuscular injection of atipamezole (1 mg/kg). Surgical spinal cord injury can also be induced under volatile anesthesia with isofluorane (5% for induction; 3% for maintenance), but, in this case, there is little analgesia after suspension of the anesthetics, and this should be preaccounted for.

2.3 Post-Surgery Care

Immediately after surgery, animals should receive 5% saline-glucose (1 mL IP) to compensate for blood loss and dehydration during procedure and placed on individual cages with hydrogel and moisturized food pellets on the bedding material to allow easy access to water and food. Post-surgery care should include saline and enrofloxacin administration, given orally to rats and via subcutaneous injection to mice (5 mg/kg), to avoid dehydration and infections, respectively. For analgesia in control animals, rats are treated with oral tramadol (0.4 mg/kg) and mice with subcutaneous buprenorphine (0.5 mg/

kg) for 3 days after surgery and whenever deemed necessary thereafter. In CNP studies, pain control in SCI animals should be discussed with the veterinary to maintain animal welfare without compromising the goals of the study. Because SCI is accompanied with little or no bladder reflex activity [95, 97], rats' bladders are manually emptied once or twice daily by abdominal compression until automatic micturition develops (approximately 14 days). In mice, because bladder emptying remains inefficient after this period [96], this procedure must be performed for longer period of times.

3 Methods

3.1 Surgical Site Preparation and Laminectomy

After confirming deep anesthesia, the first step is to shave the dorsal region to expose and disinfect the skin and prevent hair accumulation within the surgical wound. The majority of studies aims to produce a SCI at the T8/T9 segments [98–100, 101–105]. While the majority of human SCIs occur at cervical levels, low thoracic SCI in rodents remain a popular choice due to easier access to the spinal cord and animal-welfare reasons [47, 48]. During surgery, animals are placed in a heating plate to maintain body temperature at 37°C.

The skin is disinfected with povidone–iodine or *chlorhexidine* solutions. If not available, a 70% ethanol may be utilized. To expose the region of interest, after vertebrae identification by palpation, a skin incision (5 cm in the rat and 2 cm in mice) is made, and the underlying muscle and connective tissue is removed to allow for vertebrae manipulation. The use of surgical retractors is of great help during this step. After careful muscle dissection, the thoracic vertebrae are exposed. To perform a laminectomy, the vertebral spinous processes are carefully removed using micro scissors, followed by exposure of the intact spinal cord after removal of the dorsal lamina. This step must be performed carefully, to prevent any early injury to the spinal cord. The cord should be exposed at least 2–3 mm in the rat, and 1 mm in the case of mice. This space can be increased in case of contusion models. Surgical procedures are described below.

3.2 Injury Induction Methods: Complete Spinal Cord Transection

Spinal cord transection is easily performed in both rats and mice without significant need for major microscopic equipment. To perform complete transection, after exposure of the dorsal spinal cord by laminectomy (Fig. 2A), the spinal cord is fully sectioned with a scalpel or fine scissors. A small piece of sterile hemostatic sponge is placed between the retracted caudal and rostral ends of the cord to prevent excessive bleeding. Afterwards, the surgical wound is cleaned with cotton tips and closed in two layers (muscle and skin). The sutured wound and surrounding skin are then cleaned with antiseptic

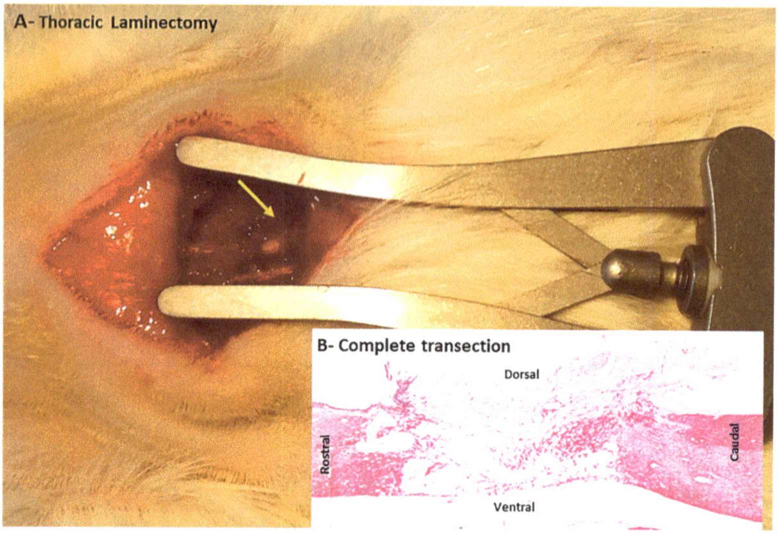

Fig. 2 Complete spinal cord transection in the rat. (**A**) Thoracic laminectomy. After removal of underlying of muscle and connective tissue, the vertebrae are identified, and a surgical retractor is inserted in the area to facilitate instruments movement. A laminectomy is performed using a fine gouge that allows a clean bone dissection. (**B**) Histological analysis by Formol-Tionin staining showing the interruption of neuronal tracts in response to a complete spinal transection. Tissue was collected 4 weeks after SCI

solution. Several types of sutures are commercially available, and silk is commonly used.

3.3 Injury Induction Methods: Incomplete Spinal Cord Transection: Hemisection

In contrast to complete transection, spinal cord hemisection preserves the neural tissue on the contralateral side of the spinal cord, which facilitates functional recovery and neuroplasticity and can be advantageous in studying CNP. To perform this partial transection, the dorsal midline of the spinal cord must be identified with a surgical microscope. A surgical needle (30 G) is then inserted dorsoventrally through the midline into the spinal cord and the needle track is used to insert one tip of sharp microsurgical scissors, while the other tip is placed along the lateral surface of one of the hemicords. A unilateral defect on the spinal cord is performed by removing approximately 2–3 mm of the neuronal tissue of the chosen side of the hemicord. To confirm the hemisection, the previously used needle should be passed through the injury several times. Finally, the lesion is closed in layers with individual sutures. To confirm the completeness of this type of lesion, the unilateral hemisection stepping (UHS) test may be applied, as a measure of the capacity of SCI animals to use their ipsilesional hindlimb in an open field environment [106]. This model is usually used to test the therapeutic approach of tissue engineering approaches [107–110], but it is also employed to examine neuropathic pain [111, 112].

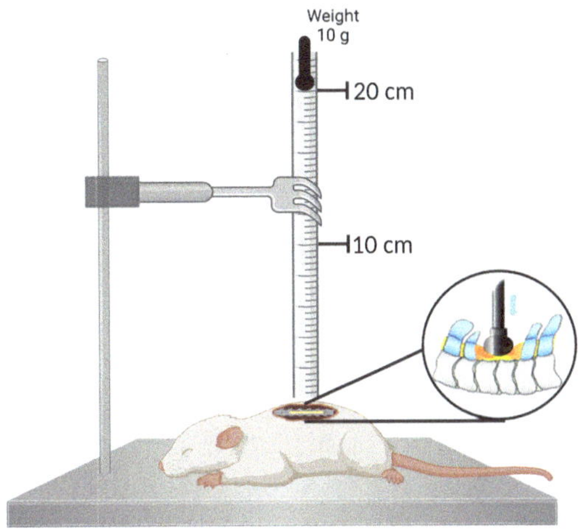

Fig. 3 Classical weight drop apparatus. The contusion method consists of a cylindrical tube with 6 mm in diameter positioned in the core of the laminectomy with the help of a flask holder, while the animal lies in a prone position. The weight consists of a steel weight (10 g, 4.5 cm long) with a 2.5 mm diameter tip. The severity of the injury is controlled by the distance the weight is dropped

3.4 Injury Induction Methods: Classical Spinal Cord Contusion by Weight Drop

Several methods are available for modeling the human contusion spinal cord injury. The weight-drop (WD) protocol, introduced by Allen in 1911 [69], is an inexpensive, consistent, and widely used model where a calibrated weight is released from predetermined heights through a guide tube, aiming to impact the surgically exposed spinal cord to induce the impact injury [113, 114]. The 20 cm cylindrical tube, usually adapted from a 10 mL plastic pipette, is attached to a manipulator's arm to allow the adjustment of the setup to the animal's position-the tube should be positioned perpendicularly to the center of the spinal cord (Fig. 3). The weight is a small metal rod weighing 10 g, with a 2.5 mm diameter tip and 4.5 cm long. Dropping the weight from specific heights will produce lesions of distinct severities. For instance, dropping the weigh from a 10 or 20 cm distance from the exposed cord, will produce a mild or severe contusion, respectively.

Before performing the method, it is important to verify if the laminectomy created enough space for the weight to reach the spinal cord and not hit the bone during impact. This would cause changes in the impact force, would affect lesion outcome, and increase variability between animals. After checking, the animal's vertebral processes are secured with the help of forceps, and the weight is positioned in the desired mark and dropped. After trauma, both the muscles and the skin are sutured, and the incision site disinfected with an antiseptic solution.

3.5 Injury Induction Methods: Improved Contusion After Adaptation from the Classical Weight Drop Method

The following approach was constructed based on previously used setups [115, 116]. The contusion device is placed on a stable rigid platform with attached forceps, which allow vertebrae stabilization during impact, as spinal trauma causes severe involuntary reflexes (Fig. 4A, B). Most materials of this device are similar to the ones of the classical method (20 cm cylindrical tube and a metal rod weighing 10 g, with a 2.5 mm diameter tip and 4.5 cm long, that can be dropped from different heights producing distinct lesion severities). However, this device has the advantage of including a braking system that allows to control the dwell time and, therefore, increasing the reproducibility of spinal contusion. The brake is a small piece of plastic 2.8 cm long, adapted with a 0.5 cm lock in the center and placed at the end of the conductive cylinder. A counterweight of 30 g is needed to retrieve the weight after the brake induction. The three components (brake, weight, and counterweight) are linked by a series of cables that are lifted in a pivot at the top of the system (Fig. 4C), so that the cables can be freely manipulated.

After laminectomy, the forceps are attached to vertebral processes (Fig. 5A), and the system is ready to be activated. Before proceeding, assessing if the laminectomy created enough space for the weight not to hit the bone during impact is also necessary. The weight is then

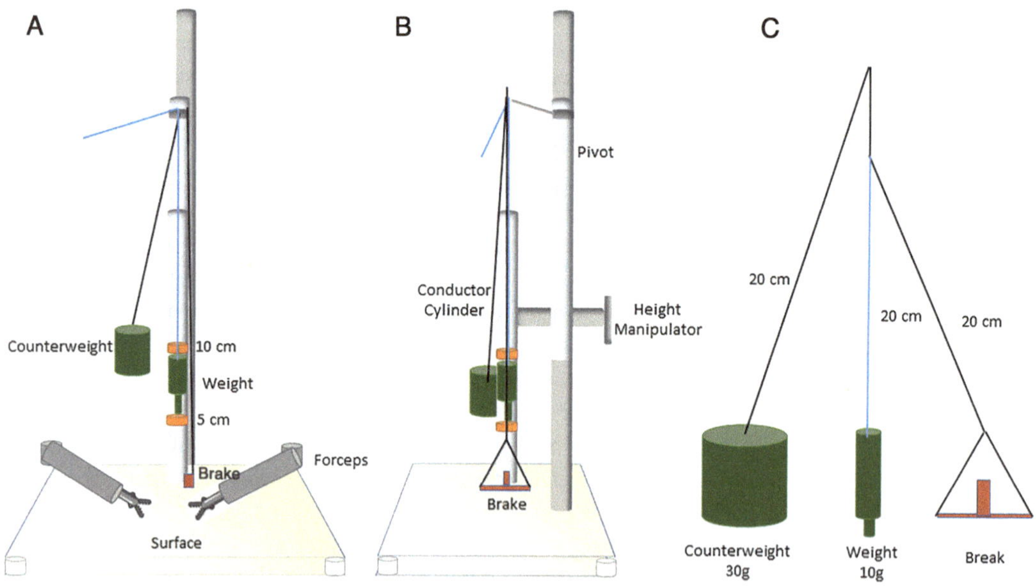

Fig. 4 Adapted contusion apparatus. (**A, B**) The contusion system consists of a rigid surface with attached forceps to stabilize the vertebral column of the animal during impact. The device is attached to a manipulator that enables adjustments of the position of the impactor tip in relation to the animal's exposed spinal cord. The weight drop device consists of a 20 cm cylindrical tube with 6 mm in diameter attached to the manipulator's arm. (**C**) The weight consists of a steel weight (10 g, 4.5 cm long) with a 2.5 mm diameter tip. The braking system is composed of a brake (a small piece of plastic 2.8 cm long, adapted with a 0.5 cm lock in the center) placed at the end of the conductor cylinder, which is linked to the impactor weight and a counterweight (30 g)

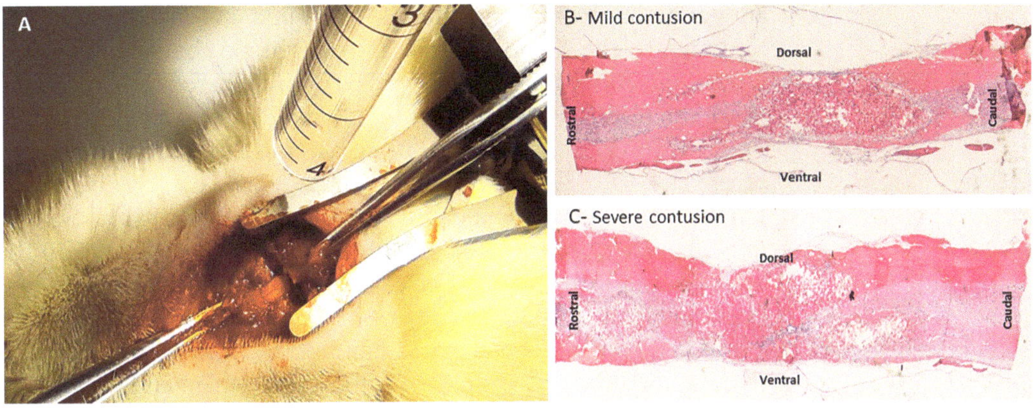

Fig. 5 Spinal cord contusion in the rat. (**A**) After laminectomy, the vertebral processes are secured with the help of forceps, and the contusion device is ready to be initiated. The spinal cord should be exposed at least 5 mm to proceed. The weight is dropped from the desired height and the contusive lesion produced. The braking system immediately retrieves the weight, making the dwell time consistent across experiments. (**B**) Histological analysis by Formol-Tionin staining showing the consequences of a mild contusion (**B**) and a severe contusion (**C**). Tissue was obtained 4 weeks after SCI

positioned at the desired height (e.g., 10 cm for severe contusion or 5 m to mild contusion, Fig. 5B, C) and released. The impactor tip touches the spinal cord and the edges of the weight release the blocking piece, which quickly results in the lift of the counterweight, retrieving the weight from the impact area. Thus, the dwell time is controlled and maintained constant between trials.

3.6 Injury Induction Methods: Automated Contusion with Impactor

The Infinite Horizon (IH) Impactor is one of the most widely used, sophisticated, and automated devices to induce a contusion spinal cord injury. It is a well-established method with standardized protocols, which facilitates experimental reproducibility and allows for the comparison of results across different laboratories. IH impactor can generate different magnitudes of injury by changing the desired force to be applied to the rodents' spinal cords, through computer software [117]. This device consists of a stepping motor that propels an impactor rack with an attached impactor tip (2.5 mm diameter for rats or 1.3 mm for mice) to the exposed spinal cord. It also includes a force sensor, to determine when the desired force has been applied, a removable base plate including stabilizing forceps and metal armatures, to stabilize the spinal column, and controls for the X, Y, and Z axis, for the correct alignment of the impactor tip over the exposed spinal cord.

The animals, which undergo a previous laminectomy, are positioned in the middle of the surgical plate (located in the surgical staging area), and the "forceps-armatures-screws" adjusted to allow the forceps to be approximately 1 mm rostral and 1 mm caudal to the laminectomy site. The spinal cord should be rigidly secured and stabilized. The animals are then carefully moved to the IH impactor and after aligning the impactor tip over the middle of the laminectomy, resorting to the X and Y axial controls, the tip should be lowered until 4 mm above the dura mater. The desired forced is then defined using the interfacing computer—100 kilo dynes (kdyne) for a mild injury, 150–175 kdyne for a moderate and 200–225 kdyne for a severe one [118]—and the impactor triggered by pressing "Start Experiment" on the program. After the injury, the impactor tip is moved away from the injury site, the animal is removed from the apparatus and the incision is cleaned with cotton tips and sutured into separate layers—muscle and skin—before being disinfected with antiseptic solution. The injury parameters, as the actual force applied to the spinal cord, the amount of tissue displacement and the velocity of the impactor tip, are directly displayed on the computer screen.

3.7 Injury Induction Methods: Spinal Cord Compression

The prevailing mechanism of SCI in humans typically involves a dual process of initial acute impact followed by prolonged compression [118]. In this model, the spinal cord is completely encircled and squeezed for the desired amount of time. This can be done with vascular clips or forceps. This method is typically applied in mice (smaller size) to avoid contusion models, which present higher probability of failure in small rodents.

Clip Compression

The first step in the clip compression method, firstly described by Rivlin and Tator in 1978 [119], is the dissection of the extradural plane between the dura mater and the adjacent vertebrae. This is typically done with a dissection hook with similar curvature and thickness as the vascular clip. The clip is then held in its fully opened applicator. Its lower blade is passed in the extradural space anterior to the cord, taking care to not damage the adjacent nerve roots and with the minimum cord displacement possible. Immediately afterwards, the clip is quickly released from the applicator and left compressing the spinal cord for the predefined time, (usually 1 min), to produce acute impact-compression. After that, the clip is opened and removed with its applicator and without any further cord manipulation. Bleeding can occur and must be controlled with a sterile hemostatic sponge to ensure a clear vision during clip placement. After injury, the muscles and the skin are sutured separately (the skin sutures may be replaced by autoclips).

Forceps Compression

Using forceps to model spinal cord compression follows the same principles and procedures as using vascular clips. The clips and applicators used in the clip compression model are expensive and manufactured to vascular specification. Using forceps, compression allows for a more reasonably priced model maintaining its reproducibility [120–122]. In this case, the spinal cord is completely compressed using fine forceps for a defined period of time—2 s to mimic a slight insult, 5 s for a moderate lesion, or 10 s to mimic a severe one (Fig.

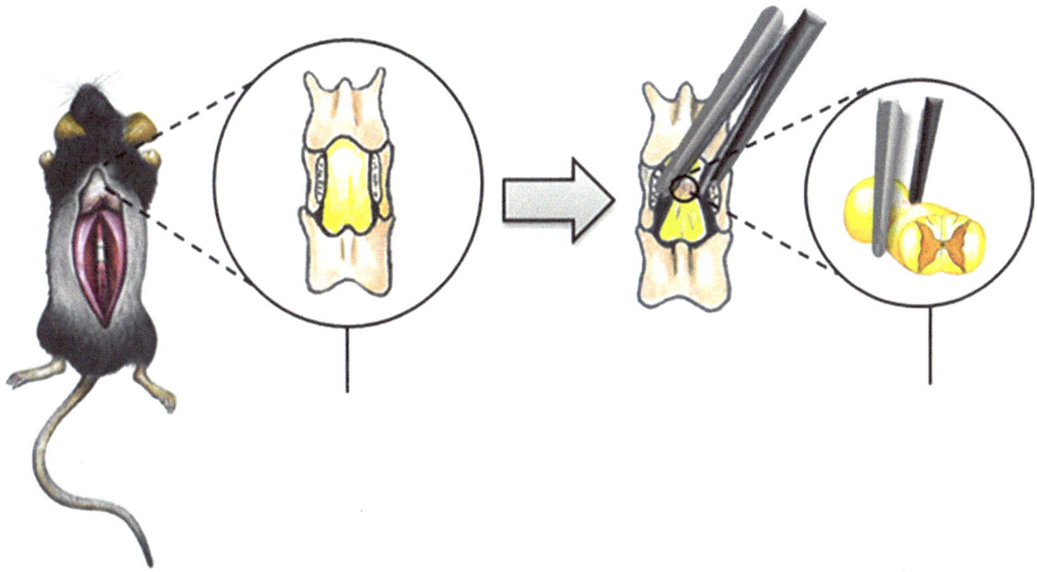

Fig. 6 Forceps compression. This method of spinal cord compression is an affordable and easy to implement alternative to the clip compression model. After a dorsal midline incision, the paravertebral muscles are retracted and the spinous and laminar arc of the vertebra column removed, exposing the spinal cord. Using the forceps Dumont #5 with the tip angled 45°, the spinal cord is completely compressed. The severity of the injury is controlled by the compression time. Lesions should be performed always with the same forceps and always by the same experimenter to reduce variability

6). It is important to make sure that the lesions are performed consistently by the same experimenter and using the same materials to reduce variability in the applied force.

3.8 Behavioral Assessment of Chronic Neuropathic Pain

The behavioral tests described below are suitable for both mice and rats, but adaptations might be required for use in smaller rodents. Moreover, in traumatic SCI animals, it is important to consider that motor function will be impaired. As responses to thermal and mechanical stimulation require a motor reaction, results should be appreciated with care, particularly in tests, in which a full plantar placement of the paw on the testing platform is necessary.

Mechanical Sensitivity: Von Frey Test

Developed in 1896 by Maximilian von Frey, this test assesses punctuating mechanical allodynia in rodents by applying filaments in specific body areas and observing the required force to elicit a withdrawal reflex to the stimulus [84]. The manual von Frey "up-down" plantar protocol stands out as the most frequently employed approach. However, the assay can also be applied to other regions of the rodent's body or conducted electronically.

The "up-down" plantar von Frey method relies on the stimulation of the plantar surface of the animal's hind paw with von Frey monofilaments with different gauges. The calculated mechanical

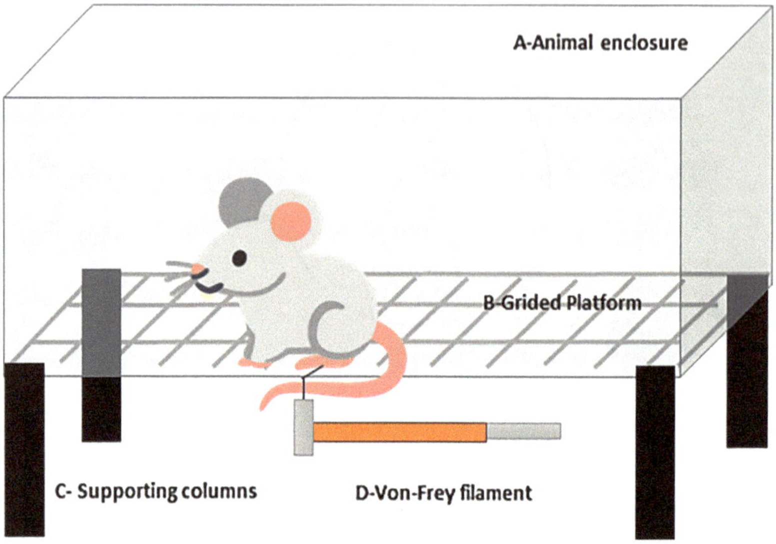

Fig. 7 "Up and down" von Frey. The apparatus includes (**A**) an animal enclosure of glass or plastic to permit animal clear observation; and (**B**) a gridded platform in which the animal is positioned. (**C**) The system must be supported with enough height to permit access to the plantar surface of the limb. (**D**) von Frey filaments. The mechanical force required to produce a hind paw withdrawal response is inversely proportional to the mechanical allodynia severity

force required to produce a hind paw withdrawal response on at least 50% of its applications (50% response threshold), serves as an indication of the tactile sensitivity to mechanical stimuli [84, 123]. The required apparatus to conduct the "up-down" plantar von Frey test includes an individual clear observation cage (Fig. 7A) positioned on top of a penetrable floor (e.g., a wire grid) (Fig. 7B) that allows the access to the plantar surface of the subject's limbs, and a set of monofilaments with different lengths and diameters that allows the application of specific reproducible forces (Fig. 7C). The von Frey setup should be suspended above eye level to allow the handling of the fibers around all sides of the platform. This test is amenable to being performed in both rats and mice. The size of the restrainer and the range of monofilaments to be applied should be adjusted depending on whether we are working with rats or mice.

To ensure greater data accuracy, all the animals should be tested under similar baseline levels of comfort and neglecting other sensory inputs such as stimulus visualization [123]. Therefore, prior to conducting any trial, it is crucial to achieve environmental habituation by acclimatizing the animals to the experimental room and the testing area. For that purpose, the animals should be placed in the testing chamber and offered some food reward for 10 min for at least four habituation sessions, or until presenting acclimatized behavior. The process also involves acclimatization to hand movements beneath the animal's paw (during the third session) and to the monofilament application to the paw, by ten presentations of a subthreshold monofilament in the fourth session.

On test day, before placing the first animal into the von Frey observation setup, and between subjects, the mesh and the testing arenas are sanitized to remove any remaining animal odors, urine, or feces. Still prior to the trials, the animals undergo an acclimatization period of 15–30 min [124, 125] to enable the cessation of cage exploration and grooming activities and are checked for weight bearing and proper plantar placement. To prevent responses based on the visual input of the monofilaments poking the skin, evaluators should check if the animals are eating the provided rewards before applying the filament. The von Frey hair is applied perpendicularly to the midplantar surface of the hind paw (Fig. 7) with increasing pressure until it bends for about 1–2 s. The testing sessions begins by testing the response to a monofilament expected to be near the 50% response threshold—15 g for rats and 0,4 g for mice [123, 126]. If there is no response, the next filament with a higher force is tested; if there is a response, the next lower force filament is tested. It is important to always wait around 30 s before retesting, allowing sensory receptors to return to baseline firing. The test must continue with the sequential appliance of higher or lower filaments based on the animal's response, until at least four trials after the first shift of direction are recorded and the sequence of results obtained. During all the trials, the experimenter should carefully monitor animal behavior to void and reperform any in which the animal retreats from the stimulus prior to the filament bending or in which the fiber directly lifts the paw (to avoid responses due to proprioceptive input).

For optimal calculation of the 50% withdrawal threshold, at least six responses in the propinquity of the estimated threshold are needed (the two responses contiguous to the threshold and the four subsequent trials) [127]. The 50% threshold is then calculated using the formula: $50\%\text{threshold}(g) = \frac{10^{(X+kd)}}{10^4}$, where X = logarithmic value of the last used filament, k = tabular value for the pattern of responses (see Appendix 1 in 124), and d = mean difference (in log units) between von Frey filaments [124, 127]. The lower the tactile threshold, the higher the mechanical hypersensitivity on the hind paws.

The "up-down" plantar von Frey test exhibits significant sensitivity and specificity, particularly when applied chronically across various SCI severities [123, 126]. This assay enables to determine tactile thresholds in unrestrained animals, eliminating the risk of handling-induced stress [84]. However, this test cannot be implemented acutely after SCI or when severe motor impairments exist since it depends on a proper weight support to avoid the monofilaments to move the limbs before the hair bends [123, 126]. Another drawback of this method is the variability in the number of measurements per animal, as well as the necessity for repeated, time-intensive measurements. This repetitive process may potentially induce sensitization or learned responses [84]. Finally, in some cases, the need for the first applied filament to be close to the expected threshold may be a problem since this value may be divergent from what we are expecting [128].

Mechanical Sensitivity: Dorsal Von Frey Test

The dorsal von Frey test is an ascending test that constitutes an alternative to determine tactile sensory thresholds during the acute period of SCI. Data from testing sensation on the dorsal surface of the hind paw can be collected in animals without weight support or trunk control, that cannot be accurately tested resorting to the typical "up-down" plantar method (122, 125).

Contrary to above, this method does not require a specific apparatus beyond the set of von Frey monofilaments. Therefore, the animals need to be well accustomed to handling and to the testing environment to avoid stress during the trials. For that purpose, the animals should be gentled for at least four sessions of 10 min prior to the test, having their head and forequarters wrapped loosely in a clean towel and the hind paws lightly rested on a table, in a way that prevents them to visualize the stimulus application (Fig. 8), while provides stability and a soothing effect. By the end of the habituation sessions, the animals should not resist to be held in the towel for several minutes. This test has only been described in rats, likely because mice are less prone to become docile enough for this type of handling.

On test day, while holding the animal as described above and after confirming it is calm for at least 10 s, the tactile stimuli can be applied. Starting with the 6.0 g, each filament is applied three times (with a 30 s interval) in an ascending order perpendicularly to the

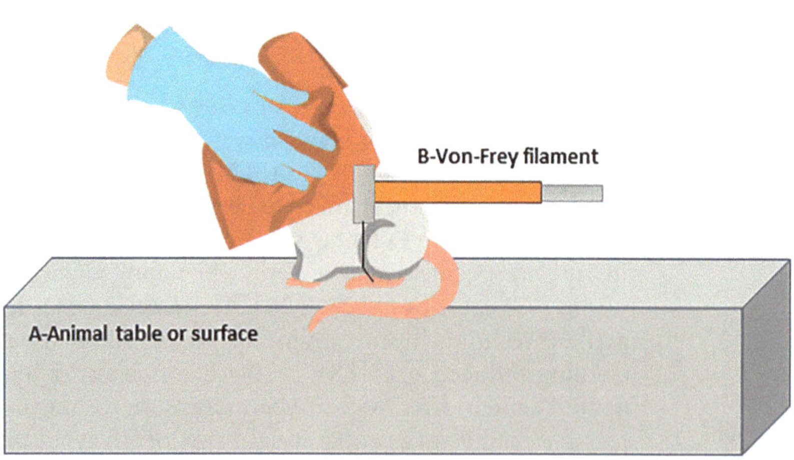

Fig. 8 Dorsal von Frey. The animal is placed in an (**A**) table or surface and restrained using a clean towel. (**B**) Von Frey filaments are applied to the desired limb. The mechanical force required to produce a hind paw withdrawal response is inversely proportional to the mechanical allodynia severity

dorsal surface of the hind paw in between the first and second metatarsal, approximately 1 cm proximal to the joint (Fig. 8). A positive response is recorded when, at the time the monofilament is removed, the animal quickly moves its paw posteriorly or externally and flexes the hind limb. While recording the number of hind paw withdrawals for each filament, the application of the next stiffer filament continues unless a positive response occurred in at least two of the trials or until the largest filament of a series of six is reached. The mechanical sensory threshold is defined as the lowest filament that induces at least two paw withdrawal responses out of three trials.

Mechanical Sensitivity: Electronic Von Frey Test

Evaluation of the sensory response can also be performed resorting to an electronic von Frey system. The electronic protocol applies the same underlying principles and procedures as the manual von Frey methods with the distinction that the aesthesiometer includes a single unbending filament that is progressively applied until a paw withdrawal is evoked, automatically recording the mechanical threshold (the exact force applied at the time the withdrawal response). After some trials, the highest and lowest values are discarded, and the middle values are averaged to obtain the final score.

The electronic system requires a reduced number of trials (typically 3–4) to determine the threshold, which avoids repetitive noxious stimuli and saves time, and generates continuous and linear data ([84, 129] Deuis et al., 2014, 2015). However, in addition to being more expensive, this technique may not be as sensitive as the monofilaments when the animals are responding in the hypersensitive range.

Thermal Sensitivity: Hot Plate Test

The hot plate test, first reported in 1944 by Woolfe and Macdonald [129], is one of the most widely employed methods to assess acute pain sensitivity to a heat stimulus in both rats and mice. The latency to respond to a heat stimulus is determined by placing an animal in an enclosed hot plate and measuring the time it takes the rodent to withdraw or lick its hind paw.

The hot plate apparatus is composed of: (1) a metal temperature plate as a surface to place the animal (Fig. 9A); (2) an adjustable temperature control (Fig. 9B); (3) a lidded transparent cylindrical restrainer (made from glass or acrylic) to limit the animal to the apparatus while allowing its clear observation (Fig. 9C); and (4) a stopwatch or a timer (if the equipment is not equipped with one). The restrainer measurements should be adjusted depending on whether we are working with rats or mice. Although the equipment is not excessively sophisticated, and as in any other behavioral test, to avoid the animals to present signs of anxiety or stress on test day, it is crucial to habituate them to the environmental factors, related both to the experimental room and the setup, until they show acclimatized behavior. Providing food rewards may also positively influence the habituation process.

Fig. 9 Hot plate. The apparatus includes (**A**) A metal temperature plate as a surface to place the animal, (**B**) A controller with adjustable temperature control and timer, and (**C**) an acrylic-lidded transparent cylindrical restrainer. The time that the animal takes to rapidly flick the tail is indicative of the heat sensitivity score

On test day, the animals are transported to the experimental room and let undisturbed for at least 15 min prior to the test. During this period, the metal plate and the cylindric cover should be cleaned with 70% ethanol (this needs to be repeated between subjects), and the hot plate can be switched on and set for the chosen temperature—usually 52.5°C for rats and 55°C for mice [130]. After reaching that temperature, a single animal is placed on the hot plate, simultaneously with the start of the timer to measure the withdrawal latency. The timer should be stopped, and the animal removed from the apparatus immediately after it shows a nociceptive reaction to heat in its hind paws (e.g., paw licking, flicking, jumping or when the defined cutoff—usually 30 s—is reached (in case of no reaction, this value is considered as latency by default). Note that front paws are commonly used for grooming and exploration and, therefore, hind paw withdrawal or licking are more reliable indicators of nociception (125; Woolfe & Macdonald, 1944). After completing all the animal testing, the average latency time is calculated. A shorter limb withdrawal latency indicates a higher severity of heat sensitization.

The hot plate test is simple, objective, reproducible, and quantitative [131], being one of the most sensitive tests for the detection of sensorimotor deficiencies following several SCI severities [132]. In contrast to other thermal sensitivity tests, like the tail flick test, the hot plate test assesses supraspinal-organized responses to a noxious stimulus [131, 132], as animals with spinal transection do not exhibit hind limb withdrawal during the trials. Therefore, the choice of this test to determine sensitivity to heat should consider the completeness and severity of SCI. Nevertheless, it is essential to be aware when selecting the temperature and duration for heat exposure to ensure that it does not cause tissue damage, especially in these animals that are unable to withdraw their limbs.

Thermal Sensitivity: Hargreaves Test

The Hargreaves test was introduced in 1988 by Hargreaves and colleagues [133]. The protocol is based on the application of a radiant or infrared heat stimulus to the plantar surface of one hindpaw. The time that an animal takes to demonstrate a nociceptive response (limb withdrawal accompanied by checking or licking) is inversely proportional to the heat sensitization severity. The adjustment of heat source intensity allows the identification of thermal allodynia or hypoalgesia.

The complete Hargreaves apparatus should include (1) a controller screen to control the intensity of stimulation, the input port, and display the reaction time (Fig. 10A); (2) an emitter/detector vessel that includes the beam controller and the movement sensor that detects the animal's reaction (Fig. 10B); (3) a large polyethene black platform (Fig. 3C); (4) a frame glass surface (Fig. 5C) and (5) individual spaces for each animal, the size of which depends on the rodent chosen for experiments (Fig. 10E). The apparatus is supported by four columns. The Hargreaves test can be performed in both

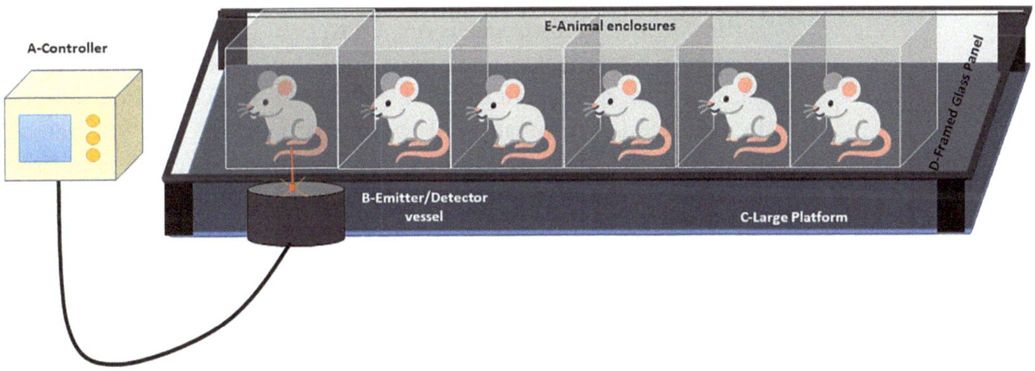

Fig. 10 Hargreaves apparatus. The system includes (**A**) a Controller unit that controls beam intensity and displays the animal reaction time; (**B**) an Emitter/detector vessel with an attached heat source and movement sensor; (**C**) Glass polyethene black surface and supporting columns; (**D**) Glass surface that supports the (**E**) animal enclosures. The time that the animal delays to demonstrate a nociceptive response is inversely proportional to the heat sensitization score

rats and mice. The species and strain can affect the test result, as full plantar placement varies across strains and species [134]. Usually, the apparatuses are designed to test six rats or ten mice simultaneously.

All procedures must be done in a quiet room, suitable for behavioral assessment. Before starting any trial, testing chambers and glass panel must be cleaned (H_2O or 70% ethanol) to eliminate any animal odor, urine, and feces from previous experiments. Before testing, all animals should be habituated to the experimenter and the setup to avoid stress-induced responses. During habituation, all electronic equipment should be turned on, to acclimatize the animals to the noise. Habituation sessions must be repeated on different days, with increasing times of 30–60 min. During this process, animals are rewarded with treats (usually corn). Habituation should also include handling by the experimenter close to the apparatus and to the moving of the emitter/detector vessel under the animals. When the habituation period is complete, baseline values should be recorded prior to any procedures, including surgery.

Experimental trials are initiated by a small acclimatization period of 10–15 min. In mice, this period should be increased to 30 min. During this period, the conditions should be set to beam intensity (50 units) and maximum reaction time (18 s-cutoff time to prevent tissue damage). These parameters may require optimization according to the specie and strain used, as well as the model used to induce SCI. The vessel is positioned underneath the center of the hind paw being tested (the vessel exhibits guiding lines that facilitate the process). Before trial initiation, proper plantar placement must be checked. To start the trial, the vessel is initiated by the experimenter and the time will automatically start to count. When the animal retrieves the paw, the system detects its movement and records the time spent since initiation (latency time), which is registered by the experimenter. Even though the process is automatic, it is important that the experimenter closely observes animal behavior to confirm if the withdrawal response is related to pain unset (normally is coupled with paw checking or licking) or if it is a voluntary movement, for instance, to reach a treat. If the latter is the case, the trial must be excluded. Importantly, if the motor responses in SCI remain impaired, the front paw might be considered for testing and other behavioral responses recorded (vocalization, moving the whiskers, etc.). The test must include five trials per animal, with at least a 5-min interval before repetition, to avoid nociceptor sensitization. The values obtained are typically presented as the average reaction of each animal in a group, and the lowest and higher values are not considered to determine the average. The shorter the withdrawal latency, the highest the animal's level of pain.

An advantage of the Hargreaves test is that it allows the separate testing of each paw, allowing the comparison between the ipsi- and contralateral heat thresholds, which might be useful in of unilateral

SCI to evaluate one-sided CNP and confirm if it spreads to the contralateral side [31, 84]. Furthermore and important in SCI animals, submitted to a severe procedure, this test for cutaneous heat sensitivity is less harsh and the classical hot plate test [84]. Importantly, investigators need to take into consideration that the Hargreaves test requires that the animal paw is fully placed on the apparatus's surface. Since SCI animals may have difficulties to perform this, the test needs to be delayed until the animals recover and become capable of correctly placing hind paws on the test surface. This means that early changes in thermal sensitivity after SCI may not be possible to evaluate. Another limitation is that the paw withdrawal time is recorded, rather than measuring the exact temperature that produces a response [84]. While this can be inferred, a thermal probe applied to the animal skin would produce a more accurate measure. To overcome this limitation, a modified version of the Hargreaves has already been proposed already perform this [135], but, to our knowledge, it is not commercially available and the testing period is significantly longer, which could increase stress levels in animals being tested.

Thermal Sensitivity: Tail Flick Test

The tail flick test evaluates the acute response to heat stimuli, using a high-intensity radiant light applied to a rodent tail. The time that the animal takes to demonstrate a withdrawal behavior (rapid flick of the tail) is recorded and indicative of heat sensitivity. It was first described by D'Amour and Smith in 1941 [136] and is useful, e.g., to test the analgesic effect of pharmacological drugs, by testing the withdrawal response before and after drug administration [84]. This heat sensitivity test can be used in rats and mice, eventually adjusting the beam to the size of the animal's tail.

The tail-flick system includes: (1) a controller screen to adjust heat intensity and an automatic timer (Fig. 11A); (2) a metal surface

Fig. 11 Tail-flick apparatus. The system includes (**A**) a Controller unit that controls light intensity and displays the animal reaction time; (**B**) a Metal platform in which the animal performs the test; (**C**) a Heat source. The time that the animal takes to rapidly flick the tail is indicative of the heat sensitivity score

in which the animal is placed (Fig. 11B); and (3) a coupled heat (Fig. 11C). As with other behavioral tests, animals should be habituated and acclimatized to the room and the setup for 30 min for at least 5 days, or until the animal no longer displays anxiety and stress. As in the Hargreaves protocol, the habituation period variable across species and strains. Factors that will affect the habituation process include the experimenter, the animal model of disease in study and the expected associated stress. Positive rewarding, e.g., with corn kernels, has a positive impact in habituation.

Once animals have been fully habituated to the experimenter and the procedure, testing can begin. Before starting the recording, animals should be placed in the room and let to acclimatize for 30 min. To record the baseline latency, the heat intensity needs to be adjusted to produce a tail-flick latency response of 3–4 s. The light source should be focused 50 mm from the tip of the tail. For mice, this value is decreased to 15 mm. To avoid tissue damage, the maximum reaction time is settled to 10 s. For the second and third stimulation, the stimulated zone should vary 10 mm in both directions from the first stimulated site. When the animal retrieves the tail, the system detects its movement and records the time spent since initiation (latency time). This is also registered by the experimenter. Once all animal testing has been concluded, the average latency time is determined. Typically, the shorter the tail withdrawal latency, the highest the animal level of pain.

The tail flick test is a relatively easy and rapid test to perform. Additionally, this paradigm does not require the full plantar support such as the Hargreaves test [133]. This might come as an advantage for SCI animals, but care should be taken when analyzing results as it might be difficult to differentiate between SCI-associated spasticity or a true positive response to tail stimulation [84]. It has been tested before in SCI rats without the need of major modifications [137].

Thermal Sensitivity: Acetone Evaporation Test

The acetone evaporation test, described for the first time in 1994 [138, 139], consists in applying a small amount of acetone to the animal paw, which evaporates and cools the skin at an estimated temperature of 15°C–21°C, considered to be innocuous in normal conditions. Positive responses to the acetone test include flicking and/or linking the stimulated hind paw and the number and duration of positive responses is recorded as a measure of cold allodynia [84]. This test can easily be applied to rats and mice.

For this test, an individual transparent acrylic chamber (Fig. 12A) placed on a wire mesh floor (Fig. 12B) is needed. The size of the chamber depends on the size of the animal being tested (rats or mice). The setup is supported by columns, allowing acetone administration with a syringe, spray, or pipette via the mesh floor (Fig. 12C). For habituation, animals should be acclimatized to the chamber for 45 min at least 3 days, or until the animal show acclimatized

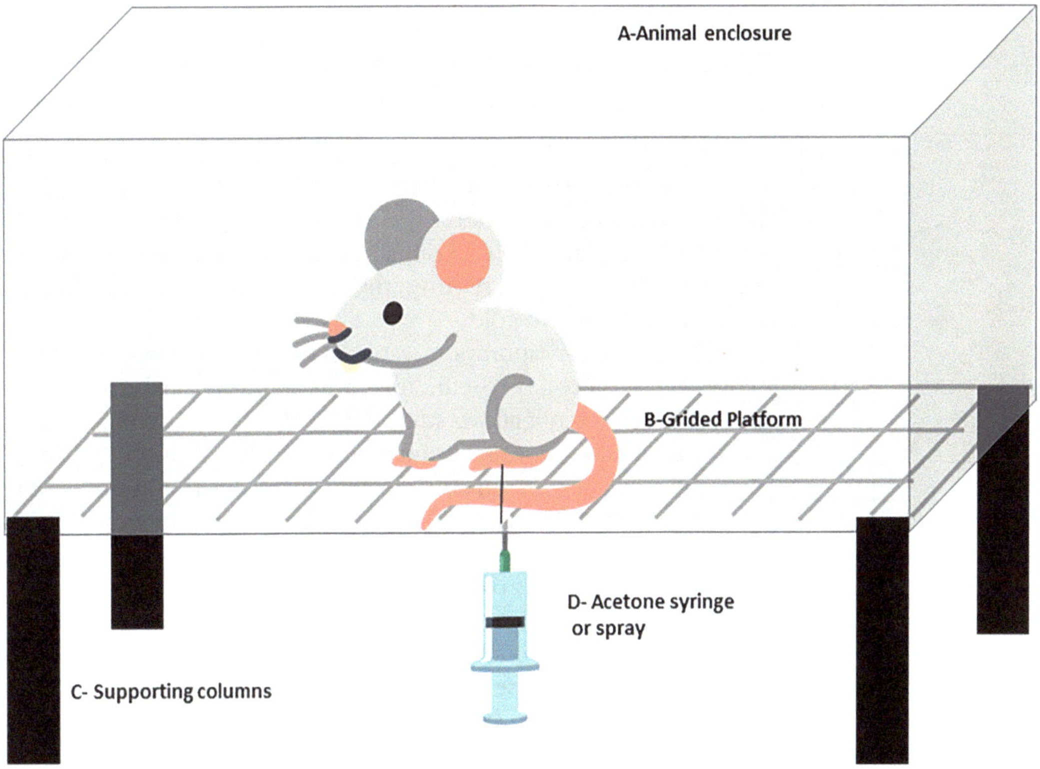

Fig. 12 Acetone-evaporating test apparatus. The acetone-evaporating apparatus consists in (**A**) a transparent acrylic chamber, placed on a (**B**) wire mesh floor; (**C**) Supporting columns that support the apparatus higher enough to provide space for the movement of the acetone syringe or spray. (**D**) The acetone is applied to the animal limb through the wire mesh floor. The amount of time that an animal takes to demonstrate a pain behavior is inversely proportional to the cold sensitization score

behavior. During this process, animals are rewarded with treats (usually corn). Habituation also includes acclimatization for the presence of the experimenter near the apparatus, and to approximation of the syringe or acetone spray. When the habituation period is complete, baseline values should be recorded.

On test day, animals should be placed in the chamber 45 min before the stimulation. An initial spray of water heated to 30°C is used as control [138, 139]. The test consists in the application of approximately 100 μL of acetone to the plantar surface of the rat's hind paw, from a distance of 2 cm. The animals are monitored for a response for 0-30 s following this application. The test is repeated 5 times with at least a 5-min intertrial interval. The outcome is measured as the total number of hind paw withdrawals per trial, or, alternatively, by scoring the reaction of the animal as following [140].:

- no reaction (score 0)
- startle response without paw withdrawal (score 1)

- brief withdrawal of the paw (score 2)
- prolonged withdrawal of 5–30 s (score 3)
- prolonged and repetitive withdrawal of 30 s with flinching and/or licking (score 4)

For each animal, the average latency time or the score observed by the experimenter is determined. The shorter the latency time, the highest the animal level of pain, indicating the presence of cold allodynia. In the case of scoring, the value is proportionally directed to hypersensitivity to cold.

The acetone-evaporation test is an easy and inexpensive test, as it does not require any specific apparatus. An additional advantage of this protocol is that it allows stimulation, enabling the comparison to the contralateral side. Furthermore, it is considered a more ethical test than the cold plate, as animals can actively escape the stimulus [84].

There are, however, some drawbacks. In spinal intact animals, the response to acetone application might be too quick for the experimenter to quantify in real-time and the video recording of the test may be necessary for subsequent slow-motion visualization. Moreover, the experimenter needs to consider that topical acetone application may course with simultaneous auditory (e.g., if using a spray) and olfactory (the acetone smell) stimulation, which may obscure the true cause of the animal's response [84, 141]. The application of acetone in itself is also challenging, once it has a lower surface tension than water, making it difficult to form uniform drops with syringes or pipettes [84, 141]. Finally, in SCI animals, motor response might be impaired at the time of testing, possibly making it difficult to detect cold allodynia at early stages of disease progression and/or not be able to correctly assess hypersensitivity to cold stimulation at chronic stages.

Thermal Sensitivity: Cold Hyperalgesia Test

The cold hyperalgesia test constitutes a reliable alternative to measure thermal hyperalgesia to a cold stimulus, in both rats and mice. The test shares all the cited advantages for the acetone-evaporation test (and some of the limitations). Yet, this test has the advantage of removing auditory and olfactory cues and is effortless to apply.

The protocol for the cold hyperalgesia test relies on the same procedures as the one for von Frey testing, regarding the apparatus, the acclimatization process, and most experimental steps. A stopwatch is necessary to measure the time required for a withdrawal response to be elicited. Ice probes are made by freezing water in small Eppendorf tubes (1.8 mL size) and applying a wooden applicator stick inside for a handle (Fig. 13A). With the animal calmly resting on the compartment over the grid (Fig. 13B and C), the tip of the probe is applied to the hairless skin of the hind paws for up to 10 s (defined cutoff) or until a withdrawal response is observed, always recording the latency times. A full cold hyperalgesia test consists in ten trials per paw, with

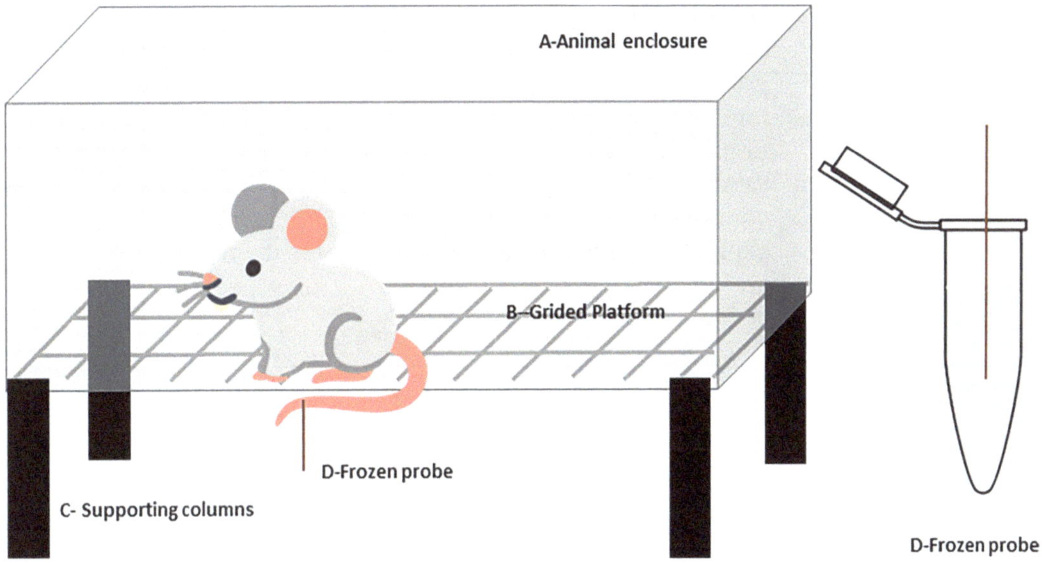

Fig. 13 Setup for detection of cold hyperalgesia. This apparatus consists in (**A**) a wooden handler inside an Eppendorf tube with frozen water as a probe, and (**B**) a transparent chamber, placed on a (**C**) wire mesh floor. The ice probe is applied to the animal limb through the wire mesh floor. The amount of time that an animal takes to demonstrate a pain behavior is inversely proportional to the cold sensitization score

30 s between them to allow the paw to return to baseline temperature prior to reapplying the ice. The results should be reported as latency time ± average reaction time (determined by measuring the elapsed time between switching the stopwatch on and off). It is inversely proportional to cold hypersensitivity. To guarantee that the recorded responses are not due to touch or wetness sensations, but actually to the thermal stimulation, an empty tube attached to a handle and a water-soaked cotton swab can be applied similar to the ice probe and the latency times compared [142].

4 Conclusions

Animal models of traumatic spinal cord injury play a crucial role in studying neuropathic pain of central origin. These animal models provide researchers with a controlled and reproducible system to investigate the underlying mechanisms and therapeutic interventions. Rats are more commonly used, but mice models are also gaining popularity, especially for genetically modified models. It is important to point out that when choosing the model, age and sex should be considered align with SCI clinical characteristics. Sexual dimorphism in pain processing and pathophysiology has been observed, emphasizing the importance of studying both males and females. Overall, animal models of SCI-induced neuropathic pain have significantly contributed to our understanding of the molecular changes in the peripheral and central nervous system. They have helped identify potential therapeutic targets and evaluate the efficacy of different interventions. However, it is important to acknowledge the limitations of animal models and the need for careful interpretation and validation of findings in human clinical trials. Future research should continue to refine animal models, consider gender differences, and explore novel therapeutic approaches for the management of pain arising after SCI.

Acknowledgments

Ana Ferreira is supported by a PhD fellowship from i3S and FCT-Fundação para a Ciência Tecnologia (UI/BD/151547/2021). Sílvia Chambel is supported by an PhD fellowship by FCT-Fundação para a Ciência Tecnologia (SFRH/BD/135868/2018). Nuno Silva is supported by the Scientific Employment Stimulus Program from FCT (CEECIND/04794/2017). We would like to thank José Graça for producing Fig. 6. We would like also to acknowledge FCT for support through projects UIDB/50026/2020 and UIDP/50026/2020.

Conflict of Interest The authors declare they have no conflict of interest to declare.

References

1. Raja SN, Carr DB, Cohen M, Finnerup NB, Flor H, Gibson S et al (2020) The revised international association for the study of pain definition of pain: concepts, challenges, and compromises. Pain 161(9):1976–1982. https://doi.org/10.1097/j.pain.0000000000001939
2. Athnaiel O, Cantillo S, Paredes S, Knezevic NN (2023) The role of sex hormones in pain-related conditions. Int J Mol Sci 24(3):1866. https://doi.org/10.3390/ijms24031866
3. St John Smith E (2018) Advances in understanding nociception and neuropathic pain. J Neurol 265(2):231–238. https://doi.org/10.1007/s00415-017-8641-6
4. Jensen TS, Baron R, Haanpaa M, Kalso E, Loeser JD, Rice ASC et al (2011) A new definition of neuropathic pain. Pain 152(10):2204–2205. https://doi.org/10.1016/j.pain.2011.06.017
5. Colloca L, Ludman T, Bouhassira D, Baron R, Dickenson AH, Yarnitsky D et al (2017) Neuropathic pain. Nat Rev Dis Primers 3:17002. https://doi.org/10.1038/nrdp.2017.2
6. O'Connor AB (2009) Neuropathic pain: quality-of-life impact, costs and cost effectiveness of therapy. Pharmacoeconomics 27(2):95–112. https://doi.org/10.2165/00019053-200927020-00002
7. O'Brien T, Breivik H (2012) The impact of chronic pain-European patients' perspective over 12 months. Scand J Pain 3(1):23–29. https://doi.org/10.1016/j.sjpain.2011.11.004
8. Watson JC, Sandroni P (2016) Central neuropathic pain syndromes. Mayo Clin Proc 91(3):372–385. https://doi.org/10.1016/j.mayocp.2016.01.017
9. James SL, Theadom A, Ellenbogen RG, Bannick MS, Montjoy-Venning W, Lucchesi LR et al (2019) Global, regional, and national burden of traumatic brain injury and spinal cord injury, 1990–2016: a systematic analysis for the Global Burden of Disease Study 2016. Lancet Neurol 18(1):56–87
10. Singh A, Tetreault L, Kalsi-Ryan S, Nouri A, Fehlings MG (2014) Global prevalence and incidence of traumatic spinal cord injury. Clin Epidemiol 6:309–331. https://doi.org/10.2147/CLEP.S68889

11. Chen Y, He Y, DeVivo MJ (2016) Changing demographics and injury profile of new traumatic spinal cord injuries in the United States, 1972–2014. Arch Phys Med Rehabil 97(10):1610–1619. https://doi.org/10.1016/j.apmr.2016.03.017
12. Devivo MJ (2012) Epidemiology of traumatic spinal cord injury: trends and future implications. Spinal Cord 50(5):365–372. https://doi.org/10.1038/sc.2011.178
13. DeVivo MJ, Chen Y (2011) Trends in new injuries, prevalent cases, and aging with spinal cord injury. Arch Phys Med Rehabil 92(3):332–338. https://doi.org/10.1016/j.apmr.2010.08.031
14. Finnerup NB, Norrbrink C, Trok K, Piehl F, Johannesen IL, Sorensen JC et al (2014) Phenotypes and predictors of pain following traumatic spinal cord injury: a prospective study. J Pain 15(1):40–48. https://doi.org/10.1016/j.jpain.2013.09.008
15. Burke D, Fullen BM, Stokes D, Lennon O (2017) Neuropathic pain prevalence following spinal cord injury: a systematic review and meta-analysis. Eur J Pain 21(1):29–44. https://doi.org/10.1002/ejp.905
16. Finnerup NB (2013) Pain in patients with spinal cord injury. Pain 154(Suppl 1):S71–S76. https://doi.org/10.1016/j.pain.2012.12.007
17. Siddall PJ, McClelland JM, Rutkowski SB, Cousins MJ (2003) A longitudinal study of the prevalence and characteristics of pain in the first 5 years following spinal cord injury. Pain 103(3):249–257
18. Bryce TN, Biering-Sørensen F, Finnerup NB, Cardenas DD, Defrin R, Lundeberg T et al (2012) International spinal cord injury pain classification: part I. Background and description. March 6–7, 2009. Spinal Cord 50(6):413–417. https://doi.org/10.1038/sc.2011.156
19. Widerström-Noga E (2017) Neuropathic pain and spinal cord injury: phenotypes and pharmacological management. Drugs 77(9):967–984. https://doi.org/10.1007/s40265-017-0747-8
20. Nakae A, Nakai K, Yano K, Hosokawa K, Shibata M, Mashimo T (2011) The animal model of spinal cord injury as an experimental pain model. J Biomed Biotechnol 2011:939023
21. Alshami AM (2019) Pain: is it all in the brain or the heart?. Curr Pain Headache Rep 23:1–4
22. Bushnell MC, Čeko M, Low LA (2013) Cognitive and emotional control of pain and its disruption in chronic pain. Nat Rev Neurosci 14(7):502–511
23. Malfliet A, Coppieters I, Van Wilgen P, Kregel J, De Pauw R, Dolphens M et al (2017) Brain changes associated with cognitive and emotional factors in chronic pain: a systematic review. Eur J Pain 21(5):769–786
24. Coull JA, Beggs S, Boudreau D, Boivin D, Tsuda M, Inoue K et al (2005) BDNF from microglia causes the shift in neuronal anion gradient underlying neuropathic pain. Nature 438(7070):1017–1021
25. Detloff MR, Fisher LC, McGaughy V, Longbrake EE, Popovich PG, Basso DM (2008) Remote activation of microglia and pro-inflammatory cytokines predict the onset and severity of below-level neuropathic pain after spinal cord injury in rats. Exp Neurol 212(2):337–347
26. Gwak YS, Hulsebosch C (2009) Remote astrocytic and microglial activation modulates neuronal hyperexcitability and below-level neuropathic pain after spinal injury in rat. Neuroscience 161(3):895–903
27. Gwak YS, Kang J, Unabia GC, Hulsebosch CE (2012) Spatial and temporal activation of spinal glial cells: role of gliopathy in central neuropathic pain following spinal cord injury in rats. Exp Neurol 234(2):362–372
28. Kato K, Koda M, Takahashi H, Sakuma T, Inada T, Kamiya K et al (2015) Granulocyte colony-stimulating factor attenuates spinal cord injury-induced mechanical allodynia in adult rats. J Neurol Sci 355(1–2):79–83
29. Watson JL, Hala TJ, Putatunda R, Sannie D, Lepore AC (2014) Persistent at-level thermal hyperalgesia and tactile allodynia accompany chronic neuronal and astrocyte activation in superficial dorsal horn following mouse cervical contusion spinal cord injury. PLOS One 9(9):e109099
30. Kramer JL, Minhas NK, Jutzeler CR, Erskine EL, Liu LJ, Ramer MS (2017) Neuropathic pain following traumatic spinal cord injury: models, measurement, and mechanisms. J Neurosci Res 95(6):1295–1306
31. Silva NA, Sousa N, Reis RL, Salgado AJ (2014) From basics to clinical: a comprehensive review on spinal cord injury. Prog Neurobiol 114:25–57
32. Kwon BK, Hillyer J, Tetzlaff W (2010) Translational research in spinal cord injury: a survey of opinion from the SCI community. J Neurotrauma 27(1):21–33
33. Wilson SG, Mogil JS (2001) Measuring pain in the (knockout) mouse: big challenges in a small mammal. Behav Brain Res 125(1–2):65–73
34. Kerr BJ, David S (2007) Pain behaviors after spinal cord contusion injury in two

commonly used mouse strains. Exp Neurol 206(2):240–247
35. Larson CM, Wilcox GL, Fairbanks CA (2019) The study of pain in rats and mice. Comp Med 69(6):555–570
36. Mogil JS, Chesler E, Wilson S, Juraska J, Sternberg W (2000) Sex differences in thermal nociception and morphine antinociception in rodents depend on genotype. Neurosci Biobehav Rev 24(3):375–389
37. Mogil JS (2019) The translatability of pain across species. Philos Trans R Soc B 374(1785):20190286
38. Datto JP, Yang J, Dietrich WD, Pearse DD (2015) Does being female provide a neuroprotective advantage following spinal cord injury?. Neural Regen Res 10(10):1533
39. Swartz KR, Fee DB, Joy KM, Roberts KN, Sun S, Scheff NN et al (2007) Gender differences in spinal cord injury are not estrogen-dependent. J Neurotrauma 24(3):473–480
40. Hubscher CH, Fell JD, Gupta DS (2010) Sex and hormonal variations in the development of at-level allodynia in a rat chronic spinal cord injury model. Neurosci Lett 477(3):153–156
41. Gaudet AD, Ayala MT, Schleicher WE, Smith EJ, Bateman EM, Maier SF et al (2017) Exploring acute-to-chronic neuropathic pain in rats after contusion spinal cord injury. Exp Neurol 295:46–54
42. Cardenas DD, Bryce TN, Shem K, Richards JS, Elhefni H (2004) Gender and minority differences in the pain experience of people with spinal cord injury. Arch Phys Med Rehabil 85(11):1774–1781
43. Mapplebeck JC, Dalgarno R, Tu Y, Moriarty O, Beggs S, Kwok CH et al (2018) Microglial P2X4R-evoked pain hypersensitivity is sexually dimorphic in rats. Pain 159(9):1752–1763
44. Sorge RE, LaCroix-Fralish ML, Tuttle AH, Sotocinal SG, Austin J-S, Ritchie J et al (2011) Spinal cord toll-like receptor 4 mediates inflammatory and neuropathic hypersensitivity in male but not female mice. J Neurosci 31(43):15450–15454
45. Sorge RE, Mapplebeck JC, Rosen S, Beggs S, Taves S, Alexander JK et al (2015) Different immune cells mediate mechanical pain hypersensitivity in male and female mice. Nature Neurosci 18(8):1081–1083
46. Cañete T, Giménez-Llort L (2021) Preserved thermal pain in 3xTg-AD mice with increased sensory-discriminative pain sensitivity in females but affective-emotional dimension in males as early sex-specific AD-phenotype biomarkers. Front Aging Neurosci 13:683412
47. Lane MA, Fuller DD, White TE, Reier PJ (2008) Respiratory neuroplasticity and cervical spinal cord injury: translational perspectives. Trends Neurosci 31(10):538–547
48. Sharif-Alhoseini M, Khormali M, Rezaei M, Safdarian M, Hajighadery A, Khalatbari M et al (2017) Animal models of spinal cord injury: a systematic review. Spinal Cord 55(8):714–721
49. Jeffries MA, Tom VJ (2021) Peripheral immune dysfunction: a problem of central importance after spinal cord injury. Biology (Basel) 10(9):928. https://doi.org/10.3390/biology10090928
50. Rodgers KA, Kigerl KA, Schwab JM, Popovich PG (2022) Immune dysfunction after spinal cord injury–a review of autonomic and neuroendocrine mechanisms. Curr Opin Pharmacol 64:102230. https://doi.org/10.1016/j.coph.2022.102230
51. Sekhon LH, Fehlings MG (2001) Epidemiology, demographics, and pathophysiology of acute spinal cord injury. Spine 26(24S):S2–S12
52. Blight A (2000) Animal models of spinal cord injury. Top Spinal Cord Inj Rehabil 6(2):1–13
53. Lee BB, Cripps RA, Fitzharris M, Wing PC (2014) The global map for traumatic spinal cord injury epidemiology: update 2011, global incidence rate. Spinal Cord 52(2):110
54. Yezierski RP (2005) Spinal cord injury: a model of central neuropathic pain. Neurosignals 14(4):182–193
55. Reier PJ, Lane MA, Hall ED, Teng Y, Howland DR (2012) Translational spinal cord injury research: preclinical guidelines and challenges. Handb Clin Neurol 109:411–433
56. Verma R, Virdi JK, Singh N, Jaggi AS (2019) Animals models of spinal cord contusion injury. Korean J Pain 32(1):12
57. Fouad K, Bennett DJ, Vavrek R, Blesch A (2013) Long-term viral brain-derived neurotrophic factor delivery promotes spasticity in rats with a cervical spinal cord hemisection. Front Neurol 4:187
58. Hains BC, Johnson KM, McAdoo DJ, Eaton MJ, Hulsebosch CE (2001) Engraftment of serotonergic precursors enhances locomotor function and attenuates chronic central pain behavior following spinal hemisection injury in the rat. Exp Neurol 171(2):361–378
59. Cheriyan T, Ryan D, Weinreb J, Cheriyan J, Paul J, Lafage V et al (2014) Spinal cord injury models: a review. Spinal Cord 52(8):588–595
60. Bennett AD, Everhart AW, Hulsebosch CE (2000) Intrathecal administration of an NMDA or a non-NMDA receptor antagonist reduces mechanical but not thermal allodynia in a rodent model of chronic central pain after spinal cord injury. Brain Res 859(1):72–82
61. Hains BC, Chastain KM, Everhart AW, McAdoo DJ, Hulsebosch CE (2000)

Transplants of adrenal medullary chromaffin cells reduce forelimb and hindlimb allodynia in a rodent model of chronic central pain after spinal cord hemisection injury. Exp Neurol 164(2):426–437
62. Kim J, Jung J-I, Na HS, Hong SK, Yoon YW (2003) Effects of morphine on mechanical allodynia in a rat model of central neuropathic pain. Neuroreport 14(7):1017–1020
63. Tanabe M, Ono K, Honda M, Ono H (2009) Gabapentin and pregabalin ameliorate mechanical hypersensitivity after spinal cord injury in mice. Eur J Pharmacol 609(1–3):65–68
64. Hoschouer EL, Yin FQ, Jakeman LB (2009) L1 cell adhesion molecule is essential for the maintenance of hyperalgesia after spinal cord injury. Exp Neurol 216(1):22–34
65. Kong F, Sun K, Zhu J, Li F, Lin F, Sun X et al (2021) PD-L1 improves motor function and alleviates neuropathic pain in male mice after spinal cord injury by inhibiting MAPK pathway. Front Immunol 12:670646
66. Masri R, Keller A (2012) Chronic pain following spinal cord injury. Spinal Cord 760:74–88
67. McDonald JW, Sadowsky C (2002) Spinal-cord injury. The Lancet 359(9304):417–425
68. Ramer LM, Ramer MS, Steeves JD (2005) Setting the stage for functional repair of spinal cord injuries: a cast of thousands. Spinal Cord 43(3):134–161. https://doi.org/10.1038/sj.sc.3101715 3101715 [pii]
69. Allen A (1911) Surgery of experimental lesion of spinal cord equivalent to crush injury of fracture dislocation of spinal column: a preliminary report. JAMA LVII(11):878–880. https://doi.org/10.1001/jama.1911.04260090100008
70. Cheriyan T, Ryan DJ, Weinreb JH, Cheriyan J, Paul JC, Lafage V et al (2014) Spinal cord injury models: a review. Spinal Cord 52(8):588–595. https://doi.org/10.1038/sc.2014.91
71. Beattie M, Bresnahan J, Komon J, Tovar C, Van Meter M, Anderson D et al (1997) Endogenous repair after spinal cord contusion injuries in the rat. Exp Neurol 148(2):453–463
72. Yoon YW, Dong H, Arends JJ, Jacquin MF (2004) Mechanical and cold allodynia in a rat spinal cord contusion model. Somatosens Mot Res 21(1):25–31
73. Hulsebosch CE, Hains BC, Crown ED, Carlton SM (2009) Mechanisms of chronic central neuropathic pain after spinal cord injury. Brain Res Rev 60(1):202–213
74. Fiford R, Bilston L, Waite P, Lu J (2004) A vertebral dislocation model of spinal cord injury in rats. J Neurotrauma 21(4):451–458
75. Kloos AD, Fisher LC, Detloff MR, Hassenzahl DL, Basso DM (2005) Stepwise motor and all-or-none sensory recovery is associated with nonlinear sparing after incremental spinal cord injury in rats. Exp Neurol 191(2):251–265
76. Marques SA, Garcez VF, Del Bel EA, Martinez AMB (2009) A simple, inexpensive and easily reproducible model of spinal cord injury in mice: morphological and functional assessment. J Neurosci Methods 177(1):183–193
77. Nashmi R, Fehlings M (2001) Changes in axonal physiology and morphology after chronic compressive injury of the rat thoracic spinal cord. Neuroscience 104(1):235–251
78. Bruce JC, Oatway MA, Weaver LC (2002) Chronic pain after clip-compression injury of the rat spinal cord. Exp Neurol 178(1):33–48
79. Shiao R, Lee-Kubli CA (2018) Neuropathic pain after spinal cord injury: challenges and research perspectives. Neurotherapeutics 15(3):635–653
80. Baastrup C, Maersk-Moller CC, Nyengaard JR, Jensen TS, Finnerup NB (2010) Spinal-, brainstem-and cerebrally mediated responses at-and below-level of a spinal cord contusion in rats: evaluation of pain-like behavior. Pain 151(3):670–679
81. Fakhri S, Sabouri S, Kiani A, Farzaei MH, Rashidi K, Mohammadi-Farani A et al (2022) Intrathecal administration of naringenin improves motor dysfunction and neuropathic pain following compression spinal cord injury in rats: relevance to its antioxidant and anti-inflammatory activities. Korean J Pain 35(3):291–302
82. M'Dahoma S, Bourgoin S, Kayser V, Barthélémy S, Chevarin C, Chali F et al (2014) Spinal cord transection-induced allodynia in rats–behavioral, physiopathological and pharmacological characterization. PLOS One 9(7):e102027
83. Kim HT, Kim T, Novotny B, Khan N, Aksamit J, Siegel S, Miranpuri GS, Resnick DK (2014) Thermal hyperalgesia assessment for rats after spinal cord injury: developing a valid and useful pain index. Spine J 14(6):984–989. https://doi.org/10.1016/j.spinee.2013.09.051. Epub 2013 Oct 24. PMID: 24361359
84. Deuis JR, Dvorakova LS, Vetter I. Methods Used to Evaluate Pain Behaviors in Rodents. Front Mol Neurosci. 2017 Sep 6;10:284. https://doi.org/10.3389/fnmol.2017.00284. PMID: 28932184; PMCID: PMC5592204.
85. Manresa JAB, Finnerup NSB, Johannesen IL, Biering-Sørensen F, Jensen TS, Arendt-Nielsen L et al (2014) Central sensitization in spinal cord injured humans assessed by reflex receptive fields. Clin Neurophysiol 125(2):352–362

86. Christensen MD, Hulsebosch CE (1997) Chronic central pain after spinal cord injury. J Neurotrauma 14(8):517–537
87. Walters ET (2018) How is chronic pain related to sympathetic dysfunction and autonomic dysreflexia following spinal cord injury. Auton Neurosci 209:79–89
88. Jakeman LB, Guan Z, Wei P, Ponnappan R, Dzwonczyk R, Popovich PG, Stokes BT (2000) Traumatic spinal cord injury produced by controlled contusion in mouse. J Neurotrauma 17(4):299–319. https://doi.org/10.1089/neu.2000.17.299. PMID: 10776914
89. Ahuja CS, Nori S, Tetreault L, Wilson J, Kwon B, Harrop J et al (2017) Traumatic spinal cord injury-repair and regeneration. Neurosurgery 80(3S):S9–S22. https://doi.org/10.1093/neuros/nyw080
90. Anderson KD (2004) Targeting recovery: priorities of the spinal cord-injured population. J Neurotrauma 21(10):1371–1383. https://doi.org/10.1089/neu.2004.21.1371
91. Collinger JL, Boninger ML, Bruns TM, Curley K, Wang W, Weber DJ (2013) Functional priorities, assistive technology, and brain-computer interfaces after spinal cord injury. J Rehabil Res Dev 50(2):145–160
92. Dave S, Cho JJ (2019) Neurogenic Shock. [Updated 2024]. In: StatPearls [Internet]. StatPearls Publishing, Treasure Island (FL), 2026 Jan. Available from: https://www.ncbi.nlm.nih.gov/books/NBK459361/
93. Silver JR (2005) Spinal shock revisited: a four-phase model. Spinal Cord 43(7):450. https://doi.org/10.1038/sj.sc.3101739
94. Anderson CE, Kozomara M, Birkhauser V, Bywater M, Gross O, Kiss S et al (2023) Temporal development of unfavourable urodynamic parameters during the first year after spinal cord injury. BJU Int 131(4):503–512. https://doi.org/10.1111/bju.15918
95. Bywater M, Tornic J, Mehnert U, Kessler TM (2018) Detrusor acontractility after acute spinal cord injury-myth or reality?. J Urol 199(6):1565–1570. https://doi.org/10.1016/j.juro.2018.01.046
96. Wada N, Shimizu T, Takai S, Shimizu N, Kanai AJ, Tyagi P et al (2017) Post-injury bladder management strategy influences lower urinary tract dysfunction in the mouse model of spinal cord injury 36(5):1301–1305. https://doi.org/10.1002/nau.23120
97. Ruiz IA, Squair JW, Phillips AA, Lukac CD, Huang D, Oxciano P et al (2018) Incidence and natural progression of neurogenic shock after traumatic spinal cord injury. J Neurotrauma 35(3):461–466. https://doi.org/10.1089/neu.2016.4947
98. Chambel SS, Ferreira A, Oliveira R, Miranda R, Vale L, Reguenga C et al (2022) Development of neurogenic detrusor overactivity after thoracic spinal cord injury is accompanied by time-dependent changes in lumbosacral expression of axonal growth regulators. Int J Mol Sci 23(15):8667
99. Ferreira A, Chambel SS, Avelino A, Cruz CD (2022) Spinal cord injury causes marked tissue rearrangement in the urethra—experimental study in the rat. Int J Mol Sci 23(24):15951
100. Oliveira R, Coelho A, Franquinho F, Sousa MM, Cruz F, Cruz CD (2019) Effects of early intravesical administration of resiniferatoxin to spinal cord-injured rats in neurogenic detrusor overactivity. Neurourol Urodyn 38(6):1540–1550. https://doi.org/10.1002/nau.24032
101. Cruz CD, McMahon SB, Cruz F (2006) Spinal ERK activation contributes to the regulation of bladder function in spinal cord injured rats. Exp Neurol 200(1):66–73
102. de Groat WC, Kawatani M, Hisamitsu T, Cheng CL, Ma CP, Thor K et al (1990) Mechanisms underlying the recovery of urinary bladder function following spinal cord injury. J Auton Nerv Syst 30(Suppl:):S71–S77
103. Seki S, Sasaki K, Fraser MO, Igawa Y, Nishizawa O, Chancellor MB et al (2002) Immunoneutralization of nerve growth factor in lumbosacral spinal cord reduces bladder hyperreflexia in spinal cord injured rats. J Urol 168(5):2269–2274. https://doi.org/10.1097/01.ju.0000025338.65642.09S0022-5347(05)64369-8
104. Yoshimura N, Bennett NE, Hayashi Y, Ogawa T, Nishizawa O, Chancellor MB et al (2006) Bladder overactivity and hyperexcitability of bladder afferent neurons after intrathecal delivery of nerve growth factor in rats. J Neurosci 26(42):10847–10855. https://doi.org/10.1523/JNEUROSCI.3023-06.2006
105. Kadekawa K, Majima T, Shimizu T, Wada N, de Groat WC, Kanai AJ et al (2017) The role of capsaicin-sensitive C-fiber afferent pathways in the control of micturition in spinal-intact and spinal cord-injured mice. Am J Physiol Renal Physiol 313(3):F796–F804. https://doi.org/10.1152/ajprenal.00097.2017
106. Lin XJ, Wen S, Deng LX, Dai H, Du X, Chen C, Walker MJ, Zhao TB, Xu XM (2020) Spinal cord lateral hemisection and asymmetric behavioral assessments in adult rats. J Vis Exp 157. https://doi.org/10.3791/57126. PMID: 32281968
107. Cerqueira SR, Oliveira JM, Silva NA, Leite-Almeida H, Ribeiro-Samy S, Almeida A et al (2013) Microglia response and in vivo therapeutic potential of methylprednisolone-loaded

dendrimer nanoparticles in spinal cord injury. Small 9(5):738–749
108. Gomes ED, Mendes SS, Assunção-Silva RC, Teixeira FG, Pires AO, Anjo SI et al (2018) Co-transplantation of adipose tissue-derived stromal cells and olfactory ensheathing cells for spinal cord injury repair. Stem Cells 36(5):696–708
109. Gomes ED, Mendes SS, Leite-Almeida H, Gimble JM, Tam RY, Shoichet MS et al (2016) Combination of a peptide-modified gellan gum hydrogel with cell therapy in a lumbar spinal cord injury animal model. Biomaterials 105:38–51
110. Silva NA, Salgado AJ, Sousa RA, Oliveira JT, Pedro AJ, Leite-Almeida H et al (2010) Development and characterization of a novel hybrid tissue engineering–based scaffold for spinal cord injury repair. Tissue Eng Part A 16(1):45–54
111. Gomes ED, Ghosh B, Lima R, Goulão M, Moreira-Gomes T, Martins-Macedo J et al (2020) Combination of a gellan gum-based hydrogel with cell therapy for the treatment of cervical spinal cord injury. Front Bioeng Biotechnol 8:984
112. Moon HC, Lee YJ, Cho CB, Park YS (2016) Suppressed GABAergic signaling in the zona incerta causes neuropathic pain in a thoracic hemisection spinal cord injury rat model. Neurosci Lett 632:55–61
113. Lima R, Monteiro S, Lopes JP, Barradas P, Vasconcelos NL, Gomes ED et al (2017) Systemic interleukin-4 administration after spinal cord injury modulates inflammation and promotes neuroprotection. Pharmaceuticals (Basel) 10(4). https://doi.org/10.3390/ph10040083
114. Lima R, Monteiro S, Gomes ED, Vasconcelos NL, Assuncao-Silva R, Morais M et al (2020) Citalopram administration does not promote function or histological recovery after spinal cord injury. Int J Mol Sci 21(14):5062. https://doi.org/10.3390/ijms21145062
115. Vasconcelos NL, Gomes ED, Oliveira EP, Silva CJ, Lima R, Sousa N et al (2016) Combining neuroprotective agents: effect of riluzole and magnesium in a rat model of thoracic spinal cord injury 16(8):1015–1024
116. Lima R, Gomes ED, Cibrao JR, Rocha LA, Assuncao-Silva RC, Rodrigues CS et al (2021) Levetiracetam treatment leads to functional recovery after thoracic or cervical injuries of the spinal cord. NPJ Regen Med 6(1):11. https://doi.org/10.1038/s41536-021-00121-7
117. Park JH, Kim JH, Oh S-K, Baek SR, Min J, Kim YW, Kim ST, Woo C-W, Jeon SR (2016) Analysis of equivalent parameters of two spinal cord injury devices: the New York University impactor versus the Infinite Horizon impactor. Spine J 16(11):1392–1403. https://doi.org/10.1016/j.spinee.2016.06.018. Epub 2016 Jun 24. PMID: 27349631
118. Puckett AH, Nunn CD, Onifer SM (2009) Veterinary care methods for rats and mice in experimental spinal cord injury studies. In: Jun C, Zao CX, Xiao-Ming X, John HZ (eds) Animal models of acute neurological injuries, Humana Press, American Society of Neuroradiology, pp 47–60
119. Rivlin A, Tator C (1978) Effect of duration of acute spinal cord compression in a new acute cord injury model in the rat. Surg Neurol 10(1):38–43
120. de Sousa N, Pinho AG, Monteiro S, Liberato V, Santos DJ, Campos J et al (2023) Acute baclofen administration promotes functional recovery after spinal cord injury. Spine J 23(3):379–391
121. Monteiro S, Pinho AG, Macieira M, Serre-Miranda C, Cibrão JR, Lima R et al (2020) Splenic sympathetic signaling contributes to acute neutrophil infiltration of the injured spinal cord. J Neuroinflammation 17(1):1–14
122. Pinho AG, Cibrão JR, Lima R, Gomes ED, Serra SC, Lentilhas-Graça J et al (2022) Immunomodulatory and regenerative effects of the full and fractioned adipose tissue derived stem cells secretome in spinal cord injury. Exp Neurol 351:113989
123. Detloff MR, Fisher LC, Deibert RJ, Basso DM (2012) Acute and chronic tactile sensory testing after spinal cord injury in rats. J Vis Exp: Jove 62:e3247
124. Chaplan SR, Bach FW, Pogrel J, Chung J, Yaksh T (1994) Quantitative assessment of tactile allodynia in the rat paw. J Neurosci Methods 53(1):55–63
125. Minett MS, Quick K (2011) Wood JNJCpimb. Behavioral measures of pain thresholds. Curr Protoc Mouse Biol 1(3):383–412
126. Detloff MR, Clark LM, Hutchinson KJ, Kloos AD, Fisher LC, Basso DM (2010) Validity of acute and chronic tactile sensory testing after spinal cord injury in rats. Exp Neurol 225(2):366–376
127. Dixon WJJArop (1980) Efficient analysis of experimental observations. Annu Rev Pharmacol Toxicol 20(1):441–462
128. Bradman MJ, Ferrini F, Salio C, Merighi A (2015) Practical mechanical threshold estimation in rodents using von frey hairs/Semmes–Weinstein monofilaments: towards a rational method. J Neurosci Methods 255:92–103
129. Woolfe G, MacDonald A (1944) The evaluation of the analgesic action of pethidine

hydrochloride (Demerol). J Pharmacol Exp Ther 80:300–307
130. Bannon AW, Malmberg AB (2007) Models of nociception: hot-plate, tail-flick, and formalin tests in rodents. Curr Protoc Neurosci Chapter 8:Unit 8.9–Unit 8.9. https://doi.org/10.1002/0471142301.ns0809s41PMID: 18428666
131. Gunn A, Bobeck EN, Weber C, Morgan MM (2011) The influence of non-nociceptive factors on hot-plate latency in rats. J Pain 12(2):222–7. https://doi.org/10.1016/j.jpain.2010.06.011
132. Giglio C, Defino HLA, Da-Silva C, De-Souza A, Del Bel EA (2006) Behavioral and physiological methods for early quantitative assessment of spinal cord injury and prognosis in rats. Braz J Med Biol Res 39:1613–1623
133. Hargreaves K, Dubner R, Brown F, Flores C, Joris J (1988) A new and sensitive method for measuring thermal nociception in cutaneous hyperalgesia. Pain 32(1):77–88
134. Rigaud M, Gemes G, Barabas M-E, Chernoff DI, Abram SE, Stucky CL et al (2008) Species and strain differences in rodent sciatic nerve anatomy: implications for studies of neuropathic pain. Pain 136(1-2):188–201
135. Banik RK, Kabadi RA (2013) A modified Hargreaves' method for assessing threshold temperatures for heat nociception. J Neurosci Methods 219(1):41–51
136. D'Amour FE, Smith DLJ (1941) A method for determining loss of pain sensation. Pharmacol Exp Ther 72(1):74–79
137. Hook MA, Woller SA, Bancroft E, Aceves M, Funk MK, Hartman J et al (2017) Neurobiological effects of morphine after spinal cord injury. J Neurotrauma 34(3):632–644. https://doi.org/10.1089/neu.2016.4507
138. Carlton SM, Lekan HA, Kim SH, Chung JM (1994) Behavioral manifestations of an experimental model for peripheral neuropathy produced by spinal nerve ligation in the primate. Pain 56(2):155–166
139. Yoon C, Wook YY, Sik NH, Ho KS, Mo CJ (1994) Behavioral signs of ongoing pain and cold allodynia in a rat model of neuropathic pain. Pain 59(3):369–376
140. Fakhri S, Abbaszadeh F, Pouriran R, Jorjani MJP (2020) Pharmacology. The effects of intrathecal ketamine on improving sensory-motor function in a rat model of compression spinal cord injury. Physiol Pharmacol 24(2):101–110
141. Yamamoto K, Tsuboi M, Kambe T, Abe K, Nakatani Y, Kawakami K et al (2016) Oxaliplatin administration increases expression of the voltage-dependent calcium channel α2δ-1 subunit in the rat spinal cord. J Pharmacol Sci 130(2):117–122
142. Lindsey AE, LoVerso RL, Tovar CA, Hill CE, Beattie MS, Bresnahan JC (2000) An analysis of changes in sensory thresholds to mild tactile and cold stimuli after experimental spinal cord injury in the rat. NNR 14(4):287–300
143. Oliveira R, Coelho A, Franquinho F, Sousa MM, Cruz F, Cruz CD (2019) Effects of early intravesical administration of resiniferatoxin to spinal cord-injured rats in neurogenic detrusor overactivity 38(6):1540–1550. https://doi.org/10.1002/nau.24032

Chapter 4

Animal Models for Neuropathic Pain Research in Type 1 and Type 2 Diabetes

Stéphanie A. Eid, Emily J. Koubek and Eva L. Feldman

Abstract

Painful diabetic neuropathy is a disabling complication of prediabetes and diabetes, associated with a significant burden on both the affected patients and society. Given the lack of disease-modifying therapies, there is an urgent need to better understand the pathogenesis of neuropathic pain. Painful diabetic neuropathy can be modeled in rodents, which provide a powerful tool for dissecting the molecular mechanisms underlying disease development and testing the therapeutic efficacy of novel targets. In this chapter, we first provide an overview of the commonly used tests to characterize pain behaviors in animals and cover the challenges associated with these tests. We provide the reader with guidelines on how to apply two of these tests, Hargreaves and von Frey tests, to assess pain behaviors in mice. We then discuss the status of rodent models of painful DPN, their main phenotypic features, limitations, and mechanistic insights gained. Here as well, we describe two protocols for establishing animal models of painful DPN: a single high-dose streptozotocin-induced model of type 1 diabetes and a high-fat diet- and streptozotocin-induced model of type 2 diabetes. Furthermore, we discuss the promise of emerging uncharacterized models that may better capture the human condition and improve our knowledge of pain processing. We contend that selecting the appropriate model for exploration of painful diabetic neuropathy and careful data interpretation are critical for advancing our understanding of the disease, its progression, and treatment.

Keywords Neuropathic pain, Obesity, Painful diabetic neuropathy, Prediabetes, Rodent models, Type 1 diabetes, Type 2 diabetes

Abbreviations

DPN	diabetic peripheral neuropathy,
HFD	high-fat diet,
IENFD	intraepidermal nerve fiber density,
MLD	multiple low dose,
NC	nerve conduction velocity,
QST	quantitative sensory testing,
SHD	single high dose,

Isaura Tavares and Hugo Leite-Almeida (eds.), *Translational Research Methods for Neuropathic Pain*, Neuromethods, vol. 235, https://doi.org/10.1007/978-1-0716-5312-8_4,

STZ	streptozotocin,
T1D	type 1 diabetes,
T2D	type 2 diabetes

1 Introduction: Painful Diabetic Peripheral Neuropathy

The prevalence of prediabetes and type 2 diabetes (T2D) is increasing worldwide due in part to the overconsumption of energy-dense foods and a sedentary lifestyle. Indeed, 541 million people have prediabetes, and over 537 million are affected with T2D worldwide [1]. Prediabetes is characterized by impaired glucose tolerance and often progresses to T2D. T2D accounts for ~90%–95% of all diabetes cases and is characterized by hyperglycemia and insulin resistance [2]. Diabetic neuropathy is a common and severe complication affecting over 50% of all diabetic patients, including those with the less prevalent insulin-deficient type 1 diabetes (T1D). Over 20 million Americans currently suffer from diabetic neuropathy caused by either prediabetes, T1D, or T2D [3]. This number is expected to increase in parallel with the growing obesity epidemic, an established risk factor for prediabetes and T2D [4]. While diabetes-induced nerve damage can present in multiple forms, the most common form is diabetic peripheral neuropathy (DPN), a length-dependent and symmetric axonal degeneration, which progresses in a distal-to-proximal, stocking-glove fashion [5]. Prediabetes results in a similar pattern of nerve injury, indicating that DPN occurs along a continuum of metabolic dysfunction [6]. The main clinical consequences of DPN include foot fractures, foot ulcers, and lower-limb amputations [7, 8], which substantially impair patients' quality of life.

Additionally, 25%–50% of patients with DPN experience neuropathic pain, which by definition, "arises from a lesion or disease of the peripheral somatosensory system resulting from diabetes" [9]. Neuropathic pain symptoms are often described as burning, tingling, electric-shock sensations, or deep aching [10, 11]. Patients may also experience allodynia (painful sensation to non-painful stimuli) and hyperalgesia (increased sensitivity to painful stimuli) [12]. Furthermore, over 50% of patients with painful DPN report severe pain [13], which worsens at night and results in sleep disturbances [14]. Together, these symptoms may limit a patient's ability to perform daily activities and affect work productivity, ultimately contributing to withdrawal from social life, anxiety, and depression [15].

Unfortunately, despite extensive research and the substantial burden associated with painful DPN, there remain several challenges: (1) Current therapies for neuropathic pain only marginally provide relief and are frequently accompanied by multiple adverse drug side effects [16]. (2) Pain mechanisms in prediabetes, T1D, and T2D are poorly understood. (3) Only a subset of DPN individuals develop painful symptoms, suggesting contributions from both genetic and epigenetic factors increase pain risk. Associations between genotype and painful DPN susceptibility have been reported [17], but have yet to be validated. (4) The lack of robust rodent models that consistently replicate key features of human painful DPN. This chapter will first provide an overview of the commonly used tests to characterize pain behaviors in animals. We will then discuss the most widely used animal models of painful DPN, including the main phenotypic features of each model, mechanistic insights gained, and important considerations when selecting the appropriate model. Furthermore, we discuss the promise of emerging uncharacterized models that may improve our knowledge of pain processing and accelerate development of novel treatments for neuropathic pain in prediabetes and diabetes.

1.1 Clinical Characteristics of Painful DPN

There are multiple types of axons that constitute a peripheral nerve. These axons, frequently referred to as fibers, are commonly divided into two types: small and large. Small fibers are unmyelinated or thinly myelinated fibers and convey temperature, pain, itch and regulate select autonomic functions. Large fibers are myelinated and convey proprioception and vibratory thresholds.

Small fiber damage is one of the earliest events in DPN and may be accompanied by continuous or intermittent pain located within the region of sensory damage [18, 19]. Painful sensation typically begins in the foot and lower-limb area, and progresses in a symmetrical, distal-to-proximal, stocking-glove pattern [20]. Some patients may also suffer from "gain-of-function" symptoms such as allodynia and hyperalgesia [12], as mentioned above. Moreover, the same patient may concomitantly report painful as well as negative symptoms such as numbness and loss of sensation, which impact the accurate diagnosis of painful DPN. Patients may occasionally experience proximal pain, including in the trunk area, or diffuse pain, affecting several body sites, which is usually associated with autonomic dysfunction [21]. Despite the severe nature of pain, sensory loss may be moderate or absent. Additionally, large fiber function is generally preserved, and motor weakness is not observed until later in the disease course.

1.2 Utility of Animal Models of Painful DPN

Pain studies using human subjects are difficult to perform due to ethical issues, the subjective nature of pain, and the heterogeneity of the observed phenotypes [22]. In this context, animal models have been critical for elucidating the pathogenesis of painful DPN because they provide experimental strategies and enable testing of novel therapeutic agents, which are otherwise not feasible in human studies. Research on painful DPN has been carried out in multiple species, including rats, which were commonly used in the 1980s. Indeed, extensive pain research using the streptozotocin (STZ)-induced T1D rat model identified several pathogenic mechanisms centered on the dysregulation of glucose metabolism in T1D [23].

However, poor translation to patient clinical care led to major reconsiderations of how closely this model recapitulates essential features of the human disease.

Only recently have studies begun to address pain processing in prediabetes and T2D, reflecting the dominant prevalence of these disorders. These studies are also shedding light on potential differences in pain mechanisms between T1D versus T2D [24]. Attractive features of mice, including cost-effectiveness, short breeding cycles, and ease of genetic manipulation, have made them the preferred species for recent painful DPN research. In this regard, the Diabetes Complications Consortium (DiaComp; www.diacomp.org) of the National Institutes of Health has refined and standardized neuropathy phenotyping protocols and identified rodent models that better replicate human DPN. Together with the Juvenile Diabetes Research Foundation and the Diabetic Neuropathy Study Group of the European Foundation for the Study of Diabetes Pain, DiaComp recommends assessing painful DPN in diabetic rodents through evoked behavioral responses to sensory stimuli [25]. However, these tests should be considered with caution as they do not capture the ongoing, spontaneous pain that diabetic patients commonly experience [26]. Rather, they measure pain-like behaviors, including limb withdrawal, licking, and agitation in response to a stimulus, which are thought to reproduce aspects of the quantitative sensory testing (QST) that measures peripheral sensory function in diabetic patients [27]. Despite these limitations, these behavioral tests are reproducible and sensitive [28]. An overview of DiaComp recommended behavioral tests is discussed below.

2 Evaluating Pain Behaviors in Diabetic Rodents

While there is no gold standard test to characterize painful DPN in rodents, commonly used tests measure nociception in response to thermal, mechanical, or chemical stimuli applied to the tail or paw (Fig. 1). Current recommendations advise using multiple behavioral tests per study and minimizing approaches that require animal restraining, a potential stressor that can induce behavioral changes independent of diabetes [25, 29]. In this section, we provide an overview of behavioral tests designed to assess pain behaviors (Table 1) and describe in detail two protocols for the study of painful DPN in the laboratory mouse. The first protocol, the Hargreaves method, details the procedure for detecting responses to a thermal stimulus. The second protocol, the Von Frey test, describes the methodology for assessing sensitivity to mechanical stimuli.

2.1 Evoked Response to Thermal Stimuli

Like diabetic patients [30], heat hyperalgesia develops early in the disease course, i.e., within weeks of diabetes onset in T1D and T2D rodents, and progresses to hypoalgesia at later disease stages [31–36].

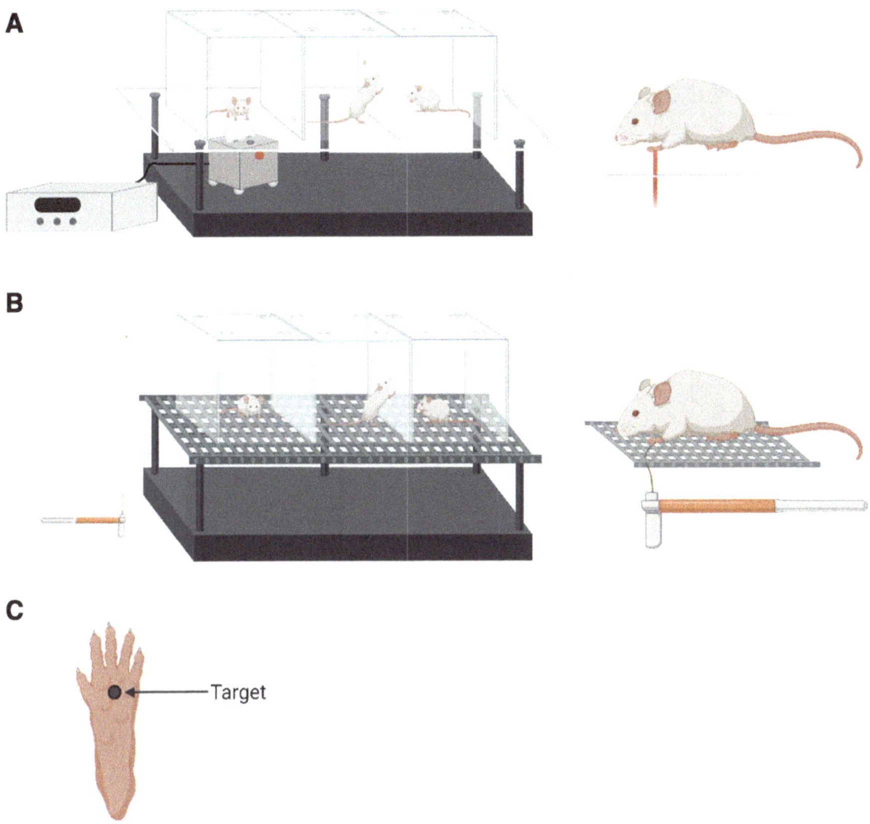

Fig. 1 Commonly used methods to assess pain behaviors in diabetic rodents. (**A**) The Hargreaves method quantifies heat thresholds in the plantar surface of the hind paw following stimulation by a radiant or infrared light source. (**B**) Mechanical sensitivity in diabetic rodents is assessed using the von Frey filaments. A positive response is recorded when the animal withdraws, licks, or shakes its hind paw after probing. (**C**) Midplantar testing area of mouse hindpaw. (Created with BioRender.com)

Thermal sensitivity is evaluated by quantifying hind paw withdrawal or tail flick latencies in response to a thermal stimulus.

The hot-plate test measures withdrawal latencies in all limbs as well as the tail [37]. This may complicate data interpretation and result in variable behavioral responses. However, an advantage of this test is that it allows the distinction between two types of small fibers, C- and Aδ-fibers, and their firing in response to graded rates of noxious skin heating [38–40]. Particularly, Aδ-fibers appear to mediate nociceptive responses evoked by high rates of noxious radiant heating, while C-fibers seem to be preferentially activated by lower temperatures and slower heating [38–40].

The time it takes for the tail to flick or twitch in response to a thermal stimulus is another method used to measure heat sensitivity in diabetic rodents [41]. However, this test has several limitations. The tail flick may be a spinal reflex, involving motor neuron and effector function, rather than an indication of nociception and sensory function [42]. Additionally, the translation potential of this test is

Table 1
Common tests to assess pain behaviors in diabetic rodents

Stimulus	Behavioral test	Description	Measurement	Limitation(s)	References
Thermal	Hargreaves method	Apply a radiant or infrared light source to paw	Paw withdrawal latencies	Repetition may stimulate learning behaviors in rats	[43, 44]
	Hot plate	Place rodent on hot plate	Withdrawal latencies in all limbs as well as the tail	Variable behavioral responses due to simultaneous testing in all limbs	[37]
	Tail flick	Subject the tail to radiant heat or immerse it in hot water	Time it takes for the tail to flick or twitch	Response may be a spinal reflex; requires animal restraining; poor translation to clinic	[41, 42]
Mechanical	von Frey	Apply pressure using von Frey filaments with predetermined forces to the point of bending	Paw withdrawal latencies	Protocol variability; subjective	[28, 58]
	Randall–Selitto paw pressure test	Apply increasing mechanical force of a restrained animal	Withdrawal latencies, vocalization, or escape response	Requires animal restraining; poor translation to clinic	[25, 51]
Chemical	Formalin test	Inject 0.2%–5% formalin into paw	Voluntary biphasic responses, such as flinching, guarding, or biting	Subjective; poor translation to clinic	[59, 62]

questionable because there is no anatomical equivalent for the tail in diabetic patients. Although this test is relatively quick, it may require animal restraining, which may induce stress.

The Hargreaves method quantifies heat thresholds in the plantar surface of the hind paw following stimulation by a radiant or infrared light source (Fig. 1A) [43]. Animals are first allowed to habituate to the testing apparatus to minimize movement and limit stress-induced responses. Habituation time differs between mice and rats, with mice typically requiring longer habituation time (30–45 min vs. 5 min in rats) [28]. Once the light source is positioned under the hind paw being tested, the time between beam activation and paw withdrawal is recorded. To prevent tissue injury, a cutoff time (10–20 s) is preset, at the end of which the light source is switched off automatically even in the absence of a response. Rodents that display heat hyperalgesia have significantly lower paw withdrawal latencies compared to healthy controls. While this test closely mimics QST, its repetitive nature may stimulate learning behaviors in rats, but not mice, potentially confounding withdrawal latencies [44]. Variations in skin thickness may also impact response time.

Protocol 1: Hargreaves Method for Assessing Evoked Response to Thermal Stimuli

Materials

(1) Mice. (2) IITC Model 336 Plantar/Tail Analgesic Meter with True Tail Temp and Heated Glass. (3) Plastic mouse chamber which can hold up to 12 mice at a time. (4) Instrument controlling heat output and timing and that includes cutoff safety features. (5) Test head that emits radiant heat light source and includes start, stop, and restart button.

Procedure

(1) Perform this test in a quiet room as external noise may affect mouse behavior. (2) Turn the heat of the glass plate warmer to medium (roughly equivalent to 32°C) and allow to warm for 10 min. (3) Set the active intensity during the test for 25% intensity and the cutoff timer for 20 s to prevent injury to the mouse. (4) Acclimate mice to the apparatus by placing them in the plastic mouse chamber on the plate for 30–45 min. Mice need to be completely calm for this test and data are invalid for agitated mice. (5) To begin the experimental test, move the test head under the mouse and use the mirror to direct the beam under the center of the hind paw. (6) Start the test by pressing the red button twice. The first press will initialize the light source and the second press will turn on the light and start the test. Following the second press, the temperature will rise to the desired temperature. (7) When the mouse moves its paw, press the red button to stop the test. (8) At test completion, the screen will flash the elapsed time. (9) Manually

record the elapsed time of the test in a logbook or attach a printer for printed records. (10) Record six measurements for each mouse, alternating between left and right hind paws, with a 10-min break between each individual stimulus. (11) Use the six measurements to calculate an average latency time for each mouse. (12) Gently return the animals to their cages. (13) Clean the apparatus and make sure you leave no smell after each use.

2.2 Evoked Response to Mechanical Stimuli

Mechanical allodynia is a painful sensation to light touch frequently experienced by diabetic patients and mediated by thinly myelinated small fibers [45]. It markedly reduces patients' quality of life in that the simple touch of clothes against their skin provokes excruciating pain [46]. Similarly, T1D and T2D rodents develop mechanical allodynia within weeks of disease onset [47, 48]. More recent data show that diet-induced obese and prediabetic mice also develop tactile allodynia, which will be discussed in greater detail below [49, 50].

The Randall–Selitto paw pressure test, which is considered a measure of mechanical hyperalgesia, determines paw withdrawal thresholds in response to increasing mechanical force of a restrained animal [51]. It is better suited for use in rats than mice that are unable to tolerate the high level of restraint required to carry out the assay [52]. In fact, although STZ-induced T1D rats exhibit mechanical hyperalgesia using this test [53, 54], handling-induced stress may interfere with the behavioral response observed [29]. Additionally, it is unclear what component of the QST is mimicked by the Randall–Selitto test, which in turn impacts its clinical relevance [25, 55]. Given these limitations, it is currently recommended to use this test in conjunction with one or more of the tests summarized in Table 1 [25].

Mechanical sensitivity in diabetic rodents can also be assessed by von Frey filaments [28], which are used in the clinic [56]. While this test can be performed both manually and electronically, the manual test is considered more reliable across experimental models of pain (Fig. 1B) [28, 57]. For the manual von Frey test, animals are first placed in a plexiglass cage with mesh bottom and allowed to acclimate. Mechanical sensitivity is then determined using different methods, including the "up-down method" and "percent response" [28].

The "up-down" method quantifies the threshold for a non-noxious mechanical stimulus applied to the midplantar surface of the hind paw through a logarithmic series of calibrated von Frey filaments with predetermined forces that are pressed to the point of bending [47, 49]. A positive response is noted when rodents elicit a brisk paw withdrawal, at which point a lighter filament is tested; if no response is observed, a larger filament is applied. Once an initial change in response is elicited, four additional filaments are applied. Animals are given a 5–10 min interval to recover between tests.

The 50% withdrawal threshold is calculated using the formula described by Chaplan et al. [58]. Diabetic rodents with mechanical allodynia display a significant decrease in mechanical thresholds compared with non-diabetic controls. In the "percent response" method, the investigator applies multiple von Frey filaments with ascending order of strength five to ten times on the region of interest. The number of positive responses to each filament is recorded and then used to calculate a percent response [58].

The use of the von Frey test for quantifying mechanical allodynia in diabetic animals provides several advantages, including its analogy to aspects of the QST and the fact that no animal restraint is required [25]. Furthermore, this test is selective to A-fiber-mediated allodynia [25]. Repeated measurements do not prompt learning behaviors, thus allowing longitudinal behavioral evaluations in the presence or absence of pharmacological agents [23, 47]. However, disadvantages of this test do exist. There is extensive variability in the protocols used, including pressure applied with von Frey filaments and the number of measurements per animal [28], which in turn affect test outcomes. Additionally, paw withdrawal is entirely dependent on the investigator's judgment, and thus is subjective. For instance, one must determine whether animal response was elicited due to aversive stimuli or because of grooming or agitation, which may introduce further variability. Hence, standardization of approaches is essential to enhance experimental reproducibility and collection of clinically relevant and translatable data.

Protocol 2: "Up-Down" Von Frey Test to Assess Evoked Response to Mechanical Stimuli

Materials

(1) Mice older than 4 weeks of age. (2) Plexiglass cage with mesh flooring that can hold six mice at a time. (3) Logarithmic series of calibrated von Frey monofilaments ranging from 1 to 6 g. (4) Logbook.

Procedure

(1) Place experimental mice in a plexiglass cage. (2) Allow mice to acclimate in the plexiglass cage for a minimum of 60 min prior to testing. It is important that mice are not active during testing. (3) Once animals are still, begin testing the left hind paw with the 1 g filament. Place the filament onto the midplantar surface of the hind paw and press until the filament bends. A brisk withdrawal of the stimulated paw is considered a positive response. (4) If a response is observed with the initial stimulus, record the response in the logbook with an "X" and continue testing with the next smallest filament (0.6 g) in the series. If no response is observed, record the response in the logbook with an "O" and continue testing with the next largest filament (1.4 g) in the series. Allow mice 10 min to recover between tests. (5) The lightest fiber to elicit a paw withdrawal is considered

the response threshold. Note all responses in the logbook. However, once the response threshold is crossed, begin recording six critical data points. The two responses on either side of the threshold can be designated as the first two responses of the six critical points. (6) Test four additional stimuli that vary sequentially up or down based on the mouse's response and record the responses. (7) Calculate the 50% g threshold using the following equation:

$$\mathrm{PR}_{\mathrm{min}} = \frac{\mathrm{PR}_t}{T} \times 60 \tag{1}$$

where X_f = value (in log units) of the last von Frey filament tested; κ = tabular value for the pattern of positive and negative responses; and δ = mean difference (in log units) between stimuli.

2.3 Evoked Response to Chemical Stimuli

The formalin test assesses voluntary responses, such as flinching, guarding, or biting, to an inflammatory compound [59], which is thought to model spontaneous pain behavior [60]. Formalin injection (0.2%–5%) into the hind paw produces a biphasic response: the first phase, which lasts up to 10 min after injection, results from the activation of primary afferent fibers, such as small unmyelinated C-fibers, and low-threshold mechanoreceptors at the injection site [61, 62]. The second phase ranges between 20 and 30 min and results from the inflammatory response and the subsequent central sensitization in the dorsal horn in addition to the ongoing afferent input [61, 62]. This pain test requires no animal restraint and provides insight into both the central and peripheral components of pain, which are considered main advantages. However, it is not analogous to aspects of the QST, and thus is not clinically relevant [25]. Additionally, this test relies on subjective judgment to differentiate between flinching, guarding, and biting, which calls for caution when interpreting data.

3 Animal Models of Painful DPN

There are several approaches for modeling diabetes and painful DPN in rodents. For instance, chemical induction with toxins, such as STZ, is cost-effective and provides a simple way to induce diabetes at different disease stages or in transgenic models with multiple transgenes. Genetic models circumvent potential confounding effects of toxins [63]. Diet manipulation allows investigating obesity and prediabetes independent of hyperglycemia [34]. Selecting the appropriate model will therefore depend on study objectives, including considerations such as metabolic abnormalities in T1D versus T2D, breeding scheme and complexity of the model, and how closely the model replicates human neuropathic pain. This section summarizes the status of the leading rodent models of painful DPN (Table 2), including advantages, limitations, and main considerations when selecting the

Table 2
Animal models of type 1 and type 2 diabetes are commonly used in pain research

Disease	Paradigm/Model	DPN phenotype			Limitations	References
		Behavior	NCV	Anatomy		
T1D	SHD-STZ (150 mg/kg in mice and 40–80 mg/kg in rats)	Tactile allodynia and thermal hypoalgesia in mice up to 20 wk Mechanical allodynia and heat hyperalgesia in rats within 8 wk of diabetes onset	↓ NCVs	↓ IENFDs	Paradigm-dependent inconsistencies in pain behavior; nonspecific toxicity; high mortality rates	[65, 69, 70]
	MLD-STZ (50 mg/kg/day for 5 days)	Thermal and mechanical hypoalgesia in C57BL/6 mice Thermal hypoalgesia in NOD mice Tactile allodynia and thermal hyperalgesia in Swiss albino mice	↓ NCVs	↓ IENFDs	Strain-dependent inconsistencies in pain behavior; reverts back to normoglycemia over time	[80, 82, 142]
	NOD	Thermal hyperalgesia at 8 wk Thermal Hypoalgesia by 12 wk	N/A	N/A	Diabetic phenotype affected by housing, diet, and sex	[102, 103]
	B6INS2^{Akita}	Tactile allodynia at 12 wk Thermal and tactile hypoalgesia at 16 wk	N/A	N/A	Sex-dependent differences in diabetic severity	[105–107]
	BB (rat)	Thermal hyperalgesia as early as 6 wk	N/A	↓ IENFDs	Require daily insulin administration for survival	[109, 110]
Prediabetes	HFD (C57BL/6 mice)	Mechanical allodynia (54% fat diet for 8 wk) Thermal and mechanical hypersensitivity (42% fat diet for 12 wk) Thermal hypoalgesia and mechanical allodynia (60% fat diet up to 31 wk)	↓ NCVs	↓ IENFDs	Strain, diet, and fat source-dependent differences in severity of prediabetic painful neuropathy	[50, 113, 114, 116]

(continued)

Table 2
(continued)

Disease	Paradigm/Model	DPN phenotype			Limitations	References
		Behavior	NCV	Anatomy		
	HFD with low-dose STZ	Mechanical allodynia and hyperalgesia	↓ NCVs	↓ IENFDs	Experimental variation due to lack of standardized protocol for HFD regimens and STZ administration	[35, 133]
T2D	*db/db* (C57BKS background)	Mechanical allodynia and thermal hyperalgesia from 8 to 12 wk of age	↓ NCVs	↓ IENFDs	Strain-dependent differences in diabetic severity, develop diabetes quickly, often infertile	[75, 140, 141]
	ob/ob	Thermal hypoalgesia and tactile allodynia by 11 wk	↓ NCVs	↓ IENFDs	Develop diabetes quickly, often infertile	[146–148]

BB biobreeding, *HFD* high-fat diet, *IENFDs* intraepidermal nerve fiber densities, *MLD-STZ* multiple low-dose streptozotocin, *NCVs* nerve conduction velocities, *NOD* nonobese diabetic; *SHD-STZ* single high-dose streptozotocin, *T1D* type 1 diabetes, *T2D* type 2 diabetes, *wk* weeks appropriate model for painful DPN studies, and describes in detail two protocols for experimental mouse models of DPN. The first protocol describes a T1D mouse model of DPN induced by administration of a single high dose (SHD) of streptozotocin (STZ). The second protocol details the methodology to produce a high-fat diet (HFD)- and STZ-induced T2D mouse model of DPN.

3.1 STZ-Induced T1D Models

T1D is characterized by the autoimmune destruction of insulin-producing pancreatic β-cells, leading to insulin deficiency and systemic hyperglycemia. These metabolic disturbances can be reproduced in rodents through multiple methods, which range from chemical depletion of β-cells with STZ to using rodents that spontaneously develop T1D. The STZ-induced T1D rat model is by far the best studied and most widely used model of painful DPN [23].

STZ induces β-cell destruction, resulting in hypoinsulinemia, hyperglycemia, and weight loss [64]. The main advantage of this model is the ease of diabetes induction in essentially any pre-existing rodent model. Based on DiaComp recommendations, mice are rendered T1D using one of two STZ paradigms: a single high dose (SHD; approximately 150 mg/kg) or consecutive multiple low doses (MLD; approximately 50 mg/kg/day for 5 days). While both paradigms produce a neuropathy phenotype, differences in the severity of nerve pathology and painful behaviors have been reported and should be carefully considered when designing painful DPN studies [63].

Reports have consistently demonstrated that the SHD-STZ model develops a robust and early neuropathy phenotype, with reduced nerve conduction velocities (NCVs) and intraepidermal nerve fiber densities (IENFDs) [65–68]. However, the effect of SHD-STZ on painful behaviors is less clear-cut. Several groups observed tactile allodynia and thermal hypoalgesia up to 20 weeks after SHD-STZ administration [65, 69]. In contrast, other studies showed that SHD-STZ mice develop hyposensitivity to mechanical, thermal, and chemical stimuli as early as 4 weeks post STZ [66, 70–72]. Factors contributing to these discrepancies may include differences in the testing paradigms used, handling-associated stress, mouse background strain, or sex dimorphism. Importantly, concerns have been raised about the nonspecific toxicity of STZ and high mortality rate in mice cohorts due to SHD-STZ injection [73].

Alternatively, the MLD-STZ paradigm is less toxic and induces a repetitive low-grade β-cell death and lymphocytic infiltration of the pancreatic islets, resulting in gradual increases in glycemia and better survival among STZ-treated mice [74]. Like the SHD-STZ model, MLD-STZ mice develop a neuropathic phenotype with delayed NCVs and IENFD loss [75–79]. However, there are still inconsistencies with respect to painful behaviors, which appear to be strain dependent. For example, a longitudinal analysis found that although MLD-STZ induced early metabolic disturbances in C57BL/6 mice (12 weeks post STZ), neuropathic symptoms were only evident after 22 weeks of STZ injection, with thermal and mechanical hypoalgesia and IENFD loss [80]. While nonobese diabetic (NOD) mice displayed reduced sensitivity to thermal stimuli 7 weeks post MLD-STZ injection [81], Swiss albino mice developed tactile allodynia and thermal hyperalgesia after 10 weeks [82]. Of note, this paradigm is complicated by the potential to revert back to normoglycemia over time, which limits long-term follow-up studies [83]. For both SHD and MLD inductions, different STZ induction paradigms can result in differential degrees of β-cell depletion and subsequent nerve damage, which in turn increases experimental variability and confounds data interpretation.

Compared to mice, STZ-induced T1D rats demonstrate more consistent pain behaviors, with mechanical allodynia and heat hyperalgesia observed up to 8 weeks after diabetes onset [23, 48, 84]. Moreover, STZ-induced T1D rats develop formalin hyperalgesia, with flinching observed during phase 2 of the formalin test [85, 86]. Heat hyperalgesia may progress to loss of sensation 8 weeks post STZ injection [87]. This regimen typically calls for a single dose of STZ (40–80 mg/kg), which is determined based on rat age and body weight [25, 88].

The exact pathogenesis of painful DPN remains largely unknown. However, the use of STZ-induced T1D model has provided a better understanding of glucose-dependent pain mechanisms in DPN [3, 89]. Indeed, studies have shown that chronic hyperglycemia contributes to thermal hyperalgesia in T1D rats by alterations in key pathways, including nerve growth factor signaling [90], Toll-like receptor (TLR) 4-dependent inflammatory response [91], increased polyol pathway flux [32], and protein kinase C (PKC) [92]. Both the polyol pathway and PKC have been implicated in the progression to thermal hypoalgesia at later disease stages [32, 92]. More recent data show that thermal hypoalgesia is also associated with mitochondrial dysfunction and an impaired AMP-activated protein kinase (AMPK)–peroxisome proliferator-activated receptor-gamma coactivator (PGC)-1α axis in the hyperglycemic milieu [87]. On the other hand, tactile allodynia in rodent models has been attributed to upregulation of sodium channel (Nav) 1.3 in conjunction with microglial activation [93], as well as impaired insulin signaling [94, 95]. Formalin hyperalgesia has been linked to increased spinal cyclooxygenase (Cox)-2 expression and polyol pathway flux in oligodendrocytes, which suggests a central rather than peripheral basis for this pain [96]. Targeting many of these pathways, namely aldose reductase and PKC, reduced neuropathic pain in vivo [97], but failed to demonstrate neuroprotection in human clinical trials. These results suggest that while glycemia and glycemic control are key pathogenic factors in T1D [98], factors besides hyperglycemia may be involved in pain processing in human DPN [99]. Accordingly, the way forward is to move away from the STZ-induced T1D rodent model centered on glucose metabolism only toward models that better capture the human pathology and may help identify nonglycemic drivers of painful DPN.

Protocol 3: SHD-STZ-Induced T1D Model of DPN

Materials

(1) C57BL/6 mice. (2) Mouse weighing scale. (3) Na citrate. (4) Citric acid. (5) ddH_2O. (6) STZ covered in aluminum foil. (7) 18 g needle on a 5.0 mL tuberculin syringe (8) 26–28½ g needle on a 1.0 mL tuberculin syringe. (9) pH meter. (10) 10% sucrose water. (11) Glucometer and glucose test strips.

Procedure

(1) Randomize mice into nondiabetic control and SHD-STZ groups. (2) Fast mice for 4 h prior to STZ induction. Drinking water should still be provided. (3) Weigh mice and calculate the amount of STZ buffer solution required for each mouse so that the final dosage is 150 mg/kg. (4) Prepare fresh STZ-Na citrate buffer solution immediately before injection as STZ degrades after 15–20 min in the Na citrate buffer. To prepare the buffer, dissolve 1.47 g of Na citrate in 50 mL of ddH_2O in a conical tube. Test the pH with a pH meter and adjust the buffer to pH 4.5 with monohydrate Na-citrate or citric acid if necessary. Sterile filter the Na citrate buffer. Weigh the appropriate amount of STZ (112.5 mg) into an Eppendorf tube for a concentration of 7.5 mg/mL in the Na citrate buffer. Cover the Eppendorf tube with aluminum foil as STZ is light sensitive. Add the STZ from the Eppendorf tube to the conical tube containing Na citrate solution and mix well. Use a 5 mL syringe with an 18 g needle to mix and move the STZ-Na citrate solution to a sterile vile. (5) Inject the first mouse with the appropriate amount of STZ solution via intraperitoneal (IP) injection using a 26–28½ g needle on a 1.0 mL tuberculin syringe. (6) Repeat the procedure for the remaining mice. For control mice, inject with citrate buffer. (7) Provide mice with 10% sucrose water for up to 48 h to avoid sudden, potentially fatal hypoglycemia. (8) Test blood glucose levels 48 h after STZ injection. If mice are hyperglycemic, they can be returned to standard drinking water. If mice on 10% sucrose water are showing signs of hypoglycemia (lethargy and slowed responsiveness), they can be given a 1.0 mL bolus of 10% sucrose water by gavage.

3.2 Spontaneous and Genetic T1D Models

Genetic T1D rodent models spontaneously develop hyperglycemia, bypassing the potential confounding effects of tissue toxicity associated with STZ. These include NOD mice, $B6Ins2^{Akita}$ mice, and the Bio-Breeding (BB) rat.

NOD mice develop spontaneous autoimmune destruction of β-cells and attain serum glucose levels in the frank diabetic range around 16–20 weeks of age [100, 101], which is somewhat variable and impacted by factors such as housing and diet [101]. Timeline for hyperglycemia onset is also affected by sex, with female NOD mice showing more susceptibility to T1D relative to males [101]. Given these limitations, the extent of nerve injury in NOD mice has not been thoroughly studied. Nonetheless, two reports showed evidence of thermal hyperalgesia at 8 weeks of age [102] and hypoalgesia by 12 weeks of age [103].

B6Ins2^{Akita} mice develop spontaneous T1D secondary to insulin-2 misfolding, leading to toxic injury to β-cells, impaired insulin secretion, and hyperglycemia [104]. B6Ins2^{Akita} mice exist on the C57BL/6J background and develop hyperglycemia and hypoinsulinemia at approximately 4 weeks of age, with worse diabetic symptoms in male mice than in female mice [104]. These mice display early symptoms of tactile allodynia at 12 weeks of age [105], followed by thermal and tactile hypoalgesia at 16 weeks of age [106], an effect that modestly, but still significantly, persists until 24 weeks of age [107].

The BB rat develops spontaneous T1D at equal frequency in males and females, secondary to pancreatic infiltration with β-cell failure between 8 and 16 weeks, and requires daily insulin administration for survival [108]. These rats develop thermal hyperalgesia starting as early as 6 weeks of age [109] and continues up to 24 weeks of age with parallel progressive C-fiber loss [110].

Compared with the STZ-induced T1D model, cellular and molecular mechanisms are relatively understudied in genetic T1D models. Nevertheless, studies in NOD mice have shown a role for the inducible bradykinin B1 receptor in mediating hyperalgesia [102] and for nitrosative stress in inducing hypoalgesia [103], effects that are conserved in the STZ model. On the other hand, the B6Ins2^{Akita} model highlighted the role of the soluble epoxide hydrolase enzyme (sEH), a regulator of the arachidonic acid pathway [105] as well as Poly(ADP-ribose) polymerase (PARP) activation [106] in neuropathic symptoms. Lastly, Kamiya et al. demonstrated that hyperalgesia in the BB rat is associated with loss of neurotrophic support and progressive degeneration of the Golgi apparatus [110]. Future studies across different clinically relevant animal models with or without pharmacological inhibitors are needed to confirm whether these proteins and pathways are suitable for therapeutic targeting in T1D painful DPN. Additionally, further neurological phenotyping will determine how closely these models capture clinical features of the human disease.

3.3 Diet-Induced Models Of Prediabetes and Type 2 Diabetes

T2D is characterized by insulin resistance and the inability of β-cells to respond to glucose, ultimately leading to hyperglycemia. Prediabetes, a state of impaired glucose tolerance and insulin resistance, often progresses to T2D [6]. These conditions are associated with painful DPN in humans [111, 112] and are modeled in rodents through dietary manipulation and spontaneous mutations in leptin or the leptin receptor.

High-fat feeding provides a simple and commonly used approach to induce painful neuropathy in rodents, as they develop obesity, insulin resistance, and prediabetes in the absence of hyperglycemia [34–36, 49, 50, 113, 114]. Sprague–Dawley rats and C57BL/6 mice are the preferred animals for diet-induced obesity and prediabetes studies, owing to their increased susceptibility to metabolic dysfunction [115]. Indeed, we confirmed and extended this finding in a comparative study that assessed metabolic and neuropathy phenotypes in commonly used mouse strains in peripheral neuropathy research: BKS, BTBR, and C57BL/6J [114]. We found that C57BL/6J mice fed a HFD (54% kcal fat) develop the most robust metabolic and neuropathy phenotypes, with increased body weight, impaired glucose tolerance, hyperinsulinemia, dyslipidemia, as well as reduced NCVs and IENFDs at 36 weeks of age [114]. These results support the C57BL/6J mouse as the model of choice for diet-induced painful neuropathy studies independent of hyperglycemia. On the other hand, the Jackson Laboratory reported that NON/ShiLtJ mice are quite sensitive to diet-induced obesity and obesity-driven hyperglycemia as they rapidly attain a hyperglycemic state when fed HFD, unlike C57BL/6 mice that develop glucose levels in the range of prediabetes, but not T2D (https://www.jax.org/news-and-insights/2006/october/non-shiltj-males-a-new-model-of-diet-induced-obesity-and-diabesity). Depending on the study goals and objectives, neuropathy phenotyping may be warranted in these models.

HFDs used in painful neuropathy research are commercially available and typically provide 42%–60% of calories from fat, which differentially impacts the severity of nerve damage and pain behaviors. As an example, Guilford et al. reported that a 54% HFD for 8 weeks reduces motor NCVs and induces mechanical allodynia but does not affect sensory NCVs, thermal sensitivity, or IENFD [116]. Gavini et al. found that a 42% HFD for 12 weeks induces thermal and mechanical hypersensitivity [50]. Changes in NCV measures or IENFDs were not reported in this study [50]. On the other hand, we showed that a 45% HFD for a longer duration of 34 weeks induces thermal hypoalgesia but does not affect mechanical responses [113]. It also delays motor and sensory NCVs and reduces IENFDs [113]. More recently, we and others have demonstrated that a 60% HFD for a duration between 11 and 31 weeks triggers more severe metabolic and neuropathy profiles, with thermal hypoalgesia, mechanical allodynia, and reduced NCVs and IENFDs [35, 117, 118]. Additional research evaluating the longitudinal effect of various fat percentages on pain behaviors, NCVs, and IENFDs, particularly early after HFD consumption, warrants consideration when designing painful neuropathy studies.

Nerve pathology and the extent of nociception also vary with fat source (lard versus plant) and saturation degree [119, 120]. Rodents placed on a lard-based HFD consisting of 42%–60% saturated fats develop obesity, prediabetes, painful symptoms, and large fiber dysfunction [118, 121]. Conversely, 54% HFD from hydrogenated vegetable shortening for a 12-week study duration induced mechanical allodynia, but had no effect on thermal sensitivity, NCVs, or IENFDs [122]. Notably, switching a lard-based HFD to diets consisting of unsaturated fats, from plant sources or fish oil, ameliorates painful neuropathy, including thermal sensitivity [119, 123]. Also of interest is the ketogenic diet, a high-fat (79%–90% kcal from plant source) and low-carbohydrate (0.3%–3% kcal) diet, which has emerged as a promising nonpharmacological approach for the treatment of metabolic disorders [124]. It triggers ketone body utilization as a main energy source, shifting energy metabolism from carbohydrates to fats [125]. With respect to neuropathic pain, Cooper et al. recently reported that replacing an HFD with a ketogenic diet reversed mechanical allodynia and promoted axonal growth [126]. Together, these data suggest that fat saturation degree may differentially impact nociception and that diets rich in unsaturated fats or ketogenic diets may offer an effective lifestyle intervention for the treatment of painful DPN in humans.

Another important consideration is the role of sex in nociception. Human studies show sex differences in neuropathic pain, with T2D female patients experiencing a higher pain intensity relative to males [127]. However, studies investigating sex dimorphism in diet-induced rodent models are almost completely absent and are therefore essential to help design sex-specific therapies for the treatment of painful neuropathy.

Mechanistically, diet-induced rodent studies have uncovered candidate targets and pathways that are dysregulated in conjunction with electrophysiological, behavioral, and anatomical deficits. Gavini et al. identified a role for liver X receptor (LXR), a master regulator of lipid homeostasis, in modulating ER stress and allodynia in HFD mice [50]. We have previously shown that NADPH oxidase–dependent oxidative stress and oxidized LDLs may be contributing factors in the development of HFD-induced hypoalgesia and prediabetic painful neuropathy [113, 128]. Our more recent findings point to a key role for mitochondrial abnormalities and reduced axonal calcium levels in the pathogenesis of painful DPN [129]. Additionally, our data suggest that beyond C-fibers, nonmyelinating Schwann cells might mediate nerve pathology in diet-induced prediabetes and warrant further research [130]. Overall, these preclinical findings advanced our understanding of pain mechanisms in prediabetes and shed light on novel players likely implicated in early peripheral neuropathy in humans well before individuals progress to frank T2D.

3.4 HFD Combined with Low-Dose STZ as a Model of T2D

Researchers have recently incorporated STZ treatment with high-fat feeding in wild-type animals to induce hyperglycemia and a T2D model (HFD-STZ). The protocol involves treating HFD-fed prediabetic mice with a low dose of STZ to deplete a portion of pancreatic β-cells, resulting in mildly impaired insulin secretion and hyperglycemia [131, 132]. Although less is known about the susceptibility of this model to diabetic complications, we and others have recently shown that HFD-STZ rodents develop DPN, with reports showing mechanical allodynia and hyperalgesia [35, 117, 118, 133, 134]. Despite having a worse glycemic status, HFD-STZ mice develop similar levels of large and small nerve fiber damage relative to HFD mice [35, 117, 118]. These results suggest that glucotoxicity alone does not drive nerve injury, which likely occurs along a continuum of metabolic dysfunction also involving dyslipidemia [89, 135].

One of the major limitations of the HFD-STZ model is the lack of a standard protocol for high-fat feeding regimens and STZ administration paradigms used to induce T2D, as well as animal age and strain. This results in experimental variation in DPN presentation and pain behaviors, potentially leading to conflicting or noncomparable findings. Additionally, data have shown that young HFD-STZ animals are not hyperinsulinemic but exhibit hypoinsulinemia [117, 136, 137]. In addition to mimicking a later disease stage, this feature has been attributed to rodent age with the view that young T2D mice, like young T2D patients, tend to display insulin secretion deficiency rather than insulin resistance [117, 137]. Choosing the appropriate rodent age, as well as the disease stage, is therefore an important consideration when inducing T2D in HFD-STZ models to study pain.

Protocol 4: HFD-STZ-Induced Mouse Model of T2D

Materials

(1) C57BL/6 J mice. (2) Na citrate. (3) Citric acid. (4) ddH_2O. (5) STZ covered in aluminum foil. (6) 18 g needle on a 5.0 mL tuberculin syringe. (7) 26–28½ g needle on a 1.0 mL tuberculin syringe. (8) 10% sucrose water. (9) HFD containing 60% energy from fat, 20% energy from carbohydrates, and 20% energy from protein (D12492, Research Diets, New Brunswick, NJ).

Procedure

(1) Randomize 4-week-old C57BL/6 J mice into standard diet (SD) and HFD + STZ groups and provide water and SD to both groups ad libitum. House mice five animals per cage. (2) After 1 week of acclimation, switch mice in the HFD + STZ group to HFD and allow mice to feed ad libitum. (3) When mice are 12 weeks of age, administer a 75 mg/kg body weight IP dose of STZ to the HFD-STZ group according to the protocol described in Sect. 3.1.1,

Protocol 3. Briefly, fast mice for 4 h prior to dosing. Weigh mice to determine appropriate dosing (75 mg/kg body weight). Prepare fresh STZ-Na citrate buffer solution immediately before injection. Inject each mouse with STZ solution using a 26–28½ g needle on a 1.0 mL tuberculin syringe. Inject SD mice with citrate buffer. Provide mice with 10% sucrose drinking water for 2 days to prevent hypoglycemic shock. (4) 72 h after the first dose, administer a second dose of STZ at 50 mg/kg body weight.

3.5 Spontaneous and Genetic T2D Models

Mutations in genes involved in carbohydrate or lipid metabolism can result in impaired insulin sensitivity, as well as increased body fat, plasma lipids, and glucose levels in rodents. Particularly, leptin is a hormone produced by adipocytes that regulates satiety, energy homeostasis, and lipid and glucose metabolism [138]. Mutations in leptin (*ob/ob* mice) or the leptin receptor (*db/db* mice) induce diabetes secondary to hyperphagia and subsequent obesity, insulin resistance, and hyperglycemia. These models have been studied extensively in T2D research as they closely resemble the human disease state.

The *db/db* mouse on the C57BKS background develops hyperinsulinemia within days and overt T2D by 4 weeks of age [139]. Similar to the human pathology, this mouse model displays mechanical allodynia and thermal hyperalgesia from 8 to 12 weeks of age [47, 140], which often progress to hypoalgesia at later disease stages [107, 141]. Delays in motor and sensory NCV are also present, as is IENFD loss, characteristic of DPN [76, 77, 142, 143]. The severity of the diabetic phenotype in *db/db* mice is strain-dependent, with evidence of less persistent hyperglycemia on the C57BL/6 background [144, 145]. Because of this limitation, most DPN studies used the C57BKS *db/db* model. When challenged with HFD, C57BL/6 *db/db* mice, however, do demonstrate indices of neurologic dysfunction, including thermal hypoalgesia [107, 114].

C57BL/6 *ob/ob* mice are considered a model of obesity and mild T2D, with clear signs of obesity by 4 weeks of age and moderate hyperglycemia between 8 and 12 weeks of age. Compared with the *db/db* model, much less is known about the susceptibility of *ob/ob* mice to painful DPN. Studies have shown that C57BL/6 *ob/ob* mice have painful DPN by 11 weeks of age, with decreased motor and sensory NCVs, thermal hypoalgesia and tactile allodynia, and small fiber degeneration [146]. Our own research echoed these findings in the BTBR *ob/ob* mice between 9 and 13 weeks of age [147], suggesting that neurological deficits are conserved across different strains. Moreover, we found that female BTBR *ob/ob* mice develop a similar degree of nerve damage as males, with thermal hypoalgesia, reduced NCVs and IENFDs [148].

Genetic T2D models are considered robust models of painful DPN and capture essential features of the human condition. One disadvantage of this model, however, is how quickly and early *db/*

db and *ob/ob* mice attain the diabetic state, relative to the gradual progression from prediabetes to full-blown T2D in humans. Second, leptin-based models are often infertile, and generating sufficient animal numbers is difficult [149]. Lastly, deficient leptin signaling may affect glucose and lipid metabolism independent of T2D, confounding data interpretation [149].

Nonetheless, these models, specifically *db/db* mice, have advanced our knowledge of cellular and molecular mechanisms involved in painful DPN pathogenesis. We previously found that elevated nerve growth factor and substance P expression in *db/db* mice could contribute to mechanical allodynia [47]. Il-10, an anti-inflammatory cytokine, is neuroprotective during painful conditions in the same model, suggesting that increasing Il-10 levels could offer an effective strategy for alleviating painful DPN [140]. Long-term treatment with the antioxidant Coenzyme Q10 prevented hypoalgesia in *db/db* mice by maintaining neuronal levels of phospholipase C (PLC) β3, which is involved in pain perception [141]. On the other hand, we found that hypoalgesia in both male and female *ob/ob* mice may be associated with alterations in genes and pathways related to inflammation and the immune response [147, 148], which is similarly reported in HFD mice with DPN [35] and human DPN [150–152]. These results suggest that ameliorating the inflammatory response may provide a viable therapeutic benefit for painful DPN across species, an avenue worthy of further investigation.

3.6 Emerging Diabetic Models

Although the available T1D and T2D rodent models capture various aspects of painful DPN, there remains a critical need for models that more adequately mirror the human condition to accelerate progress toward a much-needed treatment for this disease. For T1D, the generation of an immunodeficient mouse line engrafted with human β-cells and components of the human immune system [153] offers a promising approach to enhance translatability, but has yet to be characterized for painful DPN.

For T2D, polygenic models generated by the Jackson Laboratories, including NON/ShiLtJ, NONcNZO/LtJ, and TALLYHO/JngJ mice, have recently gained attention in the complications field as they progressively develop metabolic dysfunction and overt hyperglycemia, characteristic of the human condition (www.jax.org). Another advantage of these models is that they retain their leptin signaling, inducing a less severe and adult-onset metabolic dysfunction compared to leptin-based models. NON/ShiLtJ mice placed on a HFD develop insulin resistance and hyperglycemia and a more severe obesity phenotype compared to C57BL/6 mice. When crossed with the T2D-prone NZO/HILtJ strain, the resulting recombinant NONcNZO/LtJ mice demonstrate moderate obesity and T2D [154]. Lastly, TALLYHO/JngJ mice are characterized by moderate obesity and maturity-onset T2D as they develop hyperinsulinemia around

6 weeksof age and hyperglycemia by 12–14 weeks of age [155]. Overall, these models provide an alternative strategy to overcome the shortcomings of leptin-based models and merit investigation in painful DPN.

4 Conclusions

Neuropathic pain is a severe consequence of prediabetes and diabetes, which significantly reduces quality of life of afflicted patients. However, treatment options remain limited, and a better understanding of pain pathogenesis is needed to develop mechanism-based therapies. A barrier to progress has been a lack of reliable testing approaches to capture the multifaceted and diverse presentation of pain. Current methods used to assess nociception in rodents are centered on recording behavioral responses to painful and nonpainful stimuli, which do not capture the full spectrum of nociception. Accordingly, further refinement and standardization of these approaches will help obtain more reliable and replicable scientific data. Moreover, available rodent models mimic several clinical features of painful DPN and provide useful mechanistic insight into disease pathogenesis in T1D and T2D. However, the choice of the specific mouse model must be carefully considered based on study objectives as phenotype severity and response can differ based on model paradigm, strain, diet, and sex as described in detail above (Sect. 3) and Table 2. Therefore, the advantages and limitations of each model as well as behavioral tests used should be carefully considered to ensure collection of meaningful and translatable data that will significantly enhance our understanding of painful DPN and facilitate our progress toward a treatment for pain.

Acknowledgments

The authors would like to thank Dr. Bhumsoo Kim for creating the initial figure. Funding was provided by the National Institutes of Health (NIH) (P30DK020572 and K01DK135799 to S.A.E. and R01DK130913, 1R24082841 to E.L.F.), Novo Nordisk Foundation (NNF14OC0011633 to E.L.F.), the Nathan and Rose Milstein Research Fund (to S.A.E), Neuronetwork for Emerging Therapies at the University of Michigan (to S.A.E. and E.L.F.), and the A. Alfred Taubman Medical Research Institute (to S.A.E. and E.L.F.).

Conflict of Interest The authors declare no competing interest in connection with this chapter.

References

1. Sun H et al (2022) IDF Diabetes Atlas: global, regional and country-level diabetes prevalence estimates for 2021 and projections for 2045 Diabetes Res Clin Pract 183:109119
2. Zaccardi F et al (2016) Pathophysiology of type 1 and type 2 diabetes mellitus: a 90-year perspective Postgrad Med J 92(1084):63–69
3. Feldman EL et al (2017) New horizons in diabetic neuropathy: mechanisms, bioenergetics, and pain. Neuron 93(6):1296–1313
4. O'Brien PD et al (2017) Neurological consequences of obesity. Lancet Neurol 16(6):465–477
5. Feldman EL et al (2019) Diabetic neuropathy. Nat Rev Dis Primers 5(1):41
6. Tabak AG et al (2012) Prediabetes: a high-risk state for diabetes development. Lancet 379(9833):2279–2290
7. Timar B et al (2016) The impact of diabetic neuropathy on balance and on the risk of falls in patients with type 2 diabetes mellitus: a cross-sectional study. PLOS One 11(4):e0154654
8. Margolis DJ, Jeffcoate W (2013) Epidemiology of foot ulceration and amputation: can global variation be explained? Med Clin North Am 97(5):791–805
9. Finnerup NB et al (2016) Neuropathic pain: an updated grading system for research and clinical practice. Pain 157(8):1599–1606
10. Abbott CA et al (2011) Prevalence and characteristics of painful diabetic neuropathy in a large community-based diabetic population in the U.K. Diabetes Care 34(10):2220–2224
11. Alleman CJ et al (2015) Humanistic and economic burden of painful diabetic peripheral neuropathy in Europe: a review of the literature. Diabetes Res Clin Pract 109(2):215–225
12. Jensen TS, Finnerup NB (2014) Allodynia and hyperalgesia in neuropathic pain: clinical manifestations and mechanisms. Lancet Neurol 13(9):924–935
13. Sadosky A et al (2015) Healthcare utilization and costs in diabetes relative to the clinical spectrum of painful diabetic peripheral neuropathy. J Diabetes Complications 29(2):212–217
14. Naranjo C et al (2019) Anxiety, depression and sleep disorders in patients with diabetic neuropathic pain: a systematic review. Expert Rev Neurother 19(12):1201–1209
15. Twiddy H, Frank B, Alam U (2021) A consideration of the psychological aspects to managing patients with painful diabetic neuropathy: an insight into pain management services at a tertiary centre in the UK. Diabetes Ther 12(2):487–498
16. Cavalli E et al (2019) The neuropathic pain: an overview of the current treatment and future therapeutic approaches. Int J Immunopathol Pharmacol 33:2058738419838383
17. Meng W et al (2015) A genome-wide association study provides evidence of sex-specific involvement of Chr1p35.1 (ZSCAN20-TLR12P) and Chr8p23.1 (HMGB1P46) with diabetic neuropathic pain. EBioMedicine 2(10):1386–1393
18. Petropoulos IN et al (2018) Diagnosing diabetic neuropathy: something old, something new. Diabetes Metab J 42(4):255–269
19. Gylfadottir SS et al (2019) Painful and non-painful diabetic polyneuropathy: clinical characteristics and diagnostic issues. J Diabetes Investig 10(5):1148–1157
20. Pop-Busui R et al (2022) Diagnosis and treatment of painful diabetic peripheral neuropathy. American Diabetes Association, Arlington, VA
21. Pop-Busui R et al (2017) Diabetic neuropathy: a position statement by the American Diabetes Association. Diabetes Care 40(1):136–154
22. Finnerup NB, Kuner R, Jensen TS (2021) Neuropathic pain: from mechanisms to treatment. Physiol Rev 101(1):259–301
23. Lee-Kubli CA et al (2014) Animal models of diabetes-induced neuropathic pain. Curr Top Behav Neurosci 20:147–170
24. Middleton SJ et al (2021) Studying human nociceptors: from fundamentals to clinic. Brain 144(5):1312–1335
25. Biessels GJ et al (2014) Phenotyping animal models of diabetic neuropathy: a consensus statement of the diabetic neuropathy study group of the EASD (Neurodiab). J Peripher Nerv Syst 19(2):77–87
26. Gold MS, Gebhart GF (2010) Nociceptor sensitization in pain pathogenesis. Nat Med 16(11):1248–1257
27. Backonja MM et al (2013) Value of quantitative sensory testing in neurological and pain disorders: NeuPSIG consensus. Pain 154(9):1807–1819
28. Deuis JR, Dvorakova LS, Vetter I (2017) Methods used to evaluate pain behaviors in rodents. Front Mol Neurosci 10:284
29. Butler RK, Finn DP (2009) Stress-induced analgesia. Prog Neurobiol 88(3):184–202
30. Dyck PJ et al (2000) Patterns of quantitative sensation testing of hypoesthesia and hyperalgesia are predictive of diabetic polyneuropathy: a study of three cohorts. Nerve growth factor study group. Diabetes Care 23(4):510–517

31. Sharma S et al (2006) Curcumin attenuates thermal hyperalgesia in a diabetic mouse model of neuropathic pain. Eur J Pharmacol 536(3):256–261
32. Calcutt NA, Freshwater JD, Mizisin AP (2004) Prevention of sensory disorders in diabetic Sprague-Dawley rats by aldose reductase inhibition or treatment with ciliary neurotrophic factor. Diabetologia 47(4):718–724
33. Wang R et al (2020) Quercetin attenuates diabetic neuropathic pain by inhibiting mTOR/p70S6K pathway-mediated changes of synaptic morphology and synaptic protein levels in spinal dorsal horn of db/db mice. Eur J Pharmacol 882:173266
34. Eid SA, Feldman EL (2021) Advances in diet-induced rodent models of metabolically acquired peripheral neuropathy. Dis Model Mech 14(11):dmm049337. Epub 2021 Nov 11. https://doi.org/10.1242/dmm.049337
35. O'Brien PD et al (2020) Integrated lipidomic and transcriptomic analyses identify altered nerve triglycerides in mouse models of prediabetes and type 2 diabetes. Dis Model Mech 13(2):dmm042101
36. Davidson EP et al (2010) Diet-induced obesity in Sprague-Dawley rats causes microvascular and neural dysfunction. Diabetes Metab Res Rev 26(4):306–318
37. Forman LJ et al (1986) Streptozocin diabetes alters immunoreactive beta-endorphin levels and pain perception after 8 wk in female rats. Diabetes 35(12):1309–1313
38. Yeomans DC, Pirec V, Proudfit HK (1996) Nociceptive responses to high and low rates of noxious cutaneous heating are mediated by different nociceptors in the rat: behavioral evidence. Pain 68(1):133–140
39. Yeomans DC, Proudfit HK (1994) Characterization of the foot withdrawal response to noxious radiant heat in the rat. Pain 59(1):85–94
40. Yeomans DC, Proudfit HK (1996) Nociceptive responses to high and low rates of noxious cutaneous heating are mediated by different nociceptors in the rat: electrophysiological evidence. Pain 68(1):141–150
41. Lee JH, McCarty R (1992) Pain threshold in diabetic rats: effects of good versus poor diabetic control. Pain 50(2):231–236
42. King TE, Joynes RL, Grau JW (1997) Tail-flick test: II. The role of supraspinal systems and avoidance learning. Behav Neurosci 111(4):754–767
43. Hargreaves K et al (1988) A new and sensitive method for measuring thermal nociception in cutaneous hyperalgesia. Pain 32(1):77–88
44. Kocevski D, Tvrdeic A (2008) The effect of repeated daily measurements on paw withdrawal latencies in Hargreaves test. Coll Antropol 32(Suppl 1):93–97
45. Freeman R et al (2014) Sensory profiles of patients with neuropathic pain based on the neuropathic pain symptoms and signs. Pain 155(2):367–376
46. Costigan M, Scholz J, Woolf CJ (2009) Neuropathic pain: a maladaptive response of the nervous system to damage. Annu Rev Neurosci 32:1–32
47. Cheng HT et al (2009) Nerve growth factor mediates mechanical allodynia in a mouse model of type 2 diabetes. J Neuropathol Exp Neurol 68(11):1229–1243
48. Jolivalt CG et al (2008) Allodynia and hyperalgesia in diabetic rats are mediated by GABA and depletion of spinal potassium-chloride co-transporters. Pain 140(1):48–57
49. Cooper MA et al (2018) Intrinsic activity of C57BL/6 substrains associates with high-fat diet-induced mechanical sensitivity in mice. J Pain 19(11):1285–1295
50. Gavini CK et al (2018) Liver X receptors protect dorsal root ganglia from obesity-induced endoplasmic reticulum stress and mechanical allodynia. Cell Rep 25(2):271.e4–277.e4
51. Randall LO, Selitto JJ (1957) A method for measurement of analgesic activity on inflamed tissue. Arch Int Pharmacodyn Ther 111(4):409–419
52. Minett MS, Quick K, Wood JN (2011) Behavioral measures of pain thresholds. Curr Protoc Mouse Biol 1(3):383–412
53. Inceoglu B et al (2012) Acute augmentation of epoxygenated fatty acid levels rapidly reduces pain-related behavior in a rat model of type I diabetes. Proc Natl Acad Sci USA 109(28):11390–11395
54. Morgado C et al (2011) Minocycline completely reverses mechanical hyperalgesia in diabetic rats through microglia-induced changes in the expression of the potassium chloride co-transporter 2 (KCC2) at the spinal cord. Diabetes Obes Metab 13(2):150–159
55. Gruener G, Dyck PJ (1994) Quantitative sensory testing: methodology, applications, and future directions. J Clin Neurophysiol 11(6):568–583
56. Cruccu G et al (2010) EFNS guidelines on neuropathic pain assessment: revised 2009. Eur J Neurol 17(8):1010–1018
57. Nirogi R et al (2012) Comparison of manual and automated filaments for evaluation of neuropathic pain behavior in rats. J Pharmacol Toxicol Methods 66(1):8–13
58. Chaplan SR et al (1994) Quantitative assessment of tactile allodynia in the rat paw. J Neurosci Methods 53(1):55–63
59. Dubuisson D, Dennis SG (1977) The formalin test: a quantitative study of the analgesic effects of morphine, meperidine, and brain stem stimulation in rats and cats. Pain 4(2):161–174

60. Gregory NS et al (2013) An overview of animal models of pain: disease models and outcome measures. J Pain 14(11):1255–1269
61. Rosland JH et al (1990) The formalin test in mice: effect of formalin concentration. Pain 42(2):235–242
62. McNamara CR et al (2007) TRPA1 mediates formalin-induced pain. Proc Natl Acad Sci USA 104(33):13525–13530
63. O'Brien PD, Sakowski SA, Feldman EL (2014) Mouse models of diabetic neuropathy. ILAR J 54(3):259–272
64. Furman BL (2015) Streptozotocin-induced diabetic models in mice and rats. Curr Protoc Pharmacol 70:5.47.1–5.47.20
65. Yorek MS et al (2014) Effect of glycemic control on corneal nerves and peripheral neuropathy in streptozotocin-induced diabetic C57Bl/6J mice. J Peripher Nerv Syst 19(3):205–217
66. Johnson MS, Ryals JM, Wright DE (2008) Early loss of peptidergic intraepidermal nerve fibers in an STZ-induced mouse model of insensate diabetic neuropathy. Pain 140(1):35–47
67. Yamakawa I et al (2011) Inactivation of TNF-alpha ameliorates diabetic neuropathy in mice. Am J Physiol Endocrinol Metab 301(5):E844–E852
68. Jolivalt CG et al (2011) GLP-1 signals via ERK in peripheral nerve and prevents nerve dysfunction in diabetic mice. Diabetes Obes Metab 13(11):990–1000
69. Dogrul A et al (2011) Systemic and spinal administration of etanercept, a tumor necrosis factor alpha inhibitor, blocks tactile allodynia in diabetic mice. Acta Diabetol 48(2):135–142
70. Lennertz RC et al (2011) Impaired sensory nerve function and axon morphology in mice with diabetic neuropathy. J Neurophysiol 106(2):905–914
71. Johnson MS, Ryals JM, Wright DE (2007) Diabetes-induced chemogenic hypoalgesia is paralleled by attenuated stimulus-induced fos expression in the spinal cord of diabetic mice. J Pain 8(8):637–649
72. Drel VR et al (2007) Evaluation of the peroxynitrite decomposition catalyst Fe(III) tetra-mesitylporphyrin octasulfonate on peripheral neuropathy in a mouse model of type 1 diabetes. Int J Mol Med 20(6):783–792
73. Ventura-Sobrevilla J et al (2011) Effect of varying dose and administration of streptozotocin on blood sugar in male CD1 mice. Proc West Pharmacol Soc 54:5–9
74. McEvoy RC et al (1984) Multiple low-dose streptozotocin-induced diabetes in the mouse. Evidence for stimulation of a cytotoxic cellular immune response against an insulin-producing beta cell line. J Clin Invest 74(3):715–722
75. Eid SA et al (2020) Targeting the NADPH oxidase-4 and liver X receptor pathway preserves Schwann cell integrity in diabetic mice. Diabetes 69(3):448–464
76. Eid SA et al (2021) Differential effects of minocycline on microvascular complications in murine models of type 1 and type 2 diabetes. J Transl Sci 7(1):10–15761
77. Eid SA et al (2020) Differential effects of empagliflozin on microvascular complications in murine models of type 1 and type 2 diabetes. Biology (Basel) 9(11):347
78. Wiggin TD et al (2008) Rosiglitazone treatment reduces diabetic neuropathy in streptozotocin-treated DBA/2J mice. Endocrinology 149(10):4928–4937
79. Hu X et al (2022) Identification and quantification of nociceptive Schwann cells in mice with and without streptozotocin-induced diabetes. J Chem Neuroanat 123:102118
80. Rojas DR, Kuner R, Agarwal N (2019) Metabolomic signature of type 1 diabetes-induced sensory loss and nerve damage in diabetic neuropathy. J Mol Med (Berl) 97(6):845–854
81. Homs J et al (2011) Comparative study of peripheral neuropathy and nerve regeneration in NOD and ICR diabetic mice. J Peripher Nerv Syst 16(3):213–227
82. Atef MM et al (2019) Donepezil improves neuropathy through activation of AMPK signalling pathway in streptozotocin-induced diabetic mice. Biochem Pharmacol 159:1–10
83. Kennedy JM, Zochodne DW (2005) Experimental diabetic neuropathy with spontaneous recovery: is there irreparable damage? Diabetes 54(3):830–837
84. Heng LJ et al (2015) Docosahexaenoic acid inhibits mechanical allodynia and thermal hyperalgesia in diabetic rats by decreasing the excitability of DRG neurons. Exp Neurol 271:291–300
85. Cesena RM, Calcutt NA (1999) Gabapentin prevents hyperalgesia during the formalin test in diabetic rats. Neurosci Lett 262(2):101–104
86. Ismail CAN et al (2021) Ifenprodil reduced expression of activated microglia, BDNF and DREAM proteins in the spinal cord following formalin injection during the early stage of painful diabetic neuropathy in rats. J Mol Neurosci 71(2):379–393
87. Roy Chowdhury SK et al (2012) Impaired adenosine monophosphate-activated protein kinase signalling in dorsal root ganglia neurons is linked to mitochondrial dysfunction and peripheral neuropathy in diabetes. Brain 135(Pt 6):1751–1766
88. Calcutt NA (2004) Modeling diabetic sensory neuropathy in rats. Methods Mol Med 99:55–65

89. Eid SA et al (2023) New perspectives in diabetic neuropathy. Neuron 111(17):2623–2641
90. Manni L, Florenzano F, Aloe L (2011) Electroacupuncture counteracts the development of thermal hyperalgesia and the alteration of nerve growth factor and sensory neuromodulators induced by streptozotocin in adult rats. Diabetologia 54(7):1900–1908
91. Yan JE et al (2012) Streptozotocin-induced diabetic hyperalgesia in rats is associated with upregulation of Toll-like receptor 4 expression. Neurosci Lett 526(1):54–58
92. Sasase T et al (2005) Novel protein kinase C-beta isoform selective inhibitor JTT-010 ameliorates both hyper- and hypoalgesia in streptozotocin-induced diabetic rats. Diabetes Obes Metab 7(5):586–594
93. Cheng KI et al (2014) Persistent mechanical allodynia positively correlates with an increase in activated microglia and increased P-p38 mitogen-activated protein kinase activation in streptozotocin-induced diabetic rats. Eur J Pain 18(2):162–173
94. Sugimoto K et al (2008) Early changes in insulin receptor signaling and pain sensation in streptozotocin-induced diabetic neuropathy in rats. J Pain 9(3):237–245
95. Hoybergs YM, Meert TF (2007) The effect of low-dose insulin on mechanical sensitivity and allodynia in type I diabetes neuropathy. Neurosci Lett 417(2):149–154
96. Ramos KM et al (2007) Pathogenesis of spinally mediated hyperalgesia in diabetes. Diabetes 56(6):1569–1576
97. Singh R, Kishore L, Kaur N (2014) Diabetic peripheral neuropathy: current perspective and future directions. Pharmacol Res 80:21–35
98. Callaghan BC et al (2012) Enhanced glucose control for preventing and treating diabetic neuropathy. Cochrane Database Syst Rev 6:CD007543
99. Jensen TS et al (2021) Painful and non-painful diabetic neuropathy, diagnostic challenges and implications for future management. Brain 144(6):1632–1645
100. Ize-Ludlow D et al (2011) Progressive erosion of beta-cell function precedes the onset of hyperglycemia in the NOD mouse model of type 1 diabetes. Diabetes 60(8):2086–2091
101. Leiter EH (2001) The NOD mouse: a model for insulin-dependent diabetes mellitus. Curr Protoc Immunol Chapter 15:15.9.1–15.9.23
102. Gabra BH, Sirois P (2005) Hyperalgesia in non-obese diabetic (NOD) mice: a role for the inducible bradykinin B1 receptor. Eur J Pharmacol 514(1):61–67
103. Obrosova IG et al (2005) Role for nitrosative stress in diabetic neuropathy: evidence from studies with a peroxynitrite decomposition catalyst. FASEB J 19(3):401–403
104. Wang J et al (1999) A mutation in the insulin 2 gene induces diabetes with severe pancreatic beta-cell dysfunction in the Mody mouse. J Clin Invest 103(1):27–37
105. Wagner K et al (2017) Soluble epoxide hydrolase inhibition alleviates neuropathy in Akita (Ins2 Akita) mice. Behav Brain Res 326:69–76
106. Drel VR et al (2011) Poly(ADP-ribose)polymerase inhibition counteracts renal hypertrophy and multiple manifestations of peripheral neuropathy in diabetic Akita mice. Int J Mol Med 28(4):629–635
107. Sullivan KA et al (2007) Mouse models of diabetic neuropathy. Neurobiol Dis 28(3):276–285
108. Like AA et al (1982) Spontaneous autoimmune diabetes mellitus in the BB rat. Diabetes 31(Suppl 1 Pt 2):7–13
109. Gabra BH et al (2005) Inhibition of type 1 diabetic hyperalgesia in streptozotocin-induced Wistar versus spontaneous gene-prone BB/Worchester rats: efficacy of a selective bradykinin B1 receptor antagonist. J Neuropathol Exp Neurol 64(9):782–789
110. Kamiya H, Zhang W, Sima AA (2006) Degeneration of the Golgi and neuronal loss in dorsal root ganglia in diabetic BioBreeding/Worcester rats. Diabetologia 49(11):2763–2774
111. Callaghan B, Feldman E (2013) The metabolic syndrome and neuropathy: therapeutic challenges and opportunities. Ann Neurol 74(3):397–403
112. Stino AM, Smith AG (2017) Peripheral neuropathy in prediabetes and the metabolic syndrome. J Diabetes Investig 8(5):646–655
113. Vincent AM et al (2009) Dyslipidemia-induced neuropathy in mice: the role of oxLDL/LOX-1. Diabetes 58(10):2376–2385
114. Hinder LM et al (2017) Dietary reversal of neuropathy in a murine model of prediabetes and metabolic syndrome. Dis Model Mech 10(6):717–725
115. Buettner R, Scholmerich J, Bollheimer LC (2007) High-fat diets: modeling the metabolic disorders of human obesity in rodents. Obesity (Silver Spring) 15(4):798–808
116. Guilford BL, Ryals JM, Wright DE (2011) Phenotypic changes in diabetic neuropathy induced by a high-fat diet in diabetic C57BL/6 mice. Exp Diabetes Res 2011:848307
117. O'Brien PD et al (2018) Juvenile murine models of prediabetes and type 2 diabetes develop neuropathy. Dis Model Mech 11(12):dmm037374
118. Yorek MS et al (2015) Effect of diet-induced obesity or type 1 or type 2 diabetes on corneal nerves and peripheral neuropathy in C57Bl/6J mice. J Peripher Nerv Syst 20(1):24–31

119. Rumora AE et al (2019) The divergent roles of dietary saturated and monounsaturated fatty acids on nerve function in murine models of obesity. J Neurosci 39(19):3770–3781
120. Yorek MS et al (2017) Early vs. late intervention of high fat/low dose streptozotocin treated C57Bl/6J mice with enalapril, alpha-lipoic acid, menhaden oil or their combination: effect on diabetic neuropathy related endpoints. Neuropharmacology 116:122–131
121. Anderson NJ et al (2014) Role of insulin signaling impairment, adiponectin and dyslipidemia in peripheral and central neuropathy in mice. Dis Model Mech 7(6):625–633
122. Groover AL et al (2013) Exercise-mediated improvements in painful neuropathy associated with prediabetes in mice. Pain 154(12):2658–2667
123. Coppey L et al (2018) Effect of dietary oils on peripheral neuropathy-related endpoints in dietary obese rats. Diabetes Metab Syndr Obes 11:117–127
124. McDonald TJW, Cervenka MC (2019) Lessons learned from recent clinical trials of ketogenic diet therapies in adults. Curr Opin Clin Nutr Metab Care 22(6):418–424
125. Masino SA, Rho JM (2012) Mechanisms of ketogenic diet action. In: Noebels JL, Avoli M, Rogawski MA, Olsen RW, Delgado-Escueta AV (eds) Jasper's basic mechanisms of the epilepsies. National Center for Biotechnology Information (US), Bethesda, MD
126. Cooper MA et al (2018) A ketogenic diet reduces metabolic syndrome-induced allodynia and promotes peripheral nerve growth in mice. Exp Neurol 306:149–157
127. Abraham A et al (2018) Sex differences in neuropathic pain intensity in diabetes. J Neurol Sci 388:103–106
128. Eid SA et al (2022) Nox, Nox, are you there? The role of NADPH oxidases in the peripheral nervous system. Antioxid Redox Signal 37(7-9):613–630
129. Sajic M et al (2021) High dietary fat consumption impairs axonal mitochondrial function in vivo. J Neurosci 41(19):4321–4334
130. Eid SA et al (2023) Single-cell RNA-seq uncovers novel metabolic functions of Schwann cells beyond myelination. J Neurochem 166(2):367–388
131. Luo J et al (1998) Nongenetic mouse models of non-insulin-dependent diabetes mellitus. Metabolism 47(6):663–668
132. Gilbert ER, Fu Z, Liu D (2011) Development of a nongenetic mouse model of type 2 diabetes. Exp Diabetes Res 2011:416254
133. Barriere DA et al (2018) Combination of high-fat/high-fructose diet and low-dose streptozotocin to model long-term type-2 diabetes complications. Sci Rep 8(1):424
134. Byrne FM et al (2015) Characterisation of pain responses in the high fat diet/streptozotocin model of diabetes and the analgesic effects of antidiabetic treatments. J Diabetes Res 2015:752481
135. Eid S et al (2019) New insights into the mechanisms of diabetic complications: role of lipids and lipid metabolism. Diabetologia 62(9):1539–1549
136. Zhang M et al (2008) The characterization of high-fat diet and multiple low-dose streptozotocin induced type 2 diabetes rat model. Exp Diabetes Res 2008:704045
137. Skovso S (2014) Modeling type 2 diabetes in rats using high fat diet and streptozotocin. J Diabetes Investig 5(4):349–358
138. Munzberg H, Morrison CD (2015) Structure, production and signaling of leptin. Metabolism 64(1):13–23
139. Kleinert M et al (2018) Animal models of obesity and diabetes mellitus. Nat Rev Endocrinol 14(3):140–162
140. Yanik BM, Dauch JR, Cheng HT (2020) Interleukin-10 reduces neurogenic inflammation and pain behavior in a mouse model of type 2 diabetes. J Pain Res 13:3499–3512
141. Shi TJ et al (2013) Coenzyme Q10 prevents peripheral neuropathy and attenuates neuron loss in the db-/db-mouse, a type 2 diabetes model. Proc Natl Acad Sci USA 110(2):690–695
142. Eid SA et al (2021) Gene expression profiles of diabetic kidney disease and neuropathy in eNOS knockout mice: predictors of pathology and RAS blockade effects. FASEB J 35(5):e21467
143. Hinder LM et al (2019) Mitochondrial uncoupling has no effect on microvascular complications in type 2 diabetes. Sci Rep 9(1):881
144. Clee SM, Attie AD (2007) The genetic landscape of type 2 diabetes in mice. Endocr Rev 28(1):48–83
145. Coleman DL (1992) The influence of genetic background on the expression of mutations at the diabetes (db) locus in the mouse. VI: hepatic malic enzyme activity is associated with diabetes severity. Metabolism 41(10):1134–1136
146. Drel VR et al (2006) The leptin-deficient (ob/ob) mouse: a new animal model of peripheral neuropathy of type 2 diabetes and obesity. Diabetes 55(12):3335–3343
147. O'Brien PD et al (2015) BTBR ob/ob mice as a novel diabetic neuropathy model: neurological characterization and gene expression analyses. Neurobiol Dis 73:348–355
148. O'Brien PD et al (2016) Gender-specific differences in diabetic neuropathy in BTBR ob/ob mice. J Diabetes Complications 30(1):30–37

149. Wang B, Chandrasekera PC, Pippin JJ (2014) Leptin- and leptin receptor-deficient rodent models: relevance for human type 2 diabetes. Curr Diabetes Rev 10(2):131–145
150. Guo K et al (2019) Genome-wide DNA methylation profiling of human diabetic peripheral neuropathy in subjects with type 2 diabetes mellitus. Epigenetics 14(8):766–779
151. Guo K et al (2020) Genome-wide profiling of DNA methylation and gene expression identifies candidate genes for human diabetic neuropathy. Clin Epigenetics 12(1):123
152. McGregor BA et al (2018) Conserved transcriptional signatures in human and murine diabetic peripheral neuropathy. Sci Rep 8(1):17678
153. Brehm MA et al (2012) Advancing animal models of human type 1 diabetes by engraftment of functional human tissues in immunodeficient mice. Cold Spring Harb Perspect Med 2(5):a007757
154. Hirata T et al (2017) Pathological and gene expression analysis of a polygenic diabetes model, NONcNZO10/LtJ mice. Gene 629:52–58
155. Leiter EH et al (2013) Comparison of two new mouse models of polygenic type 2 diabetes at the Jackson Laboratory, NONcNZO10Lt/J and TALLYHO/JngJ. J Diabetes Res 2013:165327

Chapter 5

Animal Models of Chemotherapy-Induced Neuropathy: Nociceptive Behavioral Assessment, Metabolic, and Functional Evaluation of the Central Nervous System of Paclitaxel-Injected Rats

José Tiago da Costa-Pereira, Nuria Arias-Ramos, Pilar López-Larrubia and Isaura Tavares

Abstract

Chemotherapy-induced neuropathy (CIN) is one of the most prevalent side effects of cancer treatment with sensory dysfunctions including pain. The peripheral mechanisms of paclitaxel-induced neuropathy are widely understood, whereas the central processes are starting to be uncovered. Paclitaxel-induced neuropathy can be modeled in experimental conditions using rats injected with low doses of paclitaxel, which mimics chemotherapy cycles, being relevant to the clinical translation. The first part of this chapter describes behavioral technique used to validate the mechanical allodynia in animal models of paclitaxel-induced neuropathy using the von Frey test. In the second part, a protocol is detailed to study the central sensitization of paclitaxel-injected rats through the immunodetection of the c-fos, a neuronal marker, in the spinal cord. In the last part of this chapter, protocols of groundbreaking magnetic resonance imaging (MRI) techniques, manganese-enhanced MRI (MEMRI) and ex vivo spectroscopy (HRMAS), are described to evaluate the central nervous system of animal models of the paclitaxel-induced neuropathy functionally and metabolically.

Keywords Chemotherapy-induced neuropathy, Paclitaxel, Neuropathic pain, Von Frey test, c-Fos, MEMRI, Ex vivo spectroscopy, HRMAS, Rat

1 Introduction

With improved rates of cancer remission, cancer patients are living longer [1]. Chemotherapy is the most common approach in cancer treatment, but it induces several side effects. Although incidences are higher with platinum-based agents and taxanes [2], a common side effect of chemotherapy is peripheral neuropathy, which may occur during the period of cancer treatment or even subsist after its complete remission [3–5]. This condition is known as chemotherapy-induced neuropathy (CIN). Clinical studies have stated the approximate incidence of CIN after cessation of treatment is around

Isaura Tavares and Hugo Leite-Almeida (eds.), *Translational Research Methods for Neuropathic Pain*, Neuromethods, vol. 235, https://doi.org/10.1007/978-1-0716-5312-8_5,

68% at 1 month, 60% at 3 months, and 30% at 6 months. The risk of developing CIN depends on the type of chemotherapeutic agent and cumulative dose [6, 7]. Frequently, CIN imposes a reduction in the chemotherapy dose and may even induce discontinuation of the planned treatment [6, 8].

The cytostatic agents known to induce CIN include several drugs, namely anti-tubulin agents (paclitaxel, docetaxel, and vincristine), platinum agents (cisplatin, carboplatin, and oxaliplatin), and proteosome inhibitors (bortezomib). Taking into account that these drugs act through different mechanisms, although inducing neuronal toxicity at the peripheral fibers, cytostatic drugs may cause heterogeneous degrees of damage with putatively distinct effects in the nociceptive input that travels from the periphery to the central nervous system (CNS) [9].

Since paclitaxel is first-line therapeutic alone or in combination with other cytostatic drugs for solid tumors such as breast, ovarian, prostate, and non-small cell lung carcinomas [10], all procedures described in this chapter only focus on the paclitaxel treatment.

Animal models of CIN are based on the intermittent administration of the therapeutic dose of cytostatic drugs to mimic cycles of chemotherapy. They allow the study of the mechanisms prompting the progression and recovery of CIN as well as the assessment of the new pharmacological and non-pharmacological strategies for prevention and treatment of CIN. Paclitaxel-induced neuropathy can be modeled in experimental conditions using rats intraperitoneally injected with low doses of paclitaxel, which simulates chemotherapy cycles [11]. These rats exhibit several pain-like behaviors, namely mechanical and thermal hypersensitivity and, frequently, spontaneous pain [8, 12].

1.1 Methods to Study Paclitaxel-Induced Neuropathy in Rats

Preclinical studies with paclitaxel demonstrated that its cumulative administration affects the integrity of peripheral nerves triggering concomitantly behavioral manifestations such as mechanical and thermal hypersensitivity—see for instance [13–15]. The dose-dependent effects of paclitaxel treatment have also been observed in humans, where the incidence and severity of neuropathic signs and symptoms increased relative to increasing cumulative doses of paclitaxel [16]. Several parameters to evaluate CIN both in clinical and preclinical environment, with great translational potential, can be considered such as the nerve conduction velocity, the morphometry of peripheral nerves, and the presence of intraepidermal nerve fibers after paclitaxel treatment [17–20]. Nevertheless, the behavioral alterations are more consistent to validate the peripheral neuropathy. Thus, the first part of this chapter describes the protocol for the mechanical allodynia evaluation.

Along with the changes in nociceptive behavior, functional changes in the processing of sensory information by spinal cord neurons have been reported following paclitaxel treatment by electrophysiological approaches [21]. The central sensitization is a significant feature of neuropathy [22]. In conjunction with behavioral evaluation, we also present a detailed procedure to study the phenomena of central sensitization monitoring the expression of c-Fos, a specific marker of nociceptive activity at the spinal dorsal horn [23] during paclitaxel-induced neuropathy.

Besides the classical approaches to validate the animal model of CIN, neuroimaging is also emerging as a valuable tool to explore the activity of the brain with the possibility to perform longitudinal studies using the same animal noninvasively, especially greatly magnetic resonance imaging (MRI) [24, 25].

This chapter presents a comprehensive protocol to measure the brain function and regional metabolomics in animals with paclitaxel-induced neuropathy using magnetic resonance approaches. Manganese-enhanced magnetic resonance imaging (MEMRI) has been used for high-contrast imaging in animal models of neuropathic pain, since MEMRI changes spin-lattice relaxation times (T1), which is a direct measure of neuronal activation [26]. Manganese ions (Mn^{2+}) compete with calcium ions (Ca^{2+}) for the voltage-dependent calcium ion channels, which are activated during neuronal activity, permitting the accumulation of Mn^{2+} in the neurons [27]. Furthermore, astrocytes are able to uptake manganese ions located in the neural surroundings by the gap junctions present in these cells [28]. This manganese accumulation can be detected as hyperintensity in T1W images due to the selective induced contrast in those regions of higher activation. The extension of the effect can be also quantified by measuring T1 in the different cerebral regions. However, although MEMRI is a straightforward method to measure neural activity by MRI, excessive accumulation of manganese results in toxic effects causing neural damage [29], hampering MEMRI translations to the clinical routine and being exclusively employed in preclinical studies [30, 31].

Along with brain activity variations, chronic pain also causes neurochemical and metabolic changes linked to neuroinflammatory and neurodegenerative processes. Magnetic resonance spectroscopy (MRS) is a powerful analytical technique that can measure qualitatively relevant brain metabolites both in vivo and ex vivo, which reflect neuronal and glial densities. Moreover, in vivo MRS is widely used in clinical scenarios to study some chronic pain conditions, highlighting changes in metabolites such as N-acetylaspartate, glutamate, and myoinositol [32]. Additionally, ex vivo spectroscopy using high-resolution magic angle spinning (HRMAS) technique allows to obtain local spectra from dissected brain regions in which high resolution is achieved due to the sample inclination with respect to the main direction of the magnetic field and high spinning, both canceling dipolar coupling and chemical shift anisotropy, respectively [33]. The proper selection of a euthanasia method which minimizes the influence of the *post mortem* metabolism of the brain, such as the Focus Microwave Method, enables to obtain accurate metabolomic information very proximate to in vivo metabolism [34].

Taking this into account, this chapter ends with a procedure to estimate the metabolic profile of the several dissected brain areas involved in pain modulation/processing from paclitaxel-induced neuropathy animal model using ex vivo quantitative HRMAS.

2 Materials

2.1 Animals

Adult male Wistar rats (weighing 150–175 g) are housed in a 12/12 h light/dark environment at 22 ± 2°C and receive food and water ad libitum. The animals must be acclimated to the housing facility for at least 1 week before the initiation of experiments. All procedures should be approved by the institutional animal ethics committee and the national authority for animal health. Moreover, the experiments should follow the ethical guidelines of the International Association for the Study of Pain (IASP) in conscious animals.

2.2 CIN Induction

- Paclitaxel (Taxol®, Tocris Bioscience, UK).
- Dimethyl sulfoxide (DMSO, Sigma-Aldrich®, USA).
- Saline solution.
- Syringes (1 ml).
- Syringe needles 26 G.
- Thermoblock at 37°C.

2.3 Nociceptive Behavior Testing Apparatus

- Transparent Plexiglas cages with metallic mesh floor.
- Calibrated von Frey test monofilaments (Stoelting, USA).

2.4 Immunohistochemistry

Vascular Perfusion

- Cold plate apparatus (Bioseb®, USA/Canada).
- Anesthesia: sodium pentobarbital.
- Calcium-free Tyrode's solution.
- Paraformaldehyde (PFA).
- Phosphate buffer (pH = 7.2) (PBS).
- Sucrose.
- Cryostat (Leica CM3050).
- Cryoprotectant solution.

Immunoreaction

- 12-well plate.
- PBS.
- Hydrogen peroxide.

- Triton X-100 (Sigma-Aldrich®, USA).
- Normal swine serum (NSS).
- Anti-c-Fos primary antibody (Oncogene®, USA).
- Swine biotinylated anti-rabbit secondary antibody (Dako, Denmark).
- Avidin–Biotin Complex (ABC) solution (Vector, USA).
- 3,3′-diamino-benzidine (DAB, Sigma-Aldrich®, USA).
- Gelatine-coated slides.
- Xylol.
- Mounting medium (Eukitt®, Fluka Analytical, Germany).

Visualization

- Light microscope (Axioskop 40 model, Zeiss®, Switzerland).
- High-resolution digital camera (Leica EC3 model).
- LAS 4.6.0. software (Leica Microsystems®).

2.5 Manganese-Enhanced Magnetic Resonance Imaging (MEMRI)

MRI System

- MR high-field magnet (>7 T)
- Gradient coil strength > 360 mT/m with ramp times < 200 µs.
- Radio frequency (RF) coils. Several combinations can be used depending on the animal to be studied.
  - **Transceiver coil setup**: The excitation and reception of the MR signal are performed with the same volume coil of around 35–40 mm inner diameter for rats and 25–30 inner diameter for mice.
  - **Active decoupled transmit and receive coil setup**: Excitation and reception of the signal are typically made by a volume transmit coil and a brain surface coil (rat and/or mouse dedicated). In that case, coils must be actively detuned from each other.
- Other equipment, devices, and materials for MRI studies:
  - **Anesthesia equipment**: Small rodent inhalation anesthesia equipment formed by an anesthetic vaporizer, which dissolves the anesthetic liquid (isoflurane or similar) in a carrier gas (oxygen or medical air), a flowmeter (depending on the carrier gas), poly(methyl methacrylate) chambers, nose cap, circulating tubes, and an anesthesia filter.
  - **Animal holders**: Suitable animal beds made of MR-compatible material (e.g. acrylate) are used to

immobilized the animal inside the MR magnet. The holder will be slid to localize the region of interest in the isocenter of the magnet.

- **Warm blanket**: Anesthetized animals are placed in water heated beds (or some other system) to maintain the body temperature at 37°C during the imaging session.
- **Small animal physiological monitoring system:** It continuously reports data of body temperature (measured with a rectal probe) and respiratory rate (using a pneumatic pillow placed in the animal abdomen) to monitor the state of the anesthetized animal during the MRI studies.

Manganese Infusion

- A solution of manganese chloride ($MnCl_2$) 100 mM in distilled water, pH = 7.4.
- Anesthesia equipment.
- Infusion pumps to administrate the $MnCl_2$ solution at a fixed infusion rate. Also necessary: syringe, catheter needle/needle, polyethylene tubing, and physiological serum.
- Eye ointment.

Software

- Acquisition software form the MRI scanner vendor.
- **Post-processing**: MEMRI studies are based on the acquisition of T1 maps that can be computed with different software options (including commercial or home-made) to perform a pixel-by-pixel fitting of the MRI signal to Eq. 1 (where S is the MR signal value for a given repetition time (TR) and S_0 the MR signal for TR = 0). This procedure will generate a parametric map of the T1 value in each pixel. Finally, regions of interest are manually selected (using the brain atlas "Paxinos" [35] as a guide) with image analysis software such as Image J, and data are statistically analyzed.

$$S = S_0(1 - e^{-TR/T1}) \qquad (1)$$

2.6 Ex Vivo Spectroscopy

Tissue Sample Collection

- High-power (5 kW) focused microwave for experimental animal sacrifice (or any other appropriate euthanasia procedure).
- Surgical instruments for tissue dissection: scissors, scalpels, razors, and forceps.
- Cryotubes.
- Liquid nitrogen.

HRMAS Spectroscopy

- NMR vertical spectrometer (>400 MHz) equipped with an HRMAS probe.
- Zirconia HRMAS rotor with an insert and rotor cap.
- Digital balance.
- Deuterium oxide (D_2O) as solvent.
- Black marker.
- Petri dish.
- Dry ice.
- Sodium hypochlorite.
- Milli-Q water.

Software

- Acquisition software from the NMR spectrometer vendor.
- Processing of spectra can be done with different software, including: LcModel, MestReNova, MetaboAnalysit, or any other program or application that allows metabolite's quantification.

3 Methods

3.1 CIN Induction

Stock solution of paclitaxel is prepared by dissolving 50 mg of paclitaxel in 1 ml of DMSO with gentle agitation at 37°C. The stock solutions are aliquoted and stored at −20°C.

To inject the paclitaxel solution as fresh as possible, on the day of administration of paclitaxel, the aliquot is thawed and successively diluted in a saline solution (0.9% NaCl) at 37°C with gentle agitation until a final concentration of 2.0 mg/ml of paclitaxel in 4% of DMSO.

The CIN model is induced in animals weighing 175–190 g (*see* **Note 1**) by injecting 2.0 mg/kg (intraperitoneal; i.p.) of paclitaxel on four alternating days (1, 3, 5, and 7) as previously described [11].

Control group of animals is injected with vehicle solution (4% of DMSO) dissolved in saline solution (0.9% NaCl).

3.2 Mechanical Allodynia Evaluation in CIN Animals Using von Frey Test

The mechanical allodynia is assessed using the von Frey test. To perform the von Frey test, the animals are placed in an elevated transparent Plexiglas cages with metallic mesh floor and allowed to acclimate to the environment for 15–20 min (*see* **Note 2**).

After acclimatization, the von Frey test is performed by the "up and down" method which consists of the application of monofilaments between 0.4 and 26.0 g, starting with the 2.0 g monofilament. Each monofilament is consecutively applied to the mid-plantar region of the rat's hind paw until it bends for 2–3 s. In the absence of the paw withdrawal, the next higher monofilament is applied, whereas in the presence of paw withdrawal, the next lower monofilament is applied. A brisk withdrawal or hind paw elevation is recognized as a

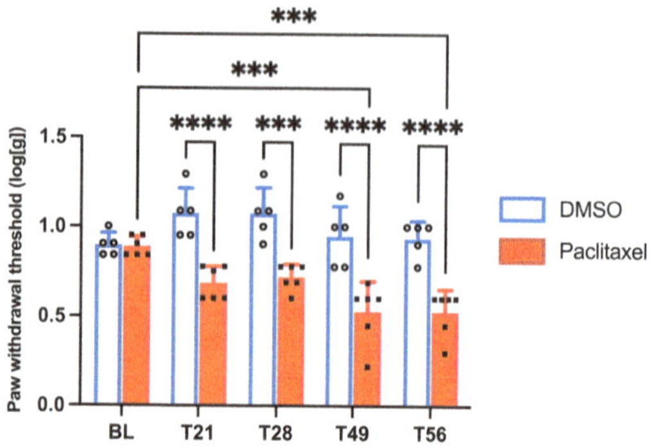

Fig. 1 Time-course effects of systemic paclitaxel administration on mechanical sensitivity assessed by the von Frey test. The test was performed before (baseline; BL), at days 21 (T21), 28 (T28), 49 (T49), and 56 (T56) after the first injection of paclitaxel or DMSO. Data are presented as mean ± SD. ***$P < 0.001$; ****$P < 0.0001$. (Reproduced from Costa-Pereira et al. [15] with permission from Elsevier)

positive response. Two trials are performed on the left or right hind paw, and each value obtained is transformed into logarithms and averaged (Fig. 1).

3.3 c-Fos Immunodetection in Spinal Dorsal Horn by Immunohistochemistry (IHC)

Vascular Perfusion and Spinal Cord Processing for IHC

At 1 month after CIN induction, the animals are placed on the cold plate surface, which is set at 0°C, three times for 50 s each, at an interval of 5 min. Two hours after the first cold stimulation, the animals were perfused (*see* **Note 3**).

The vascular perfusion is performed in animals deeply anesthetized with 65 mg/kg of sodium pentobarbital via i.p. injections. The animals are perfused with 100 ml of calcium-free Tyrode's solution followed by 750 ml of 4% PFA in 0.1 M phosphate buffer (pH = 7.2).

After perfusion, the spinal cords of the animals are dissected and immersed in 4% PFA for 4 h and then cryopreserved in 30% sucrose solution. The spinal L4 segments are then transversally sliced into 30 μm thick sections in a cryostat and then preserved in cryoprotectant solution at −20°C.

Immunohistochemical Detection

Free-floating spinal L4 sections are first washed with 0.1 M PBS four times for 10 min. Then, they are incubated for 20 min with a 30% hydrogen peroxide solution to block endogenous peroxidase activity, followed by a 10 min with PBS-T. Sections are then incubated for 2 h with a blocking solution of glycine and 10% NSS diluted in PBS-T, followed by incubation for two overnights with a rabbit-raised anti-c-Fos primary antibody diluted at 1:2500 in PBS-T and 2% NSS at 4°C. After further washing with PBS-T, the sections are incubated for 1 h with a swine biotinylated anti-rabbit secondary antibody diluted at 1:200 in PBS-T and 2% NSS at room temperature. Sections are again washed before incubation for 1 h with the ABC solution. The detection step occurs by the incubation of DAB. The sections are mounted on gelatine-coated slides, cleared in xylol, and coverslipped with mounting medium.

Photomicrographs are taken from the left and right sides of the spinal dorsal horn of each animal. A total of five random non-contiguous L4 sections are acquired under the same exposure and lighting settings from a light microscope coupled to a high-resolution digital camera and the LAS 4.6.0. software.

Quantification of c-Fos Immunoreactive Nuclei

The quantification of c-Fos immunoreactive nuclei is performed using the ImageJ® software. The Cell Counter tool is used to manually count the number of c-Fos nuclei on laminae I–II, III, IV, and V of the spinal dorsal horn (Fig. 2). The spinal laminae are carefully delimited by superimposing a digital diagram of the L4 segment taken from the atlas of the rat's brain [35] on each section of the spinal cord.

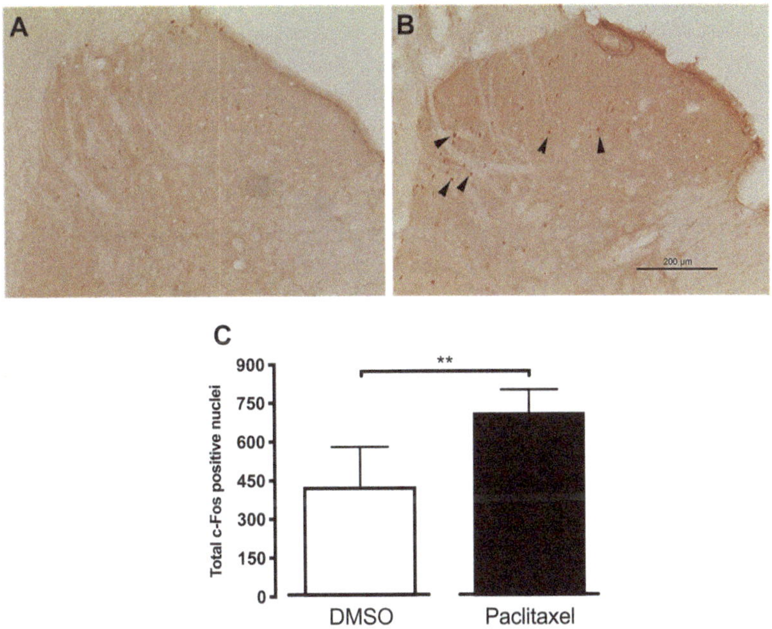

Fig. 2 Expression of c-Fos after noxious cold stimulation in the spinal dorsal horn. Representative photomicrographs of c-Fos expression in the spinal dorsal horn of animals injected with DMSO (**A**) and paclitaxel (**B**). Some c-Fos-positive nuclei are marked with arrowheads in (**B**). Scale bar in (**B**): 200 μm (**A** and **B** are at the same magnification). The total number of c-Fos positive nuclei counted at the spinal dorsal horn is shown in (**C**). Data are presented as mean ± SD $^{**}P < 0.01$ vs DMSO

3.4 Functional Activity in Brain Areas Involved in Pain Modulation with MEMRI

The functional activity in brain areas involved in pain modulation can be evaluated in vivo with MRI methodologies. To that, at the interested time point after the paclitaxel injection, the CIN-inducted animals are intraperitoneally (i.p.) infused with a manganese solution and the MEMRI studies are performed 24 h later (*see* **Note 4**, step 1).

Manganese Intraperitoneal Administration

The $MnCl_2$ is infused i.p through a custom-made canula/catheter, introduced in the ventral zone of the abdomen, employing a needle and polyethylene (PE) tubing of the appropriate length and diameter (*see* **Note 4**, step 2).

Preparation and Administration of $MnCl_2$ Solution

To acquire MEMRI studies, Mn^{2+} is used as a contrast medium. A stock solution of 100 mM $MnCl_2$ in distilled water is freshly prepared and pH adjusted to a physiological value (7.2–7.4) by adding 100 mM NaOH.

The $MnCl_2$ is infused i.p through a custom-made canula/catheter employing a needle and polyethylene (PE) tubing of the appropriate length and diameter (*see* **Note 4**, step 2). The canula is introduced in the ventral zone of the abdomen in the anesthetized animal (induced with 2%–2.5% isoflurane/oxygen in a box and maintained with 1%–1.5% isoflurane/oxygen with nose mask during the infusion) at a dose of 1 mmol/kg body weight and a rate of 0.1 mL/min.

MRI Acquisition Protocols

The MRI protocols described in this section are optimized and performed in a Bruker Biospec 7T/16 cm bore MRI system equipped with a ^1H circular polarized transmit/receive birdcage volume coil (40 mm inner diameter) and a shielded gradient system of 360 mT/m maximum amplitude (90 mm diameter). Data are acquired with Paravision 6.0.1 software (in a Linux environment) and with a Bruker AVANCE III console. Any other suitable combination of commercial or custom radiofrequency coil configuration can be employed. Slice orientation in the MRI acquisitions is selected according to the needs of the study. Before starting MRI studies, make sure to observe all the health and safety rules in a high magnetic field room. After that, place the animal in the scanner following these steps:

1. Place the animal in an anesthesia induction box (2%–2.5% isoflurane/O_2 at a flow of 1 L/min) and perform a pinprick stimulation in the left hind paw with a safety pin at an intensity sufficient to indent the skin.
2. Move the animal to the appropriate MRI bed/holder in the prone position, over a heated blanket (or any other system to maintain the body temperature) and maintain the anesthesia through the whole study (1%–1.5% isoflurane/O_2, 1 L/min). The body temperature and respiration rate need to be controlled with the physiological monitoring system during the imaging session (*see* **Note 4**, step 3). To that, a thermometer is inserted in the rectum, and a breathing sensor pillow is

allocated under the animal chest to monitor the animal status and modulate the plane of anesthesia. The animal head is immobilized during imaging acquisition, with ears and tooth bars, to avoid processing errors due to motion.

3. Introduce the animal in the scanner, either manually or using an automatic positioning system, matching the volume coil with the isocenter of the magnet.
4. Identify the study in the operating software of the system with all the required information. Prior to the first scan, an automatic or manual adjustment of some parameters must be achieved:
   a. Tuning and matching the coil to adjust the coil frequency of resonance and impedance according to the animal.
   b. Shimming to improve the magnetic field homogeneity.
   c. Calculate the RF power for the 90° excitation pulse.
   d. Adjust the basic frequency of the system.
   e. Determine the optimal receiver gain.
5. Acquire fast localization images of the three orthogonal spatial orientations of the animal (axial, coronal, and sagittal) and check its correct position inside the magnet.
6. Obtain structural images of the brain to be used as a reference. Normally, a RARE (rapid acquisition with relaxation enhancement) T2W sequence is employed for obtaining anatomical information that will be used to better identify the cerebral areas in the regions of interests (ROIs) selection.
7. Generate a T1 map acquisition through a saturation-recovery experiment with a RAREVTR (rapid acquisition with relaxation enhancement at variable TR) sequence (*see* **Note 4**, step 4). Ideally, a study should be acquired prior to the manganese administration, as basal conditions, and another 24 h after the infusion of the Mn^{2+} (MEMRI).

The MRI parameters used in the previous acquisitions are summarized in Table 1.

Processing and Data Analysis

1. Images can be processed and analyzed with a dedicated tool incorporated into the software of the MRI scanner. Data, as raw-data or in DICOM format (Digital Imaging and Communication in Medicine), are also read with different analysis software such as MATLAB (The MathWorks Inc., Natick, MA, 2000) or IDL (Iterative Data Language, Research System, Boulder, CO, USA).
2. T1 parametric maps are generated by fitting the MRI signal to Eq. 1 (Section: Software). In our experiments, we use the homemade software application (MyMapAnalyzer) developed

Table 1
MRI parameters used in the MEMRI experiment

	Localization	T2W	T1 map
Method	FLASH	RARE	RAREVTR
Repetition time (TR), ms	100	2500	150, 200, 400, 800, 1600, 3500, 6000
Echo time/Effective echo time (TE), ms	3	44	10
Number of averages	1	3	1
Acquisition time	12.8 s	3 min	20 min
RARE factor	–	8	–
Matrix size	256 × 256	256 × 256	128 × 128
Field of view (mm^2)	50	35	35
In-plane resolution (mm^2/pixel)	0.195	0.137	0.273
Orientation	Axial, coronal, and sagittal	Axial	Axial
Slice thickness (mm)	1	1.5	1.5
Number of slices	3	6	6

in MATLAB (The MathWorks Inc., Natick, MA, 2000). These parametric images are analyzed with Image J (U.S. National Institutes of Health, Bethesda, MD, USA) [19], or any other adequate software, allowing the measurement of mean T1 values from the manually or automatically selected ROIs of the brain.

3. The statistical evaluation of data can be done with different software, being typically employed: SPSS (IBM, Armonk, NY, USA), Excel (Microsoft, Redmond, WA, USA), R [20], or GraphPad Prism (GraphPad Software, San Diego, CA, USA).

To illustrate the protocol, an example of T1 maps obtained in MEMRI studies of a rat treated with paclitaxel and a rat injected with DMSO as control is shown in Fig. 3.

3.5 HRMAS Spectroscopy

Metabolomic profiles can be obtained ex vivo from HRMAS spectra of tissue collected from different brain areas involved, or not, in pain modulation.

Animals' Euthanasia

1. The animal is euthanized right after finishing the MRI session while still anesthetized. To that, it is placed in a water-refrigerated specific holder and quickly introduced in the high-power (5 kW) focused microwave system, which causes the death in

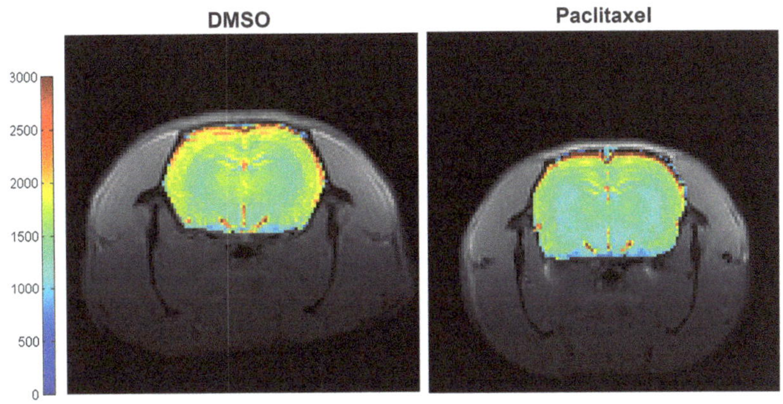

Fig. 3 Representative T1 maps of the same slice in axial orientation from DMSO- and paclitaxel-injected rats. The acquisition parameters of the T1 images are shown in Table 1

2.7 s triggering a brain metabolism arrest by heat inactivation (*see* **Note 5**, step 1).

2. After removing the rat from the MW system, the brain is excised from the skull to collect the brain regions of interest (in our case: hypothalamus, periaqueductal gray (PAG), anterior cingulate cortex (ACC), and amygdala); after microdissection, these are then snap-frozen in liquid nitrogen and stored at –80°C until the spectra acquisitions.

Sample Preparation for HRMAS Spectroscopy

1. The sample preparation must be performed at low temperature to avoid sample thawing.
2. A small piece of the cerebral ROI (10–15 mg) is resected with a scalpel and introduced into the zirconium rotor (4 mm of outer diameter) and filled with 50 µl of D_2O.
3. The rotor with the sample is quickly introduced into the pre-cooled spectrometer (at 4°C) to acquire the spectra.

Acquisition of Ex Vivo HRMAS Spectra

Ex vivo spectra explained in this section are acquired in a 11.7 T Bruker AVANCE WB spectrometer operating at 500.13 MHz for ^1H, equipped with a triple-nuclei high-resolution magic angle spinning (HR-MAS) probe and using Topspin 4.1.1 software.

1. Create the study and load a one-pulse protocol to do the adjustments.
2. Set the spinning rate at 5 kHz and the temperature at 277 K.
3. Tuning and matching the HRMAS probe to adjust the coil frequency of resonance and impedance according to the sample.
4. Acquire a ^1H one-pulse spectrum to check the field homogeneity and measure the water frequency.

5. Shim the sample either automatically or manually, trying to achieve the highest magnetic field homogeneity (*see* **Note 5**, step 3).
6. Load a water-suppression ^{1}H one-pulse protocol matching the pulse suppression frequency with the water frequency previously measured.
7. Optimize the water signal suppression by adjusting the length and power of the suppression pulse (*see* **Note 5**, step 4).
8. Load a Carr–Purcell–Meiboom–Gill (CPMG) sequence with the following acquisition parameters: water suppression pulse: 2 s (can be modified within 1–3 s), relaxation delay: 5 s, data points: 32k, number of scans: 128, echo time: 36 and/or 144 ms (*see* **Note 5**, step 5).
9. When the spectrum acquisition is completed, remove the rotor from the probe and clean it by soaking it in sodium hypochlorite for 30 min, rinse it with Milli-Q water and dry it.
10. Processing the spectra and metabolite quantification can be achieved with different programs (*see* **Note 5**, step 6). An example of a spectrum processed and quantified with the program LCModel [Provencher SW, Magn Reson Med. 1993, 30(6):672–9] can be observed in Fig. 4.

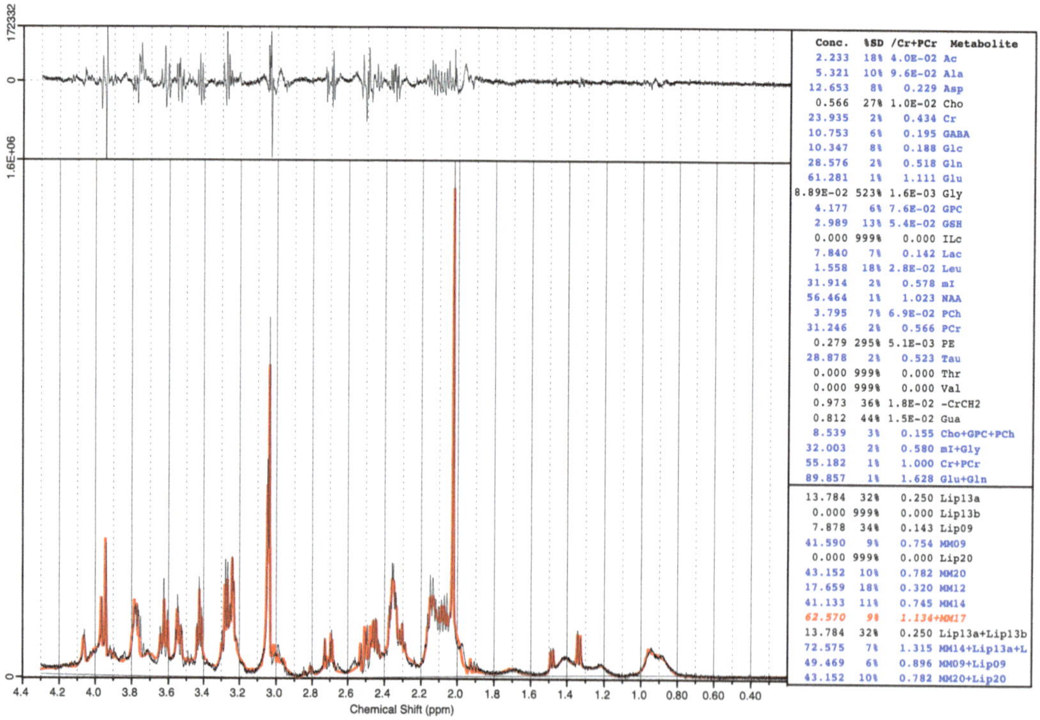

Fig. 4 Representative ex vivo HRMAS spectra and metabolite quantification processed by LCModel program

4 Notes

1. **Animals:**
For the induction of CIN, the weight of animals is determined taking into account that, weeks after, we may have to perform stereotaxic and intrathecal surgeries, which are executed at specific weights. However, it is scientifically relevant to study the effects of paclitaxel treatment on older animals, since advanced age is a risk factor for developing CIN [36].

2. **Behavioral assessment:**
The tests are performed after a period of habituation of 1 week, during which the animals were handled by the experimenter in the test room for 30 min and placed in the testing apparatus for another 30 min every day.

3. **Immunohistochemistry:**
c-Fos is a validated marker for neuronal activation in spinal cord under noxious stimulation [37]. The animals should be perfused 2 h after the first cold stimulation since the 2-h interval is required to ensure maximal expression of the proto-oncogene c-fos in the spinal dorsal horn after noxious stimulation [38, 39].

4. **MEMRI protocol:**
    1. In order to compare the functional activity of the brain before and after the CIN induction, MEMRI studies need to be carried out in control animals. The effects of Mn^{2+} in the MRI signal are highly appreciated at 24 h, but they can also be detected at other time points up to 4 days after the manganese administration.
    2. The canula/catheter can be prepared with a needle and polyethylene (PE) tubing of the appropriate length and diameter. In our studies, we use PE20 tubing (0.38 mm of inner diameter × 1.1 mm of outer diameter) with a 26 G needle, detached from the hub, placed at the end of the tube. Any other tailored needle combination can work properly. Other routes of administration, such as intravenous or intracerebral, can be used with the precaution of previously determining the optimal dose given the high neurotoxicity of manganese. A safe route of administration consists of intraperitoneal implantation of osmotic infusion pumps of adequate volume and flow rate.
    3. To perform accurate and reliable parametric MRI studies, it is very important to maintain a constant body temperature and respiration rate [40]. This is achieved by increasing/decreasing the level of isoflurane in oxygen and the temperature of the heated blanket or the system employed to warm the animal. The body temperature should be

maintained at ~37°C and the respiratory signal at 40–60 breaths per minute to minimize motion artifacts in image. In our studies, we use a physiological monitoring system (Small Animal Instruments, Inc., Stony Brook, USA).

4. T1 values in brain areas will identify those of different accumulation dependent on the different brain activity. The T1 maps allow quantifying these effects and can be generated by using different methodologies, such as saturation-recovery or inversion-recovery, either based on spin-echo or gradient-echo acquisitions. In any case, to obtain a good fitting of the signal to the mathematical equation, at least six different values of T1 contrast should be imaged. Besides, T1W images can be acquired and just measure the effects of Mn^{2+} on signal intensity. Slice orientation in the MRI studies.

5. **HRMAS:**

   1. The sacrifice of animals by heat inactivation of the metabolism is the most optimal method to obtain ex vivo the metabolomic profile because it avoids later metabolite alterations due to potential metabolism activation. Nevertheless, if the system is not available in the laboratory, other methods of euthanasia can be employed, selecting the one that is expected not to induce alterations in those metabolites of higher interest in our study. In any case, death of the animal must occur in the shortest time possible, and tissue samples must be frozen quickly after collection.
   2. Frozen tissue HRMAS spectra are acquired and ideally should not be thawed during the sample preparation process. This is best achieved by working in a cold room, at 4°C. If not, the tissue must be handled on a Petri dish over dry ice during the whole procedure of sample preparation.
   3. Having a good shim is crucial for subsequent accuracy in metabolite quantification. Automatic shim adjustment may work properly, but it can usually be improved manually by checking the signals: the taller and narrower, the better. This can be easily assessed by monitoring the lactate signal, a doublet at 1.23 ppm with two peaks 7 Hz apart. The homogeneity of the magnetic field can be varied and assessed by viewing these signals in real time. The optimal measure of a good shim is the width at half-height of a signal: a value of 1.5–2 Hz is considered good enough. Optimize the water suppression pulse by increasing or decreasing both the length and strength of the pulse in such a way that the more water signal intensity is suppressed, the better, but avoiding inverting the signal

because it will alter the actual intensity of nearby signals from other metabolites.

4. Different sequences can be employed to determine and measure metabolites with HRMAS, also 1D and 2D methodologies. The most used with biopsies or tissue samples are those named CPMG, in which signals coming from macromolecules are eliminated by acquiring an echo at a specific echo time (TE). The echo time can be selected according to the signals to be observed: with TE = 36 or 144 ms, all metabolites with T2 > 36 ms or > 144 ms respectively, will not appear in the spectrum.
5. The metabolite analysis can be made with several quantification programs, such as LCModel (http://s-provencher.com/lcmodel.shtml), MestReNova (https://mestrelab.com/download/mnova/), or MetabolAnalyst (https://www.metaboanalyst.ca). In our studies, we process and quantify the spectra with LCModel, a prior knowledge spectral fit software. This program fits the sample spectrum as a linear combination of model spectra included in a home-designed database of cerebral metabolites and contributions for lipids and macromolecules, yielding values of metabolite concentration and estimated standard deviation [Righi V. et al., J Proteome Res. 2018, 17(9):2953–2962]. Only metabolites with SD smaller than 20% were included in the final analysis of the data.

Conflict of Interest

The authors have no conflicts of interest to declare that are relevant to the content of this chapter.

References

1. Arnold M, Rutherford MJ, Bardot A, Ferlay J, Andersson TM, Myklebust TA, Tervonen H, Thursfield V, Ransom D, Shack L, Woods RR, Turner D, Leonfellner S, Ryan S, Saint-Jacques N, De P, McClure C, Ramanakumar AV, Stuart-Panko H, Engholm G, Walsh PM, Jackson C, Vernon S, Morgan E, Gavin A, Morrison DS, Huws DW, Porter G, Butler J, Bryant H, Currow DC, Hiom S, Parkin DM, Sasieni P, Lambert PC, Moller B, Soerjomataram I, Bray F (2019) Progress in cancer survival, mortality, and incidence in seven high-income countries 1995-2014 (ICBP SURVMARK-2): a population-based study. Lancet Oncol 20(11):1493–1505. https://doi.org/10.1016/S1470-2045(19)30456-5
2. Banach M, Juranek JK, Zygulska AL (2017) Chemotherapy-induced neuropathies – a growing problem for patients and health care providers. Brain Behav 7(1):e00558. https://doi.org/10.1002/brb3.558
3. Park SB, Goldstein D, Krishnan AV, Lin CS, Friedlander ML, Cassidy J, Koltzenburg M, Kiernan MC (2013) Chemotherapy-induced peripheral neurotoxicity: a critical analysis. CA Cancer J Clin 63(6):419–437. https://doi.org/10.3322/caac.21204
4. Hershman DL, Lacchetti C, Dworkin RH, Lavoie Smith EM, Bleeker J, Cavaletti G,

Chauhan C, Gavin P, Lavino A, Lustberg MB, Paice J, Schneider B, Smith ML, Smith T, Terstriep S, Wagner-Johnston N, Bak K, Loprinzi CL , American Society of Clinical Oncology (2014) Prevention and management of chemotherapy-induced peripheral neuropathy in survivors of adult cancers: American Society of Clinical Oncology clinical practice guideline. J Clin Oncol 32(18):1941–1967. https://doi.org/10.1200/JCO.2013.54.0914

5. Flatters SJL, Dougherty PM, Colvin LA (2017) Clinical and preclinical perspectives on chemotherapy-induced peripheral neuropathy (CIPN): a narrative review. Br J Anaesth 119(4):737–749. https://doi.org/10.1093/bja/aex229
6. Colvin LA (2019) Chemotherapy-induced peripheral neuropathy: where are we now?. Pain 160(Suppl 1):S1–S10. https://doi.org/10.1097/j.pain.0000000000001540
7. Seretny M, Currie GL, Sena ES, Ramnarine S, Grant R, MacLeod MR, Colvin LA, Fallon M (2014) Incidence, prevalence, and predictors of chemotherapy-induced peripheral neuropathy: a systematic review and meta-analysis. Pain 155(12):2461–2470. https://doi.org/10.1016/j.pain.2014.09.020
8. Sisignano M, Baron R, Scholich K, Geisslinger G (2014) Mechanism-based treatment for chemotherapy-induced peripheral neuropathic pain. Nat Rev Neurol 10(12):694–707. https://doi.org/10.1038/nrneurol.2014.211
9. Cavaletti G, Marmiroli P (2010) Chemotherapy-induced peripheral neurotoxicity. Nat Rev Neurol 6(12):657–666. https://doi.org/10.1038/nrneurol.2010.160
10. Staff NP, Fehrenbacher JC, Caillaud M, Damaj MI, Segal RA, Rieger S (2020) Pathogenesis of paclitaxel-induced peripheral neuropathy: a current review of in vitro and in vivo findings using rodent and human model systems. Exp Neurol 324:113121. https://doi.org/10.1016/j.expneurol.2019.113121
11. Polomano RC, Mannes AJ, Clark US, Bennett GJ (2001) A painful peripheral neuropathy in the rat produced by the chemotherapeutic drug, paclitaxel. Pain 94(3):293–304
12. Griffiths LA, Duggett NA, Pitcher AL, Flatters SJL (2018) Evoked and ongoing pain-like behaviours in a rat model of paclitaxel-induced peripheral neuropathy. Pain Res Manag 2018:8217613. https://doi.org/10.1155/2018/8217613
13. Costa-Pereira JT, Serrao P, Martins I, Tavares I (2020) Serotoninergic pain modulation from the rostral ventromedial medulla (RVM) in chemotherapy-induced neuropathy: the role of spinal 5-HT3 receptors. Eur J Neurosci 51(8):1756–1769. https://doi.org/10.1111/ejn.14614
14. Costa-Pereira JT, Ribeiro J, Martins I, Tavares I (2019) Role of spinal cord alpha(2)-adrenoreceptors in noradrenergic inhibition of nociceptive transmission during chemotherapy-induced peripheral neuropathy. Front Neurosci 13:1413. https://doi.org/10.3389/fnins.2019.01413
15. Costa-Pereira JT, Oliveira R, Guadilla I, Guillen MJ, Tavares I, Lopez-Larrubia P (2023) Neuroimaging uncovers neuronal and metabolic changes in pain modulatory brain areas in a rat model of chemotherapy-induced neuropathy - MEMRI and ex vivo spectroscopy studies. Brain Res Bull 192:12–20. https://doi.org/10.1016/j.brainresbull.2022.10.018
16. Postma TJ, Vermorken JB, Liefting AJ, Pinedo HM, Heimans JJ (1995) Paclitaxel-induced neuropathy. Ann Oncol 6(5):489–494. https://doi.org/10.1093/oxfordjournals.annonc.a059220
17. Cavaletti G, Bogliun G, Marzorati L, Zincone A, Marzola M, Colombo N, Tredici G (1995) Peripheral neurotoxicity of taxol in patients previously treated with cisplatin. Cancer 75(5):1141–1150. https://doi.org/10.1002/1097-0142(19950301)75:5<1141::aid-cncr2820750514>3.0.co;2-u
18. Boyette-Davis J, Xin W, Zhang H, Dougherty PM (2011) Intraepidermal nerve fiber loss corresponds to the development of taxol-induced hyperalgesia and can be prevented by treatment with minocycline. Pain 152(2):308–313. https://doi.org/10.1016/j.pain.2010.10.030
19. Boehmerle W, Huehnchen P, Peruzzaro S, Balkaya M, Endres M (2014) Electrophysiological, behavioral and histological characterization of paclitaxel, cisplatin, vincristine and bortezomib-induced neuropathy in C57Bl/6 mice. Sci Rep 4: 6370. https://doi.org/10.1038/srep06370
20. Toma W, Kyte SL, Bagdas D, Alkhlaif Y, Alsharari SD, Lichtman AH, Chen ZJ, Del Fabbro E, Bigbee JW, Gewirtz DA, Damaj MI (2017) Effects of paclitaxel on the development of neuropathy and affective behaviors in the mouse. Neuropharmacology 117:305–315. https://doi.org/10.1016/j.neuropharm.2017.02.020
21. Cata JP, Weng HR, Chen JH, Dougherty PM (2006) Altered discharges of spinal wide dynamic range neurons and down-regulation of glutamate transporter expression in rats with paclitaxel-induced hyperalgesia. Neuroscience 138(1):329–338. https://doi.org/10.1016/j.neuroscience.2005.11.009
22. Campbell JN, Meyer RA (2006) Mechanisms of neuropathic pain. Neuron 52(1):77–92.

https://doi.org/10.1016/j.neuron.2006.09.021
23. Abbadie C, Honore P, Besson JM (1994) Intense cold noxious stimulation of the rat hindpaw induces c-fos expression in lumbar spinal cord neurons. Neuroscience 59(2):457–468. https://doi.org/10.1016/0306-4522(94)90609-2
24. Thompson SJ, Bushnell MC (2012) Rodent functional and anatomical imaging of pain. Neurosci Lett 520(2):131–139. https://doi.org/10.1016/j.neulet.2012.03.015
25. Martucci KT, Ng P, Mackey S (2014) Neuroimaging chronic pain: what have we learned and where are we going?. Future Neurol 9(6):615–626. https://doi.org/10.2217/FNL.14.57
26. McIlwrath SL, Montera MA, Gott KM, Yang Y, Wilson CM, Selwyn R, Westlund KN (2020) Manganese-enhanced MRI reveals changes within brain anxiety and aversion circuitry in rats with chronic neuropathic pain- and anxiety-like behaviors. Neuroimage 223:117343. https://doi.org/10.1016/j.neuroimage.2020.117343
27. Koretsky AP, Silva AC (2004) Manganese-enhanced magnetic resonance imaging (MEMRI). NMR Biomed 17(8):527–531. https://doi.org/10.1002/nbm.940
28. Andrew RD, MacVicar BA, Dudek FE, Hatton GI (1981) Dye transfer through gap junctions between neuroendocrine cells of rat hypothalamus. Science 211(4487):1187–1189. https://doi.org/10.1126/science.7466393
29. Zwingmann C, Leibfritz D, Hazell AS (2003) Energy metabolism in astrocytes and neurons treated with manganese: relation among cell-specific energy failure, glucose metabolism, and intercellular trafficking using multinuclear NMR-spectroscopic analysis. J Cereb Blood Flow Metab 23(6):756–771. https://doi.org/10.1097/01.WCB.0000056062.25434.4D
30. Yang J, Li Q (2020) Manganese-enhanced magnetic resonance imaging: application in central nervous system diseases. Front Neurol 11:143. https://doi.org/10.3389/fneur.2020.00143
31. Hu TC, Pautler RG, MacGowan GA, Koretsky AP (2001) Manganese-enhanced MRI of mouse heart during changes in inotropy. Magn Reson Med 46(5):884–890. https://doi.org/10.1002/mrm.1273
32. Chang L, Munsaka SM, Kraft-Terry S, Ernst T (2013) Magnetic resonance spectroscopy to assess neuroinflammation and neuropathic pain. J Neuroimmune Pharmacol 8(3):576–593. https://doi.org/10.1007/s11481-013-9460-x
33. Beckonert O, Coen M, Keun HC, Wang Y, Ebbels TM, Holmes E, Lindon JC, Nicholson JK (2010) High-resolution magic-angle-spinning NMR spectroscopy for metabolic profiling of intact tissues. Nat Protoc 5(6):1019–1032. https://doi.org/10.1038/nprot.2010.45
34. O'Callaghan JP, Sriram K (2004) Focused microwave irradiation of the brain preserves in vivo protein phosphorylation: comparison with other methods of sacrifice and analysis of multiple phosphoproteins. J Neurosci Methods 135(1-2):159–168. https://doi.org/10.1016/j.jneumeth.2003.12.006
35. Paxinos G, Watson C (2007) The rat brain in stereotaxic coordinates, 6th edn. Elsevier Academic, Amsterdam; Oxford
36. Bulls HW, Hoogland AI, Kennedy B, James BW, Arboleda BL, Apte S, Chon HS, Small BJ, Gonzalez BD, Jim HSL (2019) A longitudinal examination of associations between age and chemotherapy-induced peripheral neuropathy in patients with gynecologic cancer. Gynecol Oncol 152(2):310–315. https://doi.org/10.1016/j.ygyno.2018.12.002
37. Lima D, Avelino A, Coimbra A (1993) Differential activation of c-fos in spinal neurones by distinct classes of noxious stimuli. Neuroreport 4(6):747–750. https://doi.org/10.1097/00001756-199306000-00037
38. Dai Y, Iwata K, Kondo E, Morimoto T, Noguchi K (2001) A selective increase in Fos expression in spinal dorsal horn neurons following graded thermal stimulation in rats with experimental mononeuropathy. Pain 90(3):287–296. https://doi.org/10.1016/S0304-3959(00)00411-5
39. Hunt SP, Pini A, Evan G (1987) Induction of c-fos-like protein in spinal cord neurons following sensory stimulation. Nature 328(6131):632–634. https://doi.org/10.1038/328632a0
40. Colby LA, Morenko BJ (2004) Clinical considerations in rodent bioimaging. Comp Med 54(6):623–630

Check for updates

Chapter 6

Incisional Model of Postoperative Pain in Rodents

Katherine Gates and Cholawat Pacharinsak

Abstract

Postoperative pain is common in rodents. The plantar or thigh incision models have been used to determine analgesia effectiveness. Under general anesthesia, the plantar or thigh incision that includes skin, fascia, and deep muscle tissue at the plantar side of rodent's hind paw or thigh, respectively. Following the incision, animals exhibit guarding or nonweight bearing behaviors. An incision in rodents also induce (mechanical and thermal) hypersensitivity. Hypersensitivity starts as early as 2–4 h after surgery and resolves within 2–7 days. These models are currently used to study behavioral changes, efficacy of analgesics, and mechanisms of an injury. We aim to provide researchers with a simple, fast, and reliable method that is ideal for modeling incisional pain. Specifically, this model can be applicable to several life stages (juvenile, adult, neonates) and in a variety of rodents (mice, rats). Two of the most common models to induce postoperative pain in rats and mice, the plantar or thigh incision model, are reviewed here. Rodent models of pain have been useful to assess the efficacy of drug therapies and the effects of differential signaling pathways. Utilization of rodent incisional pain models requires assessment of behavioral responses and an underlying understanding of the pathophysiological mechanisms. Additionally, the behavioral assays used to quantify pain and any alterations to it from therapeutic interventions will be explained.

Keywords Incisional pain model, Mechanical hypersensitivity, Thermal hypersensitivity, Behavioral nociceptive assays

Acronyms

NSAIDs	non-steroidal anti-inflammatory drugs

1 Introduction

Given the multitude of surgical models used in rodent species, an important consideration is the impact of incisional pain on rodent responses. Investigation of rodent models of postoperative pain and analgesic therapies was initially guided by Brennan et al. [1] They discussed the importance of understanding the mechanisms behind mechanical vs thermal sensitization with a rat model and how they

Isaura Tavares and Hugo Leite-Almeida (eds.), *Translational Research Methods for Neuropathic Pain*, Neuromethods, vol. 235, https://doi.org/10.1007/978-1-0716-5312-8_6,

can be modified via pharmacology. Pogatzki et al. subsequently characterized a pain model in the mouse [2]. The utility of these models stem from their applicability to model neonatal or adult stages, as well as the simplicity and speed of the surgery. Surgical models may not be inherently painful or involve the same kind of pain (neuropathic, visceral, etc.), but all invariably involve skin incisions. Within this chapter, two types of incisional models will be discussed—plantar and thigh incision. These cover both the adult and neonatal life stages and the mouse and the rat model. They are well characterized and frequently used methodologies of inducing a consistent pain response.

Beyond investigating the underlying pathways and mechanisms of injury, these models are frequently used to determine the efficacy of analgesics. Pain pathways are complex, and several targets can be considered for different drugs to modify the manifestation of pain. Classically, the pathway is separated into four phases: transduction, transmission, modulation, and perception. Optimal analgesic therapy takes advantage of this knowledge to intervein at multiple phases. This is the foundation that underlies pre-emptive analgesia and multimodal analgesia. If the noxious stimulus is modified or halted before it can complete the full pathway, phenomena like hypersensitivity or allodynia [3] can be mitigated [4].

The types of hypersensitivity that will be assessed in this chapter will be roughly categorized as mechanical and thermal [5]. Hypersensitivity to noxious stimuli or even hyperalgesia have been discussed at length in a multitude of review papers. In brief, an event (in this case an incisional surgery) induces sensitization, either peripherally or centrally. This sensitization is expressed as a lowering of the pain threshold to stimuli, with either hyperreactivity to a painful response or a painful reaction to a nonpainful input. It's been suggested that local nociceptors can be sensitized to thermal stimulation. However, the sensitization to mechanical may be more influenced by central mechanisms. This has been seen as increased pain response in dorsal horn neurons by mechanical stimuli outside the surgical area. The response to this type of noxious stimuli has been hypothesized as being primarily driven by the spinal cord. Utilizing methods that probe sensitization toward mechanical and thermal input may allow for a broader investigation of all possible hypersensitivity pathways.

Two of the most used analgesics in rodents are opioids or nonsteroidal anti-inflammatory drugs (NSAIDs). Opioids work on several receptor subtypes (mu, delta, and kappa) and activation of each will lead to different physiological effects. Classically, the mu receptor is known for its central analgesic ability, even though both delta and kappa can have analgesic properties as well, they are generally less potent [6–8]. The most frequently employed opioid is Buprenorphine. It is a partial mu agonist with a strong binding affinity to the mu receptor and both a weak kappa antagonist and delta receptor agonist [9]. The standard formulation is buprenorphine-HCl, which

Table 1
Summary for analgesic dosages tested in different animal pain models

Paper	Species	Surgical method	Pain assessment	Drugs	Drug dosages
Arthur [14]	Male and female NSG	Plantar incision	Von Frey Thermal	Saline Bup-HCl Bup-ER Bup-XR	0.1 mg/kg BID 2 days 1 mg/kg once 3.25 mg/kg once
Zhang [17]	One-day-old male and female Sprague Dawley rat pups [(Crl: CD (SD) IGS	Thigh incision	Thermal	Saline Bup-SR XR-Lo XR-Hi	1 mg/kg 0.65 mg/kg 1.3 mg/kg
Alamaw [24]	Three-month-old male Sprague–Dawley rats	Plantar incision	Von Frey Thermal	Saline Bup-SR XR-Lo XR-Hi	1.2 mg/kg 0.65 mg/kg 1.3 mg/kg
Brennan	Male Sprague-Dawley rats	Plantar incision	Von Frey	Saline Morphine	3 mg/kg
Pogatzki	Adult male C3H/He mice	Plantar incision	Von Frey Radiant heat	Vehicle Morphine-lo Morphine-hi	3 mg/kg 10 mg/kg
Weber	Female Sprague–Dawley	Dermal tail incision	Thermal—water bath Bar algometer—mechanical	None	None
Whiteside	Male Sprague–Dawley rats	Plantar incision	Analgesy-meter Von Frey Incapacitance meter	Saline Morphine Gapabentin Celecoxib Naproxen Indomethacin etoricoxib	SC 1–20 mg/kg IP 10–100 mg/kg C—3–30 mg/kg N—3–30 mg/kg I—1–30 mg/kg E—3–30 mg/kg

(continued)

Table 1
(continued)

Paper	Species	Surgical method	Pain assessment	Drugs	Drug dosages
Lu [25]	Male C57BL/6JJcl mice	Superficial incision Deep incision Orthopedic pain	Spontaneous behaviors—DWB Von Frey test	Diclofenac Buprenorphine or saline	(0.25 mg/250 μL), (1.67×10^{-3} mg/250 μL) TID POD -> POD 3
Navarro [26]	Adult male C57BL/6J mice	Superficial plantar	Von Frey Thermal Plasma drug	Saline Bup-SR Ethiqa (hi, lo)	1 mg/kg 3.25 mg/kg 6.5 mg/kg
Blaney [27]	One-day-old Sprague Dawley rat pups [(Crl: CD (SD) IGS)	Thigh incision	Thermal	Saline Bup-HCl Low-SR High-SR	0.05 mg/kg 0.5 mg/kg 1 mg/kg
Park [28]	Male BALB/c mice	Plantar incision	Dynamic plantar aesthesio-meter	Saline Pregabalin	30 mg/kg

has a duration of action approximately 6–12 h depending on the species, in the case of mice—strain, and route of administration. Commonly, this drug is administered two to three times a day, which in companion animal medicine can be managed, but repeated handling is known to increase stress in rodents [10]. As such, there has been an effort to develop longer acting formulations. The two formulation that have been tested with incisional pain methodologies include the compounded Buprenorphine-ER (extended release) and the FDA-indexed Ethiqa. Both employ a method to delay the metabolism of buprenorphine. Ethiqa contains lipid bound buprenorphine in a medium chain fatty acid triglyceride oil, and ER has the Buprenorphine HCl in a patented polymeric delivery system.

Alternatively, NSAIDs are frequently employed in studies that are confounded by the use of opioids. Both carprofen [11] and meloxicam [12] are employed as SQ injections. Carprofen inhibits cyclooxygenase activity with selective inhibition of COX-2 vs COX-1 pathways. COX-2 generates the prostaglandins that promote inflammation vs the COX-1 prostaglandins need the gastrointestinal and renal to function, which makes carprofen a more targeted analgesic drug. Meloxicam also selectively inhibits the COX-2 enzyme.

While not thoroughly researched in rodent models, local anesthetics are almost always employed in larger species as a line block before surgical incisions. Based on the goals of your study, consideration needs to be given to choosing appropriate timepoints for behavioral or hypersensitivity testing, as well as drug administration. Additionally, it has already been illustrated that there are variable responses toward certain drugs due to species (mice vs rats), age (neonatal vs adult), and strain (BALB/c, B6, NSG). As such, the model organism needs to be chosen wisely to appropriately reflect the aim of the study. While it won't be covered in this chapter, blood sampling may be required in certain studies to more accurately track serum drug levels. The timepoints for serum sampling will vary by study objections, some starting as soon as 5 min [13] and others out to 4 h [14]. The goal of the study will dictate the appropriate interval and Table 1 can be referenced for published studies.

2 Plantar Incision Model

2.1 General Consideration

When performing any type of nonterminal rodent anesthetic surgery, there are several components that need to be considered ahead of time. These components are equally applicable to both incisional models. A separate area should be dedicated for prepping the animal, surgery, and recovery. All areas require a heat source for the rodent due to their propensity for hypothermia, as well as an appropriate light source. For all surgeries, the surgeon should wear a surgical cap, clean gloves, clean lab coat, and face mask.

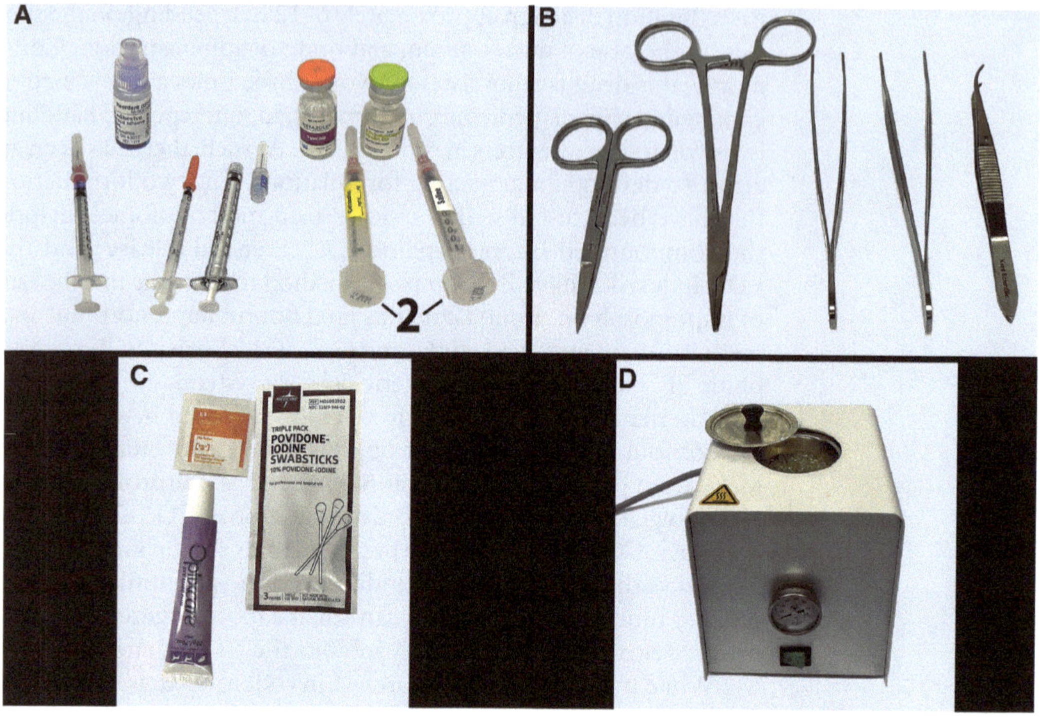

Fig. 1 Surgical Supplies: (**A**) From left to right—24 g 1 cc syringe for administration of surgical glue (VetBond), insulin syringe, and/or 1 cc syringe with 22 g needle for administration of analgesics, (2) 24 g 1 cc syringe for administration of cefazolin and saline. (**B**) Surgical instruments. (**C**) Surgical prep supplies—Ophthalmic ointment, 70% ethyl alcohol and povidone-iodine swabsticks. (**D**) Hot bead sterilizer

All surgeries discussed here will employ the tips only technique, where the tips of instruments will remain sterile, but the surgeon does not need to don a sterile gown or gloves. However, sterilize instruments via a calibrated autoclave. The tools may be sterilized between surgeries via a hot bead sterilizer (Fig. 1D), and exchanged for new autoclaved instruments every fifth animal for maximal sterility. Our recommendation is to use the inside of the surgical pack as a place to rest the sterile instrument tips. Sterile drapes may be used but alternatives that include adhesive food plastic wrap can also be used as it has been shown to be sterile when handled properly. While this procedure is only intended to cause minor pain, optimal veterinary care includes the use of highly palatable supplemental food [15].

2.2 Surgical Materials for Plantar Incision Model

The materials required for this model involves a standard nonrebreathing vaporizer anesthetic circuit (Fig. 2) with 100% oxygen supplied as the carrier gas. Isoflurane or sevoflurane can be used; sevoflurane is more often used for neonatal rodents to reduce breath holding and aversive reactions to the noxious smell of isoflurane. Gas anesthetic provides a rapid induction and recovery as well as

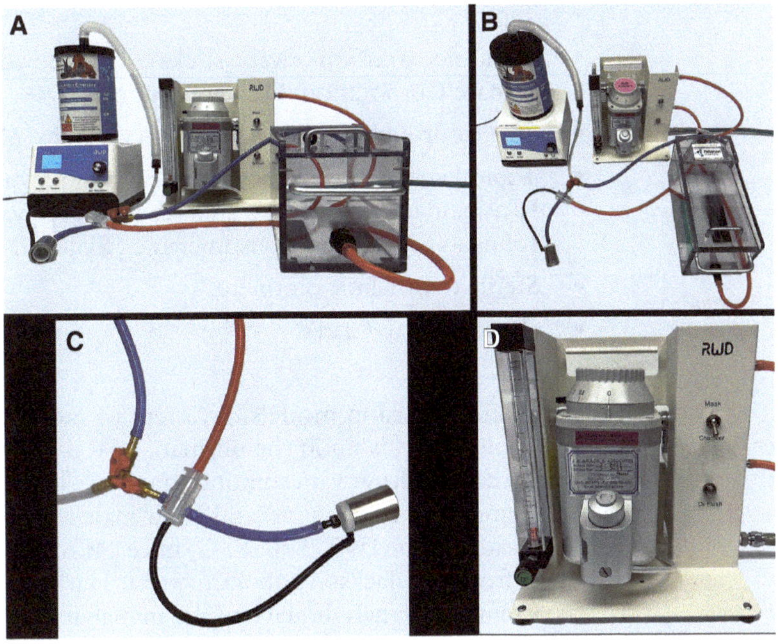

Fig. 2 Anesthetic Setup: (**A**) Front view of anesthetic set-up. (**B**) Top view of anesthetic setup. (**C**) Close-up of the recommended tubing connections from the nose-cone to induction box. (**D**) Close-up of isoflurane vaporizer

more fine-tuned intraoperative calibration of anesthetic depth. Both factors make it more appealing than the use of injectable anesthetics. Additionally, gas anesthetics do not have any analgesic properties, which could act as a confounder in pain studies.

The instruments (Fig. 2B) required include:

- Warm water circulating blanket set to 38°C (Stryker T/Pump, Portage, Michigan)
- #15 scalpel blade (VetOne®, Boise, ID)
- Curved forceps
- (2) Fine-tipped forceps
- Scissors
- Needle drivers.
- Leur lock syringes (Becton, Dickinson and Company, Franklin Lakes, NJ) fitted with 22-guage needles
- 29-gauge 0.3 mL insulin syringes (UltiMed, Inc., Excelsior, MN)
- 4-0 or silk suture
- Cefazolin and sterile water
- Sterile saline

- Tape
- Betadine solution swab sticks (10% povidone-iodine, Dynare Corporation, Orangeburg, NY)
- 70% Isopropyl Alcohol USP (Henry Schein, Melville, NY)
- Topical antibiotic ointment (triple antibiotic ointment with bacitracin zinc, neomycin sulfate, and polymyxin B sulfate, Johnson & Johnson Consumer Inc., Skillan, NJ)
- Sterile ophthalmic ointment
- (Press n' Seal, Glad)

2.3 Animal Preparation for Plantar Incision Model

For the plantar incision models, a variety of mouse strains may be used. This depends upon the ultimate goal of the study as the response to certain drugs varies among strains [14]. Both adult male C57BL/6 mice (*Mus musculus*) and adult male and female NOD. Cg-Prkdcscid Il2rgtm1Wjl/SzJ (NSG) mice (*Mus musculus*) can be purchased from The Jackson Laboratory (Bar Harbor, ME). Health reports of sentinel animals should find the animals free of minute virus of mice, mouse hepatitis virus, mouse rotavirus, Theiler's murine encephalomyelitis virus, Sendai virus, murine adenovirus 1 and 2, ectromelia virus, lymphocytic choriomeningitis virus, pneumonia virus of mice, respiratory enteric virus 3 (Reovirus 3), Mycoplasma pulmonis, endo- and ectoparasites, and pinworms. Group-housed mice in IVC caging (Innovive, San Diego, CA) on prefilled corncob or ALPHA-dri bedding, fed a commercial diet (Teklad Global 18% Protein Rodent Diet 2018, Harlan Laboratories, Madison, WI), and provided with acidified bottled water (Innovive, San Diego, CA). Maintain rooms on a 12:12-h dark:light cycle at 68°F–79°F (20°C–26°C) and 30%–70% relative humidity. Treat mice in accordance with the Guide for the Care and Use of Laboratory Animals. Acclimate mice to the facility for a minimum of 3 days prior to baseline testing. At the study's conclusion, euthanize animals by carbon dioxide asphyxiation, followed by cervical dislocation.

2.4 Surgical Procedure for Plantar Incision Model

1. Induce mice with 1%–4% isoflurane with 100% oxygen (2–4 L/min) in an induction chamber.
2. Once righting reflex is lost, and they are laterally recumbent, they can be transferred to a circulating warm-water blanket and maintained on 0.8%–2.5% isoflurane delivered in 100% oxygen (0.5 L/min) via nose cone.
3. Before starting, apply sterile ophthalmic ointment to both eyes and additional preoperative supportive of intrascapular SQ 20 mg/kg cefazolin and warmed (32–33 C) sterile saline (5 mL/kg) (Fig. 1).
4. Administer any experimental compounds:
   a. *Use a separate location from the fluids or antibiotic to avoid dilution of the compound.*

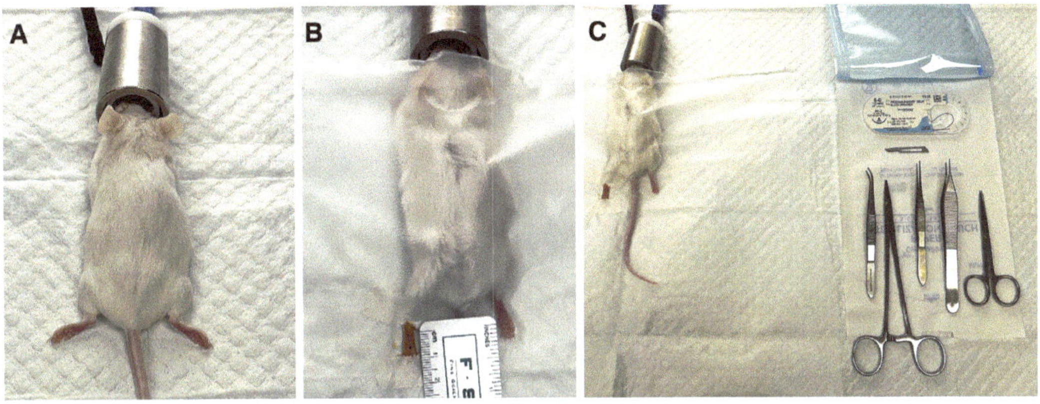

Fig. 3 Surgical Setup: (**A**) Place mouse in nose cone (**B**) Tape foot to secure surgical site and drape with sterile drape. Marking on foot indicates the recommended 0.5 cm incision length. (**C**) Prepare sterile surgical supplies adjacent to surgical area

5. Place mouse in sternal recumbency and secure (Fig. 3B).
   a. *The operated (left) hind limb should be extended and taped to the surface of the table to optimize exposure of the plantar surface.*
6. The plantar surface of the left (ipsilateral) hindpaw is aseptically prepped with three alternating betadine solution swab sticks and 70% Isopropyl Alcohol.
   a. *It is important to note that excess povidone–iodine should not be left on the skin as it is irritating and can confound later pain assessments.*
7. Cover the mouse with a sterile adherent disposable plastic drape.
8. Place the opened autoclaved instrument pack adjacent to the surgical area (Fig. 3C).
9. Drop the blade into the sterile area.
10. Cut a window into the sterile dressing to expose the surgically prepped area.
11. Confirm the depth of the mouse via toe pinch.
    a. *If there is any reaction (withdrawal, flinch,* etc.*), increase depth by 0.5% and test again after 2 min.*
    b. *During the surgery, continuously monitor anesthetic depth and record every 15 min* via *respiratory rate and effort, mucous membrane color, and withdrawal reflex.*
12. With the scalpel blade, make a 5 mm longitudinal incision 3 mm from the tibiotarsal joint through the skin and fascia on the midline of the plantar surface of the foot extending toward the digits (Fig. 4A).

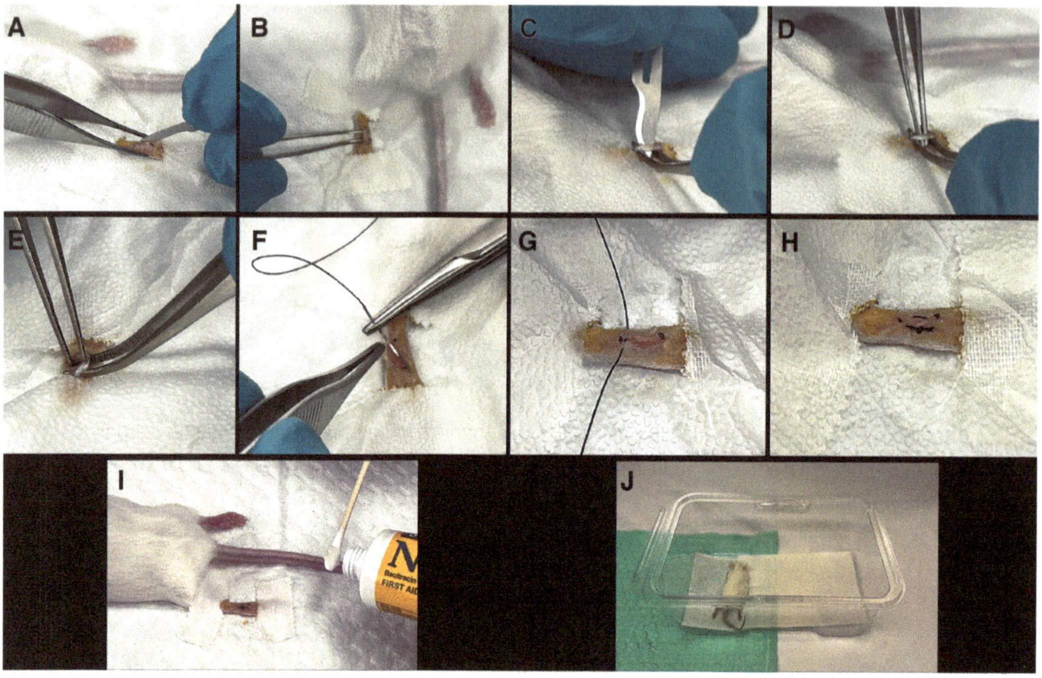

Fig. 4 Surgical model

a. *Take care not to incise the paw pad.*

13. Elevate the underlying muscle bundle using curved forceps (Fig. 4B).
14. Make a stab incision into this muscle with the point of the scalpel blade without disrupting the muscle attachments or underlying structure (Fig. 4C).
15. Using the fine-tipped forceps, apply gentle lateral traction inside the muscle incision for 10 s (Fig. 4D, E).
16. Flush the area with sterile saline and blot away excess fluid.
17. Close the skin with a single horizontal mattress pattern (Fig. 4F–H).
18. Apply triple antibiotic to the closure (Fig. 4I).
19. Transfer the mouse to the recovery cage (Fig. 4J).
    a. *All bedding should be removed from the cage and replaced with a single layer of paper towels to avoid the risk of aspiration. Half of the cage should be on a heating pad to allow the mouse to leave the heat source if desired during recovery. The mouse should be placed sternly for optimal respiration.*
20. Monitor the recovery phase until the righting reflex returns and the mouse can fully ambulate.
21. Return mouse to home cage.

3 Thigh Incision Model (Neonatal Rats)

3.1 Animal Selection

Neonatal thigh incisional pain model is preferred. One-day-old male and female Sprague Dawley rat pups can be acquired from Charles River Laboratories (Hollister, CA); house in litters of eight to ten with the dam. Assure pups are free of rat coronavirus, rat Theiler virus, Kilham rat virus, rat parvovirus, Toolan H1 virus, rat minute virus, lymphocytic choriomeningitis virus, murine adenovirus type 1 and 2, reovirus type 3, Sendai virus, pneumonia virus of mice, Mycoplasma pulmonis, mites, lice, and pinworms. House in static microisolator cages under a 12:12 light:dark cycle (lights on at 7:00 a.m.). Provide rat dams with Teklad Global 18% Protein Rodent Diet 2018 (Harlan Laboratories, Madison, WI) and water filtered by reverse osmosis ad libitum. All animals should be handled and housed according to the Guide for the Care and Use of Laboratory Animals. At the conclusion of the study, euthanize rat pups and dams with carbon dioxide asphyxiation followed by decapitation as a secondary method.

3.2 Materials for Surgical Thigh Incision Model

The material required for the thigh incision model do not vary significantly from those in the plantar model, except for the use of a sevoflurane instead of an isoflurane vaporizer. Sevoflurane is known for a more rapid induction and recovery, as well as being less aversive during induction and as such can be beneficial for use in neonatal animals [16]. However, isoflurane has been used successfully in similar procedures and is a viable alternative.

The supplies required include:

- Warm water circulating blanket set to 38°C (Stryker T/ Pump, Portage, Michigan)
- #15 scalpel blade (VetOne®, Boise, ID)
- Leur lock syringes (Becton, Dickinson and Company, Franklin Lakes, NJ) fitted with 22-gauge needles.
- 29-gauge 0.3 mL insulin syringes (UltiMed, Inc., Excelsior, MN)
- Sterile surgical tissue glue (Covetrus, Dublin, OH).
- Cefazolin
- Warmed sterile saline
- Tape
- Betadine solution swab sticks (10% povidone-iodine, Dynare Corporation, Orangeburg, NY)
- 70% Isopropyl Alcohol USP (Henry Schein, Melville, NY)

- Topical antibiotic ointment (triple antibiotic ointment with bacitracin zinc, neomycin sulfate, and polymyxin B sulfate, Johnson & Johnson Consumer Inc., Skillan, NJ)
- Sterile ophthalmic ointment
- (Press n' Seal, Glad)

3.3 Method for Thigh Incision Model Surgery [17]

1. Induce anesthesia with sevoflurane (5%–8%) in 100% O_2 (0.5–1 L/min) via induction chamber
   a. *Depth assessed by the absence of paw withdrawal reflex.*
2. Maintain with 4%–7% sevoflurane in 100% O_2 (0.5 L/min) via nose cone.
3. Place pups on a warm water circulating blanket set to 38°C.
4. Apply sterile ophthalmic ointment to both eyes.
5. Administer preoperative warmed supplemental fluids (0.9% NaCl, 5 mL/kg, SC).
6. Administer any experimental compounds.
   a. *Use a separate location from the fluids or antibiotic to avoid dilution of the compound.*
7. The surgical area is organized as described above in the plantar incision model.
8. The pups are positioned in right (contralateral) recumbency, and the left (ipsilateral) thigh is aseptically prepared with three alternating betadine solution swab sticks and 70% Isopropyl Alcohol.
9. Confirm the depth of the mouse via toe pinch.
   a. *If there is any reaction (withdrawal, flinch,* etc.*), increase depth by 0.5% and test again after 2 min.*
10. During surgery, continuously monitor anesthetic depth and record every 15 min via respiratory rate and effort, mucous membrane color and withdrawal reflex.
11. Make a 0.5-cm longitudinal skin incision on the ipsilateral thigh.
12. At 0.5-cm above the hock make a 0.3-cm longitudinal incision to the superficial biceps femoris muscle using the scalpel without disturbing the muscle attachment.
13. Close the incision with sterile surgical tissue glue.
    a. *Administer the tissue glue* via *a 1 cc syringe with 25 g needle allows for greater precision.*
14. After surgery, apply antibiotic ointment to the surgical area.
15. Recover the pups in a warm recovery cage and place back with the dam once fully recovered.

4 Hypersensitivity and Pain Testing

The primary mechanical nociceptive test discussed here will be the von Frey assay, but it should be noted that the Randall–Selitto test is another popular mechanical nociceptive test. The Randall–Selitto test is normally only used in rats due to the need for greater physical restraint. Both assays apply a mechanical stimulus that induce a physical reaction in the test subject. The von Frey hair or filament assay was initially designed in 1896 by Maximilian von Frey as a mechanical nociceptive threshold test. The filaments are made such that they will buckle elastically, but the force imparted is relatively constant compared to the degree of bending. This allows the user to determine the force at which the subject, in this case a rodent, will react due to a painful sensation. This data needs to then be interpreted by various paradigms. One of the most popular being the up-down testing paradigms popularized in rodents by Chapman et al. in 1994. Since then, there have been several advancements in the interpretation of these data for more streamlined use [18]. There are several commercially available monofilament sets for purchase (UGO BASILE Von Frey Hairs cat. No. 37450-275) and publications on how to develop home-made versions. These filaments are commonly applied to the hind paw adjacent to the surgical area and reactions such as paw withdrawal, sustained reaction, licking, or flinching are assessed as positive responses. Increasingly stronger filaments will be applied until a reaction is evoked, and this filament force dictates the pain threshold. This reactivity adjacent to a painful area indicates hyperalgesia to a mechanical stimulus.

With both the von Frey and Randall–Selitto test, there is a debate about positive threshold responses. Visual detection of response will always be a subjective measurement prone to human error and interobserver variation. Setting the threshold to vocalization explicitly indicates a conscious awareness of sensation. Waiting for vocalization brings other experimental and welfare concerns. Rodents generally do not vocalize until they experience severe pain—which is usually not applicable to the mild pain model of incisional pain. Moreover, using a method that induces unnecessarily high pain is unwarranted from a welfare stance. Vocalization in the ultrasonic range, which occurs before vocalization in the human hearing range, has been investigated as an alternative. Unfortunately, there has been poor correlation between ultrasonic vocalization and traditional evoked responses in mice and rats [19]. To this end, our methods detailed below will use withdrawal as a positive response.

The application of heat is meant to investigate an alternative pain pathway, thermal hypersensitivity. There are many experimental methods available, ranging from Hargreaves (Fig. 5), thermal probe, or hot plate (Fig. 6) to tail flick. All utilize a thermal stimulus.

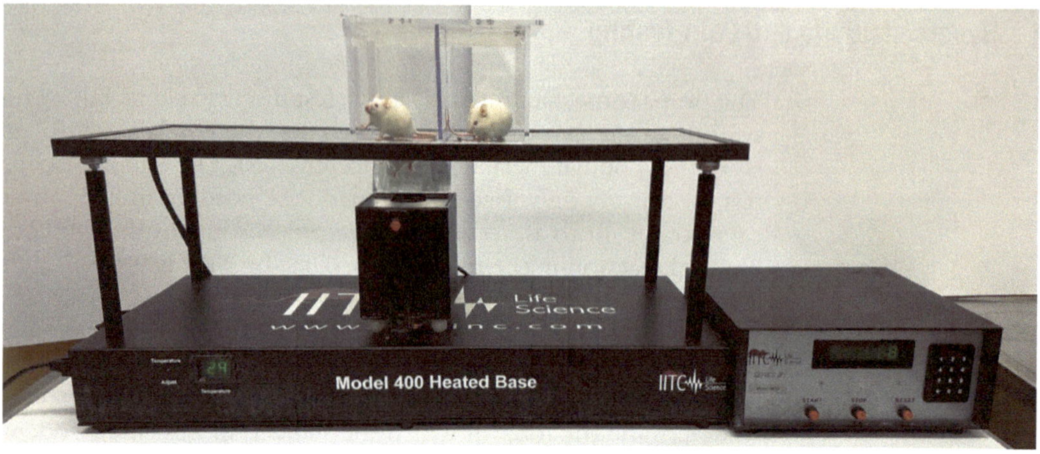

Fig. 5 Hargreaves method. Heat stimuli are produced by placing an approximately 50W light bulb in a custom-built case that allows focusing the light source (~8 mm diameter) on the plantar surface of a hindpaw of an animal. Radiant heat is applied to the plantar surface of each hindpaw and withdrawal response latencies are determined. Withdrawal latencies are measured to the nearest 0.1 s using a photocell that terminates the trial and a timer upon withdrawal

Fig. 6 Hotplate. An animal is placed into a clear chamber with a heated floor (heated to ~50–55 C). Nocifensive behaviors (paw licking, jumping, or shaking) can be measured

However, there has been debate to the appropriateness of some of these assays for assessing true hypersensitivity. The tail flick assay involves a high-intensity beam aimed at the subject's tail, but this most likely involves a peripheral withdrawal reflex that is less applicable to incisional pain. Hot plate assay is complicated by the behavioral tolerance phenomenon. With repeated assessment, a reduction in latencies in conjunction to decreased sensitivity to antinociceptive agents has been shown. This is due to the fact that the animal is barricaded on a heat conductive surface (set between 50°C and 56°C), and the time to pain-reflex behavior measured. Two alternatives to these first thermal tests are more suitable to the incisional models described here. The thermal probe test is similar to the von Frey mechanical assay where instead of a mechanical stimulus, a heated metal probe is applied directly to the hind paw of an animal atop a barred surface. Given the physical contact required, it blurs the lines between a purely thermal vs mechanical response. Alternatively, the Hargreaves test (further described below) uses only a radiant or infrared heat source that is directed at the plantar surface of the mouse. The latency in withdrawal time is measured as opposed to the heat, at which there is a response. As such, there should always be a cutoff time (usually 20 s) to avoid additional confounding tissue damage. Baseline stillness is an important component of the Hargreaves testing. The reaction to the stimulus needs to be differentiated from random movement. An appropriate acclimation time will allow the mice to relax. This can be complicated by certain experimental substances, as some have reported hyperactivity in mice given buprenorphine. Additionally, application of this assay with experimental substances that have a very short duration of action (i.e., lidocaine) is unwise given the recommended acclimation period.

4.1 General Considerations

For both assays discussed below, there are several considerations that must be considered ahead of time. These are commonalities that are applicable to most evoked behavioral assays. With any testing, and especially behavioral, there will be interobserver and intraobserver variability given the inherent subjectivity of these assays. Blinding is an incredibly important component to reduce bias in under or over interpreting evoked responses. This needs to be implemented at the start of the study and was discussed in the previous section. Acclimation and baseline readings cannot be overlooked in behavioral testing. Beyond standard institutional acclimation guidelines, acclimation to the testing rigs will reduce noise in the data. At minimum, the test subject should be left undisturbed in the von Frey and Hargreaves rigs for 15–20 min for 2 days before surgery. Anecdotal benefit has been seen with leaving them for 30–40 min. This acclimation time should remain consistent throughout the entire study and adhered to before each testing session. As much as possible, a consistent time testing during the day should also be maintained.

The timepoints selected for these experiments can vary and need to be directed by the experimental hypothesis. For most studies included in Table 1, the experimental substances were drugs that claimed efficacy up to 72 h. Frequently, the testing timepoints followed a −24, 4, 8, 24, 48, 72-h scheme for mice, or −24, 1, 4, 8, 24, and 48 h scheme for rat pups. Additionally, if serum drug levels are being tested, a corresponding sample should be taken to correlate drug concentration and efficacy.

We provide graphs from a recent publication to illustrate how the data can be represented (Figs. 5 and 6) [14]. Many groups have not found any significant differences between sexes but during the analysis, a comparison between sexes should be included to preclude sex variation.

4.2 Materials for Von Frey

Mice and rats used for these experiments should receive the incisional pain procedure or be from the naïve control group. As mentioned above, they should already be acclimatized to the von Frey rig (Fig. 7). The number of animals that can be tested will vary on

Fig. 7 Von Frey testing. An animal is placed under a clear plastic cage on an elevated plastic mesh floor. Calibrated von Frey monofilaments are used to deliver punctate mechanical stimuli to the plantar surface of the hindpaw. Withdrawal responses can be tested using thresholds or frequency. Each stimulus is applied for a duration of 1–2 s with an interstimulus interval of approximately 5 s

the setup, but the recommended amount is four. This allows the user to be able to appropriately apply a noxious stimulus to each animal equally and observe their response. Additionally, operator fatigue from using the device can occur if too many subjects are tested at once and will add unnecessary variation to the data collected. Ergonomic considerations should not be taken lightly. The user should test the setup before acclimating the mice to confirm that they are able to comfortably test for an extended period. Anchoring a wrist or hand on the rig will allow for consistent application of force with the filaments. There is a debate on the use of the contralateral foot as an internal control due to concerns over allodynia. Instead, we recommend using a saline control.

Mice

1. Clear acrylic enclosure ($10.1 \times 10.1 \times 12.5$ cm)
2. von Frey monofilaments with calibrated bending forces (0.6 g, Aesthesio, DanMic Global, San Jose, CA)

Rats

1. Clear plastic chamber ($21 \times 27 \times 15$ cm) or ($23 \times 13 \times 13$ cm)
2. Semmes-Weinstein von Frey filament (8 g)
3. Elevated mesh stand (Electronic von Frey Mesh Stand, IITC Life Science, Woodland Hills, CA) with 0.64-cm “waffle” holes

4.3 Method for Von Frey Hypersensitivity Testing

Mechanical hypersensitivity is defined as a significant increase in paw withdrawal frequency.

Mice

1. Place each mouse in a clear acrylic enclosure atop the elevated mesh stand.
2. Acclimate the mice to the testing enclosure for a minimum of 15 min.
3. Apply the von Frey monofilament for 1 s on various locations of the plantar surface of the experimental hindpaw.
4. Conduct a set number of trials per mouse—recommended ten trials per foot.
5. Predefine the threshold for a withdrawal response.
   a. *For example: a mouse lifting the paw off the mesh after von Frey stimulation during ten applications of the monofilament.*
   b. *Leave an interstimulus interval of approximately 5 s.*
6. Mechanical hypersensitivity testing should be performed prior to thermal hypersensitivity testing.

Rats

1. Place each rat in the clear plastic chamber on an elevated wire mesh platform.

2. Acclimate for a minimum of 15 min in the chamber.
3. Apply a calibrated von Frey filament to the plantar surface of the test paw at random locations.
   a. *Semmes-Weinstein (8g) works well in rats*
   b. *Avoid the heel, pads, and toes*
4. Predefine the threshold for a withdrawal response.
   a. *Lifting of a paw from the mesh for 1–2 s*
   b. *Leave an interstimulus interval of approximately 5 s*
5. Conduct a set number of trials per rat—recommended ten trials per foot
6. Throughout the entire experiment, mechanical hypersensitivity testing should be performed first, followed by thermal hypersensitivity testing.

4.4 Materials for the Hargreaves Test

Mice

1. Clear acrylic enclosure (10.1 × 10.1 × 12.5 cm) on a tempered-glass surface preheated to 29–30°C (Plantar Analgesia Meter, IITC LifeScience).
2. 50-W light bulb (Plantar Analgesia Meter, IITC Life Science, Woodland Hills, CA).

Adult Rats

1. Clear plastic chamber (23 × 13 × 13 cm) on a tempered-glass surface preheated to 29–30°C (Plantar Analgesia Meter, IITC LifeScience).
2. 50-W light bulb (Plantar Analgesia Meter, IITC Life Science, Woodland Hills, CA).

Rat Pups

1. Warm water circulating blanket (set at 38°C) (Stryker T/Pump, Portage, Michigan, USA).
2. Infrared thermometer (Extech instruments, Nashua, NH).
3. Infrared diode laser stimulator (LASMED, Mountain View, CA, USA, 490 mA, 5 mm diameter.

4.5 Method for the Hargreaves Test

Thermal hypersensitivity is defined as a significant decrease in paw withdrawal latency after the onset of focal thermal stimuli. For rat pups, purposeful paw movement away from the laser is used instead of paw withdrawal.

Adult Rats

1. Place rats in the clear plastic chamber for 15 min before testing for acclimation.
2. Measure the average temperature of the glass platform.
   a. *Should be between 29°C and 30°C*

3. Apply the thermal stimuli via the 50-W light bulb focused on the middle of the plantar surface of each hind paw.
4. A maximal threshold exposure time should be set at 20 s to prevent tissue injury.
5. Test each paw four times, with a minimum of 4 min between each trial.
6. The mean of the last three trials is used to set the withdrawal latency.

Rat Pups

1. Two to three pups can be evaluated at a time, all should be maintained on the warm water circulating blanket.
2. The baseline skin temperature is read with an infrared thermometer.
   a. *Usually between 30°C and 34°C*
3. Focus the 4 mm infrared diode laser stimulator perpendicular to the lateral thigh, 3.5 in. from the skin, with a cutoff time of 19 s to avoid tissue damage.
4. Test the thigh twice, with a 3-min intermeasure interval and average the measurements.

Mice

1. Place mice in the clear plastic chamber for 15 min before testing for acclimation.
2. Measure the average temperature of the glass platform.
   a. *Should be between 29°C and 30°C*
3. Focus the 4 × 6 mm thermal stimuli via the 50-W light bulb set at 20% on the middle of the plantar surface of each hind paw.
4. A maximal threshold exposure time should be set at 20 s to prevent tissue injury.
5. Test each paw at least three times, with a minimum of 2 min between each trial.
6. The mean of the three trials is used to set the withdrawal latency.

5 Behavioral Observational Assays

Beyond application of external stimuli, inherent mouse behaviors can be used to assess pain. While these cannot differentiate between thermal or mechanical hypersensitivity per say, they do present a more natural and potentially less biases assessment of pain given that they are nonevoked pain measures. Observation of behaviors allows for researchers to keep the animal in its home environment where it is less stressed. Several comprehensive reviews have been published on these behavioral investigations [19, 20], which can be referenced for a more in-depth analysis.

Most often these measurements come in the form of ethograms that collectively assess certain behaviors, either normal nonpainful behaviors (rearing, grooming, nest building) or painful behaviors (grimace, guarding, toe-touching). Guarding is one of the more classical behaviors that has been used in pain experiments as it can be assessed through observation alone. It is easily applied to incisional pain models, since the incision involves a limb. This statement is also true for other behaviors such as toe-touching. More advanced systems that investigate changes in locomotor activity can also be used. Several automated home-cage monitoring systems are currently on the market, such as the Laboratory Animal Behavior Observation, Registration and Analysis System (LABORAS). Ethograms have become more commonplace in laboratory animal research. The classic ethogram in mice and rats is called the Grimace scale. Organizations such as the NC3RS [21] provides a resource library where physical copies of these scales can be found. Given the stoic nature of mice and the minor pain induced by an incisional model, grimace scales may not provide sufficient differentiation between experimental groups. Alternative nonevoked and evoked protocols would be more suited in this scenario.

A common pain measurement that is applicable to many species is simply measuring weight loss. Weight loss has been used as a humane endpoint for many years and is still one of the most common humane endpoints used in rodent studies [15, 22, 23]. Most frequently these studies will use a cutoff of 15%–20% weight loss; however, recent publications have pushed for changing that number to 10%–15% depending on the severity of the model [23]. Incisional pain is a mildly painful model so having significant weight loss is unlikely; however, tracking weight loss can be a valuable nonevoked tool for investigating impact on animal comfort. Most rodent housing systems use a food hopper that requires the rodent to sit up and reach for food. The plantar incisional model will generate discomfort in that action and could reduce appetite. For the neonatal thigh incision model, impact on nursing should be confirmed by the presence of milk spots postoperatively. Recent work has proposed supplemental nutritional care as a refinement to rodent surgical models [15] in order to reduce surgical morbidity and mortality. While incisional models are only mildly painful, dietary supplementation is a recent veterinary recommended refinement to these protocols.

6 Conclusions

Incisional pain is a component of essentially all surgical procedures and has been shown to impact clinical outcomes in humans. The aim of this chapter was to provide the reader with various resources to investigate incisional pain. This is accomplished through surgical descriptions, analgesic discussion, to evoked and nonevoked measurement options. Rodents provide a valuable model due to their small size, ease of use and numerous scientifically validated assays. It is important to note that there is difficulty in differentiating nociception from conscious pain from simple spinal reflexes in any animal model. While nonevoked behavioral assays may be beneficial in moderately painful procedures, evoked assays provide a more focused assessment of thermal and mechanical hypersensitivity. Beyond the translatability to human populations, better understanding the effect of analgesics in rodents cannot be underestimated from a welfare perspective. With the increasing evidence of inter-strain variability of efficacy, these methods allow us to confirm appropriate analgesia is provided to our animal models.

Conflict of Interest

The authors have no conflicts of interest to declare that are relevant to the content of this chapter.

References

1. Brennan TJ, Vandermeulen EP, Gebhart GF (1996) Characterization of a rat model of incisional pain. Pain 64(3):493–502. https://doi.org/10.1016/0304-3959(95)01441-1
2. Pogatzki EM, Raja SN (2003) A mouse model of incisional pain. Anesthesiology 99(4):1023–1027. https://doi.org/10.1097/00000542-200310000-00041
3. He Y, Kim PY. Allodynia. (2023). In: StatPearls [Internet]. Treasure Island (FL): StatPearls Publishing; 2026 Jan-. https://www.ncbi.nlm.nih.gov/books/NBK537129/
4. Khan A, Khan S, Kim YS (2019) Insight into pain modulation: nociceptors sensitization and therapeutic targets. Curr Drug Targets 20(7):775–788
5. Modi AD, Parekh A, Pancholi YN (2023) Evaluating pain behaviours: widely used mechanical and thermal methods in rodents. Behav Brain Res 446:114417. https://doi.org/10.1016/j.bbr.2023.114417
6. Smith HS (2008) Variations in opioid responsiveness. Pain Physician 11(2):237–48
7. Smith HS (2012) Opioids and neuropathic pain. Pain Physician 15(3 Suppl):Es93–Es110
8. Gaertner DJ, Hallman TM, Hankenson FC, Batchelder MA (2008) Chapter 10—Anesthesia and analgesia for laboratory rodents. In: Fish RE, Brown MJ, Danneman PJ, Karas AZ (eds.) Anesthesia and analgesia in laboratory animals, 2nd edn. Academic Press, Amsterdam, pp. 239–297
9. Kumar R, Viswanath O, Saadabadi A. Buprenorphine (2024). In: StatPearls [Internet]. Treasure Island (FL): StatPearls Publishing; 2026 Jan-. https://www.ncbi.nlm.nih.gov/books/NBK459126/
10. Marcotte M, Bernardo A, Linga N et al (2021) Handling techniques to reduce stress in mice. J Vis Exp 175:e62593. https://doi.org/10.3791/62593
11. National Center for Biotechnology Information PubChem compound summary for CID 2581, Carprofen. https://pubchem.ncbi.nlm.nih.gov/compound/Carprofen. Accessed 21 Nov 2022
12. National Center for Biotechnology Information PubChem compound summary for CID 54677470, Meloxicam. https://pubchem.ncbi.nlm.nih.gov/compound/Meloxicam. Accessed 25 Apr 2023
13. Fitz CB, Goodroe AE, Moody DE, Fang WB, Capuano IIISV (2021) Pharmacokinetics of buprenorphine and sustained-release buprenorphine in common marmosets (*Callithrix jacchus*). J Am Assoc Lab Anim Sci 60(2):188–194. https://doi.org/10.30802/aalas-jaalas-20-000082
14. Arthur JD, Alamaw ED, Jampachairsri K et al. (2022) Efficacy of 3 buprenorphine formulations for the attenuation of hypersensitivity after plantar incision in immunodeficient NSG mice. J Am Assoc Lab Anim Sci 61(5):448–456. https://doi.org/10.30802/aalas-jaalas-22-000058
15. Gates KV, Alamaw E, Jampachairsri K, Huss MK, Pacharinsak C (2022) Efficacy of

supplemental diet gels for preventing postoperative weight loss in mice (*Mus musculus*). J Am Assoc Lab Anim Sci. 62(1):87–94. https://doi.org/10.30802/aalas-jaalas22-000030
16. Murrell JC, Mellor DJ, Johnson CB (2008) Chapter 27—Anaesthesia and analgesia in the foetus and neonate. In: Fish RE, Brown MJ, Danneman PJ, Karas AZ (eds.) Anesthesia and analgesia in laboratory animals, 2nd edn. Academic Press, London, 593–608
17. Zhang M, Alamaw E, Jampachaisri K, Huss M, Pacharinsak C (2022) Effectiveness of two extended-release buprenorphine formulations during postoperative period in neonatal rats. PLOS One 17(10):e0276327. https://doi.org/10.1371/journal.pone.0276327
18. Gonzalez-Cano R, Boivin B, Bullock D, Cornelissen L, Andrews N, Costigan M (2018) Up–down reader: an open source program for efficiently processing 50% von Frey thresholds. Front Pharmacol 9:433
19. Turner PV, Pang DS, Lofgren JL (2019) A review of pain assessment methods in laboratory rodents. Comp Med 69(6):451–467. https://doi.org/10.30802/aalas-cm-19-000042
20. Aulehner K, Leenaars C, Buchecker V et al (2022) Grimace scale, burrowing, and nest building for the assessment of post-surgical pain in mice and rats—a systematic review. Front Vet Sci 9:930005. https://doi.org/10.3389/fvets.2022.930005
21. National Centre for the Replacement Refinement & Reduction of Animals in Research Welfare assessment. Grimace scales. https://www.nc3rs.org.uk/3rs-resources/grimace-scales. Accessed 6 Nov 2022
22. Lilley E, Andrews MR, Bradbury EJ et al. (2020) Refining rodent models of spinal cord injury. Exp Neurol 328:113273. https://doi.org/10.1016/j.expneurol.2020.113273
23. Talbot SR, Biernot S, Bleich A et al. (2020) Defining body-weight reduction as a humane endpoint: a critical appraisal. Lab Anim 54(1):99–110. https://doi.org/10.1177/0023677219883319
24. Alamaw ED, Franco BD, Jampachaisri K, Huss MK, Pacharinsak C (2022) Extended-release buprenorphine, an FDAindexed analgesic, attenuates mechanical hypersensitivity in rats (*Rattus norvegicus*). J Am Assoc Lab Anim Sci 61(1):81–88. https://doi.org/10.30802/aalas-jaalas-21-000081
25. Lu F, Kato J, Toramaru T, Sugai M, Zhang M, Morisaki H (2022) Objective and quantitative evaluation of spontaneous pain-like behaviors using dynamic weight-bearing system in mouse models of postsurgical pain. J Pain Res 15:1601–1612. https://doi.org/10.2147/jpr.S359220
26. Navarro K, Jampachaisri K, Huss M, Pacharinsak C (2021) Lipid bound extended release buprenorphine (high and low doses) and sustained release buprenorphine effectively attenuate post-operative hypersensitivity in an incisional pain model in mice (*Mus musculus*). AMEM 4(2):129–137. https://doi.org/10.1002/ame2.12157
27. Blaney A, Jampachaisri K, Huss MK, Pacharinsak C (2021) Sustained release buprenorphine effectively attenuates postoperative hypersensitivity in an incisional pain model in neonatal rats (*Rattus norvegicus*). PLOS One 16(2):e0246213. https://doi.org/10.1371/journal.pone.0246213
28. Park JH, Cho SH, Kim R et al. (2021) Effect of pregabalin on nociceptive thresholds and immune responses in a mouse model of incisional pain. Korean J Pain 34(2):185–192. https://doi.org/10.3344/kjp.2021.34.2.185

Chapter 7

Rodent Models of Complex Regional Pain Syndrome Type I (CRPS-I)

Patrícia Rodrigues, Náthaly Andrighetto Ruviaro, Julia Frare and Gabriela Trevisan

Abstract

Chronic pain is a public health issue that reduces patients' quality of life and generates expenses for treatments. In this sense, we highlight the complex regional pain syndrome (CRPS), a chronic pain disorder with heterogeneous pathophysiology. The CRPS has two classifications: CRPS-I occurs due to limb trauma without initial nerve injury, while in CRPS-II, the damage is accompanied by nerve injury. However, there is still no satisfactory treatment to prevent or reduce CRPS symptoms, highlighting the need for further studies on drug development using animal models. Two main models have been studied in rodents for CRPS-I-induced nociception, including fracture/casting and ischemia/reperfusion (IR, named chronic post-ischemia pain, CPIP). These models mimic the disease symptoms by causing tissue damage, inflammatory response, and vascular injury. Thus, knowledge about the models of CRPS-I can modify the course of the syndrome, promoting the reduction of painful sensations or even pain prevention. Therefore, this book chapter will discuss the knowledge related to the pathophysiologic mechanisms involved in these two animal models of CRPS-I.

Keywords Chronic pain, Animal models, Fracture and cast, Inflammation, Ischemia, Nociception

1 Introduction

Pain is described by the International Association for the Study of Pain (IASP) as "a sensory and emotional experience associated with or resembling that associated with actual or potential tissue damage." Pain can be classified into acute and chronic pain (when the painful sensation lasts for more than 3 months). Chronic pain symptoms commonly include allodynia (pain to a stimulus that normally does not cause pain), hyperalgesia (increased pain sensation after a stimulus that causes pain), and spontaneous pain [1]. Also, chronic pain can be subdivided into primary chronic pain, found in patients with migraine and fibromyalgia, and it is not related to other chronic conditions. However, secondary chronic pain is initially a symptom of other pathologies, such as diabetic polyneuropathy and cancer

Isaura Tavares and Hugo Leite-Almeida (eds.), *Translational Research Methods for Neuropathic Pain*, Neuromethods, vol. 235, https://doi.org/10.1007/978-1-0716-5312-8_7,

pain. Complex regional pain syndrome type (CRPS) is considered a primary chronic pain syndrome associated with significant emotional distress and functional disability [1].

The CRPS has two classifications: CRPS-I occurs due to limb trauma without initial nerve injury, while in CRPS-II, the damage is accompanied by nerve injury [2]. CRPS can also be classified as nociceptive or nociplastic pain [3, 4]. Nociceptive pain results from actual or threatened damage to a non-neural tissue due to nociceptors activation. Nociplastic pain is a response to altered nociception with no clear evidence of tissue damage, similar to that observed in fibromyalgia patients [5]. In this sense, more information regarding CRPS mechanisms is required to better understand pain activation pathways helping to improve the diagnosis and treatment in the clinical setting.

Although the CRPS has two classifications (CRPS-I and CRPS-II), its distinction and diagnosis in the clinic are difficult [6]. The internationally recognized criteria for CRPS diagnosis is the Budapest criteria [7]. This criterion includes continuous pain and at least two or more signs of sensory (hyperalgesia and/or allodynia), vasomotor (temperature asymmetry and/or skin color changes), sudomotor (edema and/or sweating), motor dysfunction (weakness, tremor, dystonia), or no better diagnosis to explain it [8]. Immediately after trauma, CRPS shows an acute phase, which is accompanied by a post-traumatic inflammatory response where the innate and adaptive immune system participates and is described by the presence of edema and heat at the injured site. While the chronic phase starts about 6 months post-trauma, it is characterized by sustained pain and neurocognitive manifestations [9]. CRPS-II patients may present additional symptoms such as paresthesia, dysesthesias, and dysautonomia [10].

The first line of medicines for the treatment of CRPS acute phase is nonsteroidal anti-inflammatory drugs (NSAIDs, which inhibit cyclooxygenase enzyme and prevent the synthesis of prostaglandins, decreasing nociceptive and inflammatory processing) and corticosteroids (block the arachidonic acid cascade by inhibiting phospholipase A2) [8, 11]. For the CRPS chronic painful phase, anticonvulsants, such as gabapentin and pregabalin (inhibit the action of $\alpha 2\delta$-1 subunits, generating a selective inhibitory effect on calcium voltage-gated channels) [12], have shown efficacy in pain relief. In addition, calcium channel blockers, such as nifedipine, effectively control vasoconstriction. Also, the gamma-aminobutyric acid (GABA) agonists (baclofen) reduce dystonia and pain [13]. Besides, tricyclic antidepressants (block the reuptake of monoamines, such as norepinephrine and serotonin), can be useful for pain relief, anxiety, depression, and insomnia treatment [8]. Individuals with fractures and immobility can also be treated with the bisphosphonate class (which inhibits osteoclast activity) [8, 14]. However, most patients do not respond to pharmacological treatment, causing the chronification of pain. Besides, these compounds can lead to adverse effects (nausea, headache, xerostomia, and dizziness) [8]. Thus, there is still no satisfactory treatment to prevent or reduce CRPS symptoms, highlighting the need for further studies on drug development using animal models.

CRPS-I animal models, including fracture/casting and chronic post-ischemia pain (CPIP), mimic the disease symptoms by causing tissue damage, inflammatory response, and vascular injury [15]. In these models, the activation of peptidergic nociceptors leads to inflammatory neuropeptide release, such as calcitonin gene-related peptide (CGRP) and substance P (SP). CGRP increases peripheral blood flow through arteriolar vasodilation, while the release of SP induces plasma extravasation from venules. Both peptides activate the secretion of inflammatory cytokines such as tumor necrosis factor-alpha (TNF-α), interleukin-1 (IL-1), and interleukin-6 (IL-6), which contribute to the severity of neuronal damage [16].

Furthermore, oxidative compounds, such as hydrogen peroxide (H_2O_2) and 4-hydroxynonenal (4-HNE), can be produced in the site of injury and activate ion channels, such as transient receptor potential ankyrin 1 (TRPA1) expressed in Schwann cells, leading to prolonged recruitment of monocytes/macrophages to the site of injury and oxidative compounds formation [17]. Thus, TRPA1 activation in Schwann cells and peripheral neurons leads to the development of mechanical and cold allodynia in a CPIP model of CRPS-I caused by ischemia and reperfusion (IR) [17]. In this view, antioxidants treatment stands out since local inflammation is observed in CRPS due to oxygen-free radicals released [18]. Thus, vitamin C supplementation has been investigated in clinical trials in the first year after wrist injury to reduce the risk of developing CRPS and showed promising results [19].

In addition, another theory studied in CRPS models is the increase in autoantibodies levels in the serum and the relation with nociceptor sensitization [7]. The administration of CRPS patient's immunoglobulin G (IgG) in mice hind paw induces mechanical and cold allodynia, and heat hyperalgesia. However, no differences in the weight distribution on the injured hind limb, locomotor activity, and hind paw temperature were observed. Besides, CRPS patients with higher pain levels had an increased IgG antibody titer, suggesting a relation with painful hypersensitivity [20]. Thus, autoimmunity can also be an induction method for CRPS study, but it is not as efficient in the induction of symptoms as observed in patients.

Furthermore, peripheral nerve injury models in rodents such as chronic constriction injury (CCI), partial sciatic ligation (PSL), spinal nerve ligation (SNL), and sciatic nerve transaction (SNT) can lead to the development of CRPS type II. In addition, those models were used to study neuropathic pain, so we did not describe them in

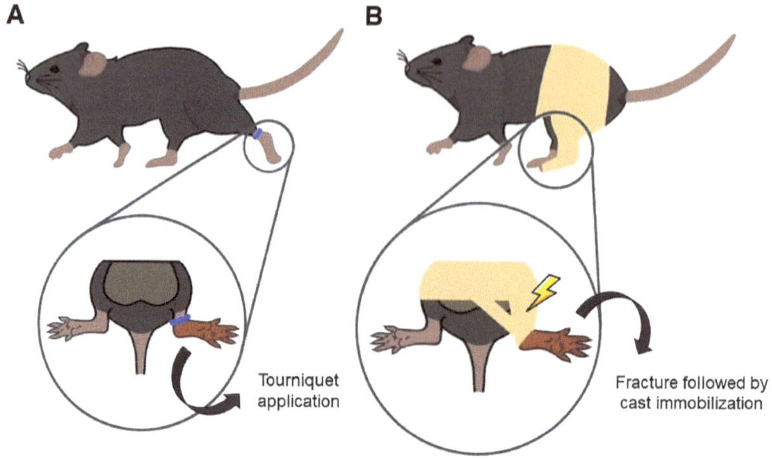

Fig. 1 The two main models of CRPS establishment in rodents are (**A**) ischemia/reperfusion (IR, named chronic post-ischemia pain, CPIP) and (**B**) fracture and cast. To induce the CPIP model, a tourniquet is placed in the hind paw of animals and remains for a period that varies between 2 and 3 h. After an ischemic period, the tourniquet is cut, allowing the blood to flow to the hind limb. In the fracture and cast model, animals endure tibia and radius fractures followed by cast immobilization. The cast remains for a period that varies between 3 and 4 weeks and is then removed. After the induction of these models, rodents showed mechanical allodynia, edema, and hind paw-enhanced temperature

this chapter, but some reviews about this topic have been recently published [7, 21, 22]. Thus, this chapter will explore two CRPS-I models (fracture/casting and CPIP) for a better understanding of the pathophysiology of this painful disease (Fig. 1).

2 Methods

2.1 Fracture and Cast Immobilization Model of CRPS-I

The tibial fracture model can be performed using rats or mice that must be housed individually with food and water ad libitum (Table 1). The distal right tibia of rats, under isoflurane anesthesia, was fractured using pliers (with an adjustable stop). Afterward, the limb was wrapped in casting tape to fix the hip, knee, and ankle in the normal standing position. The cast was applied only to the plantar surface to prevent limb constriction due to post-fracture edema and extended from the hind paw metatarsals up to a spica formed around the abdomen. Also, galvanized wire meshes were used to avoid casts chewing, and rats received postoperative hydration (saline) and opioid analgesia for 2 days. After 4 weeks, the cast was removed under isoflurane anesthesia with a vibrating cast saw, and the fracture area was manually examined [23, 24].

Table 1
The fracture and cast model induces complex regional pain syndrome (CRPS)-like behaviors in different apparatuses

Animal	Time of cast removal after fracture	Behavior	Time of evaluation	Outcomes	Reference
Sprague–Dawley rats (male, aged 10 months)	4 weeks	Mechanical allodynia (von Frey test, up-down); heat nociceptive thresholds (Hargreaves, 29°C); hind paw temperature (fine wire thermocouple); hind paw volume (laser measurement and calipter).	20 weeks	**Fracture and cast:** mechanical allodynia up to 10 weeks, heat withdrawal without difference, temperature increased up to 20 weeks, and hind paw edema increased up to 10 weeks (calipter) or up to 20 weeks (laser measurement). **Cast:** mechanical allodynia up to 6 weeks, heat withdrawal without difference, temperature increased when the cast was removed at 4 weeks, and hind paw edema increased at 4 weeks (calipter) or up to 20 weeks (laser measurement).	[23]
Sprague–Dawley rats (male, aged 10 months)	4 weeks	Mechanical allodynia (von Frey test, up-down); hind paw unweighting (incapacitance device, weight-bearing); hind paw temperature (fine wire thermocouple); hind paw volume (laser measurement).	4 weeks	**Fracture and cast:** mechanical allodynia, hind paw temperature and thickness increased and decreased the weight bearing at 4 weeks. **Control:** naive rats were used; this group did not show any alteration.	[24]
Mice C57Bl/6 (male)	3 weeks	Mechanical allodynia (von Frey test, up-down); hind paw unweighting (incapacitance device); hind paw temperature (fine wire thermocouple); hind paw volume (laser measurement).	18 weeks	**Crush:** did not show nociceptive or vascular changes at 3 weeks **Cast:** mechanical allodynia and hind paw unweighting without warmth and edema at 3 weeks. **Fracture and cast:** mechanical allodynia until 15 weeks, edema until 6 weeks, unweighting of the hind paw until 9 weeks, and warmth until 6 weeks. **Control:** naive mice were used; this group did not show any alteration.	[25]

(continued)

Table 1
continued

Animal	Time of cast removal after fracture	Behavior	Time of evaluation	Outcomes	Reference
Mice C57BL/6 (male, aged 12 weeks)	3 weeks	Open field (distance traveled and time in center); zero-maze (distance traveled and time in open quadrants); novel location recognition (time spent exploring); novel object recognition (time spent exploring); social memory (sniffing duration); mechanical allodynia (von Frey test, up-down).	7 weeks	**Fracture and cast:** without motor impairment and no difference was detected between groups for the novel location recognition test, anxiety-like behavior in the zero-maze apparatus, less time exploring the novel object and socializing with a newly introduced stranger, and mechanical allodynia (until 14 weeks). **Control:** naive mice were used; this group did not show any alteration.	[26]
Mice C57BL/6 (male and female)	3 weeks	Mechanical allodynia (von Frey test, up-down); hind paw temperature (fine wire thermocouple); hind paw volume (laser measurement); rotarod (latency to fall); capsaicin-evoked behaviors (spontaneous behaviors); conditioned place preference (time spent in the chamber); fear memory (% of freezing).	18 weeks	**Fracture and cast:** female and male displayed mechanical allodynia until 13 and 11 weeks, respectively (female mice showed difference from male with lower threshold from 7 to 12 weeks). Female and male mice showed edema (3 and 4 weeks) and temperature at 3 weeks (no difference). Female spent more time in capsaicin-evoked behaviors than males (at 18 weeks). Female and male exhibited motor disturbance at 10 weeks (no difference). Female and male presented ongoing pain (conditioned place preference test, at 9 weeks). Only female had deficits in fear memory (at 5 weeks), no difference cued memory (at 5 and 9 weeks).	[27]

Table 1
continued

Animal	Time of cast removal after fracture	Behavior	Time of evaluation	Outcomes	Reference
Mice C57BL/6 (male, aged 12–14 weeks)	3 weeks	Rotarod (latency to fall); capsaicin-evoked behaviors (spontaneous nociception); mechanical allodynia (von Frey test, up-down); hind paw temperature (fine wire thermocouple); hind paw volume (laser measurement).	18 weeks	**Fracture and cast:** without motor impairment at the 7- and 8-week and motor impairment at 18 weeks; spend more time engaging in capsaicin-evoked behaviors (at 18 weeks); mechanical allodynia (at 3 and 7 weeks); hind paw edema and temperature (at 3 weeks) **Control:** naive mice were used; this group did not show any alteration.	[28]
Mice C57BL/6 (male and female)	3 weeks	Mechanical allodynia (von Frey test, up-down).	3 weeks	**Fracture and cast:** male and female showed mechanical allodynia and unweighting (3 weeks). **Control:** naive mice were used; this group did not show any alteration.	[29]

The fracture and cast tibia model was adapted for mice using a hemostat to fracture the right medial tibia under isoflurane anesthesia. The hind limb was wrapped in casting tape from the hind paw and tied up to a spica formed around the abdomen to fix the knee and ankle, similar to the rat model. Mice received hydration (saline), antibiotics, and opioid analgesia for 2 days post-fracture, and the cast was removed after 3 weeks under isoflurane anesthesia [25–29].

2.2 CPIP Model of CRPS-I

For the CPIP model, rats or mice can be used that must be housed with food and water ad libitum (Table 2). This model can be done using rats that were first anesthetized, followed by the exposure of either the right [30] or left hind limb, and an O-ring is placed proximal to the ankle joint. The positioning of the O-ring, which remained for 3 h, allowed the blockage of blood circulation. After 3 h, the tourniquet was removed, and the blood flow was restored to the hind limb. The control group underwent the same procedure, but the O-ring was cut so that it just loosely encircled the ankle and did not cut off the blood supply to the hind paw [31–35].

The CPIP model was adapted for mice using a period of 3 h of IR. After anesthesia, animals received the placement of the O-ring elastic around either the right [36] or left side that remained for 3 h, allowing the blockage of blood circulation. After this period, the tourniquet was removed, restoring the blood flow to the hind limb. The control group underwent the same procedure, but the O-ring was cut so that it just loosely encircled the ankle and did not cut off the blood supply to the hind paw [37–39]. Another study also adapted this model using one nitrile O-ring for female mice and three nitrile O-rings for male mice during the same 3-h period. In a different study, male and female mice were used. After anesthesia, animals undergo a period of 2 h of IR using a nitrile O-ring around the left hind limb proximal to the ankle joint [40].

3 Materials

3.1 Fracture and Cast Immobilization Model of CRPS-I

Isoflurane was used as an anesthetic to conduct the fracture using pliers with an adjustable stop for rats (Visegrip, Petersen Manufacturing, Dewitt, NE) and a hemostat for mice [23–29]. The limb was wrapped in casting tape (Delta-Lite, Johnson and Johnson, Raynham, MA). Also, galvanized wire meshes were used to avoid casts chewing. Animals can receive postoperative hydration using 8 mL of subcutaneous saline, subcutaneous buprenorphine (0.3 mg/kg) analgesia, and enrofloxacin (5 mg/kg) for 2 days [23–29]. The cast was removed with a vibrating cast saw, and isoflurane is also used as an anesthetic [23–29].

3.2 CPIP Model of CRPS-I

Bolus administration (40 mg/kg, i.p.) and infusion of sodium pentobarbital (13 mg/h for the first hour, 6.5 mg/h for the second hour)

Table 2
The model of chronic post-ischemia pain (CPIP) induces complex regional pain syndrome (CRPS)-like behaviors

Animal	CPIP time	Behavior	Time of evaluation	Outcome	Reference
Long Evans hooded rats (male, 275–325 g, age not mentioned)	3 h	Mechanical allodynia (von Frey test, up-down); cold allodynia (acetone test); mechanical hyperalgesia (pin prick test); heat hyperalgesia (Hargreaves); hind paw temperature (thermocouple probe); hind paw edema (plasma extravasation—Evans Blue dye).	4 weeks	**CPIP:** mechanical allodynia and hyperalgesia, cold allodynia up to 4 weeks for ipsilateral side (maximum time of evaluation was 4 weeks), hyperemia and edema up to 2 h post-CPIP. Heat hyperalgesia was not detected. **Control:** naive rats were used; this group did not show any alteration.	[31]
Long Evans hooded rats (male, 275–325 g, age not mentioned)	3 h	Mechanical allodynia (von Frey test, up-down); evoked nociceptive behavior (norepinephrine).	1 week	**CPIP:** mechanical allodynia and norephinefrine-induced nociception (at 2 and 7 days). **Control:** naive rats were used; this group did not show any alteration.	[32]
Wistar rats (male, 200–250 g, age not mentioned)	3 h	Mechanical allodynia (von Frey test, up-down); cold allodynia (acetone test); heat hyperalgesia (Hargreaves test); hind paw edema (digital caliper).	2 weeks	**CPIP:** mechanical and cold allodynia at 1 and 14 days in the ipisilateral side. Heat hyperalgesia and edema were not detected in the ipsilateral side. **Control:** naive rats were used; this group did not show any alteration.	[33]
Sprague–Dawley rats (male, 280–320 g, age not mentioned)	3 h	Mechanical allodynia (automoted von Frey test).	3 days	**CPIP:** mechanical allodynia at 3 days in the ipsilateral and contralateral side. **Control:** naive rats were used; this group did not show any alteration.	[34]
Sprague–Dawley rats (male, 300–320 g, aged 8–10 weeks)	3 h	Mechanical allodynia (von Frey test, up-down); hind paw edema (digital caliper).	2 weeks	**CPIP:** mechanical allodynia up to 2 weeks in the ipsilateral and contralateral sides. Hind paw edema up to 2 days in the ipsilateral side. **Control:** naive rats were used; this group did not show any alteration	[30]

(continued)

Table 2
continued

Animal	CPIP time	Behavior	Time of evaluation	Outcome	Reference
Sprague–Dawley rats (male, 200–220 g, age not mentioned)	3 h	Mechanical allodynia (von Frey test, up-down); thermal hypersensitivity (heat plate 25°C).	2 weeks	**CPIP:** mechanical allodynia and thermal hypersensitivity up to 2 weeks in the ipsilateral side. **Control:** naive rats wereused; this group did not show any alteration.	[35]
Swiss CD1 mice (male, 25–30 g, aged 8 weeks)	3 h	Mechanical allodynia (von Frey test, up-down); cold allodynia (acetone test).	2 days	**CPIP:** mechanical and cold allodynia at 2 days in the ipsilateral side. **Control:** naive mice were used; this group did not show any alteration. *No baseline evaluation for mechanical and cold threshold.	[37]
Swiss CD1 mice (male, 25–30 g, aged 8 weeks)	3 h	Mechanical allodynia (von Frey test, up-down).	1 week	**CPIP:** mechanical allodynia up to 1 week in the ipsilateral side. **Control:** naive mice were not used.	[38]
Sprague–Dawley rats (male, 280–320 g, aged 7–8 weeks)	3 h	Mechanical allodynia (von Frey test, up-down); Thermal hyperalgesia (Hargreaves, 55 °C).	1 week	**CPIP:** Mechanical allodynia and heat hyperalgesia up to 1 week in the ipsilateral side. **Control:** naive mice were not used.	[39]
Swiss mice (male, 25–35 g, age not mentioned)	3 h	Mechanical allodynia (von Frey test, frequency of withdrawal method to 0.4 g filament).	15 days	**CPIP:** mechanical allodynia up to 15 days in the ipsilateral side. **Control:** naive mice were used; this group did not show any alteration.	[36]

Table 2
continued

Animal	CPIP time	Behavior	Time of evaluation	Outcome	Reference
Swiss mice (male and female, 40–60 g, aged 60–120 days)	3 h	Mechanical allodynia (von Frey test, frequency of withdrawal method to 0.4 g filament); cold allodynia (acetone test); hind paw edema (digital micrometer).	22 days	**CPIP:** mechanical allodynia up to 21 days in male and female, cold allodynia up to 22 days in male and female, and edema (up to and 5 h) in male (ipsilateral). **Control**: naive mice were used; this group did not show any alteration.	[41]
C57BL/6 mice (male and female, 20–30 g, age not mentioned)	2 h	Mechanical allodynia (von Frey test, up-down); cold allodynia (acetone test); ongoing pain-like behavior (hind paw score); spontaneous nociception; hind paw temperature (infrared thermometer); hind paw edema (digital caliper).	20 days	**CPIP:** mechanical allodynia and cold allodynia were detected up to 20 days in the ipsilateral side in male and female mice. Hind paw temperature and edema up to 3 h in the ipsilateral side in male and female mice. Ongoing pain-like behavior and spontaneous nociception at 3 h after CPIP. **Control**: naive mice were used; this group did not show any alteration. No difference was observed between male or female mice for allodynia, edema, and hind paw temperature.	[40]

was used as an anesthetic to conduct the CPIP model in Long Evans hooded rats [31, 32] or Sprague–Dawley [30, 34] using a Nitrile 70 Durometer O-ring (O-rings West, Seattle, WA) with 7/32-inch internal diameter [30–32, 34]. A mixture of ketamine (90 mg/kg) and xylazine (3 mg/kg) (i.p) was used as an anesthetic to conduct the CPIP model using a Nitrile 70 Durometer O-ring with 7/32-inch internal diameter in Wistar rats [33] or C57BL6 mice [40]. Anesthesia with an intraperitoneal injection of 50 mg/kg of sodium phenobarbital and maintained with an infusion of sodium phenobarbital 20 mg/kg/h using a Nitrile 70 Durometer O-ring with 7/32-inch internal diameter was used in Sprague–Dawley rats [30]. Intraperitoneal injection of sodium pentobarbital (60 mg/kg), followed by 20 mg/kg every 60 min during the 3-h ischemic period, was used while applying a Nitrile 70 Durometer O-ring (Kangda Chemical Company, China), with 0.56 cm of internal diameter in Sprague–Dawley rats [35]. Bolus (55 mg/kg, i.p.) of sodium pentobarbital was required for anesthesia and maintained (27.5 mg/kg, i.p) with the same anesthetic. Using Swiss CD1 mice, a Nitrile 70 Durometer O-ring (O-rings West, Seattle, WA) with a 5/64-inch internal diameter was placed in the hind limb [37]. In another study, isoflurane (1%–2%) with a Nitrile 70 Durometer O-ring, with a 5/64-inch internal diameter, was also used in Swiss CD1 mice [38, 39]. Other adaptations occur, using Swiss mice with a bolus (7%, 0.6 mL/kg, i.p.) of chloral hydrate and placement of an O-ring (Elástico Ligadura 000–1237, Uniden) with 1.2 mm internal diameter [36] or a 7/32-inch internal diameter [41].

4 Literature Review About CRPS-I Models

4.1 Fracture and Cast Immobilization Model of CRPS-I

Patients with tibia and radius fractures can develop pain, functional loss, and disability until 1-year post-trauma. These symptoms are known as CRPS-I and can result in edema, pain, weakness, contractures, and bone loss without a clear pathophysiology [16]. In this view, Guo et al. (2004) first standardized the tibial fracture model and observed that the paw temperature increased without changes in skin color. Besides, both groups showed edema and mechanical allodynia, but the fracture group remained with these symptoms for a long period. No differences in heat withdrawal were detected in this model. In addition, bone loss (periarticular regions of the femur and tibia) peaked 8 weeks post-fracture and slowly recovered over the 20 weeks of evaluation in the fractured and cast groups. Therefore, this study demonstrated that the rats' tibial fracture, a CRPS-I model, corresponds to the patient's clinical condition, but the immobilization alone also caused a brief CRPS-like syndrome [23].

Subsequently, Sabsovich et al. (2008) used a control group without fracture or cast to evaluate the CRPS-like symptoms in the same fracture and cast model in rats using the Guo et al. (2004) method.

Fracture and cast rats had mechanical allodynia, weight-bearing decrease in the fractured limb, and increased hind paw temperature and thickness compared to the control. Also, the fracture group presented bone loss in the distal femur and L4 vertebral body [24]. Therefore, using naive control animals, this study showed CRPS-like symptoms 4 weeks after fracture and cast in male rats.

Afterward, the fracture and cast tibia model was adapted for mice by Guo et al. (2012), control mice did not pass the operation and did not use a cast. Also, these authors described two other control mice groups: mice with a crush injury that did not cause fracture and mice with a cast without fracture. The mice fracture group presented mechanical allodynia, edema, unweighting, and warmth of the hind paw. The group with cast developed nociception (mechanical allodynia and hind paw unweighting) without warmth and edema of the hind paw. However, the crush group did not show any nociceptive or vascular changes. Besides, fracture mice exhibited bone loss in the ipsilateral femur compared to controls [25]. Although the cast was removed 1 week earlier in mice compared to rats, mice fracture and casting developed the same symptoms observed in rats. Thus, the mice model allows for studying the fracture-induced CPIP mechanism with wild-type and knockout animals.

CRPS patients also presented anxiety and memory dysfunction, thus a different research group induced the fracture and cast model in mice to analyze cognitive changes and anxiety-like behavior [42]. The fracture and cast mice group did not show motor impairment or anxiety-like behavior in the open field test, and no difference was detected between groups for the novel location recognition test. However, animals presented anxiety-like behavior without motor changes in the zero-maze apparatus. In addition, the fracture and cast group spend less time exploring the novel object than the familiar object and socializing with a newly introduced stranger, suggesting a memory deficit and diminished social memory, respectively. This study also detected mechanical allodynia in the hind paw [26]. Thus, anxiety-like behavior and impaired memory without motor changes were observed in the mice fracture model and can be related to nociceptive behavior and CRPS-like symptoms.

Then, the following author investigates the spontaneous behavior of mice after capsaicin injection in the right hind paw plantar surface and the forced locomotor activity. Fracture and cast spent more time responding to capsaicin-evoked nociception, but mechanical allodynia was not increased after capsaicin or saline injection. Mice did not show motor impairment in the earlier weeks, only at 18 weeks after the induction of the model [28].

Subsequently, the same author also investigated the sex difference for nociception and two different behaviors (conditioned place preference and fear memory) in the adapted fracture and cast model for mice [25]. Female mice showed a lower mechanical threshold compared to male mice after fracture and casting induction, but no difference was identified for edema and temperature of the hind paw. Also, female mice showed increased nociception time after capsaicin injection in the hind paw compared to males. In the rotarod test, both female and male mice showed signs of motor alteration. The conditioned place preference is based on unconditioned reward due to pain relief, so females and males spent more time in the morphine chamber, suggesting ongoing pain without sex difference. Finally, only female mice showed fear memory [27]. Thus, this was the first report of sex differences for nociceptive behavior of mice post-fracture and cast model. Also, those authors bring out for the first time a spontaneous pain behavior and memory evaluation in this model. However, the protocol used in this study did not use a control group without fracture.

Following, Guo et al. (2019) also investigated the sex differences and hormone effects on the 3 and 7 weeks after tibia fracture and casting, inducing nociception in mice. A sham group was used as a control group, and the fracture and cast model was performed [25]. Mice that underwent ovariectomy/castration or sham surgery developed mechanical allodynia and reduced unweighting of the hind paw post-fracture and casting. However, this study did not compare male and female mice after ovariectomy/castration for mechanical allodynia determination [29].

In summary, the novel studies using this model should preferentially focus on using C57BL/6 female and male mice because there is a sex difference in the development of nociception in this model. The control group needs to be inserted to be compared with fracture and casting animals. Also, the nociception parameters that could be evaluated are mechanical allodynia (using von Frey test, up-down) and unweighting (uncapacitance device) 7 weeks after fracture and casting because motor alteration will appear after 7 weeks of model induction. The edema and temperature of the hind paw should be measured until 6 weeks after the fracture and casting model. The anxiety-like behavior should be investigated in the fracture and cast model using the elevated plus maze test.

Thus, for the fracture and cast model of CRPS-I, both rats and mice showed nociceptive, vascular changes, and bone loss after fracture and cast immobilization, similar to patients. Although fracture and cast immobilization is one type of animal model that represents a real lesion that can occur in CRPS patients, it is an invasive model due to fracture surgery and two anesthesia procedures. Also, all studies show the need for the control group to compare with fracture and cast groups since the cast group developed a CRPS-like syndrome. Besides, there is still a need for future studies in this model that evaluate non-evoked spontaneous pain behavior (e.g., nesting behavior, burrowing activity, and grimace scale), heat hyperalgesia, and cold allodynia (Tables 1 and 3). In addition, further studies should investigate the differences in pain sensation between female and male rodents.

Table 3
Comparison of behavior responses between the fracture/cast and chronic post-ischemia pain (CPIP) model in mice and rats

Behaviors	Fracture and cast	CPIP
Mechanical allodynia (von Frey test)	Yes (rats and mice)	Yes (rats and mice)
Mechanical hyperalgesia (pin prick test)	Not evaluated (rats and mice)	Yes (rat) Not evaluated (mice)
Cold allodynia(acetone test)	Not evaluated (rats and mice)	Yes (rats and mice)
Hind paw unweighting (weight-bearing)	Yes(rats and mice)	Not evaluated (rats and mice)
Heat nociceptive thresholds (Hargreaves)	No (rats) Not evaluated (mice)	Yes (rats and mice)
Hind paw temperature (fine wire thermocouple)	Yes (rats and mice)	Yes (rats and mice)
Hind paw volume (edema caliper)	Yes (rats and mice)	Yes (rats and mice)
Spontaneous locomotor activity (open field)	Not evaluated (rats) No (mice)	Not evaluated (rats and mice)
Anxiety-like behavior (open field)	Not evaluated (rats) No (mice)	Not evaluated (rats and mice)
Anxiety-like behavior (zero-maze)	Not evaluated (rats) Yes (mice)	Not evaluated (rats and mice)
Novel location recognition (working memory)	Not evaluated (rats) No (mice)	Not evaluated (rats and mice)
Novel object recognition (working memory)	Not evaluated (rats) Yes (mice)	Not evaluated (rats and mice)
Social memory (social recognition memory)	Not evaluated (rats) Yes (mice)	Not evaluated (rats and mice)
Forced locomotor activity (rotarod)	Not evaluated (rats) Yes (mice)	Not evaluated(rats and mice)
Spontaneous nociception(capsaicin-evoked behaviors)	Not evaluated (rats) Yes (mice)	Yes (rats) Not evaluated (mice)
Spontaneous nociception (seconds)	Not evaluated (rats and mice)	Not evaluated (rats) Yes (mice)
Evoked nociceptive behavior (norepinephrine)	Not evaluated (rats and mice)	Yes (rats) Not evaluated (mice)
Ongoing pain (conditioned place preference)	Not evaluated (rats) Yes (mice)	Not evaluated (rats and mice)
Ongoing pain (score)	Not evaluated (rats and mice)	Not evaluated (rats) Yes (mice)

(continued)

Table 3
continued

Behaviors	Fracture and cast	CPIP
Fear memory (learning and memory)	Not evaluated (rats) Yes (female mice)	Not evaluated (rats and mice)
Female and male comparison	Not evaluated (rats) Yes (mice)	Not evaluated (rats) Yes (mice)

"No" is when the behavioral test was not statistically different for that model; "Yes" is when the behavioral test was statistically different for that model; "Not evaluated" is when the behavioral test was not assessed for that model

4.2 CPIP Model of CRPS-I

Besides fracture, patients undergoing ischemia and reperfusion after limb trauma show tissue injury and may develop CRPS-I. Thus, inflammation and acute pain occur during reperfusion and reoxygenation of the tissue [43]. This process can result in allodynia, hyperalgesia, edema, and motor dysfunction in patients [44]. Besides, there are occluding arteries IR models, but these models are not commonly used for preclinical research; the most used model is the O-ring method [45]. Thus, in this chapter, we will just describe the standardization of this method.

The first model of IR was standardized by Coderre et al. (2004) using male Long Evans hooded rats to establish CPIP. This model was generated through the exposure of the left hind limb and placement of the O-ring proximal to the ankle joint. Animals were first anesthetized (sodium pentobarbital), followed by the positioning of the O-ring that remained for 3 h cutting off the blood circulation. After 3 h, the tourniquet was removed, allowing the blood reperfusion to the hind limb. Animals showed mechanical allodynia and hyperalgesia, cold allodynia, hyperemia, and edema, but heat hyperalgesia was not detected. Nociception parameters were detected in the ipsilateral and contralateral hind paw after CPIP, but the contralateral side did not show similar results at all time points evaluated [31]. Thus, this was the first model that characterized nociception and vascular changes in a model of CRPS-I using CPIP.

After CPIP model standardization, other authors made a few alterations over the years, whether in the animal model itself, the nociceptive tests, or the methods used. Subsequently, another research group used the same animal model of CPIP [31] and evaluated the mechanical allodynia and evoked nociception caused by norepinephrine intraplantar injection. Rats that developed mechanical allodynia in the ipsilateral hind paw also showed norepinephrine-induced nociception [32].

Subsequently, Klafke et al. (2015) used male Wistar rats and applied the same CPIP model [31]. Animals showed mechanical and cold allodynia in the ipsilateral hind paw, but thermal hyperalgesia and edema were not detected [33].

Then, later on, Ryu et al. (2010) used a different rat strain (Sprague–Dawley) to establish the IR model induced by IR injury [31]. The authors described mechanical allodynia using an automated von Frey test in both ipsilateral and contralateral hind paws [34]. Then, Chen et al. (2020) followed the same IR model and rat strain to assess mechanical allodynia and hind paw edema. CPIP animals showed consistent mechanical allodynia in both ipsilateral and contralateral hind paws, but edema was only present in the ipsilateral hind paw compared to sham animals [30]. Another author used the same model and rat strain to determine mechanical allodynia and thermal hypersensitivity in CPIP animals. Rats presented consistent mechanical allodynia and thermal hypersensitivity in the ipsilateral hind paw compared to the control groups [35].

Millecamps et al. (2010) were the first to use mice (male Swiss CD1) to establish the CPIP model, although still following Coderre et al. (2004) footsteps. CPIP was generated by 3 h of ischemia and reperfusion with subsequent tourniquet removal. Mechanical and cold allodynia were only tested at 2 days post-IR, only observing the acute phase, but baseline evaluation of mechanical and cold threshold was not assessed [37]. Hsiao et al. (2016) then repeated the new standard model established [37] using the same mice strain (Swiss CD1) and evaluated the mechanical allodynia until 7 days after CPIP. Animals presented persistent mechanical allodynia in the ipsilateral hind paw when induced by CPIP [38]. A few years later, Hsiao et al. (2019) repeated this CPIP animal model and tested it for mechanical allodynia and thermal hyperalgesia. The nociceptive tests showed that CPIP animals presented both mechanical allodynia and thermal hyperalgesia in the ipsilateral hind paw compared to baseline [39]. These studies did not use a control group without CPIP [38, 39].

Nucci et al. (2012) used the same CPIP model of 3 h of IR but altered the mice outbred (Swiss) to measure mechanical allodynia. CPIP mice showed mechanical allodynia in the ipsilateral hind paw compared to sham groups [36]. Later, Gonçalves et al. (2020) adapted the IR model using 3 h of IR in male and female Swiss mice. After anesthesia, male mice received three nitrile O-rings around the left hind limb joint proximal to the ankle. In female mice, only one O-ring was used since three O-rings caused an exaggerated inflammatory process with tissue necrosis in a pilot test. Thus, only one O-ring was used to avoid animal loss for female mice. Female and male animals develop mechanical and cold allodynia, edema was only evaluated in male mice and showed significance [41]. This was the first study that used male and female mice in the CPIP model, but the authors did not compare the responses between both sexes.

Another article used a different period of IR (2 h) in male and female mice of the C57BL/6 strain. In this model, De Prá et al. (2019) standardized CPIP using a nitrile O-ring around the left hind limb proximal to the ankle joint to act as a tourniquet for 2 h. After the procedure, animals developed spontaneous nociception, ongoing pain-like behavior, mechanical allodynia, cold allodynia, hind paw increased temperature, and hind paw edema on the ipsilateral side. In this model, no difference was observed between male and female mice for a nociceptive response, hind paw increased temperature, or hind paw edema detection [40]. Similar to Coderre et al. (2004), this study detected mechanical and cold allodynia after CPIP. Also, this was the first study that used male and female mice to evaluate spontaneous nociception and ongoing pain-like behavior but did not evaluate heat hyperalgesia [40].

Considering the studies published until now using the CPIP model, we detected that further protocols using the CPIP model can be conducted with mice and rats, using female and male animals, since the sex difference was not completely elucidated. In addition, the control group should be inserted to be compared with CPIP animals. Also, the nociception parameters that could be evaluated are mechanical allodynia (using the von Frey test, up-down) and cold allodynia (acetone test). The edema and temperature of the hind paw should be until measured 2 h after IR induction.

Regarding the CPIP model, rats and mice showed similar nociceptive and vascular changes in the hind limbs after IR exposure, as observed in patients, but the bone loss was not evaluated. This model is a noninvasive method of CPIP injury establishment with low cost and diminished rate of animal loss. However, the tourniquet should be inserted carefully into the hind paw to avoid the reduction of O-ring elasticity for efficient installation of IR. Similarly, to the fracture and cast model, there is still a need for future studies in the IR model that evaluate non-evoked spontaneous pain behavior (e.g., nesting behavior, burrowing activity, and grimace scale), hind paw unweighting, and locomotor activity (Tables 2 and 3). Finally, more studies are needed that evaluate anxiety- and depression-like behaviors, and memory loss.

5 Conclusions

Although the CPIP model is a simple method of inducing CRPS-I-like behaviors in animals, only the fracture and cast model evaluated the bone loss closer to those observed in CRPS-I patients after fracture (Fig. 1). However, cold allodynia, a classical symptom presented by CRPS patients, was only evaluated in the CPIP model. Therefore, mechanical allodynia can be considered a standard nociceptive behavior since most studies observed this type of nociception after the induction of CRPS-I models in rodents. However, only the fracture and cast model showed a reduction in nociceptive behavior 18 weeks post-induction; for the CPIP model, there is still a need to identify the time course for nociception after IR. Consequently, future studies still need to evaluate mechanical allodynia after 4 weeks post-induction for the CPIP model. However, the CPIP model is a simple method to induce CRPS model. Thus, if a mechanism should be studied using a model of CRPS-I, initially the CPIP model is recommended. However, if fracture and bone loss are the goals of the study, the fracture and cast model should be applied. Finally, both models show limitations, but most of the studies presented nociceptive and vascular changes as observed in CRPS-I patients in the clinical setting.

Conflict of Interest

The authors have no conflicts of interest to declare that are relevant to the content of this chapter.

References

1. Treede R-D, Rief W, Barke A et al (2019) Chronic pain as a symptom or a disease: the IASP Classification of Chronic Pain for the International Classification of Diseases (ICD-11). Pain 160:19–27
2. Shim H, Rose J, Halle S, Shekane P (2019) Complex regional pain syndrome: a narrative review for the practising clinician. Br J Anaesth 123:e424–e433
3. Popkirov S, Enax-Krumova EK, Mainka T et al (2020) Functional pain disorders – more than nociplastic pain. Neurorehabilitation 47:343–353
4. Nicholas M, Vlaeyenb JWS, Riefe W et al (2019) The IASP classification of chronic pain for ICD-11: chronic primary pain. Pain 160:28–37
5. Fitzcharles MA, Cohen SP, Clauw DJ et al (2021) Nociplastic pain: towards an understanding of prevalent pain conditions. Lancet 397:2098–2110
6. Mouraux D, Lenoir C, Tuna T et al (2019) The long-term effect of complex regional pain syndrome type 1 on disability and quality of life after foot injury. Disabil Rehabil 43:967–975
7. Taylor SS, Noor N, Urits I et al (2021) Complex regional pain syndrome: a comprehensive review. Pain Ther 10:875–892
8. Harden RN, McCabe CS, Goebel A et al (2022) Complex regional pain syndrome: Practical Diagnostic and Treatment Guidelines, 5th Edition. Pain Med 23:S1–S53
9. Baronio M, Sadia H, Paolacci S et al (2020) Molecular aspects of regional pain syndrome. Pain Res Manag 2020:7697214
10. Goebel A, Birklein F, Brunner F et al (2021) The Valencia consensus-based adaptation of the IASP complex regional pain syndrome diagnostic criteria. Pain 162:2346–2348
11. Williams DM (2018) Clinical pharmacology of corticosteroids. Respir Care 63:655–670
12. Sills GJ (2006) The mechanisms of action of gabapentin and pregabalin. Curr Opin Pharmacol 6:108–113
13. Goh EL, Chidambaram S, Ma D (2017) Complex regional pain syndrome: A recent update. Burn Trauma 5:1–11
14. Varenna M, Manara M, Rovelli F et al (2017) Predictors of responsiveness to bisphosphonate treatment in patients with complex regional pain syndrome type I: a retrospective chart analysis. Pain Med (United States) 18:1131–1138
15. Tajerian M, Clark JD (2016) New concepts in CRPS. Hand Clin 32:41–49
16. Birklein F, Ajit SK, Goebel A et al (2018) Complex regional pain syndrome — phenotypic characteristics and potential biomarkers. Nat Rev Neurol 14:272–284
17. De Logu F, De Prá SDT, De David Antoniazzi CT et al (2020) Macrophages and Schwann cell TRPA1 mediate chronic allodynia in a mouse model of complex regional pain syndrome type I. Brain Behav Immun 88:535–546
18. Perez RSGM, Zuurmond WWA, Bezemer PD et al (2003) The treatment of complex regional pain syndrome type I with free radical scavengers: a randomized controlled study. Pain 102:297–307
19. Aïm F, Klouche S, Frison A et al (2017) Efficacy of vitamin C in preventing complex regional pain syndrome after wrist fracture: a systematic review and meta-analysis. Orthop Traumatol Surg Res 103:465–470
20. Tékus V, Hajna Z, Borbély É et al (2014) A CRPS-IgG-transfer-trauma model reproducing inflammatory and positive sensory signs

associated with complex regional pain syndrome. Pain 155:299–308
21. Challa SR (2015) Surgical animal models of neuropathic pain: pros and cons. Int J Neurosci 125:170–174
22. Ronchi G, Morano M, Fregnan F et al (2019) The median nerve injury model in pre-clinical research – a critical review on benefits and limitations. Front Cell Neurosci 13:1–21
23. Guo TZ, Offley SC, Boyd EA et al (2004) Substance P signaling contributes to the vascular and nociceptive abnormalities observed in a tibial fracture rat model of complex regional pain syndrome type I. Pain 108:95–107
24. Sabsovich I, Guo TZ, Wei T et al (2008) TNF signaling contributes to the development of nociceptive sensitization in a tibia fracture model of complex regional pain syndrome type I. Pain 137:507–519
25. Guo TZ, Wei T, Shi X, et al. (2012) Neuropeptide deficient mice have attenuated nociceptive, vascular, and inflammatory changes in a tibia fracture model of complex regional pain syndrome. Mol Pain 8:85
26. Tajerian M, Leu D, Zou Y et al (2014) Brain neuroplastic changes accompany anxiety and memory deficits in a model of complex regional pain syndrome. Anesthesiology 121:852–865
27. Tajerian M, Sahbaie P, Sun Y et al (2015) Sex differences in a murine model of complex regional pain syndrome. Neurobiol Learn Mem 123:100–109
28. Tajerian M, David Leu, Yang P et al (2015) Differential efficacy of ketamine in the acute versus chronic stages of complex regional pain syndrome in mice. Anesthesiology 123:1435–1447
29. Guo TZ, Shi X, Li WW, et al (2019) Sex differences in the temporal development of pronociceptive immune responses in the tibia fracture mouse model. Pain 160:2013–2027
30. Chen R, Yin C, Hu Q et al (2020) Expression profiling of spinal cord dorsal horn in a rat model of complex regional pain syndrome type-I uncovers potential mechanisms mediating pain and neuroinflammation responses. J Neuroinflammation 17:1–19
31. Coderre TJ, Xanthos DN, Francis L, Bennett GJ (2004) Chronic post-ischemia pain (CPIP): a novel animal model of complex regional pain syndrome-type I (CRPS-I; reflex sympathetic dystrophy) produced by prolonged hindpaw ischemia and reperfusion in the rat. Pain 112:94–105
32. Xanthos DN, Bennett GJ, Coderre TJ (2008) Norepinephrine-induced nociception and vasoconstrictor hypersensitivity in rats with chronic post-ischemia pain. Pain 137:640–651
33. Klafke JZ, da Silva MA, Rossato MF et al (2015) Acute and chronic nociceptive phases observed in a rat hind paw ischemia/reperfusion model depend on different mechanisms. Pflugers Arch 468:229–241
34. Ryu TH, Jung KY, Ha MJ et al (2010) Superoxide and nitric oxide involvement in enhancing of N-methyl-D-aspartate receptor-mediated central sensitization in the chronic post-ischemia pain model. Korean J Pain 23:1–10
35. Xu X, Tao X, Huang P et al (2020) N-methyl-D-aspartate receptor subunit 2B on keratinocyte mediates peripheral and central sensitization in chronic post-ischemic pain in male rats. Brain Behav Immun 87:579–590
36. Nucci C, Mazzardo-Martins L, Stramosk J et al (2012) Oleaginous extract from the fruits *Pterodon pubescens* Benth induces antinociception in animal models of acute and chronic pain. J Ethnopharmacol 143:170–178
37. Millecamps M, Laferrire A, Ragavendran JV et al (2010) Role of peripheral endothelin receptors in an animal model of complex regional pain syndrome type 1 (CRPS-I). Pain 151:174–183
38. Hsiao HT, Lin YC, Wang JCF et al (2016) Hypoxia inducible factor-1α inhibition produced anti-allodynia effect and suppressed inflammatory cytokine production in early stage of mouse complex regional pain syndrome model. Clin Exp Pharmacol Physiol 43:355–359
39. Hsiao HT, Liu YY, Wang JCF et al (2019) The analgesic effect of propofol associated with the inhibition of hypoxia inducible factor and inflammasome in complex regional pain syndrome. J Biomed Sci 26:1–11
40. De Prá SDT, Antoniazzi CT, de D, Ferro PR et al (2019) Nociceptive mechanisms involved in the acute and chronic phases of a complex regional pain syndrome type 1 model in mice. Eur J Pharmacol 859:172555
41. Gonçalves ECD, Vieira G, Gonçalves TR et al (2020) Bradykinin receptors play a critical role in the chronic post-ischaemia pain model. Cell Mol Neurobiol 41:63–78
42. Jung YH, Kim H, Jeon SY et al (2018) Peripheral and central metabolites affecting depression, anxiety, suicidal ideation, and anger in complex regional pain syndrome patients using a magnetic resonance spectroscopy: a pilot study. Psychiatry Investig 15:891–899
43. Eltzschig HK, Eckle T (2011) Ischemia and reperfusion—from mechanism to translation. Nat Med 17:1391–1401
44. Neumeister MW, Romanelli MR (2020) Complex regional pain syndrome. Clin Plast Surg 47:305–310
45. Muthuraman A, Ramesh M, Sood S (2010) Development of animal model for vasculatic neuropathy: induction by ischemic-reperfusion in the rat femoral artery. J Neurosci Methods 186:215–221

Chapter 8

Arthritis Murine Models for the Study of Inflammation and Pain

Sara Martínez-Ramos and Samuel García

Abstract

Arthritis is the pathological process hallmark of the rheumatic and musculoskeletal diseases (RMD), which affects millions worldwide and leads to significant health and socioeconomic challenges. Osteoarthritis (OA), rheumatoid arthritis (RA), Spondyloarthritis (SpA), Juvenile Idiopathic Arthritis (JIA), and gout are common types of arthritis. Despite advancements in therapy, issues persist with nonresponders, side effects, and a lack of effective treatments for some conditions. Murine models, therefore, play a pivotal role in understanding arthritis pathology and pharmacological research. A range of these models are available, each designed to study different RMD from a specific focus, while offering insights into disease progression, inflammation, injury, and pain. This chapter summarizes the main murine models in arthritis—from spontaneous to induced, and including transgenic—along with their principal characteristics, providing a comprehensive overview of the field, and always advocating for an ethical and responsible approach to the use of animals in biomedical research.

Keywords Arthritis, Pain, Inflammation, Joint damage

1 Introduction

Arthritis, baldly defined by the Human Phenotype Ontology as "inflammation of a joint" (*HP:0001369*), is a condition belonging to the group of rheumatic musculoskeletal diseases (RMD). Different types of arthritis are widely described according to their etiology and pathogenesis so, depending on its pathological mechanisms, arthritis can be traumatic, septic or immune-mediated [1]. RMD constitute one of the largest groups of diseases, affecting hundreds of millions of people just in western countries [2]. The health of these patients is impaired physically and mentally, eroded by pain and disability, with a consequent loss of quality of life and an enormous socioeconomic impact [2–6]. Osteoarthritis (OA), rheumatoid arthritis (RA), Spondyloarthritis (SpA), Juvenile Idiopathic Arthritis (JIA), or gout are largely reported [7, 8].

Isaura Tavares and Hugo Leite-Almeida (eds.), *Translational Research Methods for Neuropathic Pain*, Neuromethods, vol. 235, https://doi.org/10.1007/978-1-0716-5312-8_8,

In the last decades, multiple therapies for the treatment of arthritis have emerged, all of them previously tested in animals for their approval. Despite the great advances, non-responder patients, side effects or even the lack of effective therapies for some disorders are indeed routine issues. Therefore, new alternatives and greater efficiency are still a big need. This is the reason why reproducible and refined use of animal models remains essential in arthritis pathology and drug mechanisms research.

It is worth noting that some of the traditional models are not arthritis type-distinctive and might be applied for several diseases. Fortunately, greatly precise murine models are available nowadays for reflecting particular arthritis phenotypes and pathogenesis.

This chapter discusses the available alternatives in murine models of arthritis, from traditional to most recent choices in a variety of arthritic diseases.

2 Selection of the Model and Experiment Design

The selection of the model might depend not only on the disease but the experimental aim, hence different alternatives should be chosen according to the study focus. The availability of models and strains for arthritis is broad, so multiple options need to be carefully analyzed and evaluated. Genetics, sex, age, housing, circadian rhythms, welfare, or diet may influence the disease progression and severity [9–12], and therefore, the experiment design. For example, in posttraumatic arthritis, a higher severity is described in males and intact mice, compared to females and neutered mice, respectively [13]. By the same procedure, old mice showed increased expression of arthritis-related genes compared to young mice [14]. Also differences between sexes [15] and ages [16] have been described in rats, with worse diagnosis in males and elders.

Considering behavior is important as well. Most of models can or must be used with concrete strains, but the management varies between them. Namely, DBA/1 and C57BL/6 mice, especially males, are more prone to aggression [17]. Fortunately, from our experience, environmental enrichment, early recaging (when necessary), as well as patient and caring manipulation extremely improves and eases handling over the weeks.

Pain management is key since arthritis is linked to both nociceptive and neuropathic pain [18]. It might be controlled with drugs but trying not to interfere with the research aim. In any case, humane endpoint is imperative when suffering exceeds the ethically acceptable. This, by consortium of authors, translates into presence of severe ulcers, aberrant gait sustained for more than 72 h, incapacitating paw swelling, spontaneous vocalizations, and deep weight loss [17], besides any other unjustifiable reason. Apart from that, manipulation and restraining must be careful, prioritizing cupped hands, tunnels or containers instead of traditional tail handling. The procedures must be performed gently and accurately, reducing the infliction of pain and stress in the animals.

A variety of murine models for RMD and their main characteristics in terms of inflammation and pain are presented below.

3 Murine Models in RMD

3.1 Osteoarthritis

Osteoarthritis (OA) is the most common form of arthritis [19], mainly affecting women with prominent knee and hip involvement, and a mean prevalence of 16%, increased with age [20, 21]. The pathogenesis is a complex of mechanical, biochemical, genetic, and immune processes [22] that lead to cartilage destruction and loss of bone margins [23]. OA can be classified as primary, with idiopathic no well-known origin, or secondary to biomechanics, trauma, infection, or inflammation [23]. In early OA, the thickness of cartilage increases due to chondrocyte response to joint injury [24]. The chondrocytes are not able to repair the tissue, ossification is triggered, and reactive inflammation occurs in the surrounding synovial cavity. While disease progresses, the inflammatory mediators promote cartilage damage, aberrant bone formation with osteophyte onsets and nerve irritation that causes articular pain [25]. By the moment, there is not any drug available than can control such progression [26].

The murine models of OA can be spontaneous (naturally-occurring and transgenic models) or induced (chemically, mechanically, or surgically). Spontaneous models present more similarities in the disease progression with primary human OA, but the time of progression might be slow and variable [27, 28]. Transgenic (Tg) models enable the study of a single gene role but restrain considering the multifactorial interplay intrinsic to disease pathogenesis [29]. It should be noted that therapy results in spontaneous models cannot be translated to induced models, or vice versa, since the pathogenic mechanisms differ between them [30].

When choosing which species fits better for the study, several aspects need to be reviewed. Table 1 presents condensed information about pros, cons and requirements of murine models in OA.

In this context, the optimal animal model should reflect human-like OA reliably, quickly and with a 100% penetrance; the progression should mimic all disease stages and the anatomy, biomechanics, and tissue histology should resemble that of humans with a sufficient size of structures and animal expenses and care requirements should not be high [34, 40]. Since there is not a perfect universal model, the one that best suits the requirements of the study should be chosen.

Table 1
Pros, cons, and requirements for murine models in OA

Pros	Cons	Requirements
Short lifespan—quick OA onset development of OA [31, 32] Small size of mice—full-joint histological sections [33] Larger size of rats—easier surgical approaches [34, 35] Economic management and maintenance [33, 36] Useful for toxicology assays and basic research [35, 36] Completely described genomes [34] Easy production of mouse transgenic models [36, 37]	High variability between strains—challenging interpretation among experiments [37] Incomplete ossification of the epiphyseal plate—complex evaluation of skeletal maturity [38, 39]. Simple cartilage and thicker calcification in mice—three chondrocyte strata in humans [36, 37]. More vascularized adipose tissue and muscle in the medial stifle[a] in rats [40]. Induction methods for OA-related pain differ from those of OA joint destruction—discrepancies between pain and injury induced pathways [41]. Demanding postoperative management may be in both species [36] Unavailable rat transgenic models [36] Challenging macroscopic evaluation [33]	Adults only (mice > 10 weeks, rats > 3 months)—greater capability for joint recovery and physiological discrepancies in young animals [30]. More resemblance with human OA in older animals—severity varies with age [38]. For testing therapies, transgenic models are the most appropriate—genetic engineering targeting [30]

[a]Stifle: veterinary term referred to the tibio-patello-femoral joint of animals, which corresponds to the knee in humans

Table 2
Methods in the evaluation of OA changes in murine models

Macroscopic changes	Microscopic changes	Inflammation
Visual inspection Imaging [45–47] • MicroCT—bone tissue • Magnetic resonance imaging (MRI)—soft tissue	Mouse • OARSI cartilage histopathology assessment system (OOCHAS) [48] • OOCHAS modified [49] • Mankin system, Histologic/histochemical grading system (HHGS) [50, 51] *Rat* • OOCHAS modified [52] • Mankin modified [53, 54]	• Thermography [55] • Scintigraphy [56] • Laser Doppler [57]

Assessment

Usually, the stifle is established as the reference point for the assessment of OA [42], although temporomandibular joint involvement is also reported for STR/ort [43] and Discoidin Domain Receptor 1 knock-out (Ddr1−/−) mice [44]. Table 2 summarizes different methods for the assessment of OA changes in the joints.

When evaluating discomfort in animals, multiple time-points are required. The spontaneous pain (or pain-supressed) behavior, as diminished ambulation, is different from pain-evoked behavior, so multiple tests need to be employed. Usually, nociception is not measured but hyperalgesia and allodynia, which are signs linked to neuropathic pain [58, 59]. Some of the most commonly used pain monitoring tests are the Rotarod Test, the Dynamic Weight Bearing (DWB), the gait Analysis by CatWalk®, the Hind Limb Withdrawal by electronic unit Test the mechanical sensitivity (paw withdrawal threshold) by von Frey Test, the thermal sensitivity by Hargreaves Test, the reaction to intraplantar injection of noxious substance, the spontaneous behavior with automated assessment and the Vocalizations in response to Knee extension/Pressure Test [60–63]. Biochemically, the intensity of Substance P signal from ipsilateral and contralateral lumbar Dorsal Root Ganglia (DRG) can be compared by immunohistochemistry [60].

OA Models

Given the diversity within OA models and their mechanisms, Table 3 outlines the main ones and their most relevant characteristics.

Table 3
Murine models in OA

Spontaneous models in OA		
Naturally-occurring		
Predispose strains: • STR/ort and C57BL/ OA-resistant strains (recommend controls): • CBA	OA onset at 4–6 month-old mice. Arthritis signs mainly in the medial tibial plateau. Also in stifle, talocrural, elbow, and temporomandibular joints Increased incidence is described in males.	[64–68]
Transgenic models		
Genetic engineering manipulation of: • Extracellular matrix genes • Cartilage proteinases and inhibitors • Proinflammatory mediators • OA signaling pathways	Available for spontaneous and gene-specific models. Risk of comorbidities, low generalizability, and high costs.	[29, 65, 69–72]
Chemically induced models in OA		
Mono-iodoacetate (MIA) model	Widely used in pain studies (mouse and rat). Joint injury and ache (mechanical allodynia and decreased motion) notified within the first week. High reproducibility, early onset. No pathological pathways of human OA.	[30, 42, 63, 73, 75]
Mechanically induced models in OA		
Intraarticular tibial plateau fracture (IATPF) *Cyclic articular cartilage tibial compression (CACTC)* *Tibial compression overload: anterior cruciate ligament (ACL) injury*	Developed without invasive methods, by destabilizing the joint usually in C57BL/6 mice. Allodynia, hypersensitivity, and limited mobility. Contralateral (no destabilized) limb, sham injured, and no loaded animals are recommended as controls.	[42, 76, 78]

(continued)

Table 3
(continued)

Spontaneous models in OA		
Surgically induced models in OA		
Anterior cruciate ligament transection (ACLT) *Meniscectomy (MSX)/medial meniscal transection (MNX)* *Medial meniscus tear (MMT)* *Destabilization of the medial meniscus (DMM)*	Injury is caused by joint instability and inflammation through invasive methods, resembling posttraumatic OA. Useful for the study of pain, mainly mechanical allodynia after few weeks. Sham controls are required. Risk of postoperative complication and required surgical ability.	[27, 30, 75, 79]
Other models		
Obesity	Risk factor in human OA. It causes OA in rodents by the increment of loading and systemic inflammation. It can be combined with CACTC or transgenic strains.	[80–83]
Ovariectomy	Facilitation of cartilage turnover erosion due to the lack of estrogen.	[27, 84]

3.2 Rheumatoid Arthritis

Rheumatoid arthritis (RA) is the most prevalent disorder among the immune-mediated RMD, affecting around 1% of people worldwide, especially elderly females [85, 86]. RA is an autoimmune, inflammatory, chronic condition that causes joint inflammation and destruction, affecting both bone and cartilage [87]. Besides joint-related signs, particularly swelling, pain or morning stiffness, patients also present systemic manifestations as lung disease, heart disease, arteritis, atherosclerosis, keratoconjunctivitis, or lymphoma [88, 89].

The etiology is not completely understood but genetic, metabolic, and environmental risks factors have been associated with the initiation of RA [86, 90]. Inside the synovium, the pathological mechanism involves T and B cells, fibroblast-like synoviocytes (FLS), macrophages, chondrocytes, and bone cells, among others. The process is mediated by proinflammatory cytokines, being tumor necrosis factor (TNF), interleukin (IL)-6, and IL-1 the most important [91]. These cytokines trigger endothelial and immune cells which, in turn, perpetuate the inflammatory processes, activate FLS and enhance angiogenesis and osteoclastogenesis through RANK—RANKL junction [92–94]. This results in bone erosion and pannus formation with active synovitis.

Results obtained from animal models can be troubling to translate, since pathophysiological mechanisms may differ between animal models and human disease, but they are essential for the development of therapies. Importantly, 90% of those animals engaged in RA research are rodents [95]. A variety of murine models reveal

Table 4
Pros, cons, and requirements for murine models in RA

Pros	Cons	Requirements
Lifespan, size, anatomy—saved time and optimized assessment Replication of autoantibody activity [98] Replication of multiple disease pathways [10]	Lifespan, size, anatomy—divergences between humans and rodents [99, 100] Gender variances—immunologically and in pain-like behaviors [98]	Controlled environment—influence on disease onset [101, 102] Fitting model according to the pathophysiological mechanism/ treatment response of study [103, 104]

histopathological changes quite similar to humans RA [96], suggesting that disease pathways are multiple, differentiated and, possibly, shared by humans and rodents. Additionally, meta-analyses have revealed powerful similarities in gene expression patterns between the species during inflammation [97], supporting the use of murine models in rheumatology research. The pros, cons and requirements for murine models in RA are summarized in Table 4. Thus, considering the experimental objective, the most suitable model should be elected.

Assessment

In inflammatory arthritis, the joints are mainly evaluated in vivo by macroscopic exam. Multiple scores had been established which, in summary, assign values to different grades of inflammation, swelling, and reddening [105–108]. This visual inspection might be complemented by using callipers, plethysmography [109], infrared thermal imaging (IRT), radiographic analysis [110], and MRI [111].

Locomotion can be assessed by open field test [112] and gait analysis [113]. Thermal (of preference) and mechanical sensitivity are evaluated by Hargreaves and von Frey test, respectively [18, 112, 114]. Also genetics and serum profiling are helpful when determining inflammatory cytokines. However, the establishment of reachable biochemical parameters is pending.

RA Models

Several models can be used in RA research. Spontaneous models are considered those which pathogeneses is closer to the human condition, although mechanisms might not be de same. To solve this problem, it is possible to replicate those mechanisms through genetic engineering (humanized models) and induction protocols of autoimmune-like disease (induced models). Table 5 outlines the different RA models and their most relevant characteristics.

3.3 Spondyloarthritis

Spondyloarthritis (SpA) constitute a group of chronic arthritis within RMD, mainly affecting the sacroiliac and spinal joints, although peripheral joints are usually involved. Worldwide, the prevalence is approximately 1%, affecting both sexes in the same extend. The onset occurs around the third decade of age, one of the earliest among RMD [147, 148].

Table 5
Murine models in RA.

Spontaneous models in RA		
K/BxN mice	F1 of KRN mice crossed with the NOD strain, the MHC II I-A^{g7}-expressing offspring reveals spontaneous distal polyarthritis at the age of 4–5 weeks. T cell-autoreactivity to G6PI is the main pathogenic mechanism and mechanical allodynia is evidenced.	[115–118]
SKG mice	Chronic peripheral autoimmune arthritis in 2-month-old mutant BALB/c mice, caused by T cells autoreactivity due to altered TCR transduction, an only in non-SPF conditions.	[119]
IL-1ra knocked out	5-week-old *IL1ra*$^{-/-}$ mice reveal T-lymphocyte-mediated chronic peripheral polyarthritis. Individuals express anxiety-like behavior with aging.	[120–122]
Humanized models in RA		
Human TNF transgenic model *Human Leukocyte Antigen (HLA) Class II transgenic models* *Human Autoantigen-Specific TCR transgenic models*	The induction of an RA-like disease by incorporating human proinflammatory genes in mice is considered the most accurate method to resemble RA pathogenic mechanisms. Mechanical and thermal hyperalgesia, and decreased ambulation are reported.	[123–125]
Induced models in RA		
Collagen-induced arthritis (CIA)	Immunization with heterologous collagen type II (CII), emulsified in Incomplete Freund's Adjuvant (IFA) prompts an acute/subacute Th17-mediated polyarthritis in both mice and rats. Thermal hypersensitivity, mechanical allodynia and hyperalgesia, and decreased locomotion, are detected.	[126–131]
Collagen antibody-induced arthritis (CAIA)	BALB/c or CBA mice are treated with anti-CII arthritogenic cocktail combined with LPS, triggering arthritis within the first week by activating the complement through classical and alternative pathways. Cold sensitivity predominates during the inflammatory phase. Mechanical hypersensitivity is reported even after spontaneous arthritis resolving.	[132, 133]
Complete freund's adjuvant-induced arthritis (CFA)	Intraarticular administration of CFA in C57BL/6 mice is quickly followed by monoarthritis with mechanical allodynia, while systemic CFA is rats causes peripheric polyarthritis. This is mainly macroscopically detectable, but it is set as a useful model for pain, revealing thermal hyperalgesia, mechanical allodynia, and reduced motion.	[134–137]

(continued)

Table 5
(continued)

Spontaneous models in RA		
Recombinant human PG-G1 domain-induced arthritis (GIA)	BALB/c mice immunization with the rhG1 domain of cartilage PG results in arthritis with immune complexes deposition and Th1/Th17 polarization. Mechanical hyperalgesia is displayed.	[138, 139]
Glucose-6-phosphate isomerase-induced arthritis (G6PI)/ F759 mouse	G6PI in CFA administration induces a T cell-dependent polyarthritis in DBA/1 strain, primarily. The animals exhibit mechanical and thermal hyperalgesia [140].	[140, 141]
K/BxN serum transfer (STA)	K/BxN serum injection in naïve mice induces self-resolving polyarthritis, although chronicity can be reached with repeated administrations. Maintained hypersensitivity and mechanical allodynia are evidenced [142].	[142–146]

SpA is an umbrella term that includes different diseases, such as Ankylosing spondylitis (AS), psoriatic arthritis (PsA), undifferentiated spondyloarthritis (USpA), reactive arthritis (ReA), enteropathic arthritis, axial spondyloarthritis (axSpA), and peripheral SpA (pSpA). They share genetic, clinical, and immunological features, but they differently predominate in axial or peripheral body regions [148, 149]. Articular and extraarticular signs, comorbidities and higher mortality rates are reported in these patients [150–153].

Autoinflammatory processes in SpA are mostly mediated by the HLA-B27 molecule. The exact pathway is still unknown, but it results in the activation of autoreactive T cells by $CD8^+$ and antigen presenting cells, releasers of proinflammatory mediators as IL-17, IL-23, TNF, IL-6 or RANK, among others. These cytokines activate other immune and effector cells (FLS, osteoblasts), inducing inflammation, bone erosion, aberrant bone formation and, finally, leading to joint disruption [154].

Since advantages, limitations, and assessments can be extrapolated from those already mentioned for RA models, the different available models will be presented in Table 6. Nevertheless, as we are not aware of any existing SpA-specific model for the evaluation of pain, and suggesting the use of those techniques already described, we urge the development of targeted methods, especially for axial tenderness.

3.4 Other Arthritic Etiologies

As previously mentioned, the cause of arthritis is variable and multiple. In this section, some additional but less conventional models are briefly explained.

Juvenile Idiopathic Arthritis

Juvenile idiopathic arthritis (JIA) is a group of diseases included within the RMD and which affect children under 16 years of age with

Table 6
Murine models in SpA

Spontaneous and humanized models in SpA		
DBA/1 mice	Spontaneous PsA-like disease with enthesitis, dactylitis and onychoperiostitis from the fourth week of age.	[155]
HLA-B27 Tg rats	Rats overexpressing *HLA-B27*—$CD4^+$-mediated spontaneous SpA-like signs. HLA-B27/ human β2 microglobulin (Huβ2m) Tg rats (lines 21-4 H and 33-3 and 21-3×283-2 F1)—axial and peripheral arthritis.	[156–160]
IL-23/IL-17-related models	IL-23 minicircle-induced B10.RIII mice model)—Th17/osteoclast-mediated distal arthritis and psoriasis-like disease. Fungal-derived adjuvants administration in SKG mice in pathogen-free conditions—IL-23/IL-17 axis-dependent SpA-like signs.	[161–164]
TNF Tg	$TNF^{\Delta ARE/-}$ mice—axial and peripheral chronic polyarthritis with gut inflammation. TgA86 mice—axial arthritis and new bone formation without gut inflammation.	[165–167]
Induced models in RA		
CAIA model	axSpA-like signs in C57BL/6 > 6-week-old.	[168]
Proteoglycan-induced spondylitis (PGISp)	In $IL4^{-/-}$ mice, AS-like disease.	[169]
Yersinia-induced arthritis	ReA-like disease.	[170]

an incidence of 1.6–23/100,000 incidence, particularly in girls [171]. A model of JIA consists in challenging $IFN\gamma^{-/-}$ BALB/c mice with CFA. Arthritis, systemic inflammation, skin rash, and hematological changes are described [172]. With emerging evidence, others as the IL-6 Tg mice with TLR stimulation or the *Il1RN−/−* on a BALB/c background might be of interest for the study of JIA [173].

Gout

Gout is another RMD with a 1%–6.8% prevalence caused by the accumulation around the joints of crystalized form of uric acid, the monosodium urate monohydrate (MSU) [174].

The principal and more efficient mouse model outlined for this disorder is the MSU-induced model, by intraarticular administration of MSU crystals, triggering arthritic injury and pain thought the release of cytokines. In rats, it can be combined with oral Potassium oxonate (PO), which inhibits uricase activity, shaping the Acute Gouty Arthritis with Hyperuricemia model. Finally, the Diet-induced hyperuricemia model is set off in rats feed with hypoxanthine, yeast extract and PO oxalate [175].

Septic Arthritis

Septic arthritis is an articular infectious-inflammatory disease usually boosted by *Staphylococcus aureus* or *Pseudomonas aeruginosa*. The parenteral administration of these microorganisms in NRMI, C57BL/6, 1295 V, and BALB/c strains provokes $CD4^+$ mediated arthritic changes [176–178].

4 Conclusion

In this chapter, we have summarized the most important experimental murine models for arthritis research. As stated at the beginning, is always important to consider the research aim and the scientific objectives, but also the possibilities in terms of personnel skills, facilities, or time. Above all, the principle of the 3Rs (Replacement, Reduction, and Refinement) and welfare standards must be prioritized, driven us to a rational use of the animal experimentation, an area still indispensable in Biomedicine. We can affirm that, to date, arthritis models have proven to be irreplaceable, efficient, and sufficient in the fight against the wellness-limiting condition that is arthritis. However, new insights, designs, and accuracy improvements are still to be explored and implemented in order to push scientific, ethical progress while working with sentient beings.

Conflict of Interest

The authors have no conflicts of interest to declare that are relevant to the content of this chapter.

References

1. DiCarlo EF, Klein MJ (2012) Arthritis pathology. Surg Pathol Clin 5:15–65
2. Barbour KE, Helmick CG, Boring M, Brady TJ (2017) Vital signs: prevalence of doctor-diagnosed arthritis and arthritis-attributable activity limitation—United States, 2013–2015. MMWR Morb Mortal Wkly Rep 66:246–253
3. Vanhoof J, Declerck K, Geusens P (2002) Prevalence of rheumatic diseases in a rheumatological outpatient practice. Ann Rheum Dis 61:453–455
4. Brown TD, Johnston RC, Saltzman CL, Marsh JL, Buckwalter JA (2006) Posttraumatic osteoarthritis: a first estimate of incidence, prevalence, and burden of disease. J Orthop Trauma 20:739–744
5. Kalagate R, Rivera A, Pritchard CH, Brent LH (2013) Septic arthritis: changing trends in epidemiology over two decades. Ann Rheum Dis 71(280):2–280
6. Veronese N et al (2016) Pain Increases the Risk of Developing Frailty in Older Adults with Osteoarthritis. Pain Med 18:414–427. https://doi.org/10.1093/pm/pnw163
7. Gabriel SE, Michaud K (2009) Epidemiological studies in incidence, prevalence, mortality, and comorbidity of the rheumatic diseases. Arthritis Res Ther 11:229
8. Martínez-Ramos S, Rafael-Vidal C, Pego-Reigosa JM, García S (2022) Monocytes and macrophages in spondyloarthritis: functional roles and effects of current therapies. Cells 11:515
9. Kilkenny C, Browne WJ, Cuthill IC, Emerson M, Altman DG (2010) Improving bioscience research reporting: the ARRIVE guidelines for reporting animal research. PLOS Biol 8:e1000412

10. Vincent TL, Williams RO, Maciewicz R, Silman A, Garside P (2012) Mapping pathogenesis of arthritis through small animal models. Rheumatology 51:1931–1941
11. Chang Y et al (2021) Systematic analysis reveals that colony housing aligns gait profiles and strengthens link between histological and micro-CT bone markers in rat models of osteoarthritis. FASEB J 35:e21451.
12. Griffin TM et al (2010) Diet-induced obesity differentially regulates behavioral, biomechanical, and molecular risk factors for osteoarthritis in mice. Arthritis Res Ther 12:R130
13. Ma H-L et al (2007) Osteoarthritis severity is sex dependent in a surgical mouse model. Osteoarthritis Cartilage 15:695–700
14. Loeser RF et al (2012) Microarray analysis reveals age-related differences in gene expression during the development of osteoarthritis in mice. Arthritis Rheum 64:705–717
15. Pucha KA, McKinney JM, Fuller JM, Willett NJ (2020) Characterization of OA development between sexes in the rat medial meniscal transection model. Osteoarthr Cartil Open 2:100066
16. Ferrándiz ML et al (2014) Influence of age on osteoarthritis progression after anterior cruciate ligament transection in rats. Exp Gerontol 55:44–48
17. Hawkins P et al. (2015) Applying refinement to the use of mice and rats in rheumatoid arthritis research. Inflammopharmacology 23:131–150
18. Fischer BD, Adeyemo A, O'Leary ME, Bottaro A (2017) Animal models of rheumatoid pain: experimental systems and insights. Arthritis Res Ther 19:146
19. Sakkas LI, Platsoucas CD (2007) The role of T cells in the pathogenesis of osteoarthritis. Arthritis Rheum 56:409–424
20. Cui A et al (2020) Global, regional prevalence, incidence and risk factors of knee osteoarthritis in population-based studies. EClinicalMedicine 29–30:100587
21. Prieto-Alhambra D et al (2014) Incidence and risk factors for clinically diagnosed knee, hip and hand osteoarthritis: influences of age, gender and osteoarthritis affecting other joints. Ann Rheum Dis 73:1659–1664
22. Hunter DJ, Bierma-Zeinstra S (2019) Osteoarthritis. Lancet 393:1745–1759
23. Altman R et al (1986) Development of criteria for the classification and reporting of osteoarthritis: classification of osteoarthritis of the knee. Arthritis Rheum 29:1039–1049
24. Rim YA, Nam Y, Ju JH (2020) The role of chondrocyte hypertrophy and senescence in osteoarthritis initiation and progression. Int J Mol Sci 21:2358
25. Johnson VL, Hunter DJ (2014) The epidemiology of osteoarthritis. Best Pract Res Clin Rheumatol 28:5–15
26. Hawker GA (2019) Osteoarthritis is a serious disease. Clin Exp Rheumatol 120(37 Suppl): 3–6.
27. Lampropoulou-Adamidou K et al (2014) Useful animal models for the research of osteoarthritis. Eur J Orthop Surg Traumatol 24:263–271
28. Teeple E, Jay GD, Elsaid KA, Fleming BC (2013) Animal models of osteoarthritis: challenges of model selection and analysis. AAPS J 15:438–446
29. Little CB, Hunter DJ (2013) Post-traumatic osteoarthritis: from mouse models to clinical trials. Nat Rev Rheumatol 9:485–497
30. Poole R et al (2010) Recommendations for the use of preclinical models in the study and treatment of osteoarthritis. Osteoarthritis Cartilage 18:S10–S16
31. Lorenz J, Grässel S 2014. Experimental Osteoarthritis Models in Mice, In: Mouse Genetics. Methods in Molecular Biology. Humana Press, New York, pp. 401–419
32. Fang H, Beier F (2014) Mouse models of osteoarthritis: modelling risk factors and assessing outcomes. Nat Rev Rheumatol 10:413–421
33. McNulty MA et al (2012) Histopathology of naturally occurring and surgically induced osteoarthritis in mice. Osteoarthritis Cartilage 20:949–956
34. Little C, Smith M (2008) Animal models of osteoarthritis. Curr Rheumatol Rev 4:175–182
35. Bendele AM (2001) Animal models of osteoarthritis. J Musculoskelet Neuronal Interact 1:363–376
36. Gregory MH et al (2012) A review of translational animal models for knee osteoarthritis. Arthritis 2012:1–14
37. McCoy AM (2015) Animal models of osteoarthritis. Vet Pathol 52:803–818
38. Silberberg M, Silberberg R (1941) Age changes of bones and joints in various strains of mice. Am J Anat 68:69–95
39. Cook JL et al (2014) Animal models of cartilage repair. Bone Joint Res 3:89–94
40. Cope PJ, Ourradi K, Li Y, Sharif M (2019) Models of osteoarthritis: the good, the bad and the promising. Osteoarthritis Cartilage 27:230–239
41. Little CB, Zaki S (2012) What constitutes an "animal model of osteoarthritis"—the need for consensus?. Osteoarthritis Cartilage 20:261–267
42. Kuyinu EL, Narayanan G, Nair LS, Laurencin CT (2016) Animal models of osteoarthritis:

classification, update, and measurement of outcomes. J Orthop Surg Res 11:19
43. Kumagai K et al (2015) Spontaneously developed osteoarthritis in the temporomandibular joint in STR/ort mice. Biomed Rep 3:453–456
44. Schminke B et al (2014) A discoidin domain receptor 1 knock-out mouse as a novel model for osteoarthritis of the temporomandibular joint. Cell Mol Life Sci 71:1081–1096
45. Hall ME, Wang AS, Gold GE, Levenston ME (2022) Contrast solution properties and scan parameters influence the apparent diffusivity of computed tomography contrast agents in articular cartilage. J R Soc Interface 19: 20220403.
46. Goebel JC et al (2010) In vivo high-resolution MRI (7T) of femoro-tibial cartilage changes in the rat anterior cruciate ligament transection model of osteoarthritis: a cross-sectional study. Rheumatology 49:1654–1664
47. Thote T et al (2013) Localized 3D analysis of cartilage composition and morphology in small animal models of joint degeneration. Osteoarthritis Cartilage 21:1132–1141
48. Glasson SS, Chambers MG, van den Berg WB, Little CB (2010) The OARSI histopathology initiative—recommendations for histological assessments of osteoarthritis in the mouse. Osteoarthritis Cartilage 18:S17–S23
49. Aigner T et al (2010) Histopathology atlas of animal model systems—overview of guiding principles. Osteoarthritis Cartilage 18:S2–S6
50. Mankin HJ (1971) Biochemical and metabolic aspects of osteoarthritis. Orthop Clin North Am 2:19–31
51. van der Sluijs JA et al (1992) The reliability of the mankin score for osteoarthritis. J Orthop Res 10:58–61
52. Gerwin N, Bendele AM, Glasson S, Carlson CS (2010) The OARSI histopathology initiative—recommendations for histological assessments of osteoarthritis in the rat. Osteoarthritis Cartilage 18:S24–S34
53. Afara I, Prasadam I, Crawford R, Xiao Y, Oloyede A (2012) Non-destructive evaluation of articular cartilage defects using near-infrared (NIR) spectroscopy in osteoarthritic rat models and its direct relation to Mankin score. Osteoarthritis Cartilage 20:1367–1373
54. Gibson M et al (2014) Intra-articular delivery of glucosamine for treatment of experimental osteoarthritis created by a medial meniscectomy in a rat model. J Orthop Res 32:302–309
55. Nosrati Z, Bergamo M, Rodríguez-Rodríguez C, Saatchi K, Häfeli UO (2020) Refinement and validation of infrared thermal imaging (IRT): a non-invasive technique to measure disease activity in a mouse model of rheumatoid arthritis. Arthritis Res Ther 22:281
56. Oelschlegel AM, Goldschmidt J (2020) Functional Neuroimaging in Rodents Using Cerebral Blood Flow SPECT. Front Phys 8:152.
57. Tang GL, Kim KJ (2021) Laser Doppler Perfusion Imaging in the Mouse Hindlimb. J Vis Exp 170:e62012. https://doi.org/10.3791/62012
58. Jensen TS, Finnerup NB (2014) Allodynia and hyperalgesia in neuropathic pain: clinical manifestations and mechanisms. Lancet Neurol 13:924–935
59. Jacobs BY, Allen KD (2020) Factors affecting the reliability of behavioral assessments for rodent osteoarthritis models. Lab Anim 54:317–329
60. Krug HE, Dorman C, Funkenbusch S, Frizelle S, Mahowald M (2013) Measurement of pain in an acute murine arthritis pain model using a dynamic weight bearing device and evoked pain responses: effect of intra-articular capsaicin pretreatment. Osteoarthritis Cartilage 21:S269–S270
61. Piel MJ, Kroin JS, van Wijnen AJ, Kc R, Im H-J (2014) Pain assessment in animal models of osteoarthritis. Gene 537:184–188
62. Ruan MZC, Patel RM, Dawson BC, Jiang M-M, Lee BHL (2013) Pain, motor and gait assessment of murine osteoarthritis in a cruciate ligament transection model. Osteoarthritis Cartilage 21:1355–1364
63. Malfait AM, Little CB, McDougall JJ (2013) A commentary on modelling osteoarthritis pain in small animals. Osteoarthritis Cartilage 21:1316–1326
64. Poulet B, Westerhof TAT, Hamilton RW, Shefelbine SJ, Pitsillides AA (2013) Spontaneous osteoarthritis in Str/ort mice is unlikely due to greater vulnerability to mechanical trauma. Osteoarthritis Cartilage 21:756–763
65. Bapat S, Hubbard D, Munjal A, Hunter M, Fulzele S (2018) Pros and cons of mouse models for studying osteoarthritis. Clin Transl Med 7:36.
66. Poulet B (2016) Non-invasive loading model of murine osteoarthritis. Curr Rheumatol Rep 18:40
67. Staines KA, Poulet B, Wentworth DN, Pitsillides AA (2017) The STR/ort mouse model of spontaneous osteoarthritis—an update. Osteoarthritis Cartilage 25:802–808
68. Walton M (1977) Degenerative joint disease in the mouse knee; histological observations. J Pathol 123:109–122
69. Vo N et al (2013) An overview of underlying causes and animal models for the study of

age-related degenerative disorders of the spine and synovial joints. J Orthop Res 31:831–837
70. Helminen HJ (2002) Transgenic mouse models for studying the role of cartilage macromolecules in osteoarthritis. Rheumatology 41:848–856
71. Glasson SS (2007) In vivo osteoarthritis target validation utilizing genetically-modified mice. Curr Drug Targets 8:367–376
72. Chang JC et al (2018) SOST/sclerostin improves posttraumatic osteoarthritis and inhibits MMP2/3 expression after injury. J Bone Miner Res 33:1105–1113
73. Abaei M, et al (2016) Neural correlates of hyperalgesia in the monosodium iodoacetate model of osteoarthritis pain. Mol Pain 12: 1744806916642445.
74. Guzman RE, Evans MG, Bove S, Morenko B, Kilgore K (2003) Mono-iodoacetate-induced histologic changes in subchondral bone and articular cartilage of rat femorotibial joints: an animal model of osteoarthritis. Toxicol Pathol 31:619–624
75. Hong J-I, Park IY, Kim HA (2020) Understanding the molecular mechanisms underlying the pathogenesis of arthritis pain using animal models. Int J Mol Sci 21:533
76. Lockwood KA, Chu BT, Anderson MJ, Haudenschild DR, Christiansen BA (2014) Comparison of loading rate-dependent injury modes in a murine model of post-traumatic osteoarthritis. J Orthop Res 32:79–88
77. Khorasani MS et al (2015) Effect of alendronate on post-traumatic osteoarthritis induced by anterior cruciate ligament rupture in mice. Arthritis Res Ther 17:30
78. Heegde F et al (2019) Noninvasive mechanical joint loading as an alternative model for osteoarthritic pain. Arthritis Rheumatol 71:1078–1088
79. Alves CJ et al (2020) Nociceptive mechanisms driving pain in a post-traumatic osteoarthritis mouse model. Sci Rep 10:15271
80. Blagojevic M, Jinks C, Jeffery A, Jordan KP (2010) Risk factors for onset of osteoarthritis of the knee in older adults: a systematic review and meta-analysis. Osteoarthritis Cartilage 18:24–33
81. Griffin TM, Huebner JL, Kraus VB, Yan Z, Guilak F (2012) Induction of osteoarthritis and metabolic inflammation by a very high-fat diet in mice: effects of short-term exercise. Arthritis Rheum 64:443–453
82. Louer CR et al (2012) Diet-induced obesity significantly increases the severity of post-traumatic arthritis in mice. Arthritis Rheum 64:3220–3230
83. Kozijn AE et al (2019) Human C-reactive protein aggravates osteoarthritis development in mice on a high-fat diet. Osteoarthritis Cartilage 27:118–128
84. Høegh-Andersen P et al (2004) Ovariectomized rats as a model of postmenopausal osteoarthritis: validation and application. Arthritis Res Ther 6:R169
85. Almutairi KB, Nossent JC, Preen DB, Keen HI, Inderjeeth CA (2021) The prevalence of rheumatoid arthritis: a systematic review of population-based studies. J Rheumatol 48:669–676
86. Finckh A (2022) Global epidemiology of rheumatoid arthritis. Nat Rev Rheumatol 10:591–602.
87. Atzeni F, et al (2015) Pain in systemic inflammatory rheumatic diseases. Best Pract Res Clin Rheumatol 29:42–52.
88. Lin YJ, Anzaghe M, Schülke S (2020) Update on the Pathomechanism, Diagnosis, and Treatment Options for Rheumatoid Arthritis. Cells 9:880.
89. Park E, Griffin J, Bathon JM (2021) Myocardial Dysfunction and Heart Failure in Rheumatoid Arthritis. Arthritis & Rheumatology 74:184–199.
90. Anquetil F, Clavel C, Offer G, Serre G, Sebbag M (2015) IgM and IgA rheumatoid factors purified from rheumatoid arthritis sera boost the Fc receptor– and complement-dependent effector functions of the disease-specific anti–citrullinated protein autoantibodies. J Immunol 194:3664–3674
91. Spinelli FR, Colbert RA, Gadina M (2021) JAK1: Number one in the family; Number one in inflammation? Rheumatology (United Kingdom) 60:II3–II10.
92. Smolen JS, Aletaha D, McInnes IB (2016) Rheumatoid arthritis. Lancet 388:2023–2038
93. McInnes IB, Schett G (2007) Cytokines in the pathogenesis of rheumatoid arthritis. Nat Rev Immunol 7:429–442
94. van Raemdonck K et al (2021) Interleukin-34 reprograms glycolytic and osteoclastic rheumatoid arthritis macrophages via syndecan 1 and macrophage colony-stimulating factor receptor. Arthritis Rheumatol 73:2003–2014
95. Phillips KA et al (2014) Why primate models matter. Am J Primatol 76:801–827
96. Berg W (2009) Lessons from animal models of arthritis over the past decade. Arthritis Res Ther 11:250
97. Takao K, Miyakawa T (2015) Genomic responses in mouse models greatly mimic human inflammatory diseases. Proc Natl Acad Sci 112:1167–1172
98. Krock E, Jurczak A, Svensson CI (2018) Pain pathogenesis in rheumatoid arthritis—what have we learned from animal models?. Pain 159:S98–S109

99. Vincent TL, Williams RO, Maciewicz R, Silman A, Garside P (2012) Mapping pathogenesis of arthritis through small animal models. Rheumatology 51:1931–1941
100. Dimitrijević M et al (2018) Collagen-induced arthritis in dark agouti rats as a model for study of immunological sexual dimorphisms in the human disease. Exp Mol Pathol 105:10–22
101. Bo M et al (2020) Role of infections in the pathogenesis of rheumatoid arthritis: focus on mycobacteria. Microorganisms 8:1459
102. Arleevskaya MI, Kravtsova OA, Lemerle J, Renaudineau Y, Tsibulkin AP (2016) How Rheumatoid Arthritis Can Result from Provocation of the Immune System by Microorganisms and Viruses. Front Microbiol 7:1296.
103. Zhao T et al (2022) How to Model Rheumatoid Arthritis in Animals: From Rodents to Non-Human Primates. Front Immunol 13, 887460.
104. McGonigle P, Ruggeri B (2014) Animal models of human disease: challenges in enabling translation. Biochem Pharmacol 87:162–171
105. Igea A et al (2022) Central role of semaphorin 3B in a serum-induced arthritis model and reduced levels in patients with rheumatoid arthritis. Arthritis Rheumatol 74:972–983
106. Mossiat C et al (2015) Association between arthritis score at the onset of the disease and long-term locomotor outcome in adjuvant-induced arthritis in rats. Arthritis Res Ther 17:184
107. Lim MA et al (2017) Development of the Digital Arthritis Index, a Novel Metric to Measure Disease Parameters in a Rat Model of Rheumatoid Arthritis. Front Pharmacol 8:818.
108. Brand DD, Latham KA, Rosloniec EF (2007) Collagen-induced arthritis. Nat Protoc 2:1269–1275
109. Bolon B et al (2011) Rodent preclinical models for developing novel antiarthritic molecules: comparative biology and preferred methods for evaluating efficacy. J Biomed Biotechnol 2011:1–21
110. Barck KH et al (2004) Quantification of cortical bone loss and repair for therapeutic evaluation in collagen-induced arthritis, by micro-computed tomography and automated image analysis. Arthritis Rheum 50:3377–3386
111. Rose S, Waters EA, Haney CR, Meade CTJ, Perlman H (2013) High-resolution magnetic resonance imaging of ankle joints in murine arthritis discriminates inflammation and bone destruction in a quantifiable manner. Arthritis Rheum 65:2279–2289
112. Mausset-Bonnefont A-L et al (2019) Arthritis sensory and motor scale: predicting functional deficits from the clinical score in collagen-induced arthritis. Arthritis Res Ther 21:264
113. Ängeby Möller K et al (2018) Gait analysis and weight bearing in pre-clinical joint pain research. J Neurosci Methods 300:92–102
114. Cheah M, Fawcett J, Andrews M (2017) Assessment of Thermal Pain Sensation in Rats and Mice Using the Hargreaves Test. Bio Protoc 7:e2506.
115. Kouskoff V et al (1996) Organ-specific disease provoked by systemic autoimmunity. Cell 87:811–822
116. Ditzel HJ (2004) The K/BxN mouse: a model of human inflammatory arthritis. Trends Mol Med 10:40–45
117. Kamradt T, Schubert D (2005) The role and clinical implications of G6PI in experimental models of rheumatoid arthritis. Arthritis Res Ther 7:20
118. Gonçalves dos Santos G et al (2020) The neuropathic phenotype of the K/BxN transgenic mouse with spontaneous arthritis: pain, nerve sprouting and joint remodeling. Sci Rep 10:15596
119. Sakaguchi N et al (2003) Altered thymic T-cell selection due to a mutation of the ZAP-70 gene causes autoimmune arthritis in mice. Nature 426:454–460
120. Horai R et al. (1998) Production of mice deficient in genes for interleukin (IL)-1α, IL-1β, IL-1α/β, and IL-1 receptor antagonist shows that IL-1β is crucial in turpentine-induced fever development and glucocorticoid secretion. J Exp Med 187:1463–1475
121. Horai R et al (2000) Development of chronic inflammatory arthropathy resembling rheumatoid arthritis in interleukin 1 receptor antagonist–deficient mice. J Exp Med 191:313–320
122. Wakabayashi C et al (2015) IL-1 receptor-antagonist (IL-1Ra) knockout mice show anxiety-like behavior by aging. Neurosci Lett 599:20–25
123. Schinnerling K, Rosas C, Soto L, Thomas R, Aguillón JC (2019) Humanized Mouse Models of Rheumatoid Arthritis for Studies on Immunopathogenesis and Preclinical Testing of Cell-Based Therapies. Front Immunol 10:203.
124. Andersson EC et al (1998) Definition of MHC and T cell receptor contacts in the HLA-DR4restricted immunodominant epitope in type II collagen and characterization of collagen-induced arthritis in HLA-DR4 and human CD4 transgenic mice. Proc Natl Acad Sci 95:7574–7579

125. Hess A et al (2011) Blockade of TNF-α rapidly inhibits pain responses in the central nervous system. Proc Natl Acad Sci 108:3731–3736
126. Caplazi P et al. (2015) Mouse models of rheumatoid arthritis. Vet Pathol 52:819–826
127. Hu C et al (2022) Dihydroarteannuin Ameliorates Collagen-Induced Arthritis Via Inhibiting B Cell Activation by Activating the FcγRIIb/Lyn/SHP-1 Pathway. Front Pharmacol 13:883835.
128. Brand DD, Latham KA, Rosloniec EF (2007) Collagen-induced arthritis. Nat Protoc 2:1269–1275
129. Choudhary N, Bhatt LK, Prabhavalkar KS (2018) Experimental animal models for rheumatoid arthritis. Immunopharmacol Immunotoxicol 40:193–200
130. Inglis JJ et al (2007) Collagen-induced arthritis as a model of hyperalgesia: functional and cellular analysis of the analgesic actions of tumor necrosis factor blockade. Arthritis Rheum 56:4015–4023
131. Nieto FR et al (2016) Neuron-immune mechanisms contribute to pain in early stages of arthritis. J Neuroinflammation 13:96
132. Nandakumar, KS, Holmdahl, R, 2007. Collagen Antibody Induced Arthritis, In: Methods Mol Med. Humana Press, United States, pp. 215–223
133. Bas DB et al (2012) Collagen antibody-induced arthritis evokes persistent pain with spinal glial involvement and transient prostaglandin dependency. Arthritis Rheum 64:3886–3896
134. Fernandes ES et al (2011) A distinct role for transient receptor potential ankyrin 1, in addition to transient receptor potential vanilloid 1, in tumor necrosis factor α-induced inflammatory hyperalgesia and Freund's complete adjuvant-induced monarthritis. Arthritis Rheum 63:819–829
135. Silva Rodrigues J et al (2018) Sulforaphane modulates joint inflammation in a murine model of complete freund's adjuvant-induced mono-arthritis. Molecules 23:988
136. Muley MM, Krustev E, McDougall JJ (2016) Preclinical assessment of inflammatory pain. CNS Neurosci Ther 22:88–101
137. Noh ASM et al. (2021) Effects of different doses of complete Freund's adjuvant on nociceptive behaviour and inflammatory parameters in polyarthritic rat model mimicking rheumatoid arthritis. PLoS One 16:e0260423
138. Glant TT et al (2011) Proteoglycan-induced arthritis and recombinant human proteoglycan aggrecan G1 domain-induced arthritis in BALB/c mice resembling two subtypes of rheumatoid arthritis. Arthritis Rheum 63:1312–1321
139. Horváth Á et al (2018) Regulatory role of capsaicin-sensitive peptidergic sensory nerves in the proteoglycan-induced autoimmune arthritis model of the mouse. J Neuroinflammation 15:335.
140. Ebbinghaus M et al (2019) Contribution of inflammation and bone destruction to pain in arthritis: a study in murine glucose-6-phosphate isomerase–induced arthritis. Arthritis Rheumatol 71:2016–2026
141. Schubert D, Maier B, Morawietz L, Krenn V, Kamradt T (2004) Immunization with glucose-6-phosphate isomerase induces T cell-dependent peripheral polyarthritis in genetically unaltered mice. J Immunol 172:4503–4509
142. Christianson CA et al (2010) Characterization of the acute and persistent pain state present in K/BxN serum transfer arthritis. Pain 151:394–403
143. Christensen AD, Haase C, Cook AD, Hamilton JA (2016) K/BxN Serum-Transfer Arthritis as a Model for Human Inflammatory Arthritis. Front Immunol 7:213.
144. Solomon S et al (2002) Transmission of antibody-induced arthritis is independent of complement component 4 (C4) and the complement receptors 1 and 2 (CD21/35). Eur J Immunol 32:644
145. Archer AM et al (2016) ApoE deficiency exacerbates the development and sustainment of a semi-chronic K/BxN serum transfer-induced arthritis model. J Transl Med 14: 170
146. Rose S et al (2013) A novel mouse model that develops spontaneous arthritis and is predisposed towards atherosclerosis. Ann Rheum Dis 72:89–95
147. Vanhoof J, Declerck K, Geusens P (2002) Prevalence of rheumatic diseases in a rheumatological outpatient practice. Ann Rheum Dis 61:453–455
148. Martínez-Ramos S, Rafael-Vidal C, Pego-Reigosa JM, García S (2022) Monocytes and macrophages in spondyloarthritis: functional roles and effects of current therapies. Cells 11:515
149. Sieper J, Poddubnyy D (2016) New evidence on the management of spondyloarthritis. Nat Rev Rheumatol 12:282–295
150. Mathieu S, Motreff P, Soubrier M (2010) Spondyloarthropathies: an independent cardiovascular risk factor?. Joint Bone Spine 77:542–545
151. Maguire S, Gallagher P, O'Shea F (2022) The negative impact of depression in women with axial spondyloarthropathy. Joint Bone Spine 89:105261
152. Curtis JR et al (2021) The annual diagnostic prevalence of ankylosing spondylitis and axial spondyloarthritis in the United States using

medicare and marketscan databases. ACR Open Rheumatol 3:743–752
153. Colaco K et al (2021) Trends in mortality and cause-specific mortality among patients with psoriasis and psoriatic arthritis in Ontario, Canada. J Am Acad Dermatol 84:1302–1309
154. Crotti TN et al (2002) Receptor activator NF-kappaB ligand (RANKL) expression in synovial tissue from patients with rheumatoid arthritis, spondyloarthropathy, osteoarthritis, and from normal patients: semiquantitative and quantitative analysis. Ann Rheum Dis 61:1047–1054
155. Lories RJU, Matthys P, de Vlam K, Derese I, Luyten FP (2004) Ankylosing enthesitis, dactylitis, and onychoperiostitis in male DBA/1 mice: a model of psoriatic arthritis. Ann Rheum Dis 63:595–598
156. Hammer RE, Maika SD, Richardson JA, Tang J-P, Taurog JD (1990) Spontaneous inflammatory disease in transgenic rats expressing HLA-B27 and human β2m: An animal model of HLA-B27-associated human disorders. Cell 63:1099–1112
157. Khare SD, Hansen J, Luthra HS, David CS (1996) HLA-B27 heavy chains contribute to spontaneous inflammatory disease in B27/human beta2-microglobulin (beta2m) double transgenic mice with disrupted mouse beta2m. J Clin Invest 98:2746–2755
158. Breban M et al (1996) T cells, but not thymic exposure to HLA-B27, are required for the inflammatory disease of HLA-B27 transgenic rats. J Immunol 156:794–803
159. Taurog JD, Maika SD, Simmons WA, Breban M, Hammer RE (1993) Susceptibility to inflammatory disease in HLA-B27 transgenic rat lines correlates with the level of B27 expression. J Immunol 150:4168–4178
160. Tran TM et al (2006) Additional human β 2-microglobulin curbs HLA–B27 misfolding and promotes arthritis and spondylitis without colitis in male HLA–B27–transgenic rats. Arthritis Rheum 54:1317–1327
161. Adamopoulos IE et al (2011) IL-23 Is critical for induction of arthritis, osteoclast formation, and maintenance of bone mass. J Immunol 187:951–959
162. Sherlock JP et al (2012) IL-23 induces spondyloarthropathy by acting on ROR-γt+ CD3+CD4−CD8− entheseal resident T cells. Nat Med 18:1069–1076
163. Jeong H et al (2018) Spondyloarthritis features in zymosan-induced SKG mice. Joint Bone Spine 85:583–591
164. Ruutu M et al (2012) β-glucan triggers spondylarthritis and Crohn's disease-like ileitis in SKG mice. Arthritis Rheum 64:2211–2222
165. Kontoyiannis D, Pasparakis M, Pizarro TT, Cominelli F, Kollias G (1999) Impaired on/off regulation of tnf biosynthesis in mice lacking TNF AU-rich elements. Immunity 10:387–398
166. Christodoulou-Vafeiadou E et al (2020) Ectopic bone formation and systemic bone loss in a transmembrane TNF-driven model of human spondyloarthritis. Arthritis Res Ther 22:232
167. Kaaij MH et al (2020) Transmembrane TNF drives osteoproliferative joint inflammation reminiscent of human spondyloarthritis. J Exp Med 217:e20200288.
168. Holers VM, la Rosa FG, Banda NK (2021) A Potential New Mouse Model of Axial Spondyloarthritis Involving the Complement System. Immune Netw 21, e45.
169. Bardos T et al (2005) A longitudinal study on an autoimmune murine model of ankylosing spondylitis. Ann Rheum Dis 64:981–987
170. Yong Z, Hill JL, Hirofuji T, Mander M, Yu DTY (1988) An experimental mouse model of Yersinia-induced reactive arthritis. Microb Pathog 4:305–310
171. Thierry S, Fautrel B, Lemelle I, Guillemin F (2014) Prevalence and incidence of juvenile idiopathic arthritis: a systematic review. Joint Bone Spine 81:112–117
172. Avau A et al (2014) Systemic juvenile idiopathic arthritis-like syndrome in mice following stimulation of the immune system with freund's complete adjuvant: regulation by interferon-γ. Arthritis Rheumatol 66:1340–1351
173. Inoue N, Schulert GS (2023) Mouse models of systemic juvenile idiopathic arthritis and macrophage activation syndrome. Arthritis Res Ther 25:48
174. Dehlin M, Jacobsson L, Roddy E (2020) Global epidemiology of gout: prevalence, incidence, treatment patterns and risk factors. Nat Rev Rheumatol 16:380–390
175. Patil T, Soni A, Acharya S (2021) A brief review on in vivo models for Gouty arthritis. Metab Open 11:100100
176. Bremell T, Abdelnour A, Tarkowski A (1992) Histopathological and serological progression of experimental Staphylococcus aureus arthritis. Infect Immun 60:2976–2985
177. Colavite P, Sartori A (2014) Septic arthritis: immunopathogenesis, experimental models and therapy. J Venom Anim Toxins Incl Trop Dis 20:19
178. Jin T et al (2019) A novel mouse model for septic arthritis induced by *Pseudomonas aeruginosa*. Sci Rep 9:16868

Check for updates

Chapter 9

Investigating Sex-Specific Manifestations in Models of Pain

Loren J. Martin

Abstract

Sex differences in pain are of great interest to the public and of increasing clinical and translational significance among researchers. However, the number of possible experimental designs and interpretations in animal pain research is large, and design choices can strongly influence conclusions. This chapter outlines practical strategies for identifying and investigating sex differences in preclinical pain research. I first discuss the rationale for including both sexes in pain studies, including ethical, operational, and translational considerations, as well as common concerns about sample size and estrous cycle variability. I then describe an experimental sequence for determining whether a sex difference exists in a pain phenotype, beginning with comparisons between gonadally intact adult females and males. When a sex difference is observed, subsequent experiments can test whether it depends on gonadal hormones at the time of testing, hormone actions during development or puberty, sex chromosome complement, local hormone synthesis, immune mechanisms, stress, conditioning, pain history, or sexually dimorphic afferent and analgesic pathways. Overall, this chapter emphasizes that sex differences in pain should not be treated as simple male-female comparisons, but as experimentally tractable biological phenomena whose mechanisms require careful design, appropriate controls, and cautious interpretation. A more rigorous approach to sex as a biological variable in preclinical pain research may improve translational relevance and support the development of more effective therapeutics.

Keywords Sex differences, Pain, Nociception, Analgesia, Preclinical models, Rodents, Gonadal hormones, Estrous cycle, Immune mechanisms, Stress and conditioning, Experimental design

1 Introduction

Chronic pain is one of the most prevalent and debilitating conditions affecting as many as 20% of the population worldwide and is the foremost cause of long-term sick leave and disability [1–5]. Chronic pain imposes a profound health care burden that not only affects the individual but also permeates throughout all avenues of social, work, and family life. After age, female sex is by far the strongest risk factor for chronic pain. Women report having chronic pain one and a half to two times as often as men, and two out of every three persons receiving disability pensions due to chronic pain conditions are women [6–8]. In addition to increased prevalence, women report having greater pain than men suffering from the same condition [8].

Isaura Tavares and Hugo Leite-Almeida (eds.), *Translational Research Methods for Neuropathic Pain*, Neuromethods, vol. 235, https://doi.org/10.1007/978-1-0716-5312-8_9,

Women are also more sensitive to experimental pain stimuli in the laboratory, indicating that sex differences in pain are not merely sex differences in tissue pathology but rather differences in pain transduction, modulation, and perception [9].

Among the most challenging aspects of treating a person in pain is identifying the type(s) and mechanism(s) of pain. It is becoming more evident that aside from differences in pain sensitivity, males and females possess fundamentally different biological pathways that underlie the development and maintenance of chronic pain conditions [9–12]. The lack of therapeutic options for debilitating pain conditions highlights the need for better treatments. However, the development of new analgesics has yielded many failures—mainly due to lack of efficacy in phase II clinical trials—despite robust preclinical efficacy and a detailed understanding of pain pathways, which are fundamentally different between the sexes [12]. There is also considerable evidence from animal studies that sex hormones, principally estrogens and testosterone, play a crucial role. Several studies have shown that sex differences in pain can be abolished in laboratory animals by gonadectomy and *reversed* by cross-sex administration of hormones to gonadectomized animals [13–15].

In preclinical studies, there is an inherent bias toward the use of predominantly male mice [16–18], which derives from the fear and misconception that female behavior and physiology are more variable due to estrous cycle fluctuations. However, this has proven not to be the case; female rodents are no more variable than their male counterparts, even when not accounting for the estrous cycle [19]. A male-specific bias is not limited to pain research, as a preference for male rodents is especially evident in the fields of neuroscience (5.5:1), pharmacology (5:1), and physiology (3.7:1) [20]. Directly comparing several behavioral traits of male mice with those of females at random stages of the estrous cycle revealed that for most traits, the variability within each sex was equivalent regardless of the stage of the estrous cycle in females [18, 21]. Female mice from several strains are also no more variable than males in how they displayed pain [19]. Thus, the estrous cycle itself may not contribute to significant data variability, or male mice may present their own sex-specific variability (i.e., fighting, dominance) [19]. Therefore, there needs to be a global reassessment of the widespread notion among researchers that hormonal fluctuations during the estrous cycle of female mice render their data more variable than data obtained from male mice [19].

Given the substantially greater prevalence of many clinical pain conditions in women versus men [21] and growing evidence for sex differences in sensitivity to experimental pain and analgesics [22], it is recommended that all pain researchers test their hypotheses in both sexes. It is incorrect to assume that data obtained in male animals will generalize to females. Therefore, if only males are used in experiments, investigators must provide a rationale for excluding females. The potential limitation in the generalizability of the findings must be addressed, particularly when examining a phenomenon with greater prevalence or severity in females.

1.1 Importance of Identifying Sex-Specific Manifestations in Pain

The rapidly maturing pain field is moving from discussing whether sex differences in pain exist to recognizing the importance of these differences. Focus is now heavily directed toward understanding (1) the factors that lead to the expression of sex and gender differences in the pain experience, (2) the mechanisms that underlie these differences, and (3) how these differences can inform clinical trials and management of pain [23].

When designing and interpreting experiments on sex differences in pain, it is essential to acknowledge that sex differences in several biological systems may directly or indirectly affect pain phenotypes. For example, male rodents have a higher body fat percentage than females [24], while humans have the opposite [25]. A greater body fat percentage can affect the distribution of lipophilic drugs and therefore impact analgesic drug potency, efficacy, and duration of action. Immune responses also differ between the sexes [22], leading to sex differences in response to chronic inflammatory and neuropathic pain. The activity level of female rodents varies significantly across the estrous cycle [26], which may affect pain tests that require subjects to move freely, such as the hot plate test. There are also sex differences in how rodents respond to noxious stimuli, resulting in differential outcomes for both pain threshold and tolerance tests [27]. Finally, unpleasant effects of analgesics such as respiratory depression and nausea may differ in females versus males [28–30], which is vital when comparing analgesic efficacy.

For pain research, the use of primarily male subjects [16] to model a predominantly female disorder represents an obvious mismatch between basic science and clinical epidemiology. The implications of this mismatch depend on how similar the physiology of pain in males is to that of females. Furthermore, the exclusion of female rodents in experimental testing leads to an inadequate reflection of the clinical population. It risks the translational applicability of preclinical research, which may significantly contribute to the inadequacy of current pharmacotherapies. Epidemiological considerations aside, there are many advantages to including female rodents in preclinical research. This section highlights some of these advantages from an ethical and operational standpoint.

1.2 Practical and Ethical Considerations

The first point that I would make in terms of practical considerations is one of enhanced scientific merit. The inclusion of male and female rodents in the initial design and conception of an experiment increases the likelihood that investigators will test hypotheses in an unbiased fashion. This is important for discovering natural and fundamental differences in pain neurobiology but may lead to unexpected findings. For instance, we have shown that compared to females, male mice and men exhibit more pain when returned to a location previously associated with pain [14]. We serendipitously uncovered these responses because we routinely include male and female mice in our experiments and stratify our statistical analyses by sex. However, it is quite puzzling that males and not females developed this hypersensitivity given that up to 70% of chronic pain patients are women. Whether these differences derive from evolutionary roles is one possibility that we have considered. For example, it may be more beneficial for females not to remember discrete pain episodes, as this would be disadvantageous during childbirth (and the birth of subsequent offspring). Conversely, males have historically assumed a hunter-protector role, and remembering where they encountered physical pain (i.e., from a predator attack or violent adversary) could be adaptive if their survival depends on it.

A second argument is that identifying sex differences during the early stages of preclinical research has greater societal impact. The inclusion of male and female subjects will ultimately save time, money, and resources when developing new analgesics because potential sex differences will have been identified earlier and hypotheses regarding sex-specific efficacy may be developed before clinical trials begin [31]. Clinical trials are costly and take a long time to conduct; however, there is a plethora of data showing that different medications work differently in males and females [28, 30, 32]. By not identifying potential sex differences early during the preclinical stages of research, investigators are shifting this burden to clinical trialists who may not be equipped to find sex differences in their primary outcome measure and could consider a drug a "failure" due to inadequate overall efficacy. Identifying potential sex differences early on during the research process could also prevent an even more costly and dangerous situation where approved drugs must be withdrawn from the market due to unexpected side effects in a sex not initially included in the study. Having sex as a variable in preclinical pain research is likely to save money in the long run by increasing reproducibility and minimizing the failure of clinical trials.

The use of both sexes also has practical cost-saving advantages [19]. If a lab breeds experimental animals in-house, there is no need for the wasteful and unethical practice of euthanizing half of the offspring. If a lab purchases animals from a commercial supplier, females are sold at a cheaper rate than males due to the lower demand. Females' smaller size and body weight can also be advantageous for pharmacological studies, where investigators can reduce the amount of experimental drugs administered on a mg/kg basis. The smaller female size is significant when a drug resource is scarce or expensive.

The next section of this chapter will discuss methods and strategies for identifying and understanding sex differences in preclinical pain research with a focus on rodent models. Should a researcher choose to investigate sex differences in the trait of interest, they can explore some or all the possibilities mentioned below until a satisfactory conclusion is reached. Several references have been provided for each option below, including detailed guidelines to help design and implement the experiments.

2 Methods and Strategies for Identifying Sex Differences in Pain and Their Underlying Mechanisms

2.1 The Issue of Sample Size

A common argument against including both sexes in designing new experiments is that the inclusion of males and females will necessitate using double the animals, which will ultimately cost more money. However, I argue that standard sample sizes should be employed when initially conducting exploratory studies incorporating both sexes. For instance, if your typical experiments have sample sizes ranging between 8 and 10 per group, then a simple way to include both sexes is to split this sample evenly between males and females (i.e., 5 and 5). In our experience, this approach is sufficient to infer whether your data are trending toward a potential sex difference. If data are trending toward a sex difference, then you could choose whether to follow up on this sex difference or merely report the trending effect in a publication. However, this is an essential point as many studies include both males and females but do not stratify their analysis by sex. Including a sub-analysis where sex is considered is helpful and informative—even if this information is only included in supplemental material—because it allows other researchers to decide whether potential sex differences may exist and whether they will follow up on your initial findings in their laboratories. Trending sex differences or variable effects may require a more nuanced approach such as comparing male mice with two or more groups of females where the estrous cycle is tracked. For example, a five-group design comparing males with females on each of the 4 days of the rodent estrous cycle would inform us if there was a sex difference and, if so, on which days of the estrous cycle [33]. This type of approach would inform sex hormone effects in female mice. Still, we urge caution in adopting this approach, especially for labs new to sex difference research or when investigating a novel trait. In addition, the estrous cycle is tracked in female mice using vaginal swabs, which adds another element of variability and stress to female mice. In our experience, this could be sufficient to "mask" some pain phenotypes (Fig. 1).

2.2 Is There a Sex Difference in the Trait of Interest?

It has generally been established that the first step of studying sex differences in any pain phenomenon should be a comparison of gonadally intact adult females and males. This is important because it informs us about the trait under natural conditions, with intact and

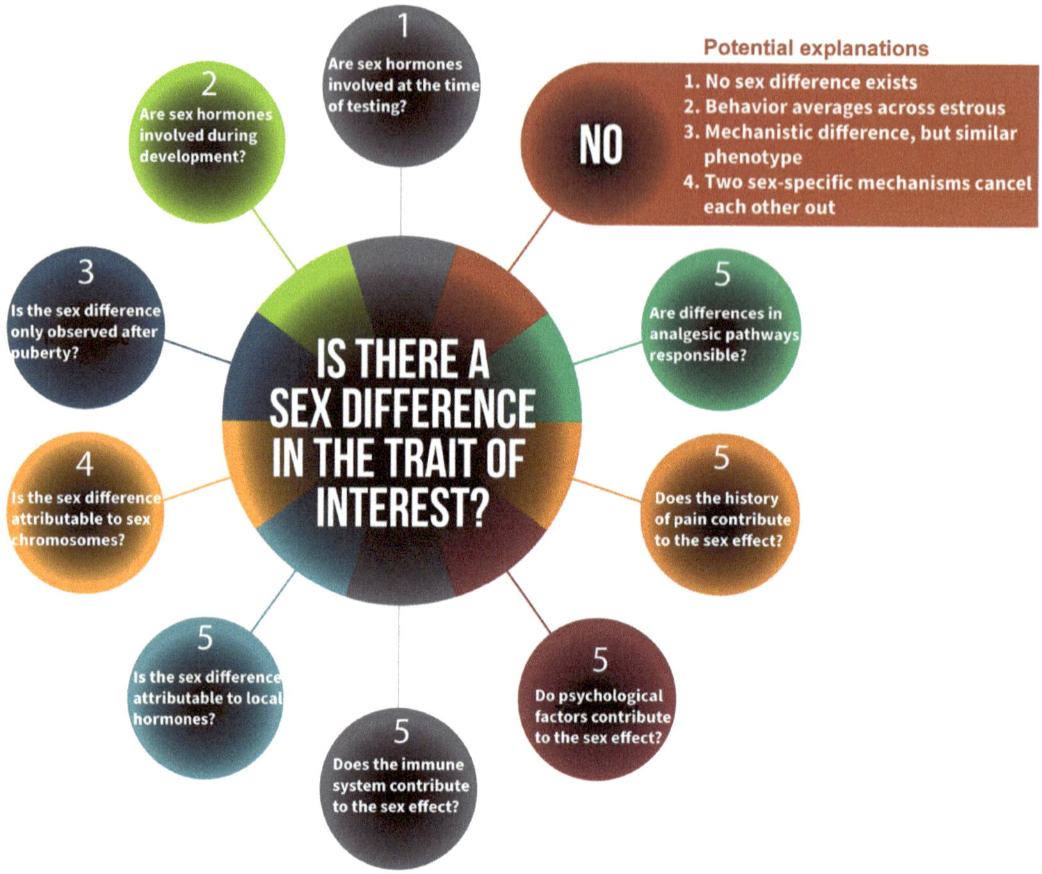

Fig. 1 Questions to help direct researchers in identifying sex differences in pain and their underlying mechanisms. This schematic depicts suggested questions that a researcher can follow to investigate potential sex differences in pain phenotypes. The investigator can start with a comparison of gonadally intact adult females and males to determine if a sex difference exists in the trait of interest. If no sex difference is observed, conclusion must be drawn with caution as there can be four different explanations. If a sex difference is observed, the next step is to investigate whether the differences are caused by gonadal hormones at the time of testing (1), during development (2), or postpuberty (3). If evidence is found at any of these steps, further experiments can be conducted to determine the role of specific hormones and their mechanisms of action. If manipulations of gonadal hormones do not completely explain the sex difference, then the investigator may test whether sex chromosomes (4) or other physiological systems are responsible (5). When considering other physiological systems, these questions may be asked in any sequential order, but should be considered within the greater context of the research questions. The numbering (1–5) represents a sequential order that an investigator can follow when investigating sex differences

fully operational physiological systems. If it has not been established that a sex difference exists, it is not necessary to track different stages of the females' estrous cycle during initial investigations. However, the absence of a sex difference should be interpreted with caution as there could be at least four possible explanations: (1) no sex difference exists in the trait of interest; (2) there is a sex difference only at a particular stage of the female estrous cycle, and behavioral responses average across the estrous cycle and therefore cannot be observed [34]; (3) there is a mechanistic sex difference that ultimately results in an equivalent magnitude of the trait in both sexes [35]; and (4) two sex-specific mechanisms, such as hormonal and genetic factors, canceled each other out resulting in no observable sex difference [36].

Moreover, it is vital to note all the biological and environmental factors around the experimental subjects. For example, researchers should always record the age of the subjects at the time of testing, the time of day of the testing, housing conditions, and testing methods. Some sex differences occur only in certain environments and at specific ages (see Table 1). Logically, it is often of interest to determine whether there is a sex difference in adults of reproductive age. If a sex difference is detected, researchers must include the conditions under which testing occurred. It would be an error to generalize the observed sex difference to other ages or conditions. Investigators must also choose the type of pain model/test carefully. For instance, the level of noxious stimuli administered is of relevance as too high of an intensity can result in a ceiling effect where all subjects or cells respond at the limit of their capacity [37], thereby potentially concealing any individual/sex differences. It is also essential to choose an appropriate rodent strain for the pain model of interest as the response to pain, and sex differences are often species- and strain-dependent [38]. Caution must also be exercised when testing male and female rodents in a laboratory setting as several aspects of the experimental design and environmental factors can affect the outcome of pain tests. See Table 2 for factors that may increase and/or decrease pain and/or analgesic response. Suppose a sex difference is observed in gonadally intact animals. In that case, the next logical step is to investigate whether the differences are caused by the activational (i.e., time of testing), organizational (i.e., during development), or another phase (i.e., postpuberty) [39] of gonadal hormones. For a more comprehensive review, refer to "Strategies and Methods for Research on Sex Differences in Brain and Behavior" by Becker et al. [39].

2.3 Do the Actions of Gonadal Hormones Cause the Observed Sex Difference at the Time of Testing?

There are two steps to answering this question. First, gonadectomy or chemical castration (via steroid receptor blockers) is performed. Gonadectomy—the removal of the testes or ovaries—is usually the first choice as this has been widely studied and performed in rodents. Following gonadectomy, it may take a few days to weeks for circulating gonadal hormones to dissipate; therefore, how long after gonadectomy testing is performed is an important factor to consider [84, 85]. If gonadectomy alters the trait in one sex, we can conclude that the gonads affect the trait in that sex [86, 87]. However, it does not implicate the effect of a specific gonadal hormone, as gonadectomy itself can disrupt other physiological systems, such as the

Table 1
Factors affecting pain sensitivity and analgesia with noted sex-dependent effects

Factor	Sex difference finding	Effect on pain/ analgesia	Reference
Age-related			
Neonatal (P0–P7)	Nociception	↑ in XX mice irrespective of gonadal status	[40]
	Morphine analgesia	↑ in male mice	[40, 41]
Young (6–8 weeks)	κ-opioid analgesia	↓ in female than male mice	[42]
	Chronic pain development	↑ in female than male rats	[43]
Adult (>2 months)	Morphine analgesia	↓ in female rats, ↑ in male rats	[44–46]
Stress-related			
Breeding season	Stress-induced analgesia	↑ in male, ↓ in female deer mice	[47]
Early-life stress	Adult pain perception	↑ in male rats	[48]
Sex	Neurochemical mechanism of stress-induced analgesia	= in males and females	[49, 50]
Social factors			
Mild social threat	Pain sensitivity	↑ in male dyads	[51]
Severe social threat	Analgesia	↑ in male dyads	[51]
Cycles			
Dark period of circadian cycle	Opioid sensitivity	↑ in male than female mice	[52]
Proestrus stage of menstrual cycle	Pain sensitivity	↑ in female than male rats	[53–55]
Other			
Gonadal hormones	Pain sensitivity	↑ in females than males	[56–60]
Sex	Neurochemical mechanism of pain hypersensitivity	= in males and females	[10, 52, 61, 62]
Strain	Pain sensitivity	↑ in females than males	[58, 59, 63–65]
Strain	Morphine analgesia	↑ in males than females	[49, 64, 66–68]

↑ indicates increased sensitivity to pain/analgesic; ↓ indicates decreased sensitivity to pain/analgesic; = indicates similar magnitude of pain/analgesia

Table 2
Experimental and environmental factors that may affect pain sensitivity and analgesia in rodents

Experimental and environmental factors	Effect on pain and/or analgesia
Stress (from various factors, see below)	Stress can "mask" the actual pain phenotype by causing stress-induced analgesia or hyperalgesia [69].
Housing or testing experimental and control groups together	Olfactory cues have been shown to mediate the transfer of hyperalgesia to bystander mice [70]. This implies that it may not be optimal to test or house experimental and control groups together in the same room/apparatus or cage [71], respectively.
Presence of unfamiliar male	Olfactory cues from unfamiliar males, rodent or human, cause stress-induced analgesia in both male and female rodents [72].
	This implies that it may not be optimal to test male or female rodents in the presence of other unfamiliar male rodents in the same room. Laboratories investigating phenomena that may be affected by stress, such as pain and analgesia, should also account for, or at least report, experimenter sex in their protocols.
Vaginal lavage for the staging of estrous cycle in female rodents	Obtaining repeated vaginal samples in female rodents can induce stress [73], and it is not clear whether daily handling of males is enough to control for this potential confound. Alternative methods such as monitoring motor activity in the home cage (which peaks during proestrus) [74] or visual assessment of the vulva [75] can be used in pain assays where stress may be a confound.
Handling	Picking up mice by the tail has been shown to induce aversion and high anxiety, whereas use of tunnels or open hand led to voluntary approach, lower anxiety, and acceptance of physical restraint [76].
	In addition, acute handling leads to a greater corticosterone response (stress) in female rats compared to males [77].
Time of day	Pain sensitivity shows seasonal and diurnal variations in both sexes in mice, with sensitivity being greater in the dark phase compared to the light phase [78].
Novel environment	Being placed in a new testing room under bright lights can produce stress-induced analgesia in mice [51] which can be ameliorated by habituating them to the test room [79].
Spatial crowding in rat cages	Crowding induces stress in male rats (likely due to inter-male aggression and the maintenance of dominance hierarchies) but calms females [80]. Female rats, in contrast, were stressed when individually housed.
Order of testing within a cage	The efficacy of opioid and non-opioid analgesic drugs is affected by order of testing within a cage, with the first mouse tested from a cage 50% more sensitive to the drug than the last mouse [81].

(continued)

Table 2
continued

Experimental and environmental factors	Effect on pain and/or analgesia
Cage/housing enrichment	Using physical objects as housing enrichment has been reported to ameliorate pain in both mice and rats [82, 83].
Humidity in test room	Sensitivity to noxious thermal stimuli is higher in conditions of high humidity [81].

neural afferents from gonads and the hypothalamic–pituitary–adrenal feedback loop [88].

Suppose a change is observed in gonadectomized subjects. In that case, the next step is to administer gonadal hormones to determine if hormone replacement reinstates the trait to its natural state, i.e., the state typical of gonadally intact animals. A common approach is a three-group design—some gonadectomized animals receive a treatment to replace the missing gonadal hormone, some gonadectomized animals receive a control treatment (i.e., oil or saline), and a gonadally intact comparison group. If estradiol treatment in ovariectomized female animals returns their phenotype to that of an intact female, this is strong evidence that estradiol from the ovaries influences the trait, directly or indirectly [23, 32]. Note that the hormone concentration to be administered must be chosen to fall within the physiological range of the subject, which avoids pharmacological side effects from the hormone itself. The interval between hormone replacement therapy and testing must also be considered as gonadal hormones can have both immediate and delayed actions [85].

Chemical castration is a common alternative to gonadectomy. This is done via administering drugs that block specific gonadal hormone production or receptors. Examples are flutamide, an androgen receptor blocker [89]; tamoxifen, an estrogen receptor antagonist [90]; RU486, an anti-progestin [91]; and fadrozole, an inhibitor of the enzyme aromatase that catalyzes 17β-estradiol synthesis from testosterone [92]. Researchers intending to use these drugs must make informed decisions based on the dual-antagonistic effect some of these may have. For example, tamoxifen, an estrogen receptor antagonist, can also act as a partial estrogen agonist in some peripheral tissues [93]; and flutamide, the androgen receptor antagonist, can affect aromatase activity that is induced by androgens [94] and act as an androgen agonist in hippocampal tissues [95].

Knockout mice that allow precise control of receptor activation/deactivation are also available. Examples are the testicular feminized (Tfm) mice that lack androgen responsiveness [96] and the estrogen receptor knockout (ERKO) mice that lack estrogen receptiveness [97, 98]. However, they do not represent animals in which all effects of a specific hormone have been eliminated. Other hormone receptors are functional in these animals, and nearly all gonadal hormones act via mechanisms other than those mediated by their classic receptors. Also, these universal knockout mice do not provide any information on the specific tissue or life stage at which the sexually dimorphic trait is observed. To regulate specific steroid receptor expression spatially and temporally, one would need a conditional knockout mouse line [99]. This will allow for determining the localized action of the hormone and identification of the life stage at which the trait is influenced by the hormone acting through specific receptors.

2.4 Is the Trait Sexually Dimorphic by the Action of Gonadal Steroids During Development?

If gonadectomy around the time of testing in adulthood does not eliminate the sex difference or if there is a sex difference when both gonadectomized males and females receive the same hormone treatment, a reasonable hypothesis is that testosterone (or its metabolites) may act during the early stages of development to produce permanent sex differences in the tissues mediating the trait. The same experimental approaches described for testosterone below are applicable for manipulating ovarian hormones during various periods of development to determine their importance in causing sex differences.

The investigator may choose females, males, or both sexes as subjects to test this hypothesis. They can increase testosterone levels in females to induce masculinization or reduce testosterone in males to prevent masculinization. In females, physiological doses of testosterone are administered during fetal or early postnatal life. The trait is measured in adulthood to see if it has changed to the masculine state. It is important to have a three-group design approach—one group of females receiving testosterone, another group of females receiving a placebo, and a control group of males receiving a placebo. Suppose testosterone treatment during female development changes the trait to a similar phenotype observed in males. In that case, we may conclude that the actions of testosterone sexually differentiate the trait during development [100, 101].

In males, testosterone exposure can be reduced before birth by administering androgen receptor blockers to pregnant mothers, which will pass through the placenta to the fetus [89]. Just after birth, masculinization can be further diminished by gonadectomy and administering estrogen antagonists or aromatase inhibitors to prevent the conversion of testosterone to 17β-estradiol (required for the masculinization of the brain). A four-group design is applicable here: one group of gonadectomized males receiving anti-testosterone treatments throughout early development, another group of gonadectomized males receiving testosterone, a group of sham gonadectomized males receiving placebo treatment, and a group of females receiving placebo as well. Suppose blocking testosterone during fetal and postnatal periods prevents the development of a masculine trait in males, and testosterone treatment in gonadectomized males produces the masculine trait. In that case, researchers can conclude that the trait is sexually dimorphic due to the actions of testosterone during the developmental period tested [101].

2.5 Is the Sex Difference Age-Dependent, i.e., Observed Only After Puberty?

If manipulating gonadal hormones during testing or early development fails to abolish the sex difference in the trait of interest, researchers may consider other critical life phases where gonadal hormones are known to cause permanent changes in adulthood. One such developmental period is puberty, during which gonadal steroids can produce persistent effects on traits measured later in life [102]. In rodents, puberty generally occurs during adolescence, beginning around postnatal day 28–42 in mice and slightly later in rats, although timing varies across sex, strain, and housing conditions. Therefore, manipulating hormones during this developmental window may abolish sex differences in the initially measured trait when assessed later in adulthood. However, if this fails to eliminate the sex difference, the investigator can begin exploring other physiological systems that are directly, indirectly, or not connected to gonadal hormones. Some of these options are discussed below.

2.6 Is the Observed Sex Difference Attributable to Genetic (Sex Chromosome) Contributions?

Some sex differences are mediated via sex chromosomes directly. This is not to say that they work entirely independently of sex hormones, as hormones can play a role in mediating the actions of the genes. Researchers can use mouse models to identify sex differences in a trait caused by sex chromosome genes when gonadal secretions are absent [103] or kept constant [104, 105]. These mice have the same gonadal sex (either males or females) but different sex chromosome complement (XX vs. XY) [104, 105]. Suppose there is a sex difference in a pain phenotype in males with XX versus males with XY sex chromosomes. We can conclude that sex chromosome complement contributed to the observed difference, either through Y chromosome genes present only in XY mice, or through X chromosome genes, which differ between XX and XY mice in genomic dosage and parental imprinting [36, 106, 107]. For instance, it has been reported that XX mice show greater thermal sensitivity compared to XY mice, irrespective of their gonadal type [43, 108]. This result does not indicate which genes are responsible or where they act. However, this type of "four core genotypes" model allows for investigating the interaction between sex chromosomes and sex hormones. It is also possible to test the effect of gonadal hormones independent of sex chromosomes. Mouse models comprising the same sex chromosome (either XX or XY) but different gonadal phenotypes (males vs. females) are also available. If a sex difference is observed between males and females with XY sex chromosomes, the gonads can be attributed to the difference [109]. For example, it has been reported that gonadal male mice show greater analgesic effects of morphine compared to gonadal females, irrespective of their sex chromosome complement [40].

2.7 Is the Observed Sex Difference Attributable to Local Hormone Effects?

Effects on pain that were initially attributed to the actions of gonadal hormones could be due to local hormone synthesis in the brain and spinal cord. Examples are progesterone [110, 111], estradiol [112, 113], and testosterone [114]. Neurons containing aromatase, the enzyme that catalyzes the conversion of testosterone to estradiol, are located in the dorsal horn of the spinal cord and medulla and form a distinct subpopulation of inhibitory interneurons [115, 116]. They are structurally positioned to receive somatic and visceral nociceptive inputs and can potentially regulate pain through local estrogen synthesis or inhibitory neurotransmitter release [115]. Experiments involving the blockade of local hormone synthesis (via specific antagonists or enzyme inhibitors) or exogenous infusion of a specific hormone in a specific part of the central nervous system can reveal its involvement in the observed sex difference. One such experiment in pain has been reported by Evrard and Balthazart where they showed that an acute blockade of the local synthesis of estrogens in the quail spinal dorsal horn significantly reduced sensitivity to a thermally painful stimulus [117]. They were also able to induce thermal sensitivity by the local infusion of estradiol [117]. Rapid estrogenic regulation of nociception at the spinal level has also been shown in rats where spinal administration of 17β-estradiol, the most potent natural estrogen, produced significant mechanical allodynia and thermal hyperalgesia [118].

2.8 Could Psychological Factors such as Stress and Conditioning Contribute to Sex Differences in Pain/Analgesia Phenotypes?

Several animal and human studies show males and females respond differentially to stress [11, 119–121]. Sometimes stress can have opposing effects in males versus females [122]. Pain itself is also stressful; therefore, any sex difference observed in pain phenotypes could result from sex differences in stress response. Both basal and stress-induced hypothalamic–pituitary–adrenal (HPA) axis secretion is more prominent in female rats than in males [123]. Estrogens facilitate HPA secretions, leading to elevated basal and stress-induced corticosterone levels in the proestrus phase of the female estrous cycle [123]. Researchers must consider this variation in stress response with estrous cyclicity when conducting experiments on sex differences in pain phenotypes.

HPA stress hormones, such as glucocorticoids, can alter/disrupt gonadal hormone secretions [124]. The HPA axis shows a pronounced circadian rhythm, with glucocorticoids peaking at the onset of the active phase of an animal's light–dark cycle [125]. Therefore, the impact of stress/glucocorticoids on behavioral and physiological sex differences may depend on the time of testing within the animal's light–dark cycle and must be considered in sex difference studies in rodents. It is also important to have proper sham controls for gonadectomy studies as surgery can have persistent effects on the HPA axis [126]. Vaginal smears for detecting the estrous cycle phase in females can also act as an acute stressor. See Table 2 for less stressful alternative methods.

2.9 Can the History of Pain Lead to Sex Differences in Future Pain Phenotypes?

Past pain experiences can result in sexually dimorphic responses to future clinical or experimentally induced pain [127]. Women are more likely to have a greater pain history and show lower pain thresholds and tolerance [127]. Therefore, the length of pain/chronicity is critical when designing preclinical experiments to investigate sex differences in pain. For example, it may be helpful to examine whether sex differences in rodents change/amplify with chronicity or repeated exposure to pain.

Past pain or repetitive painful stimuli can result in gonadal hormone-dependent adaptations in nociception, leading to sex differences in chronic pain development [128]. On the other hand, chronic pain can directly affect a rodent's circulating gonadal hormones, which can, in turn, alter its pain sensitivity [129]. Hence, it is crucial to carefully decide on the type of painful stimuli, acute or chronic, that will be utilized in an experiment as it may result in different pain phenotypes in males versus females.

2.10 Is the Immune System Responsible for the Sex Differences?

The immune system plays a significant role in the expression, development, and maintenance of chronic pain. Therefore, it should be considered when a sex difference is observed in a pain phenotype. Different immune system cell populations are activated in males versus females in response to pain [10]. Males depend on microglia, whereas females rely on T cells to mediate chronic pain. Testosterone inhibits T cells by decreasing T-helper function [130] and suppressing B cells [131] and $CD4^+$ T-cell differentiation [132] in males, possibly leading to the reliance on microglia to mediate proinflammatory responses. Estrogen and its receptors also regulate immune cells leading to sex differences in the cellular level [133–135]. Overall, high estrogen and low testosterone levels in females predispose them to greater proinflammatory cell activation, whereas in males, high testosterone levels lead to the activation of anti-inflammatory cell types [23, 136, 137].

2.11 Can Sex Differences in Pain Be Due to Sexually Dimorphic Analgesic Pathways vs. Sexually Dimorphic Afferent Pathways?

Some sex differences in pain may be due to sexually dimorphic endogenous analgesic pathways [138, 139]. These pathways are different in terms of their activation pattern in response to pain and exogenous opioids. An example is the estradiol-dependent μ-opioid receptor-mediated endogenous neurotransmission in women [140]. Females in the high-estrogen state of their menstrual cycle have greater activation of endogenous opioid neurotransmission in response to pain than when they were in the low-estrogen state. Hence, greater hyperalgesic responses were observed during the low-estrogen period. Male rodents also have a more robust spinal cord endorphin/μ-opioid receptor analgesic system than females, leading to a greater female prevalence and intensity of chronic pain syndromes [141].

3 Conclusion

Sex differences in pain are an active and growing area of research. Endocrine, nervous, and immune systems interact in complex ways to produce sex differences. This chapter has outlined some of the possible underlying mechanisms, both hormone-dependent and hormone-independent, and methods for investigating them. However, this chapter is by no means exhaustive. This chapter intends to initiate and encourage more researchers to explore sex differences in their experiments, which may ultimately reveal surprising and unexpected sex differences in pain phenotypes. Males and females of all species are inherently different and discovering the underlying mechanisms that give rise to these differences will be critical for developing new therapies and leading to a new frontier of personalized pain medicine.

Conflict of Interest

The authors have no conflicts of interest to declare that are relevant to the content of this chapter.

References

1. Goldberg DS, McGee SJ (2011) Pain as a global public health priority. BMC Public Health 11:770
2. Yong RJ, Mullins PM, Bhattacharyya N (2022) Prevalence of chronic pain among adults in the United States. Pain 163:E328–E332
3. Dahlhamer J, Lucas J, Zelaya C, Nahin R, Mackey S, DeBar L et al. (2018) Prevalence of chronic pain and high-impact chronic pain among adults - United States, 2016. MMWR Morb Mortal Wkly Rep 67:1001–1006
4. Gaskin DJ, Richard P (2012) The economic costs of pain in the United States. J Pain 13:715–724
5. Lynch ME (2011) The need for a Canadian pain strategy. Pain Res Manage 16(2):77–80
6. Morris S, Fawcett G, Brisebois L, Hughes J (2018) A demographic, employment and income profile of Canadians with disabilities aged 15 years and over, 2017. In: Canadian Survey on Disability. Statistics Canada. https://www150.statcan.gc.ca/n1/pub/89-654-x/89-654-x2018002-eng.htm. Accessed 9 Apr 2022.
7. Burlock A (2017) Women with disabilities. In: Women in Canada: a gender-based statistical report. Statistics Canada. https://www150.statcan.gc.ca/n1/pub/89-503-x/2015001/article/14695-eng.htm. Accessed 9 Apr 2022.
8. Mills SEE, Nicolson KP, Smith BH (2019) Chronic pain: a review of its epidemiology and associated factors in population-based studies. Br J Anaesth 123:e273
9. Bulls HW, Freeman EL, Anderson AJ, Robbins MT, Ness TJ, Goodin BR (2015) Sex differences in experimental measures of pain sensitivity and endogenous pain inhibition. J Pain Res 8:311–320
10. Sorge RE, Mapplebeck JCS, Rosen S, Beggs S, Taves S, Alexander JK et al. (2015) Different immune cells mediate mechanical pain hypersensitivity in male and female mice. Nat Neurosci 18(8):1081–1083
11. Quiton RL, Greenspan JD (2007) Sex differences in endogenous pain modulation by distracting and painful conditioning stimulation. Pain 132:S134–S149

12. Bartley EJ, Fillingim RB (2013) Sex differences in pain: a brief review of clinical and experimental findings. Br J Anaesth 111:52–58
13. Lisa Chanda M, Mogil JS (2006) Sex differences in the effects of amiloride on formalin test nociception in mice. Am J Physiol Regul Integr Comp Physiol 291:335–342
14. Martin LJ, Acland EL, Cho C, Gandhi W, Chen D, Corley E et al. (2019) Male-specific conditioned pain hypersensitivity in mice and humans. Curr Biol 29:192–201.e4
15. LaCroix-Fralish ML, Tawfik VL, DeLeo JA (2005) The organizational and activational effects of sex hormones on tactile and thermal hypersensitivity following lumbar nerve root injury in male and female rats. Pain 114:71–80
16. Zucker I, Beery AK (2010) Males still dominate animal studies. Nature 465(7299):690–690
17. Lee SK (2018) Sex as an important biological variable in biomedical research. BMB Rep 51:167–173
18. Prendergast BJ, Onishi KG, Zucker I (2014) Female mice liberated for inclusion in neuroscience and biomedical research. Neurosci Biobehav Rev 40:1–5
19. Mogil JS, Chanda ML (2005) The case for the inclusion of female subjects in basic science studies of pain. Pain 117:1–5
20. Beery AK, Zucker I (2011) Sex bias in neuroscience and biomedical research. Neurosci Biobehav Rev 35:565–572
21. Sorge RE, Totsch SK (2017) Sex differences in pain. J Neurosci Res 95:1271–1281
22. Berkley KJ, Zalcman SS, Simon VR (2006) Sex and gender differences in pain and inflammation: a rapidly maturing field. Am J Physiol Regul Integr Comp Physiol 291:R241–R244
23. Greenspan JD, Craft RM, LeResche L, Arendt-Nielsen L, Berkley KJ, Fillingim RB et al. (2007) Studying sex and gender differences in pain and analgesia: a consensus report. Pain 132:S26–S45
24. Leshner AI, Collier G (1973) The effects of gonadectomy on the sex differences in dietary self-selection patterns and carcass compositions of rats. Physiol Behav 11:671–676
25. Greil H (2006) Patterns of sexual dimorphism from birth to senescence. Coll Antropol 30:637–641
26. Gerall AA, Napoli AM, Cooper UC (1973) Daily and hourly estrous running in intact, spayed, and estrone implanted rats. Physiol Behav 10:225–229
27. Riley JL, Robinson ME, Wise EA, Myers CD, Fillingim RB (1998) Sex differences in the perception of noxious experimental stimuli: a meta-analysis. Pain 74:181–187
28. Cepeda MS, Farrar JT, Baumgarten M, Boston R, Carr DB, Strom BL (2003) Side effects of opioids during short-term administration: effect of age, gender, and race. Clin Pharmacol Ther 74:102–112
29. Dahan A, Sarton E, Teppema L, Olievier G (1998) Sex-related differences in the influence of morphine on ventilatory control in humans. Anesthesiology 88:903–913
30. Sarton E, Olofsen E, Romberg R, den Hartigh J, Kest B, Nieuwenhuijs D et al. (2000) Sex differences in morphine analgesia: an experimental study in healthy volunteers. Anesthesiology 93:1245–1254
31. Klein SL, Schiebinger L, Stefanick ML, Cahill L, Danska J, de Vries GJ et al. (2015) Opinion: sex inclusion in basic research drives discovery. Proc Natl Acad Sci USA 112:5257–5258
32. Yonkers KA, Kando JC, Cole JO, Blumenthal S (1992) Gender differences in pharmacokinetics and pharmacodynamics of psychotropic medication. Am J Psychiatry 149:587–595
33. Becker JB, Robinson TE, Lorenz KA (1982) Sex differences and estrous cycle variations in amphetamine-elicited rotational behavior. Eur J Pharmacol 80:65–72
34. Stoffel EC, Ulibarri CM, Craft RM (2003) Gonadal steroid hormone modulation of nociception, morphine antinociception and reproductive indices in male and female rats. Pain 103:285–302
35. Liu NJ, Gintzler AR (2000) Prolonged ovarian sex steroid treatment of male rats produces antinociception: identification of sex-based divergent analgesic mechanisms. Pain 85:273–281
36. Palaszynski KM, Smith DL, Kamrava S, Burgoyne PS, Arnold AP, Voskuhl RR (2005) A yin-yang effect between sex chromosome complement and sex hormones on the immune response. Endocrinology 146(8):3280–3285.
37. Woolf CJ, Ma Q (2007) Nociceptors—noxious stimulus detectors. Neuron 55:353–364
38. Smith JC (2019) A review of strain and sex differences in response to pain and analgesia in mice. Comp Med 69(6):490–500
39. Becker JB, Arnold AP, Berkley KJ, Blaustein JD, Eckel LA, Hampson E et al. (2005) Strategies and methods for research on sex differences in brain and behavior. Endocrinology 146:1650–1673
40. Gioiosa L, Chen X, Watkins R, Umeda EA, Arnold AP (2008) Sex Chromosome complement affects nociception and analgesia in newborn mice. J Pain 9:962–969
41. Sternberg WF, Smith L, Scorr L (2004) Nociception and antinociception during the first week of life in mice: sex differences and test dependence. J Pain 5:420–426
42. Sternberg WF, Ritchie J, Mogil JS (2004) Qualitative sex differences in κ-opioid analgesia

in mice are dependent on age. Neurosci Lett 363:178–181
43 Ferrari LF, Araldi D, Green P, Levine JD (2018) Age-dependent sexual dimorphism in susceptibility to develop chronic pain in the rat. Neuroscience 387:170–177
44. Islam AK, Cooper ML, Bodnar RJ (1993) Interactions among aging, gender, and gonadectomy effects upon morphine antinociception in rats. Physiol Behav 54:45–53
45. Baker L, Ratka A (2002) Sex-specific differences in levels of morphine, morphine-3-glucuronide, and morphine antinociception in rats. Pain 95:65–74
46. Kepler KL, Kest B, Kiefel JM, Cooper ML, Bodnar RJ (1989) Roles of gender, gonadectomy and estrous phase in the analgesic effects of intracerebroventricular morphine in rats. Pharmacol Biochem Behav 34:119–127
47. Kavaliers M, Galea LAM (1995) Sex differences in the expression and antagonism of swim stress-induced analgesia in deer mice vary with the breeding season. Pain 63:327–334
48. Prusator DK, Greenwood-Van Meerveld B (2016) Sex-related differences in pain behaviors following three early life stress paradigms. Biol Sex Differ 7:29
49. Mogil JS, Sternberg WF, Kest B, Marek P, Liebeskind JC (1993) Sex differences in the antagonism of swim stress-induced analgesia: effects of gonadectomy and estrogen replacement. Pain 53:17–25
50. Mogil JS, Belknap JK (1997) Sex and genotype determine the selective activation of neurochemically-distinct mechanisms of swim stress-induced analgesia. Pharmacol Biochem Behav 56:61–66
51. Langford DJ, Tuttle AH, Briscoe C, Harvey-Lewis C, Baran I, Gleeson P et al. (2011) Varying perceived social threat modulates pain behavior in male mice. J Pain 12:125–132
52. Kavaliers M, Innes DGL (1987) Sex and day-night differences in opiate-induced responses of insular wild deer mice, *Peromyscus maniculatus triangularis*. Pharmacol Biochem Behav 27:477–482
53. Bradshaw H, Miller J, Ling Q, Malsnee K, Ruda MA (2000) Sex differences and phases of the estrous cycle alter the response of spinal cord dynorphin neurons to peripheral inflammation and hyperalgesia. Pain 85:93–99
54. Moloney RD, Sajjad J, Foley T, Felice VD, Dinan TG, Cryan JF et al. (2016) Estrous cycle influences excitatory amino acid transport and visceral pain sensitivity in the rat: effects of early-life stress. Biol Sex Differ 7:33.
55. Kaur S, Benton WL, Tongkhuya SA, Lopez CMC, Uphouse L, Averitt DL (2018) Sex differences and estrous cycle effects of peripheral serotonin-evoked rodent pain behaviors. Neuroscience 384:87–100
56. Beatty WW, Beatty PA (1970) Hormonal determinants of sex differences in avoidance behavior and reactivity to electric shock in the rat. J Comp Physiol Psychol 73:446–455
57. Baamonde AI, Hidalgo A, Andrés-Trelles F (1989) Sex-related differences in the effects of morphine and stress on visceral pain. Neuropharmacology 28:967–970
58. Marks HE, Hobbs SH (1972) Changes in stimulus reactivity following gonadectomy in male and female rats of different ages. Physiol Behav 8:1113–1119
59. Beatty WW, Fessler RG (1977) Gonadectomy and sensitivity to electric shock in the rat. Physiol Behav 19:1–6
60. Romero MT, Kepler KL, Cooper ML, Komisaruk BR, Bodnar RJ (1987) Modulation of gender-specific effects upon swim analgesia in gonadectomized rats. Physiol Behav 40:39–45
61. Mogil JS, Wilson SG, Chesler EJ, Rankin AL, Nemmani KVS, Lariviere WR et al. (2003) The melanocortin-1 receptor gene mediates female-specific mechanisms of analgesia in mice and humans. Proc Natl Acad Sci USA 100:4867–4872
62. Juni A, Cai M, Stankova M, Waxman AR, Arout C, Klein G et al. (2010) Sex-specific mediation of opioid-induced hyperalgesia by the melanocortin-1 receptor. Anesthesiology 112:181–188
63. Deleo JA, Rutkowski MD (2000) Gender differences in rat neuropathic pain sensitivity is dependent on strain. Neurosci Lett 282:197–199
64. Kest B, Wilson S, Mogil J (1999) Sex differences in supraspinal morphine analgesia are dependent on genotype. J Pharmacol Exp Ther 289:1370–1375
65. Bon K, Lantéri-Minet M, Menétrey D, Berkley KJ (1997) Sex, time-of-day and estrous variations in behavioral and bladder histological consequences of cyclophosphamide-induced cystitis in rats. Pain 73:423–429
66. Kasson BG, George R (1984) Endocrine influences on the actions of morphine: IV. Effects of sex and strain. Life Sci 34:1627–1634
67. Krzanowska EK, Bodnar RJ (1999) Morphine antinociception elicited from the ventrolateral periaqueductal gray is sensitive to sex and gonadectomy differences in rats. Brain Res 821:224–230
68. Candido J, Lutfy K, Billings B, Sierra V, Duttaroy A, Inturrisi CE et al. (1992) Effect of adrenal and sex hormones on opioid analgesia and opioid receptor regulation. Pharmacol Biochem Behav 42:685–692

69. Jørum E (1988) Analgesia or hyperalgesia following stress correlates with emotional behavior in rats. Pain 32:341–348
70. Smith ML, Hostetler CM, Heinricher MM, Ryabinin AE (2016) Social transfer of pain in mice. Sci Adv 2:e1600855
71. Baptista-De-Souza D, Nunciato AC, Pereira BC, Fachinni G, Zaniboni CR, Canto-De-Souza A (2015) Mice undergoing neuropathic pain induce anxiogenic-like effects and hypernociception in cagemates. Behav Pharmacol 26:664–672
72. Sorge RE, Martin LJ, Isbester KA, Sotocinal SG, Rosen S, Tuttle AH et al. (2014) Olfactory exposure to males, including men, causes stress and related analgesia in rodents. Nat Methods 11(6):629–632
73. Becegato M, Meurer YSR, Paiva-Santos MA, Lima AC, Marinho GF, Bioni VS et al. (2021) Impaired discriminative avoidance and increased plasma corticosterone levels induced by vaginal lavage procedure in rats. Physiol Behav 232:113343
74. Brobeck JR, Wheatland M, Strominger JL (1947) Variations in regulation of energy exchange associated with estrus, diestrus and pseudopregnancy in rats. Endocrinology 40:65–72
75. Ajayi AF, Akhigbe RE (2020) Staging of the estrous cycle and induction of estrus in experimental rodents: an update. Fertil Res Pract 6(1):1–15
76. Hurst JL, West RS (2010) Taming anxiety in laboratory mice. Nat Methods 7:825–826
77. Wilson MA, Biscardi R (1994) Sex differences in GABA/benzodiazepine receptor changes and corticosterone release after acute stress in rats. Exp Brain Res 101(2):297–306
78. Perissin L, Facchin P, Porro CA (2003) Tonic pain response in mice: effects of sex, season and time of day. Life Sci 72:897–907
79. Abbott FV, Franklin KBJ, Connell B (1986) The stress of a novel environment reduces formalin pain: possible role of serotonin. Eur J Pharmacol 126:141–144
80. Brown KJ, Grunberg NE (1995) Effects of housing on male and female rats: crowding stresses males but calms females. Physiol Behav 58:1085–1089
81. Chesler EJ, Wilson SG, Lariviere WR, Rodriguez-Zas SL, Mogil JS (2002) Identification and ranking of genetic and laboratory environment factors influencing a behavioral trait, thermal nociception, via computational analysis of a large data archive. Neurosci Biobehav Rev 26:907–923
82. Pham TM, Hagman B, Codita A, van Loo PLP, Strömmer L, Baumans V (2010) Housing environment influences the need for pain relief during post-operative recovery in mice. Physiol Behav 99:663–668
83. Tall JM (2009) Housing supplementation decreases the magnitude of inflammation-induced nociception in rats. Behav Brain Res 197:230–233
84. Liuzzi FJ, Scoville SA, Bufton SM (1999) Effects of short-term estrogen replacement on trkA mRNA levels in axotomized dorsal root ganglion neurons. Exp Neurol 159:433–440
85. Craft RM, Ulibarri C, Leitl MD, Sumner JE (2008) Dose- and time-dependent estradiol modulation of morphine antinociception in adult female rats. Eur J Pain 12:472–479
86. Camp DM, Becker JB, Robinson TE (1986) Sex differences in the effects of gonadectomy on amphetamine-induced rotational behavior in rats. Behav Neural Biol 46:491–495
87. Gililland KR, Finn DA (2007) The impact of gonadectomy and adrenalectomy on acute withdrawal severity in male and female C57BL/6J and DBA/2J mice following a single high dose of ethanol. Alcohol Clin Exp Res 31(11):1846–1857
88. Ajika K, Krulich L, Fawcett CP, McCann SM (1972) Effects of estrogen on plasma and pituitary gonadotropins and prolactin, and on hypothalamic releasing and inhibiting factors. Neuroendocrinology 9:304–315
89. Breedlove SM, Arnold AP (1983) Hormonal control of a developing neuromuscular system. I. Complete demasculinization of the male rat spinal nucleus of the bulbocavernosus using the anti-androgen flutamide. J Neurosci 3:417–423
90. Etgen AM (1979) Antiestrogens: effects of tamoxifen, nafoxidine, and CI-628 on sexual behavior, cytoplasmic receptors, and nuclear binding of estrogen. Horm Behav 13:97–112
91. Brown TJ, Blaustein JD (1984) Inhibition of sexual behavior in female guinea pigs by a progestin receptor antagonist. Brain Res 301:343–349
92. Azcoitia I, Sierra A, Veiga S, Honda SI, Harada N, Garcia-Segura LM (2001) Brain aromatase is neuroprotective. J Neurobiol 47:318–329
93. Jordan VC (1998) Antiestrogenic action of raloxifene and tamoxifen: today and tomorrow. J Natl Cancer Inst 90:967–971
94. Beyer C, Green SJ, Hutchison JB (1994) Androgens influence sexual differentiation of embryonic mouse hypothalamic aromatase neurons *in vitro*. Endocrinology 135:1220–1226
95. MacLusky NJ, Hajszan T, Leranth C (2004) Effects of dehydroepiandrosterone and flutamide on hippocampal CA1 spine synapse density in male and female rats: implications for the role

of androgens in maintenance of hippocampal structure. Endocrinology 145:4154–4161
96. He WW, Kumar MV, Tindall DJ (1991) A frame-shift mutation in the androgen receptor gene causes complete androgen insensitivity in the testicular-feminized mouse. Nucleic Acids Res 19(9):2373–2378
97. Couse JF, Korach KS (1999) Estrogen receptor null mice: what have we learned and where will they lead us?. Endocr Rev 20:358–417
98. Korach KS, Couse JF, Curtis SW et al. (1996) Estrogen receptor gene disruption: molecular characterization and experimental and clinical phenotypes. Recent Prog Horm Res 51:159–186. https://www.researchgate.net/publication/14501395_Estrogen_receptor_gene_disruption_Molecular_characterization_and_experimental_and_clinical_phenotypes. Accessed 13 Mar 2022.
99. Hruska KS, Tilli MT, Ren S, Cotarla I, Kwong T, Li M et al. (2002) Conditional over-expression of estrogen receptor alpha in a transgenic mouse model. Transgenic Res 11(4):361–372
100. Breedlove SM, Arnold AP (1983) Hormonal control of a developing neuromuscular system. II. Sensitive periods for the androgen-induced masculinization of the rat spinal nucleus of the bulbocavernosus. J Neurosci 3:424–432
101. Nordeen EJ, Nordeen KW, Sengelaub DR, Arnold AP (1985) Androgens prevent normally occurring cell death in a sexually dimorphic spinal nucleus. Science (1979) 229:671–673
102. Romeo RD, Richardson HN, Sisk CL (2002) Puberty and the maturation of the male brain and sexual behavior: recasting a behavioral potential. Neurosci Biobehav Rev 26:381–391
103. Parker KL, Schedl A, Schimmer BP (1999) Gene interactions in gonadal development. Annu Rev Physiol 61:417–433.
104. Burgoyne PS (1993) A Y-chromosomal effect on blastocyst cell number in mice. Development 117:341–345
105. Lovell-Badge R, Robertson E (1990) XY female mice resulting from a heritable mutation in the primary testis-determining gene, Tdy. Development 109:635–646
106. Gatewood JD, Wills A, Shetty S, Xu J, Arnold AP, Burgoyne PS et al (2006) Sex chromosome complement and gonadal sex influence aggressive and parental behaviors in mice. J Neurosci 26(8):2335-2342
107. Arnold AP, Burgoyne PS (2004) Are XX and XY brain cells intrinsically different?. Trends Endocrinol Metab 15:6–11
108. Gioiosa L, Chen X, Watkins R, Klanfer N, Bryant CD, Evans CJ et al. (2008) Sex chromosome complement affects nociception in tests of acute and chronic exposure to morphine in mice. Horm Behav 53(1):124–130
109. de Vries GJ, Rissman EF, Simerly RB, Yang LY, Scordalakes EM, Auger CJ et al. (2002) A model system for study of sex chromosome effects on sexually dimorphic neural and behavioral traits. J Neurosci 22:9005–9014
110. Micevych P, Sinchak K, Mills RH, Tao L, LaPolt P, Lu JKH (2003) The luteinizing hormone surge is preceded by an estrogen-induced increase of hypothalamic progesterone in ovariectomized and adrenalectomized rats. Neuroendocrinology 78:29–35
111. Sinchak K, Mills RH, Tao L, LaPolt P, Lu JKH, Micevych P (2003) Estrogen induces de novo progesterone synthesis in astrocytes. Dev Neurosci 25:343–348
112. Amateau SK, Alt JJ, Stamps CL, McCarthy MM (2004) Brain estradiol content in newborn rats: sex differences, regional heterogeneity, and possible de novo synthesis by the female telencephalon. Endocrinology 145:2906–2917
113. Kretz O, Fester L, Wehrenberg U, Zhou L, Brauckmann S, Zhao S et al. (2004) Hippocampal synapses depend on hippocampal estrogen synthesis. J Neurosci 24:5913–5921
114. Diotel N, Charlier TD, Lefebvre d'Hellencourt C, Couret D, Trudeau VL, Nicolau JC et al. (2018) Steroid transport, local synthesis, and signaling within the brain: roles in neurogenesis, neuroprotection, and sexual behaviors. Front Neurosci 12:84
115. Tran M, Kuhn JA, Bráz JM, Basbaum AI (2017) Neuronal aromatase expression in pain processing regions of the medullary and spinal cord dorsal horn. J Comp Neurol 525(16):3414–3428
116. Evrard H, Baillien M, Agne‘ A, Foidart A, Absil P, Harada N et al. (2000) Localization and controls of aromatase in the quail spinal cord. J Comp Neurol 423:552–564
117. Evrard HC, Balthazart J (2004) Rapid regulation of pain by estrogens synthesized in spinal dorsal horn neurons. J Neurosci 24(33):7225–7229
118. Zhang Y, Lü N, Zhao ZQ, Zhang YQ (2012) Involvement of estrogen in rapid pain modulation in the rat spinal cord. Neurochem Res 37:2697–2705
119. Kitay JI (1961) Sex differences in adrenal cortical secretion in the rat. Endocrinology 68:818–824
120. Sanford SD, Kersh BC, Thorn BE, Rich MA, Ward LC (2002) Psychosocial mediators of sex differences in pain responsivity. J Pain 3:58–64
121. Kudielka BM, Kirschbaum C (2005) Sex differences in HPA axis responses to stress: a review. Biol Psychol 69:113–132

122. Shors TJ, Chua C, Falduto J (2001) Sex differences and opposite effects of stress on dendritic spine density in the male versus female hippocampus. J Neurosci 21:6292–6297
123. Carey MP, Deterd CH, de Koning J, Helmerhorst F, de Kloet ER (1995) The influence of ovarian steroids on hypothalamic-pituitary-adrenal regulation in the female rat. J Endocrinol 144:311–321
124. Karsch FJ, Battaglia DF, Breen KM, Debus N, Harris TG (2009) Mechanisms for ovarian cycle disruption by immune/inflammatory stress. Stress 5:101–112
125. Keller-Wood ME, Dallman MF (1984) Corticosteroid inhibition of ACTH secretion. Endocr Rev 5:1–24
126. Schmidt ED, Binnekade R, Janszen AWJW, Tilders FJH (1996) Short stressor induced long-lasting increases of vasopressin stores in hypothalamic corticotropin-releasing hormone (CRH) neurons in adult rats. J Neuroendocrinol 8:703–712
127. Rollman GB, Abdel-Shaheed J, Gillespie JM, Jones KS (2004) Does past pain influence current pain: biological and psychosocial models of sex differences. Eur J Pain 8:427–433
128. Aloisi AM, Ceccarelli I, Fiorenzani P (2003) Gonadectomy affects hormonal and behavioral responses to repetitive nociceptive stimulation in male rats. Ann N Y Acad Sci 1007:232–237
129. Amini H, Ahmadiani A (2005) *In vivo* evidence for an increase in 5alpha-reductase activity in the rat central nervous system following morphine exposure. Int J Dev Neurosci 23:621–626
130. Grossman CJ, Roselle GA, Mendenhall CL (1991) Sex steroid regulation of autoimmunity. J Steroid Biochem Mol Biol 40:649–659
131. Sthoeger ZM, Chiorazzi N, Lahita RG (1988) Regulation of the immune response by sex hormones. I. *In vitro* effects of estradiol and testosterone on pokeweed mitogen-induced human B cell differentiation. J Immunol 141(1):91–98
132. Kissick HT, Sanda MG, Dunn LK, Pellegrini KL, On ST, Noel JK et al. (2014) Androgens alter T-cell immunity by inhibiting T-helper 1 differentiation. Proc Natl Acad Sci USA 111:9887–9892
133. Loram LC, Sholar PW, Taylor FR, Wiesler JL, Babb JA, Strand KA et al. (2012) Sex and estradiol influence glial pro-inflammatory responses to lipopolysaccharide in rats. Psychoneuroendocrinology 37:1688–1699
134. Phiel KL, Henderson RA, Adelman SJ, Elloso MM (2005) Differential estrogen receptor gene expression in human peripheral blood mononuclear cell populations. Immunol Lett 97:107–113
135. Kovats S (2015) Estrogen receptors regulate innate immune cells and signaling pathways. Cell Immunol 294:63–69
136. Rosen S, Ham B, Mogil JS (2017) Sex differences in neuroimmunity and pain. J Neurosci Res 95:500–508
137. Calippe B, Douin-Echinard V, Delpy L, Laffargue M, Lélu K, Krust A et al. (2010) 17β-estradiol promotes TLR4-triggered proinflammatory mediator production through direct estrogen receptor α signaling in macrophages *in vivo*. J Immunol 185:1169–1176
138. Loyd DR, Murphy AZ (2006) Sex differences in the anatomical and functional organization of the periaqueductal gray-rostral ventromedial medullary pathway in the rat: a potential circuit mediating the sexually dimorphic actions of morphine. J Comp Neurol 496:723–738
139. Loyd DR, Morgan MM, Murphy AZ (2007) Morphine preferentially activates the periaqueductal gray–rostral ventromedial medullary pathway in the male rat: a potential mechanism for sex differences in antinociception. Neuroscience 147:456–468
140. Smith YR, Stohler CS, Nichols TE, Bueller JA, Koeppe RA, Zubieta JK (2006) Pronociceptive and antinociceptive effects of estradiol through endogenous opioid neurotransmission in women. J Neurosci 26:5777–5785
141. Gupta DS, von Gizycki H, Gintzler AR (2007) Sex-/ovarian steroid-dependent release of endomorphin 2 from spinal cord. J Pharmacol Exp Ther 321:635–641

Check for updates

Chapter 10

Individual Differences Using Social Interactions Testing: A Modified Resident-Intruder Test for Measuring Social Behaviors in Rats with Neuropathic Injury

James W.M. Kang and Kevin A. Keay

Abstract

Individuals suffering from chronic neuropathic pain often report severe and disabling impairments to their relationships and social interactions, in addition to signs of sensory dysfunction. These underappreciated impacts of pain, on social interactions and relationships are often reported to lead to major impacts on quality of life. Similarly, in preclinical models of neuropathic pain, disturbances in social interactions have been described, and a modified Resident–Intruder test has been developed as a robust method to define impairments in social interactions in subpopulations of rats, following neuropathic injury. The use of animal models is essential in unravelling the neurobiological mechanisms that underlie individual susceptibilities to developing complex behavioral disturbances in chronic neuropathic pain. We describe a detailed methodological approach for identifying individual differences using a modified Resident–Intruder social interactions test in rats following a neuropathic injury, and discuss the advantages and limitations of experimental approaches to studying individual differences in preclinical models.

Keywords Neuropathic pain, Resident–Intruder test, Social interactions, Individual differences, Dominance behaviors, Chronic constriction injury (CCI), Behavioral subpopulations, Preclinical model

1 Introduction

Robust interindividual differences in pain sensitivity, as well as in the susceptibility to develop chronic pain, and its comorbid psychological and social impacts have been repeatedly observed [1]. Pain *comorbidities* (e.g., loss of sleep, changes in appetite, depression) are often claimed to increase the disabling nature of sensory "pain" [2]. Comorbid changes can also include disrupted social interactions and impaired social relationships, for which many chronic pain patients can lead to isolation and lack of support. Improving our understanding of the biological bases (i.e., genetic, epigenetic, neural, glial) of the selective expression of disabling comorbidities, and their social

Isaura Tavares and Hugo Leite-Almeida (eds.), *Translational Research Methods for Neuropathic Pain*, Neuromethods, vol. 235, https://doi.org/10.1007/978-1-0716-5312-8_10,

and environmental determinants may reveal important targets of intervention to improve outcomes in chronic pain populations.

In a cross-sectional survey of postsurgical patients, persistent pain was found to be prevalent at the site of surgery in 40% of patients, and of these individuals, "51% reported chronic pain when questioned without specific reference to the surgery" [3]. Similarly, in men following surgical inguinal hernia repair, 55% of individuals reported that "*pain significantly interfered with work or leisure activity*" [4]. Disruptions to social behaviors are also observed in preclinical models of neuropathic pain, and these models provide an opportunity to identify the biological mechanisms underlying this aspect of chronic pain [5–8]. Over a number of years, we have been using a modified Resident–Intruder social interactions test to characterize the impact of nerve injury—*chronic constriction injury of the sciatic nerve (CCI)*—and the development of neuropathic pain on social behaviors in male rats [9, 10]. The Resident–Intruder test evaluates interactions of pairs of rats, where one rat is introduced into the homecage of another. This test was most often used to investigate male–male interactions, and it was usually conceptualized and described using the language and context of territorial defense and the establishment of social hierarchies. A major use of the test has been to study psychosocial stress, with a focus on the intruder rat [11]; however, we took a different approach and focused on the resident. Our modified protocol (see Table 1) differs specifically from the more usual applications of the Resident–Intruder test in three important ways: (1) the behaviors of the *Resident* rat, rather than the

Table 1
Comparison of features of the Original Resident–Intruder test[11] and the modified Resident–Intruder test [10]

Features	Original Resident–Intruder test [11]	Modified Resident–Intruder test [10
Resident Housed with female conspecifc	✓	✗
Bedding partially cleaned	✓	✓
Intruder is smaller nonaggressive strain	✓	✗
Intruder weight-matched conspecific	✗	✓
Recording in the dark	✓	✓
Expression of Violence behavior by *Resident*	✓	✗
Expression of Dominance/Defensive behaviors	✓	✓
Expression of social behaviors	✓	✓
Expression of nonsocial behaviors	✓	✓

Intruder rat are our primary focus; (2) the *Resident* rats are not housed with females, but instead they are individually-housed within visual, olfactory, and auditory proximity of other individually housed *Residents*; and (3) *the Intruders* are age, weight, size, and strain matched with the *Resident* rats. These modifications were made to significantly reduce the expression of offensive-aggressive and *violent* behaviors that characterize the classic Resident–Intruder test [11]. The overtly aggressive behaviors of *Residents* are replaced by rich repertoires of much less intense dominance behaviors, which were described in early ethological studies of laboratory rodent behavior by Grant and Mackintosh (1963) [37]. These behaviors are essential for the establishment of stable social hierarchies under laboratory housing conditions. Building on this background, the social interactions of *Resident* and *Intruder* rats can be reliably categorized into four, mutually-exclusive behavioral classes: (i) dominance behaviors, (ii) social behaviors, (iii) nonsocial behaviors, and (iv) submissive behaviors. The duration of expression of each class of behavior has been shown to be stable over repeated daily testing periods, with *Resident* rats typically performing social behaviors for ~30% of the time, dominance behaviors ~30% of the time, and nonsocial behaviors for the remaining 40% of the time. Submissive behaviors are rarely expressed by *Resident* rats.

Notwithstanding individual variance in social interactions between *Resident* rats, following sciatic nerve constriction injury ~45% of *Resident* rats show altered social interactions. These changes are characterized by reductions in the duration dominance behaviors with a corresponding increase in nonsocial behaviors. Interestingly, disassociations between sensory hypersensitivity and alterations in complex behaviors have also been observed in the clinical population [12–16]. Our preclinical rat model of neuropathic injury, similarly, show these uncoupling of sensory hypersensitivity (hyperalgesia and allodynia) and alterations in complex behaviors as seen by reductions in dominance behaviors [9, 10, 17].

Using our modified Resident–Intruder test to measure the natural variance in the expression of social interactions. We have identified subpopulations of rats with a neuropathic injury, that in addition to alterations in their social behaviors, show interindividual differences in sleep [10], the balance of activity in the hypothalamic-pituitary-thyroid (HPT) axis [18, 19], the balance of activity in the hypothalamic-pituitary-adrenal (HPA) axis [20], and the response at the site of injury, dorsal root ganglion and spinal cord [21, 22].

Dominance behaviors are coordinated by a series of subcortical and cortical neural structures including the prefrontal cortex, hypothalamus, and the periaqueductal gray (PAG), which have shown to be associated to interindividual variations in performance in the Resident–Intruder test. These include anatomically specific patterns of altered glial morphology/activity [23], cellular injury [24], glucocorticoid receptor expression, alterations in catecholaminergic signaling [25, 26], and selective changes in the expression of cholecystokinin

(CCK) in the PAG [27]. In addition, we have reported alterations of dopaminergic signaling in the nucleus accumbens [17] and significant morphological changes in the hippocampus [28]. Further, we have shown additional examples of individual variance following a peripheral nerve injury in work that revealed distinct subpopulations of rats with reduced motivation to consume a sucrose solution [29].

Taken together, we have shown that our modified Resident–Intruder test reveals individual variations in the behaviors of rats in a preclinical model of neuropathic injury, which reflects the individual variations seen in similar clinical populations.

2 Materials

Materials for performing the modified Resident–Intruder social interactions test are listed in Table 2.

Table 2

Specific equipment and consumables with suppliers details if specific, that are required to repeat the behavioral experiments described in this chapter

Items	Product number, vendor, notes
Animals	Outbred Male Sprague–Dawley rats (ARC/Ozgene Perth, WA, Australia)
Modified Resident–Intruder Testing	
• Resident Cages	Clear Perspex cages (40 × 36 × 24 cm) with open top grid lid.
• Intruder Cages	Polycarbonate cages.
• Video recording and analysis	Digital infrared camera (i.e., Sony PXW-X70 or equivalent) Tripod PowerLab 8/35 data acquisition device (AD Instruments, LabChart 7 software) connected to an analog input device.
Surgical Instruments and Consumables	Povidine Iodine 70% (v/v) ethanol Cotton tips/buds Isoflurane Quinine Vaseline Antiseptic powder Gauze Curved forceps Straight forceps Curved surgical scissors 4-0 chromic gut Michel Clips/surgical staples and applicator
Sensory Testing	Von Frey Filaments Glass plate (8 mm thick) and freezer bricks Mechanical test cage (12 × 15 × 15 cm) with mesh floor. Thermal test cage [open top Plexigalss cylinder (15cm in diameter × 30 cm in height)].

3 Methods

To examine alterations in resident–intruder behaviors following a neuropathic injury each rat is treated as an independent and continuous variable to identify individual differences in behavioral repertoires. The modified Resident–Intruder test presented here is based on the methodology first described by Monassi, Bandler, and Keay [10], who assessed alterations in social behaviors following sciatic nerve injury in male Sprague–Dawley rats and categorized them into three distinct subgroups.

The procedures described below include the details required to monitor and control external variables to quantify individual differences in social behaviors of male Sprague–Dawley rats.

3.1 Animals

Outbred, male Sprague–Dawley rats (ARC/Ozgene, Perth, WA, Australia) are used in all of our experiments. On arrival, rats are allocated arbitrarily into either "*Resident*" or "*Intruder*" experimental groups. Ensure before commencing any rodent experiments you have received approval for the procedures described below from your corresponding institutional animal ethics committee. We recommend housing *Intruder* rats in group of three to four per cage and maintaining a *12 h reverse light dark cycle* (e.g.: lights OFF at 7 a.m., lights ON at 7 p.m.). Group housing of *Intruder* rats is crucial for maintaining the stable expression of social behaviors and developing a stable social hierarchy. The *Resident* rats are individually housed in clear Perspex cages (40 × 36 × 24 cm) with open top grid lids in a room specifically dedicated to housing *Resident* rats (*12 h reverse light dark cycle*). This ensures that there is no opportunity for familiarity on exposure to an *Intruder* rat. We recommend that the *Resident* rats are used in cohorts of at least six individuals housed in the same room. This minimizes the potential impacts of socially isolating the rats; the open cage lids of the *Resident* rat's home cages allows for vocal communication with other *Resident* rats, as well as allowing relatively free access to olfactory and auditory cues.

3.2 Housing

Resident Rats

On arrival, at the laboratory, *Resident* rats are weighed and immediately singly housed in a *12 h reverse light dark cycle*. From the time of arrival, rats are considered to have entered the experimental protocol that is inclusive of the habituation period (see Fig. 1). During the entire experimental period, rats are only exposed to a single experimenter (exceptions are made for emergencies of course, or the unlikely event that animal welfare may be at risk). During the period of habituation to the new housing environment, for which we recommend a minimum of 7 days, general husbandry including weighing occurs daily at the same time. General husbandry is performed by the experimenter to minimize variations in the homecage experience that can impact behavioral outcomes. During cage cleaning for the

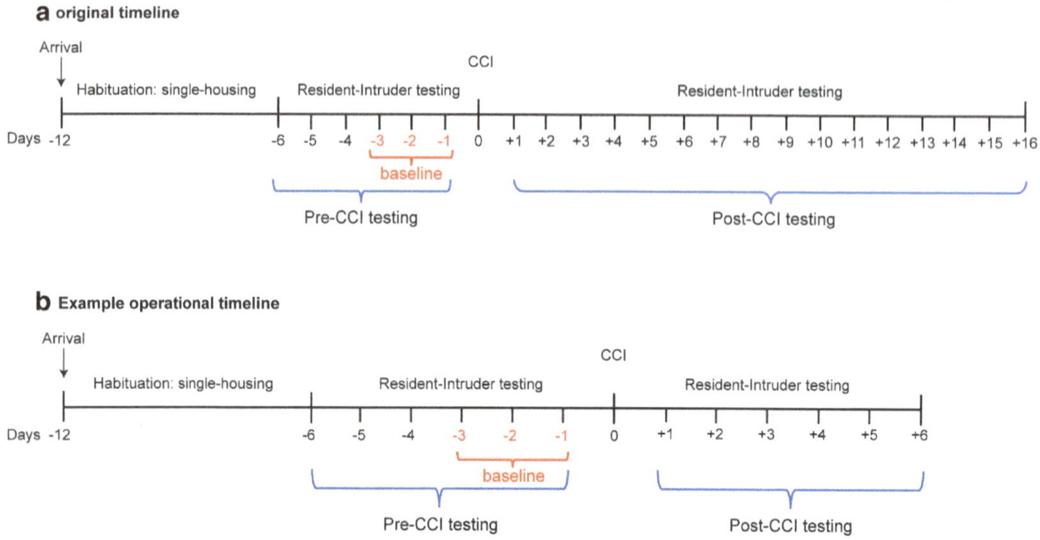

Fig. 1 (**a**) Original modified Resident–Intruder experimental timeline used by Monassi et al. [10] to characterize CCI rats into behavioral subcategories; (**b**) exhibits an example experimental timeline that can be utilized to characterize CCI rats into behavioral subcategories according to changes in dominance behavior

Resident rats, only 30%–50% of the cage bedding is removed and replaced with fresh materials, and a constant level of environmental enrichment is maintained, replacing any nonuseable objects. This partial clean of the *Resident* cage is to maintain homecage familiarity and to encourage territoriality based on the olfactory cues of the residents odor. These cues not only maintain the *Resident's* territory but signal to the *Intruder* that it is the home cage of unknown conspecific. Two days prior to sensory testing *Resident* rats are habituated twice (once a day) for 30 min to the sensory testing apparatus.

Resident–intruder, social interactions testing occurs in the *Resident* rats housing room on a bench separate to the cage racking. This minimizes changes to the *Resident's* home cage environment and enables the assessment of stable home cage behaviors of the *Resident* rat (see Fig. 2).

Intruder Rats

On arrival, *Intruder* rats are group housed three to four to a cage on a 12 h reverse light-dark cycle. *Intruder* rats are housed in a separate housing room to the *Resident* rats. General husbandry of the *Intruder* rats is performed at approximately the same time each day during their light-dark cycle.

3.3 Modified Resident–Intruder Testing

Resident rats are left to habituate to their home cage for a minimum of 7 days to establish their territory. Following the habituation period, *Resident* rats are tested daily in the resident–intruder social interactions test for a minimum of 4 days. In the modified Resident–Intruder social interactions test, the behaviors of the *Resident* rat

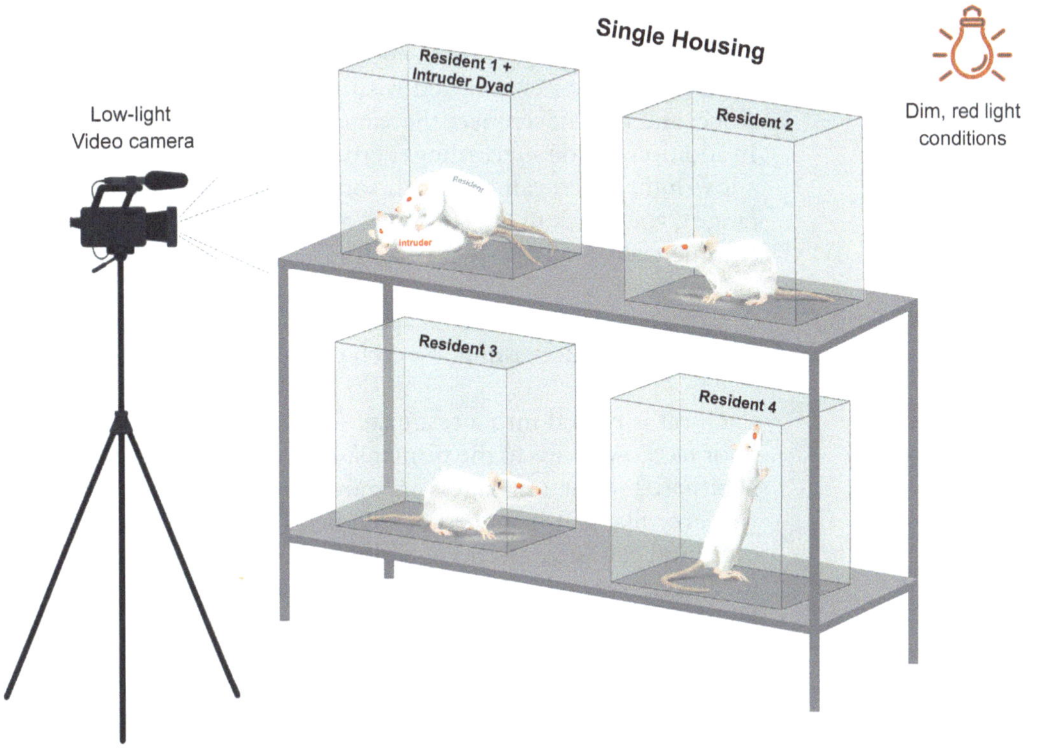

Fig. 2 A diagrammatic representation of the Resident housing/testing room. The Resident single housing home cages (right) are organized next to fellow Resident cages on the shelving to allow for visual, auditory, and olfactory detection and communication. The Intruder rat is introduced into the Resident's home-cage in situ (top left cage) and is video-recorded with a low-light video camera with the lights off for 6 min. Following the removal of the Intruder rat from the first Resident cage, the testing is repeated with the adjacent Resident rat in their home cage in situ. Note: if space does not permit for the video recording to be performed with the Resident home cages in situ. The Resident home-cage may be moved to a table that is located inside the Resident housing room to perform the video recordingImages modified from Sosa et al. [20] with permission from authors.

resulting from the introduction of an age, sex, and weight matched conspecific into their "home cage" are video-recorded with a low-light infrared camera and later quantified [9, 10]. It should be noted in our experiments that our focus is on the *Resident* rat's behavior, this is in contrast to the usual focus on the *Intruder* rat's behaviors during the resident–intruder interaction (see Fig. 3 and Table 1).

On each day, the *Resident* rat's home cage is moved to the testing table (see Fig. 2), and the rats homecage behaviors are video recorded in the dark for a period of 2 min. Immediately after this period, an uninjured, weight-matched, same-age, same-sex *Intruder* rat is placed into the *Resident* rat's home cage, and the interactions are recorded for a further period of 6 min [30–33]. The *Intruder* rat can be identified in the video records as prior to being introduced to the *Residents* cage, multiple stripes are applied to fur with an odor-free black marker pen. These marks readily distinguish the *Resident* rat from the *Intruder* rat during analysis of the video-recordings. *Resident* rat are never tested with the same *Intruder* more than twice, and they never meet the same *Intruder* on consecutive days. In addition to video-recordings of the social interactions, made using a low-light video camera, both sonic and ultrasonic (22–28 kHz range), vocalizations are recorded, and made audible by means of a commercially available "bat detector."

3.4 Sensory Testing

Sensory testing is conducted approximately 2 h after the modified Resident–Intruder testing, on each of the test days.

Tactile (Mechanical) Sensitivity Testing

Each rat is placed into a test cage (12 × 15 × 15 cm) with a mesh floor to allow access to the plantar surface of the hind paws. Rats are habituated to the test cage for a minimum of 20 min, after which a von Frey filament is applied perpendicular to the plantar surface of each hind paw in an ascending series (0.217 g–15 g). The cutoff time for a negative response is 5 s, and a positive response for the filament is if the hind paw is withdrawn at least seven out of ten trials (70%).

Thermal Sensitivity Testing

Each rat is placed on an 8 mm thick glass plate cooled to ~11 °C ± 1 °C by placing it on "freezer bricks." An open-topped Plexiglass cylinder

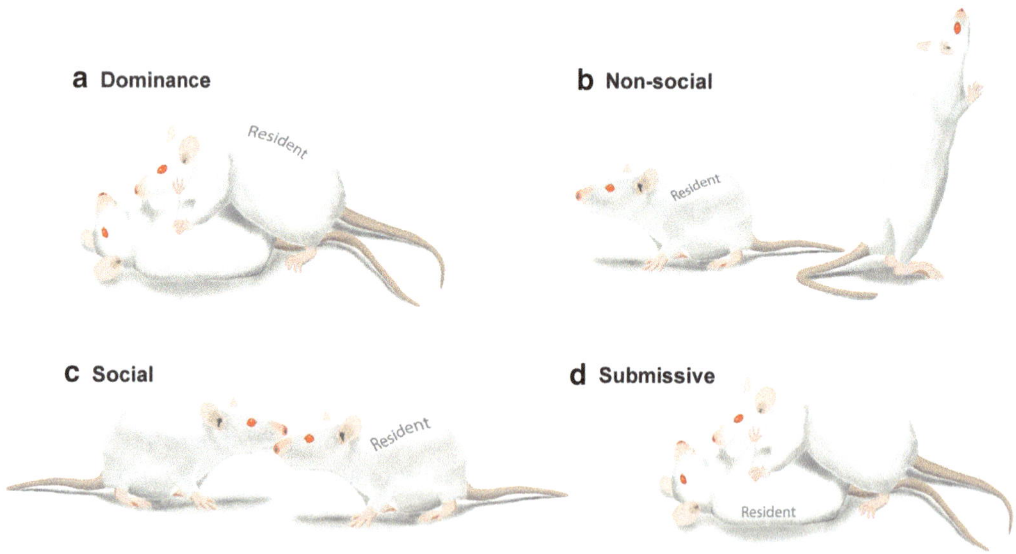

Fig. 3 Illustration of the mutually exclusive behaviors expressed by the **Resident** rat during the modified Resident–Intruder test (**a**) dominance behavior, Resident is "on top" of a submissive Intruder rat; (**b**) nonsocial behavior, Resident is exploring the cage, head is orientated away from Intruder; (**c**) social behavior, Resident is sniffing the orofacial region of the Intruder; (**d**) submissive behavior, Resident is supine underneath an Intruder rat

(15 cm diameter × 30 cm height) is placed on top of the cooled glass. The frequency of hind paw withdrawals over a 6 min test period are recorded in a low-light video camera and later scored.

Neuropathic Pain Model: Chronic Constriction Injury of the Sciatic Nerve

The protocol for nerve constriction is based on surgical procedures first described by Bennett and Xie [34]. Aseptic techniques are used for all procedures. All surfaces are sterilized prior to operation, and surgical instruments are sterilized between each procedure. Rats are initially anesthetized (5% isoflurane, in 100% oxygen) with a custom-made anesthetic induction box, which allows the rats to move freely. Once unconscious, the rats are positioned prone, and anesthesia is maintained with a nose cone (2%–2.5% isoflurane in 100% oxygen). The appropriate anesthetic depth is determined by the loss of the eyeblink and pedal reflexes. Rats are monitored for respiratory rate, heart rate, and mucosal coloring continuously throughout the procedure. The right hindlimb is placed in tension by gently taping the right hindpaw to the surgical table to maintain a 90° angle at the knee joint. The fur is shaved, and site of incision sterilized. The sciatic nerve is revealed by a blunt dissection through the biceps femoris in a plane parallel to the femur bone. The nerve is isolated by removing the surrounding connective tissue again using blunt dissection with a pair of curved forceps, four chromic gut ligatures (chromic gut 5-0 Ethicon, Johnson & Johnson Medical) are then tied loosely to surround the sciatic nerve, but not tight enough to impede the perineurial flow. Each ligature is spaced approximately 1 mm apart and tied using a square knot, just proximal to the nerves trifurcation.

Following ligation, the nerve is repositioned to its original location and the skin closed using sutures or surgical staples.

A topical mixture of antibiotic powder (Tricin®), quinine, and petroleum jelly is applied to the incision site. This combination prevents the rat from investigating and biting the surgical site and ensures rapid healing. Rats are then moved to a heated recovery cage where they are monitored continuously until ambulatory and eating, and drinking. Once recovered rats are returned to their home cage.

Procedure

1. House the *Resident* males individually in their home cage for a minimum one week prior to behavioral testing.
2. Only clean a maximum of half of the bedding from the *Resident's* home cage.
3. Test day: Leave *Resident* home cage in situ. If not possible position *Resident* home cage on a test table within the *Resident* housing/testing facility for video recording. Remove any environmental enrichment and water bottles.
4. Commence recording the behavior of the *Resident* using a low-light high definition video camera for at least 2 min

before the introduction of an *Intruder* rat to the *Resident's* home cage.

5. Bring in the *Intruder* rat by itself into the housing room. Mark the back of the *Intruder's* fur with a black permanent marker pen for identification purposes. Introduce the *Intruder* rat to the *Resident's* home cage.
6. Video record interactions for a minimum of 6 min.
7. On completion of the social interaction. Remove the *Intruder* rat and return to its home cage. Replace enrichment and water and return *Resident*'s home cage to original location.
8. Approximately 2 h after Resident–Intruder test. Rats are tested for sensory sensitivity (mechanical or thermal sensitivity testing).
9. Repeat Steps 3–8 for a minimum four exposures prior to nerve injury (*or any model of chronic pain*).
10. Surgical intervention: chronic constriction injury
11. Repeat Steps 3–8 a minimum of six tests post-CCI to observe individual differences that enable categorization into previously reported subpopulations.
    a. No Effect.
    b. Persistent Effect.
    c. Recovery.
12. Detailed analysis of social interactions.

3.5 Analysis: Behavioral Categories of Social Interactions

Video-recordings of the *Resident's* behaviors, during their 6 min interaction with the *Intruder* rat are coded into four main categories: dominant behaviors, social behaviors, nonsocial behaviors, and submissive behaviors (see Fig. 3). These categories are identical to those previously described by Depaulis and colleagues [35, 36] and which were based on the earlier descriptions of Grant and Mackintosh [37].

Dominant Behaviors—These behaviors are characterized by active offensive postures and behaviors. Standing above or leaning on the supine partner (i.e., *over-the-top* behavior), offensive sideways (*lateral attack*) posture, biting targeted at the neck or back of the partner, and rapid locomotion or hop/jumps directed forwards and away from the partner.

Social Behaviors—General investigation; involving locomotion towards, head-orienting to; crawling under or over any part of the partner's (*Intruder's*) body. Allogrooming; light nibbling or licking or any part of the partner'*s* (*Intruder's*) body or sniffing of the *Intruder* often focused around the ano-genital region.

Nonsocial Behaviors—cage exploration, self-grooming, or behaviors not directed towards the *Intruder*. This may include locomotion, rearing, sniffing, or digging oriented towards the walls or the floor of the test cage.

Submissive Behaviors—Defensive alerting or freezing, defensive sideways or supine posture to the approach/attack of the partner (often accompanied by 22–28 kHz vocalizations).

The cumulative time spent in each of the four behavioral categories, i.e., dominant behaviors, social behaviors, nonsocial behaviors, or submissive behaviors is calculated for each *Resident* rat. In the laboratory, we use a PowerLab 8/35 data acquisition device (ADInstruments: www.adinstruments.com/) connected to an analog input device with four buttons corresponding to each of the social interactions. The analog signal is converted to time using LabChart software (ADInstruments: www.adinstruments.com/). Similar devices and associated software is readily accessed and would allow identical analysis to be performed.

Baseline measurements are calculated by taking the mean expression levels of each behavior for the 3 test days prior to nerve injury (sciatic nerve CCI). All post-CCI behavioral measurements are compared to the mean pre-CCI behavioral scores within each individual rat (see Fig. 1).

Operational criteria to define behavioral effects:

- ***Persistent Effect***—an individual rat has at least 30% reduction in the duration of dominance behavior compared to their mean pre-CCI behaviors.

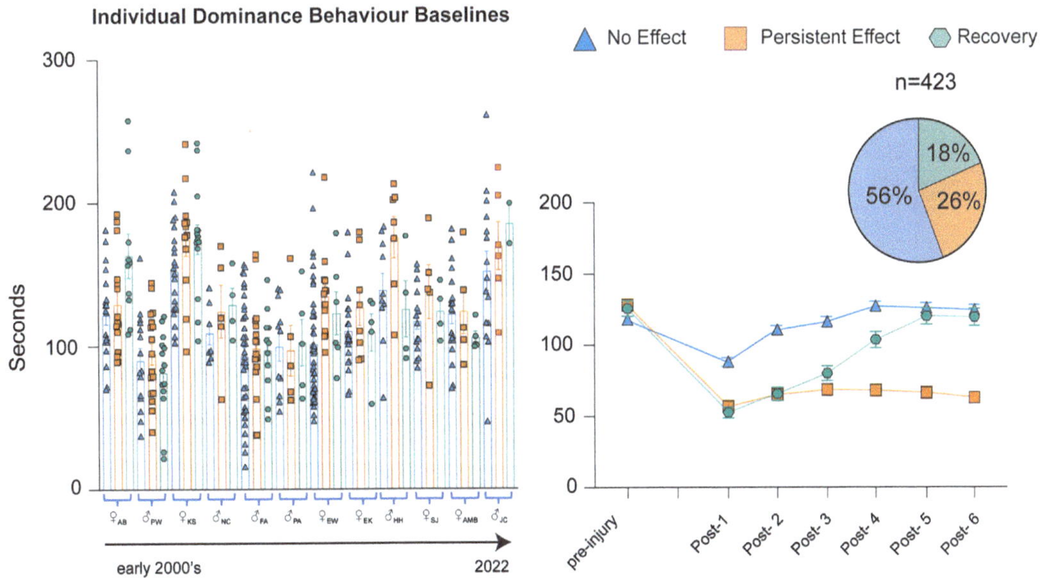

Fig. 4 Shows a bar graph of the individual dominance behavior profiles at baseline after rats were tested in the modified Resident–Intruder test by a male (♂$_{initials/experimenter}$) or female (♀$_{initials/experimenter}$) experimenters beginning these experiments in the early 2000's up until 2022. Rats were individually categorized into behavioral subgroups ***No Effect (237/423), Persistent Effect (110/423) and Recovery (76/423)*** rats according to the changes in dominance behaviors expressed in the modified Resident–Intruder test over a 6-day period post-CCI of the right sciatic nerve. The line graph depicts the mean change in dominance behaviors of all rats that underwent Resident–Intruder testing from early 2000's to 2022. The pie graph depicts the percentage distribution of all rats ($n = 423$) into their behavioral subgroups

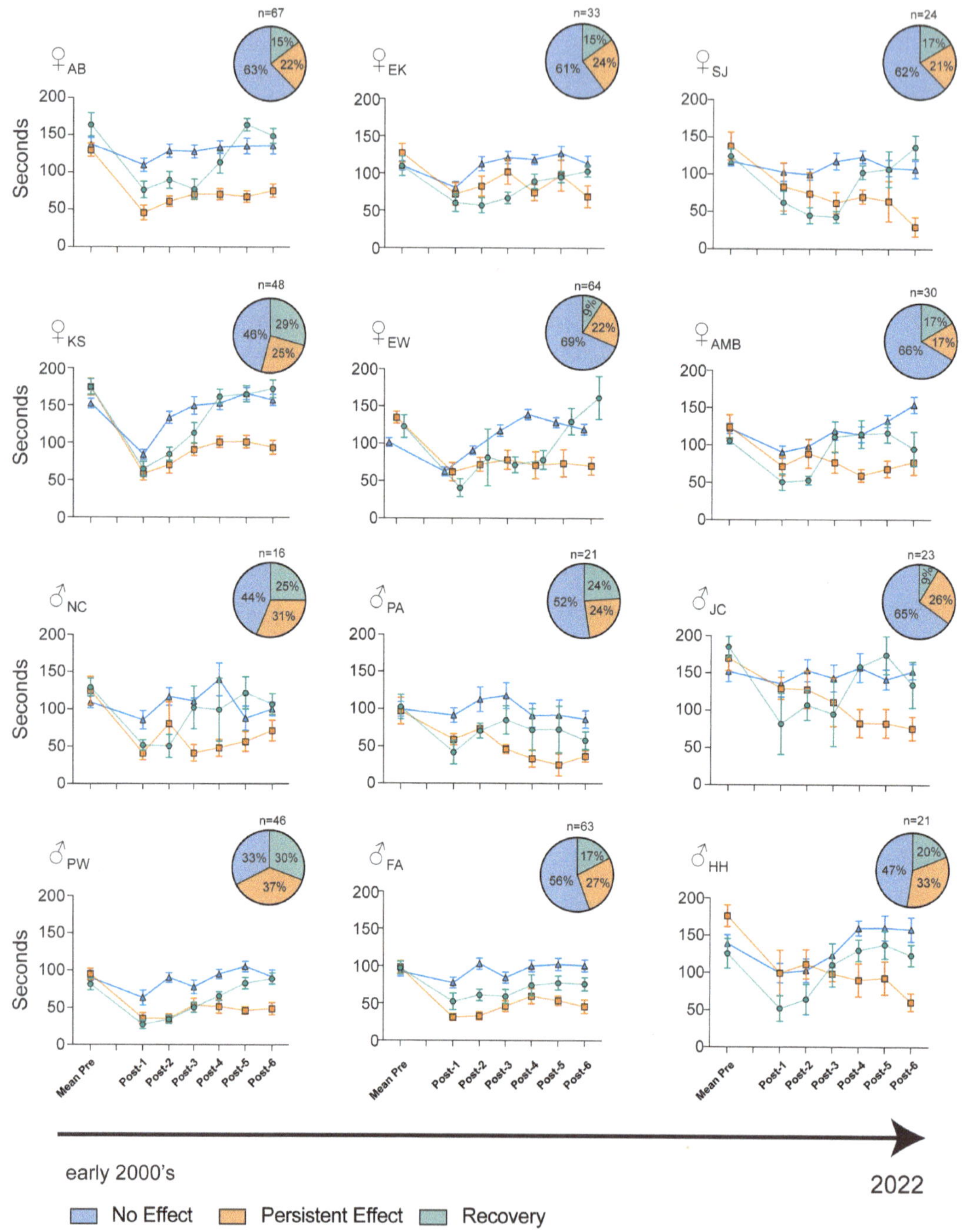

Fig. 5 The mean change in dominance behaviors of individually characterized rats following a CCI of the right sciatic nerve of each male (♂$_{initials/experimenter}$) or female (♀$_{initials/experimenter}$) experimenters beginning in the early 2000's up until 2022. The behavioral subgroups categorized by changes in dominance behavior are depicted as ***No Effect (blue), Persistent Effect (orange) and Recovery (green)***. The pie graphs show the percentage distribution into the behavioral subgroups revealed by each experimenterThe data from the individual experimenters have been previously published in the following journals. ♀$_{AB}$ [22–24, 38]; ♀$_{EK}$ [28]; ♀$_{SJ}$ (*unpublished*) ♀$_{KS}$ [20]; ♀$_{EW}$ [18, 19]; ♀$_{AMB}$ [22, 39]; ♂$_{NC}$ (*unpublished*) ♂$_{PA}$ [17]; ♂$_{JC}$ (*unpublished*) ♂$_{PW}$ [25, 27]; ♂$_{FA}$ [27]; ♂$_{HH}$ [25, 39, 40]

- ***No Effect***—No change between or less than 30% reduction in post-CCI dominance behavior when compared to their mean pre-CCI behaviors.
- ***Recovery***—*Resident* that had an initial decrease in dominance behaviors greater than 30% of the mean of their pre-CCI behaviors but returned to pre-CCI levels after post-CCI day 4.

Examples of stable changes in dominance behaviors over time and experimenter are shown in Figs. 4 and 5.

4 Conclusions

This chapter outlines steps to perform a modified Resident–Intruder social interactions test to determine individual differences in the response of rats to CCI of the right sciatic nerve. A striking feature of the behavioral changes is that they do not correlate with the usual features used to determine the presence of pain following nerve injury, namely, allodynia and hyperalgesia. The majority of studies cited in the chapter show that all CCI rats show significant mechanical allodynia and increased cold sensitivity; however, the degree of these sensory changes does not relate to the presence of changes in

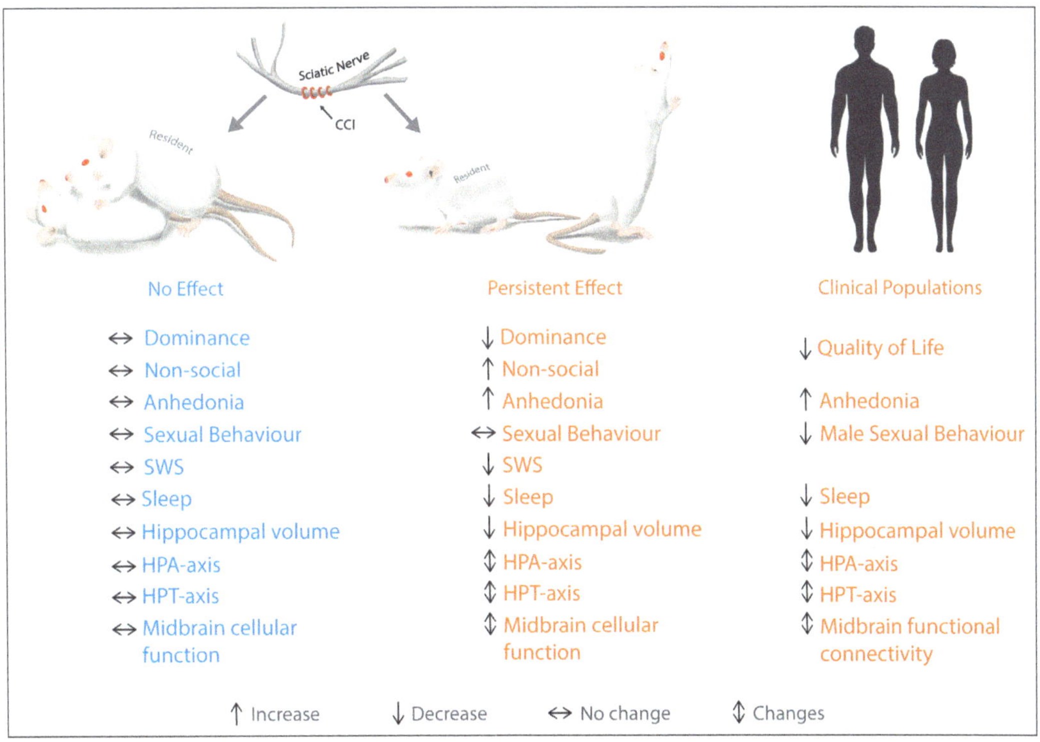

Fig. 6 Summary of changes observed between *No effect* and *Persistent effect* rats following a CCI of the sciatic nerve are shown in the top panel. A brief comparison of similarities observed in *Persistent effect* rats and the clinical population of chronic neuropathic pain individuals

resident–intruder social interactions. We have suggested that this behavioral measure might operate as a preclinical correlate of aspects of the disabilities described by human chronic pain patients. One of the advantages of the modified Resident–Intruder test that we have used is the remarkable stability of the test over time and with different operators. We have provided evidence that the expression of dominance behaviors are stable, and the frequency of nerve injury triggered behavioral responses differs little between male and female experimenters (Figs. 4 and 5). Further, not only are nerve injured rats separable into three subpopulations (i) *No Effect*, (ii) *Persistent Effect*, and (iii) *Recovery*, these groups defined by changes in social interactions also differ in a range of other behavioral, physiological, and neuroanatomical changes. These striking comorbidities are summarized in Fig. 6. The modified Resident–Intruder test provides a robust protocol that allows the identification of individual differences that are more akin to the clinical population. Further, this methodology provides a platform for the application of other powerful experimental techniques including electrophysiology, calcium imaging, cannulation studies, tract tracing, opto-, and chemo-genetic approaches, and pharmacological approaches, etc. to probe the important translational neurobiological questions that we have raised.

Conflict of Interest

The authors have no conflicts of interest to declare that are relevant to the content of this chapter.

References

1. Mogil JS (2021) Sources of individual differences in pain. Ann Rev Neurosci 44:1–25
2. Dueñas M, Ojeda B, Salazar A, Mico JA, Failde I (2016) A review of chronic pain impact on patients, their social environment and the health care system. J Pain Res 9: 457–467
3. Johansen A, Romundstad L, Nielsen CS, Schirmer H, Stubhaug A (2012) Persistent postsurgical pain in a general population: prevalence and predictors in the Tromsø study. Pain 153:1390–1396
4. Mikkelsen T, Werner MU, Lassen B, Kehlet H (2004) Pain and sensory dysfunction 6 to 12 months after inguinal herniotomy. Anesth Analg 99:146–151
5. Gregoire S, Michaud V, Chapuy E, Eschalier A, Ardid D (2012) Study of emotional and cognitive impairments in mononeuropathic rats: effect of duloxetine and gabapentin. Pain 153:1657–1663
6. Langford DJ, Crager SE, Shehzad Z, Smith SB, Sotocinal SG, Levenstadt JS, Chanda ML, Levitin DJ, Mogil JS (2006) Social modulation of pain as evidence for empathy in mice. Science 312:1967–1970
7. Martin LJ, Tuttle AH, Mogil JS (2014) The interaction between pain and social behavior in humans and rodents. Curr Top Behav Neurosci 20:233–250
8. Tansley SN, Tuttle AH, Wu N, Tohyama S, Dossett K, Gerstein L, Ham B, Austin JS, Sotocinal SG, Mogil JS (2019) Modulation of social behavior and dominance status by chronic pain in mice. Genes Brain Behav 18:e12514
9. Keay KA, Monassi CR, Levison DB, Bandler R (2004) Peripheral nerve injury evokes disabilities and sensory dysfunction in a subpopulation of rats: a closer model to human chronic neuropathic pain?. Neurosci Letts 361:188–191
10. Monassi CR, Bandler R, Keay KA (2003) A subpopulation of rats show social and sleep–waking

changes typical of chronic neuropathic pain following peripheral nerve injury. Eur J Neurosci 17:1907–1920
11. Koolhaas JM, Coppens CM, de Boer SF, Buwalda B, Meerlo P, Timmermans PJ (2013) The resident-intruder paradigm: a standardized test for aggression, violence and social stress. J Vis Exp 77:e4367
12. Harding VR, de C.W AC, Richardson PH, Nicholas MK, Jackson JL, Richardson IH, Pither CE (1994) The development of a battery of measures for assessing physical functioning of chronic pain patients. Pain 58:367–375
13. Menefee LA, Frank ED, Doghramji K, Picarello K, Park JJ, Jalali S, Perez-Schwartz L (2000) Self-reported sleep quality and quality of life for individuals with chronic pain conditions. Clin J Pain 16:290–297
14. Reuler JB, Girard DE, Nardone DA (1980) The chronic pain syndrome: misconceptions and management. Ann Intern Med 93:588–596
15. Sternbach RA (1974) Psychological aspects of pain and the selection of patients. Clin Neurosurg 21:323–333
16. Timmermans G, Sternbach RA (1974) Factors of human chronic pain: an analysis of personality and pain reaction variables. Science 184:806–808
17. Austin PJ, Beyer K, Bembrick AL, Keay KA (2010) Peripheral nerve injury differentially regulates dopaminergic pathways in the nucleus accumbens of rats with either 'pain alone' or 'pain and disability'. Neuroscience 171:329–343
18. Kilburn-Watt E, Banati RB, Keay KA (2010) Altered thyroid hormones and behavioural change in a sub-population of rats following chronic constriction injury. J Neuroendocrinol 22:960–970
19. Kilburn-Watt E, Banati RB, Keay KA (2014) Rats with altered behaviour following nerve injury show evidence of centrally altered thyroid regulation. Brain Res Bull 107:110–118
20. Sosa MK, Boorman DC, Keay KA (2022) Sciatic nerve injury rebalances the hypothalamic-pituitary-adrenal axis in rats with persistent changes to their social behaviours. J Neuroendocrinol 34:e13131
21. Austin PJ, Berglund AM, Siu S, Fiore NT, Gerke-Duncan MB, Ollerenshaw SL, Leigh SJ, Kunjan PA, Kang JW, Keay KA (2015b) Evidence for a distinct neuro-immune signature in rats that develop behavioural disability after nerve injury. J Neuroinflammation 12: 96
22. Austin PJ, Bembrick AL, Denyer GS, Keay KA (2015a) Injury-dependent and Disability-specific lumbar spinal gene regulation following sciatic nerveiInjury in the rat. PLOS One 10:e0124755
23. Mor D, Bembrick AL, Austin PJ, Wyllie PM, Creber NJ, Denyer GS, Keay KA (2010) Anatomically specific patterns of glial activation in the periaqueductal gray of the sub-population of rats showing pain and disability following chronic constriction injury of the sciatic nerve. Neuroscience 166:1167–1184
24. Mor D, Bembrick AL, Austin PJ, Keay KA (2011) Evidence for cellular injury in the midbrain of rats following chronic constriction injury of the sciatic nerve. J Chem Neuroanat 41:158–169
25. Mor D, Kang JW, Wyllie P, Thirunavukarasu V, Houlton H, Austin PJ, Keay KA (2015) Recruitment of dorsal midbrain catecholaminergic pathways in the recovery from nerve injury evoked disabilities. Mol Pain 11: 50
26. Mor D, Keay KA (2013) Differential regulation of glucocorticoid receptor expression in distinct columns of periaqueductal grey in rats with behavioural disability following nerve injury. Cell Mol Neurobiol 33:953–963
27. Keay KA, Argueta MA, Zafir DN, Wyllie PM, Michael GJ, Boorman DC (2021) Evidence that increased cholecystokinin (CCK) in the periaqueductal gray (PAG) facilitates changes in Resident-Intruder social interactions triggered by peripheral nerve injury. J Neurochem 158:1151–1171
28. Kalman E, Keay KA (2014) Different patterns of morphological changes in the hippocampus and dentate gyrus accompany the differential expression of disability following nerve injury. J Anat 225:591–603
29. Hakim JD, Chami J, Keay KA (2020) μ-Opioid and dopamine-D2 receptor expression in the nucleus accumbens of male Sprague-Dawley rats whose sucrose consumption, but not preference, decreases after nerve injury. Behav Brain Res 381:112416
30. Blanchard DC, Blanchard RJ (1990) Behavioral correlates of chronic dominance-subordination relationships of male rats in a seminatural situation. Neurosci Biobehav Rev 14:455–462
31. Blanchard RJ, Blanchard DC (1977) Aggressive behavior in the rat. Behav Biol 21:197–224
32. Blanchard RJ, Blanchard DC, Takahashi LK (1978) Pain and aggression in the rat. Behav Biol 23:291–305
33. Blanchard RJ, Fukunaga K, Blanchard DC, Kelley MJ (1975) Conspecific aggression in the laboratory rat. J Comp Physiol Psychol 89:1204–1209
34. Bennett GJ, Xie YK (1988) A peripheral mononeuropathy in rat that produces disorders of pain sensation like those seen in man. Pain 33:87–107
35. Depaulis A, Keay KA, Bandler R (1992) Longitudinal neuronal organization of defensive

reactions in the midbrain periaqueductal gray region of the rat. Exp Brain Res 90:307–318
36. Depaulis A, Keay KA, Bandler R (1994) Quiescence and hyporeactivity evoked by activation of cell bodies in the ventrolateral midbrain periaqueductal gray of the rat. Exp Brain Res 99:75–83
37. Grant EC, Mackintosh JH (1963) A comparison of the social postures of some common laboratory rodents. Behaviour 21:246–259
38. Bembrick AL, Boorman DC, Keay KA (2020) Disability-specific genes GRIN1, GRIN2 and CNR1 show injury-dependent protein expression in the lumbar spinal cord of CCI rats. Neurosci Letts 728:134982
39. Kang JWM, Keay KA, Mor D (2017) Resolving the contributions of anaesthesia, surgery, and nerve injury on brain derived neurotrophic factor expression in the medial prefrontal cortex of male rats in the CCI model of neuropathic pain. J Neurosci Res 95:2376–2390
40. Castorina A, Vogiatzis M, Kang JWM, Keay KA (2019) PACAP and VIP expression in the periaqueductal grey of the rat following sciatic nerve constriction injury. Neuropeptides 74:60–69

Chapter 11

Ultrasonic Vocalizations in the Context of Rodent Chronic Pain Research: A Protocol

Camila Pasquini de Souza, Vanessa B. P. Lejeune, Roberto Andreatini, Rainer K. W. Schwarting and Juliana Geremias Chichorro

Abstract

Pain is a multidimensional experience that involves sensory, affective, and motivational components. However, pain evaluation in the preclinical setting is mostly based on sensory tests that fail at capturing the affective pain component. Evaluation of the emotional component of pain in rats has been proposed as a way to optimize potential translational value of nonclinical studies. In this regard, the analysis of ultrasonic vocalization (USV) is considered a well-established measure of positive and negative emotional states in rats. It is well characterized that laboratory rats emit 50-kHz USV in appetitive situations, such as play behavior in juvenile rats, mating or in response to psychostimulants, whereas 22-kHz USVs may index a state of negative activation, including stressful and painful situations. Such aversive and appetitive USV seem to serve distinct communicative functions and depend on distinct brain mechanisms. There is cumulating evidence that USV correlates with acute pain in a variety of models, but data on chronic pain models is controversial. Such inconsistencies probably arise from many parameters that can affect USV recording and interpretation, such as fear, stress, novelty, rodent strain, and sex, among others. The appropriate register and interpretation of USV in rats subjected to chronic pain models may represent a new approach for understanding the brain circuitry underlying chronic pain states, and may provide a useful tool to predict the effectiveness of pharmacological therapies. Thus, the aim of this chapter is to provide: (1) USV basic principles; (2) current evidence of USV application in pain research; (3) a detailed methods description; and (4) concluding remarks that highlight the utility and limitations of this approach.

Keywords Nociception, Acute pain, Chronic pain, Affective pain component, USV, 50-kHz calls, 22-kHz calls

1 Introduction

The idea to use an ultrasound detector was originally described by Noyes and Pierce in 1938 to record ultrasound calls emitted by crickets in bioacoustics studies. The development of this technique advanced with the study of bats' echolocation in 1958. This method converts all detectable high-frequency signals to the audible range,

Isaura Tavares and Hugo Leite-Almeida (eds.), *Translational Research Methods for Neuropathic Pain*, Neuromethods, vol. 235, https://doi.org/10.1007/978-1-0716-5312-8_11,

with a wide range of the sound spectrum [1]. In rats and some other mammals, USV were first reported by Anderson in 1954 [2].

The first-generation detectors were important to define modern ultrasonic bioacoustics. They were widely used in the animals' research in the 1970 and 1980 decades and formed the base of the actual ultrasonic bioacoustics research [3]. The digital era turned the process more reliable and accessible and boosted its application in studies with rodents, bats, birds, and insects.

In rodents, USV emission occurs in several situations, such as mother–infant interactions, where they serve a protective function of maternal care, adult social interactions, copulation, agonistic interactions, reproductive behavior, external threat, and emotional expressions [4–6]. Juvenile and adult rats, e.g., emit two types of USV: (1) aversive 22-kHz USV characterized by relatively long call durations (~0.5–2.5 s) and low peak frequencies (~22–28 kHz), and (2) appetitive 50-kHz USV, which are short (~10–70 ms) and have high peak frequencies (~50–90 kHz). The 22-kHz calls are often emitted in bouts, whereas the emission of appetitive 50-kHz USV is characterized by a substantial temporal and structural complexity. Rats emit 22-kHz calls in stressful and/or painful situations to express a negative state and/or to alert conspecifics of a potential danger. On the other hand, 50-kHz calls emission are related to appetitive stimuli and social interaction. At least two main categories of appetitive calls have been described: [1] flat, which may have a social coordinating function and [2] frequency modulated (FM), which include step, mixed, and trill calls, that are mainly related to rewarding and hedonic situations [7, 8]. This section summarizes the data from initial USV studies in rat pain models and discusses the current literature.

The analysis of pain-like behaviors is fundamental for the evaluation of mechanisms and validation of new therapeutic targets. Studies focusing on audible and ultrasonic vocalizations have been performed and validated over the past decades, making it a quantitative and convenient method for measuring pain-like behaviors [9]. Some studies conducted in the 1980s [10–12] showed that rats submitted to painful or harmful stimuli emitted audible sounds referred to as pain vocalizations. In addition, they suggested that rats emit USV before and after receiving electrical stimuli. Interestingly, the authors demonstrated that animals that received mild painful stimuli did not express audible vocalizations, but USV, which occurred as single pulses or in short bouts [6].

The relation between nociception and USV was highlighted in the study of Ardid and collaborators [13]. The authors conducted experiments using USV in pain models with three different approaches: analysis of vocalizations after short-term painful stimuli, evaluation of the relationship between stimulus intensity and vocal response, and study of peripheral fibers responsible for the vocal response. Audible vocalizations were considered screams or cries that corresponded to defensive behaviors, but in addition to that, aversive USV calls (i.e., 22-kHz calls) were detected during painful stimuli. The authors demonstrated that the vocal, audible and ultrasonic components were related to activation of peripheral Aδ and C fibers, suggesting a relationship between USV and the sensory pain component [13].

The formalin test is widely used for the screening of drugs with analgesic potential [14]. Formalin injection into the hind paw or upper lip typically produces a biphasic pain response, separated by a brief quiescent period. The first phase is thought to be mediated by direct activation of nociceptors, and the second by the development of central sensitization and local inflammation [15]. Thus, the first phase is susceptible to drugs that have the capability of silencing nociceptors, such as local anesthetics and opioids, while the second phase is amenable to reduction by anti-inflammatory agents [14, 16]. Using the paw formalin test, Oliveira and Barros [17] showed an increase in the emission of aversive calls mostly in the quiescent period of the formalin test, indicating that that USV analysis may complement pain-like behavior assessment during this test. Morphine pretreatment caused a significant reduction of USV emission in a naloxone-reversible manner, suggesting that USV emission would represent an additional nociceptive measure.

Such increased emission of 22-kHz calls has been demonstrated in different acute pain models. Lim et al. [18] showed that animals subjected to plantar incision, a postoperative pain model, emitted aversive calls, which were significantly reduced by potential analgesic substances (i.e., *H. procumbens* extracts or naproxen). Additionally, Jourdan and collaborators [19] demonstrated that rats subject to electrical pulses applied to the tail also emitted aversive calls. In this pain model, aspirin had no effect on the vocal response, but morphine reduced 22-kHz calls emission, an effect that was reverted by naloxone. Moreover, Martino and collaborators [20] showed that sensitization of the trigeminal afferents by intracerebroventricular injection of lipopolysaccharide increased USV induced by an air puff focused on the head and this was significantly reduced by morphine, ketorolac, and antimigraine agents. Likewise, Bobade and collaborators [21] showed that intra-arterial administration of bradykinin, which was used to mimic acute pain related to a migraine attack, induced aversive USV emission, which was reduced with the antimigraine agent sumatriptan. Another example is the study of Han and colleagues [9], who induced acute arthritis in rats by injecting a carrageenan solution into the joint cavity, and detected increased aversive call emission after noxious mechanical stimulation of the joint. Altogether, these data indicate that acute painful stimulation is related to emission of aversive 22-kHz calls, which can be reduced by analgesic drugs.

As most of the studies have used morphine as a positive control, it is important to mention that in the dose ranges used, morphine caused a reduction in aversive calls [12, 22], but failed to increase or alter the number of appetitive calls (i.e., 50-kHz) [23, 24]. Studies conducted by Miczek et al. [6, 22] and Vivian et al. [25] showed that morphine changed both USV induced by nociceptive and non-nociceptive stimuli, suggesting that opioids not only interfere with the sensory component of pain, but also may reduce the aversiveness of the situation. Few studies have explored this aspect, but it is interesting to mention that the opioid peptides beta-endorphin, dynorphin, and encephalin were able to suppress aversive USV, but audible squeaks were not changed in studies using animal models of electric foot shock [6, 11].

The initial idea that aversive 22-kHz calls would reflect nociception itself, has been replaced by the idea that it rather represents behavioral and emotional consequences of pain. In this regard, Bardin et al. [26] assessed the effects of the antidepressants milnacipran and duloxetine and the anticonvulsant pregabalin in stress-induced ultrasonic vocalization (i.e., 22 kHz calls following presentation of a conditioned stimulus previously associated with foot-shocks) and acute/inflammatory nociception after intraplantar formalin injection. All three drugs reduced nociceptive paw elevation and licking induced by formalin, but also reduced stress-induced USV, with different potencies. Thus, it is important to keep in mind that aversive calls may be elicited by nociceptive stimuli, such as stress, and may also be inhibited by drugs that interfere with the emotional pain component.

In this regard, Miczek et al. [6] demonstrated that USV expresses the affective component of pain after a nociceptive stimulus. According to the authors, nociceptive stimulation would not be responsible for the emission of USV but rather its behavioral and emotional developments. For instance, animals that received benzodiazepine treatment in doses that did not alter motor activity presented a reflex response to the nociceptive stimulation in addition to a reduction in USV, suggesting an affective response to the painful stimuli [6]. This idea was recently discussed by Brudzynski [4], who emphasized that 22-kHz calls emission does not signal nociception itself. Even though emission of 22-kHz calls increases during nociceptive stimulation and is amenable to analgesic drugs, it is more likely that it express parameters secondary to the pain experience but not nociception itself [4].

Studies from our group have focused on the analysis of USV to address the emotional component of pain. In the study of Barroso et al. [24], we used the orofacial formalin test to investigate whether acute nociceptive stimulation was related to changes in emission of aversive and appetitive calls and how different social contexts affected nociceptive behavior (i.e., increased facial grooming) and USV emission. The main findings were: (1) orofacial formalin injection induced aversive calls in addition to nociceptive behavior in rats; (2) both responses were significantly reduced by morphine pretreatment; (3) the context altered pain-related responses, including USV emission. We observed no effect on nociceptive behavior or calls emitted by the formalin-injected rats after exposure to a conspecific cage-mate, but the observer rat showed emotional contagion of pain-like behavior (i.e., increased spontaneous facial grooming). Interestingly, observers exposed to cage-mates emitted appetitive calls, which were markedly reduced when they were exposed to formalin-injected rats. In contrast, when formalin-injected rats were exposed to non-cage-mates, nociceptive behavior decreased but call emission was not affected. These results suggest that the social context may modulate pain-like behavior and that USV analysis may represent an additional measure for the evaluation of the emotional aspect of pain and its modulation.

We have further characterized USV emission after formalin injection into the upper lip showing that one day after formalin injection, rats presented facial hyperalgesia, but increased emission of 22-kHz calls seen briefly after the injection was no longer detected. A similar observation was made 3 days after formalin injection. However, a significant change in the emission of 50-kHz calls was detected over time, a finding that will be discussed in the next section [23]. Taken together, these findings suggest that emission of 22-kHz calls may serve as alarm calls to warn conspecifics of a potential danger or aversive situation, but as the pain persists, visible nociceptive behavior and acoustic affective expression changes. In this sense, USV analysis may represent a useful tool to detect emotional changes related to pain over time.

Accumulating evidence suggests that acute nociception generates an increase in 22-kHz calls by rats and that analgesic treatment is capable of reducing aversive USV emission [17, 18, 20]. However, the advance of knowledge in the field suggests caution, since emission of 22-kHz calls may not be interpreted as a direct nociception measure, but as a signal of a negative affective state related to pain. Careful recording, analysis and interpretation of USV in acute pain models may provide valuable and additional data to basic pain research studies, as outlined in the following.

1.1 USV Analysis in Experimental Models of Chronic Pain

The establishment of USV as a measure of the emotional state of rats aroused interest in evaluating this parameter in rat models of chronic pain. The relevance of this approach lies in the fact that pain is a complex experience, which involves sensory, affective, and motivational components [27, 28]. While the last decades of preclinical studies brought significant advances in the comprehension of the sensory component of pain, the lack of assessment of the emotional pain component constituted one obstacle in the translation of efficacy from preclinical to clinical studies [27]. Thus, USV measurement may represent a tool for the assessment of the emotional pain component and the efficacy of new therapies for chronic pain with a high likelihood of translational relevance.

However, the utility of USV in models of persistent and/or chronic pain is still a matter of controversy. Induction of arthritis in rats is one of the most frequently used models to show changes in USV emission related to persistent pain [9, 29–35]. In these studies, an increase of 22-kHz USV has been demonstrated mainly in two different contexts: social interaction or after noxious stimulation of the inflamed limb or paw. For instance, it has been shown that when arthritic rats are confronted with naïve rats, they emit a high number of 22-kHz calls compared to intact rats. This increase of aversive calls emission by arthritic rats during social interaction was significantly reduced by several analgesic drugs, including morphine, ketoprofen, and aspirin, suggesting that USV responses may represent a pain-like behavior [29–31].

Likewise, noxious stimulation of the inflamed limb after induction of experimental arthritis by kaolin-carrageenan injection into the knee also resulted in increased 22-kHz call emission. In this model, a series of studies demonstrated that drugs, which interfere with the activity of the central nucleus of the amygdala (CeA) reduce the emission of aversive USV [32–35]. It is widely accepted that the amygdala plays a key role in emotional-affective aspects of pain. The CeA is the main recipient of nociceptive information through the spinoparabrachio-amygdaloid pathway, and it represents the main output of amygdala-processed nociceptive information [32]. Altogether, these studies indicate that USV analysis may be useful in the study of the emotional pain component and in the investigation of therapeutic strategies that change activity in the amygdala.

On the other hand, few studies have assessed USV in other models of chronic pain and its utility is still a matter of debate. Jordan and collaborators, e.g., did not detect changes in USV emission in rats subjected to persistent and chronic pain models [5]. The authors first assessed spontaneous USV emission of healthy rats compared to rats with persistent inflammatory pain induced by carrageenan or Freund's adjuvant (CFA model) or diabetic neuropathic pain induced by streptozotocin, but no difference was found between groups in all three models. In the CFA model, the authors also evaluated the influence of social interaction in USV emission, again with no effect. Finally, the effect of aspirin, which was previously shown to inhibit USV emission [31] was tested, but the treatment did not change USV emission by arthritic rats. These results led the authors to categorically affirm that "ultrasonic vocalization does not allow chronic pain to be evaluated in rats" [5]. However, more recent studies, including from our group, have provided consistent data supporting the idea that USV analysis may contribute to reveal the emotional aspect of chronic pain [23, 36].

The characterization of USV in the neuropathic pain context is very limited. In the model of chronic constriction injury (CCI) of the sciatic nerve, Burgdorf and collaborators [36] assessed whether 22- and 50-kHz calls could be used as measure of nonevoked nociception. Rats subject to hedonic stimulation (i.e., experimenter-induced rough-and-tumble play) 2–3 weeks after CCI emitted a lower number of 50-kHz calls, in the interval between stimulations, compared to sham animals, and both groups displayed a very low number of 22-kHz calls. When USV emission of rats was monitored in 15 min bins from 3 to 5 h, and also 2–3 weeks after surgery, CCI rats showed a marked reduction of 50-kHz calls, and an increase in the emission of 22-kHz calls compared to the sham group. In line with these observations, a recent publication from our group indicated that rats subjected to CCI of the infraorbital nerve display a reduced number of 50-kHz calls compared to the sham group 2 weeks after the surgery during a 10 min recording time. During this period, no emission of 22-kHz calls was detected [37]. It is possible that measuring UVS emission for longer periods, as performed by Burgdorf and colleagues [36], would permit a more reliable detection of the changes in USVs induced, including changes in 22-kHz calls. In spite of that, these findings suggest that rat USVs may be useful to capture the affective component of chronic neuropathic pain.

Considering that 22-kHz USV are alarm calls directed to the group to signal danger or potential danger, what would their significance in chronic pain states be? Current evidence suggests that 22-kHz calls are not a direct measure, of nociception, but they may reflect emotional components related to persistent pain, such as anxiety and dysphoria. This observation arose mainly from studies that evaluated USV emission during withdrawal from drugs of abuse. The dysphoria period was associated with high rates of aversive calls, which were interpreted to reflect anxiety and frustration associated with drug withdrawal [38–40]. Finally, it is important to mention that two subtypes of 22-kHz calls have been reported: short (10–500 ms) and long (300–3000 ms). The former might reflect a dysphoric state without the presence of an external threat, whereas the latter might signal the presence of a predator or other dangers [4, 41]. Thus, detection of 22-kHz calls during a chronic pain state may contribute in the evaluation of the emotional pain component.

Another aspect that deserves consideration is that emission of 22-kHz emission during a chronic pain state may be triggered by noxious stimulation, again reflecting pain aversiveness. This was recently demonstrated by Presto and collaborators [42] in two well-established models of chronic pain, monoarthritis induced by kaolin plus carrageenan and L5 spinal nerve ligation. The authors showed that both, audible and aversive USV were elicited by a brief noxious stimulus applied to the affected limb. Moreover, they demonstrated that female rats exhibited increased audible and ultrasonic components of vocalizations compared to males in both pain models. Interestingly, analysis of vocalizations represented the strongest indicator of sexual difference in both pain models, as such differences were not detected for mechanical hyperalgesia and anxiety-like behavior. Considering the burgeoning interest in sex dichotomy in pain, studies that assess USV in male and female rats are clearly warranted.

Although our group has not detected 22-kHz calls in conditions of persistent pain in rats [23], we and others have demonstrated that acute nociception induces a remarkable increase in the emission of 22-kHz "long" calls, which may serve as an alarm call to signal the pain aversiveness [17, 23, 24]. For instance, formalin injection into the rat's hind paw or upper lip led to aversive call emission briefly after the injection, which occurred concomitantly to the expression of the nociceptive behaviors related to each site of injection, for instance, paw licking and flinching and facial grooming, respectively [17, 23, 24]. Interestingly, less than an hour after formalin injection, rats stopped showing overt pain-like behaviors, but it was possible to assess hyperalgesia in the injected area for several days, due to peripheral and central sensitization caused by formalin. Based on these observations, we analyzed USV emission in later periods after formalin, and found that on day 1 after injection, rats presented facial heat hyperalgesia, but there was no difference in the emission of USV compared to the control group. On day 3 after formalin, facial heat hyperalgesia was still evident but now accompanied by a decrease in the emission of 50-kHz calls compared to the control group. To reinforce the idea that persistent pain is associated with reduction in the emission of appetitive calls, we also assessed USV emission in a model of postoperative pain induced by intraoral incision. Facial heat hyperalgesia was present on days 1 and 3 after surgery, which were the same time points for detection of reduced emission of 50-kHz calls by operated rats compared to sham, whereas 22-kHz calls were not detected in any of the groups and assessment periods in both models. Moreover, the decrease of 50-kHz calls emission coincided with the appearance of anxiety-like behavior in both models [23]. Likewise, CCI of the infraorbital nerve results in reduction of 50-kHz calls emission in the same period that animals present facial mechanical allodynia and anxiety-like behavior [43, 44]. Therefore, a question that arose was whether the reduction in USV would be related to pain- or anxiety-like behavior. In this regard, our group demonstrated that the reduction of 50-kHz calls emission after CCI of the infraorbital nerve was restored by drugs with analgesic (i.e., carbamazepine and lidocaine) but not anxiolytic (i.e., midazolam) properties [45]. These results reinforce the idea that reduction in appetitive calls emission during chronic pain states is a valuable tool to assess the emotional pain component.

It is important to highlight that the flat subtype of 50-kHz calls was the only one reduced in animals with persistent inflammatory or neuropathic pain [23, 37, 45]. While FM calls are mainly related to rewarding and hedonic situations and may reflect a positive affective state, flat calls are considered to have a social-coordinating function [4, 7]. Thus, the full implication of this finding remains to be clarified, but it is possible that the chronic pain state causes impairment in social interaction, e.g., social withdrawal. On the other hand, Burgdorf and colleagues showed that FM, but not flat calls, were reduced in rats subjected to CCI of the sciatic nerve [36]. Again, it is important to emphasize that capture and interpretation of USV may be influenced by several parameters, and one crucial difference between our studies and that of Burgdorf is that we evaluated spontaneous USV emission, while they measure USV triggered by a hedonic stimulus. Altogether, these finding suggest that the chronic pain state may be associated with the reduction of both hedonic and nonhedonic 50-kHz USV and reinforce the idea that USV analysis can be used to capture pain-induced changes in the positive emotional arousal expressed by 50 kHz USV emission.

The neural circuits of aversive and appetitive USV, as well as its pharmacological modulation, have been well-characterized. For instance, emission of appetitive calls, to both social and nonsocial contexts, involves activation of ascending mesolimbic system, where it has been suggested to occur an increase dopamine in the nucleus accumbens [4]. The chronic pain state is associated with a decrease in the dopaminergic tone in the nucleus accumbens, while pain relief may have the opposite effect [27]. Thus, altered function of the mesolimbic system during persistent pain may have a role in pain chronification, but may also account for the reduction in the emission of appetitive calls [23], which may represent a change in the emotional perception of pain. Table 1 summarizes the main findings related to USV analysis in pain studies.

In conclusion, we believe that inconsistent data regarding USV emission in chronic pain models may be, at least in part, due to the lack of standardized and reproducible techniques to induce and measure these rather complex vocalizations (i.e., number, frequency, amplitude, duration). It is also noteworthy that the detection equipment and methods of USV analysis have been rapidly and substantially evolved in the past years, which allow deeper and more detailed data collection and interpretation. As will be pointed out in the next section, USV emission is influenced by the test environment, social context and may differ depending on the rats' strain and sex. Regarding pain studies, emission may also differ according to stimulus application, response measurement, and pain type (i.e., acute or chronic). Thus, using standardized methods and minimizing variability are key factors for the use of USV as a complimentary measure of pain. A detailed description of the method used by our group can be found in the next section.

Table 1
Summary of the main findings related to the analysis of USV in pain models

Acute pain models	USVs characteristics	References
Thermal or electrical stimuli applied to the tail	22-kHz calls emission, related to activation of nociceptors, reduced by morphine	[13, 22]
Paw formalin test	Increased emission of 22-kHz calls, attenuated by morphine	[17, 26]
Orofacial formalin test	Increased emission of 22-kHz calls, attenuated by morphine. 50-kHz calls were reduced 3 days after formalin injection	[23, 24]
Postoperative pain model (plantar incision)	Increased emission of aversive calls (22-kHz calls)	[18]
Postoperative pain (intraoral incision)	22-kHz calls were not detected, but 50-kHz calls were reduced 3 days after surgery	[23]
Sensitization of trigeminal afferents (by LPS or bradykinin)	Increased 22-kHz calls emission, reduced by anti-migraine agents	[20, 21]
Acute arthritis by carrageenan	Increased 22-kHz calls emission	[9]
Chronic pain models		
Arthritis induced by CFA	Increased emission of 22-kHz calls, reduced by morphine and anti-inflammatory agents	[29–31]
Arthritis induced by kaolin-carrageenan	Increased emission of 22-kHz calls 6 h after the induction, reduced by drugs that interfere with the activity of the central nucleus of the amygdala	[33–35]
Neuropathic pain (CCI of the sciatic nerve)	Decrease in 50-kHz calls emission; increased emission of 22-kHz calls	[36]
Neuropathic pain (L5 spinal nerve ligation)	Increased emission of 22-kHz calls elicited by a brief noxious stimulus applied to the affected limb. Greater effect in female compared to male rats	[42]
Neuropathic pain (CCI of the infraorbital nerve	Decrease in 50-kHz calls emission; which was restored by carbamazepine and lidocaine. 22-kHz calls emission was not detected	[37, 45]
Main findings		
- Acute painful stimulation is related to emission of aversive 22-kHz calls, which can be reduced by analgesic drugs. - Emission of 22-kHz calls may not be interpreted as a direct pain measure, but as a signal of a negative affective state related to pain. - USV analysis may represent a useful tool to detect emotional changes related to pain over time. - Neuropathic pain seems to be related to a decrease in the 50 kHz calls emission. - USV analysis can be used to capture pain-induced changes in the positive emotional arousal expressed by 50 kHz calls emission.		

2 Methods

2.1 Animals: Strain, Housing, and Handling

Rat strain and sex: Choose the appropriate rat strain for the model. Studies with USV in pain models have employed Wistar [24], Sprague–Dawley [31, 46], and Lewis [29] strains. It is recommended to use both male and female rats due to demonstration of sexual dimorphism in the rat larynx, as well as in all three major categories of USVs, which are differentially affected by experimental conditions (for review see [47]).

Housing conditions: The housing conditions should be standardized. Rats should be housed together for at least 1 week before testing, in groups of four, under controlled conditions of light (12:12 h light: dark cycle) and temperature (22°C ± 1°C), and with chow and water ad libitum.

Handling: Evaluation of USV is very sensitive to the animal's anxiety and stress levels. Therefore, all efforts to minimize stress should be made, including habituation to the experimenter. A handling protocol of 5 min per animal for 3 consecutive days should be applied before any USV testing. If possible, sessions of handling should be done in the test room. If not, animals must be habituated to the testing room for an hour to avoid additional stress during the experiments.

2.2 Test Room, Apparatus, and Equipment

General arrangement: Different experimental room arrangements can be used, but the general idea is that the subject should be tested in a quiet, sound attenuated room with dim illumination. As an example, a test room with USV microphone and video camera to record the call emissions and behavior can be used and a recording room where the experimenter records the experiment. USV data can be stored for posterior analysis through Avisoft SASLabPro, DeepSqueak, or similar software (Fig. 1).

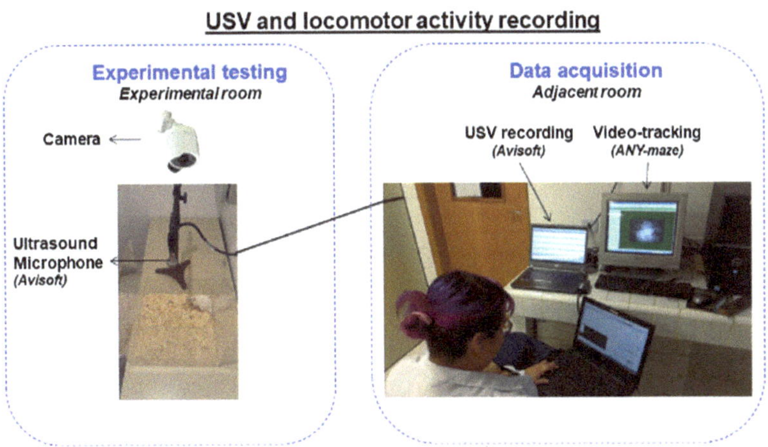

Fig. 1 Example of experimental and data acquisition room arrangement

Lux Meter: The testing room must be not aversive; therefore, it should be lit by a dim light placed centrally above the experimental apparatus (recommended lux values between 10 and 40; e.g., [23, 24]). A good strategy is to control light with a lux meter to maintain adequate lighting inside the room and the apparatus, between subjects and across experiments.

Apparatus: Different kinds of apparatus can be used, depending on which type of pain model is being used to evoke vocalizations (e.g., spontaneous pain or pain induced by pressure or thermal stimuli). It is important to take into consideration the acoustics of high frequency vocalizations and be aware that the shape and size of the apparatus might influence the capturing of vocalizations. As an example, one can use an activity box (acrylic, 40 × 40 × 40 cm) to observe and evaluate the rat throughout the experiment for both ultrasonic vocalization and pain behaviors following a painful stimulus, see [24] (Fig. 2A).

Microphone position: Ultrasonic signals are shorter in wavelength than audible sounds, so correct microphone placement for recording and soundproofing is paramount for optimal ultrasonic recording and to reduce reflections and echoes. Large acoustically reflective surfaces surrounding the microphone may induce reverberation of the sound, and one way to prevent this problem is to place the animal cage in a soundproof chamber that absorbs ultrasound and prevents the signals from bouncing back to the microphone. It is recommended to run background noise measurements prior to recording to determine additional possible sound sources that can also be captured by the microphone, hence interfering with the signal/noise ratio. Overhead fluorescent lighting and adjacent computer hardware often generate frequencies in the ultrasonic band and can interfere with USV recording. Removing noise sources or repositioning equipment can greatly improve measurements. Special attention should be given in behavioral experiments in which more than one animal is evaluated in the same room to avoid cross talk between recording channels. This can be done by using insulating materials between cages or positioning the cages and microphones at a distance that reduces interference. The receiving angle of the microphone depends on the frequency of the sound being captured. The distance between the floor of the arena and the microphone must be adapted to each apparatus it will be mounted onto. The Avisoft microphone (CM16/CMPA), e.g., is mounted centrally 45 cm above the floor of the acrylic box (40 × 40 × 40 cm; see Fig. 2B). All of these elements should be tested and standardized in each lab since they vary greatly between lab environments and brands of equipment being used.

2.3 USV Recording and Analysis

Ultrasonic Vocalization Recording: The emission of USV is monitored through a dedicated microphone sensible to the ultrasonic range (e.g., Avisoft CM16/CMPA, Pettersson M500-384, etc.). As an example, we use the Avisoft system (Avisoft Bioacoustics, Berlin, Germany), with a condenser microphone CM16, sensitive to

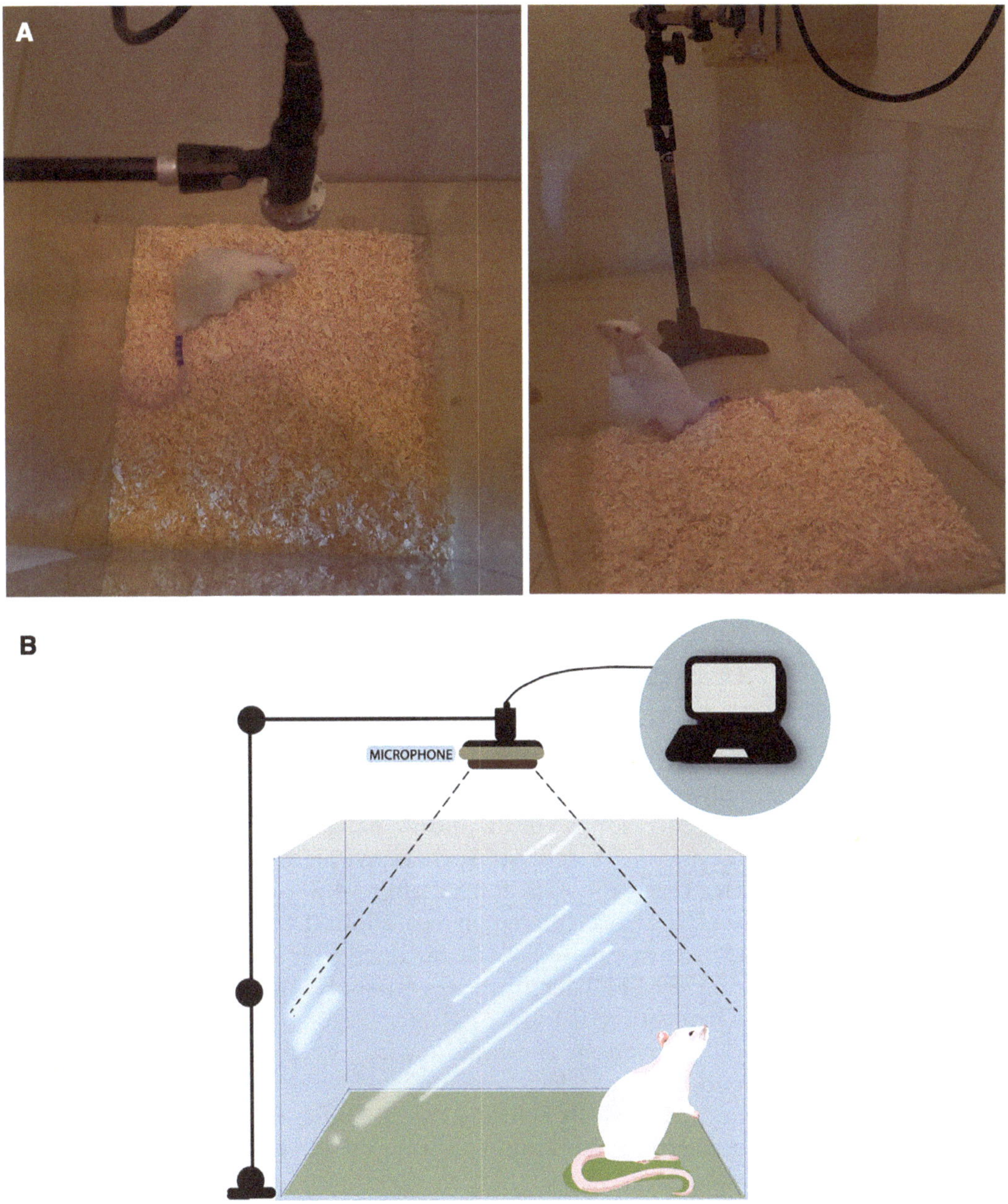

Fig. 2 Example of observation arena and microphone placement

frequencies between 15 and 180 kHz, connected to an UltraSound-Gate (116 H or equivalent; Avisoft Bioacoustics, Berlin, Germany) via an extension cable (Fig. 1). The microphone is mounted 45 cm above the floor of the observation cage (Fig. 2) and connected to a computer with the recording software Avisoft Recorder USGH (version 2.95, Avisoft Bioacoustics) set at a sampling rate of 214,285 Hz and 16-bit format. Standardization of recording parameters is important

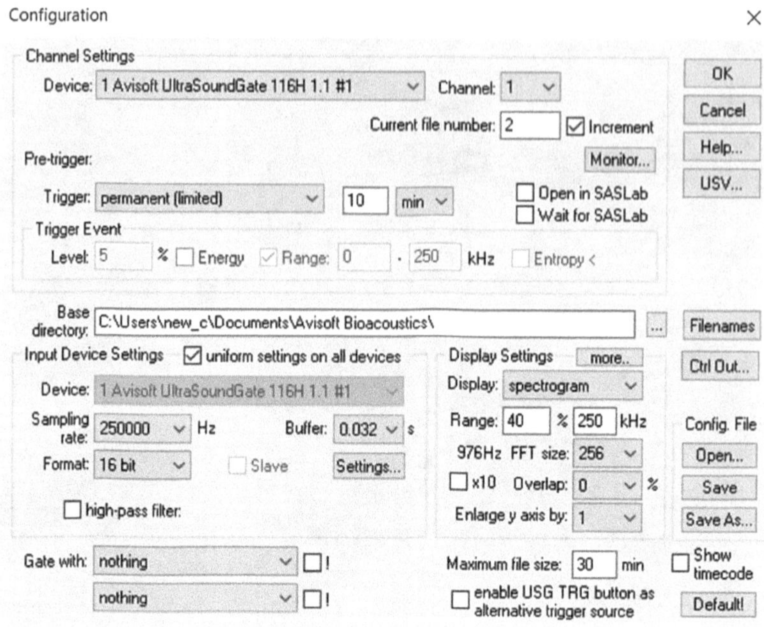

Fig. 3 Settings for Recording for Avisoft Recorder USGH (version 2.95, Avisoft Bioacoustics)

for later analysis of acquired data (see Fig. 3 for Avisoft Recorder software).

Ultrasonic Vocalization Analysis: Different software can be used for USV analysis, such as Avisoft SASLabPro, DeepSqueak, etc. Using the Avisoft SASLabPro 4.38 software, high-resolution spectrograms (frequency resolution: 0.488 kHz, time resolution: 0.512 ms; Fig. 4A) can be obtained after a fast Fourier transformation (512 FFT-length, 100% frame, Hamming Window, 75%-time window overlap).

Conversely, the audio recordings stored as.WAV files can be transferred to a deep learning-based system, an example being DeepSqueak [48]. By splitting the audio files into short segments and converting them into sonograms (Fig. 4B), DeepSqueak detects vocalizations by running the generated images in a Faster-RCNN (regional convolutional neural network) object detector.

Based on its frequency band, USV can be categorized into low (22-kHz) or high frequency vocalizations (50-kHz).

22-kHz USV: 22-kHz USVs are observed in the frequency band between 20 and 35 kHz (Fig. 5) and have long call durations in comparison to 50-kHz ones. They are predominantly flat with little to no frequency variation [8, 49, 50]. The 22-kHz vocalizations appear in spectrograms as rather straight flat lines with a narrow bandwidth of 1–4 kHz. They can be divided into short calls and long calls.

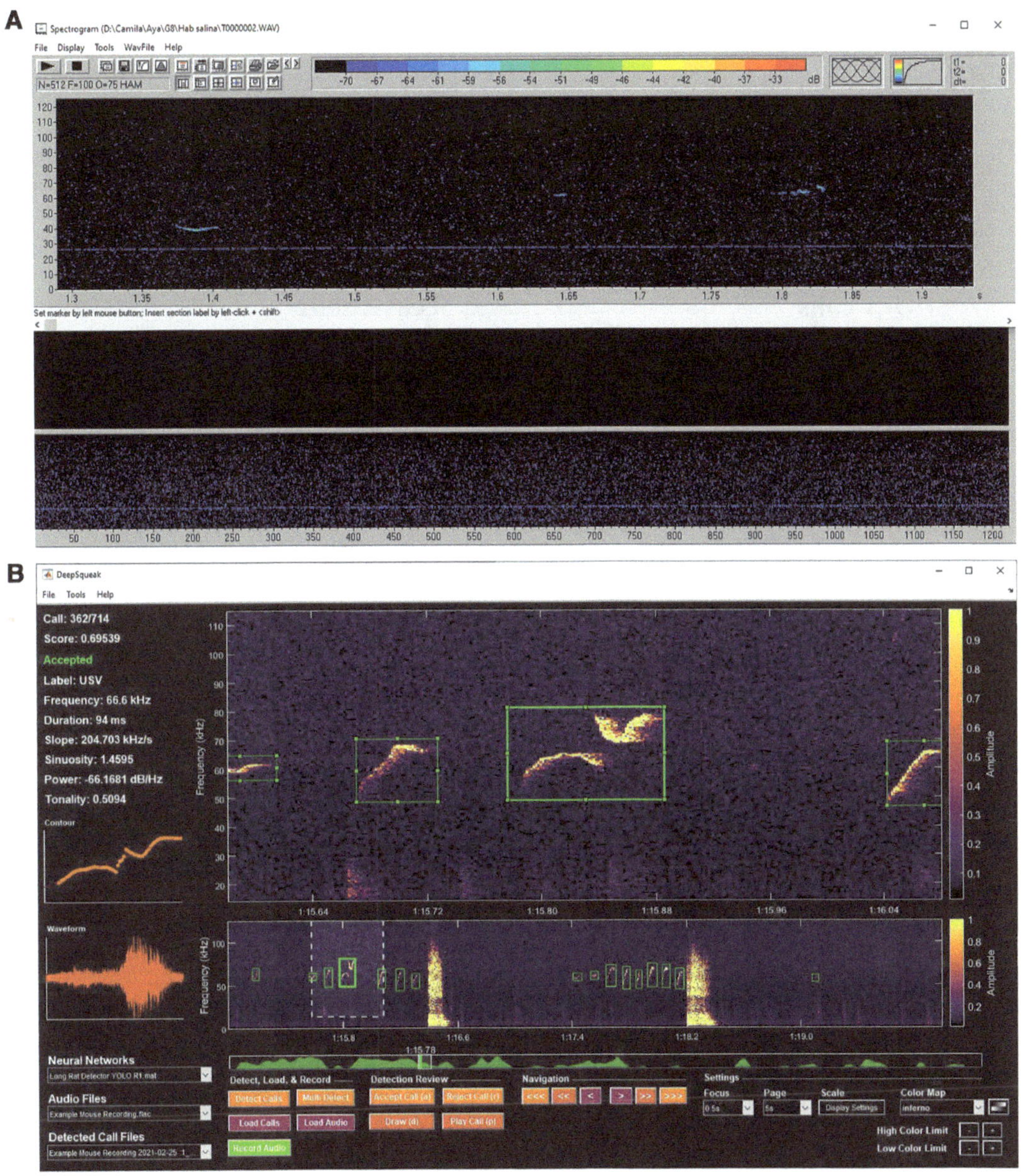

Fig. 4 Example of USV call spectrograms analyzed via AVISOFT SASLABPRO 4.38 (**A**) and DEEP SQUEAK v. 3 (**B**, courtesy of Dr. Kevin Coffey)

50-kHz USV: 50-kHz USV are diverse and their shape can be used to further classify them into subcategories (Fig. 5). Flat calls are defined as those with a peak frequency change up to 5-kHz. Modulated calls can be subdivided into trills (a call element with one or more peak frequency change higher than 5-kHz or with two or more peak frequency changes in opposite directions with at least 5-kHz apart), step call (when a flat call has at least one short flat element adjoining in the beginning or at the end of the call; one of these calls

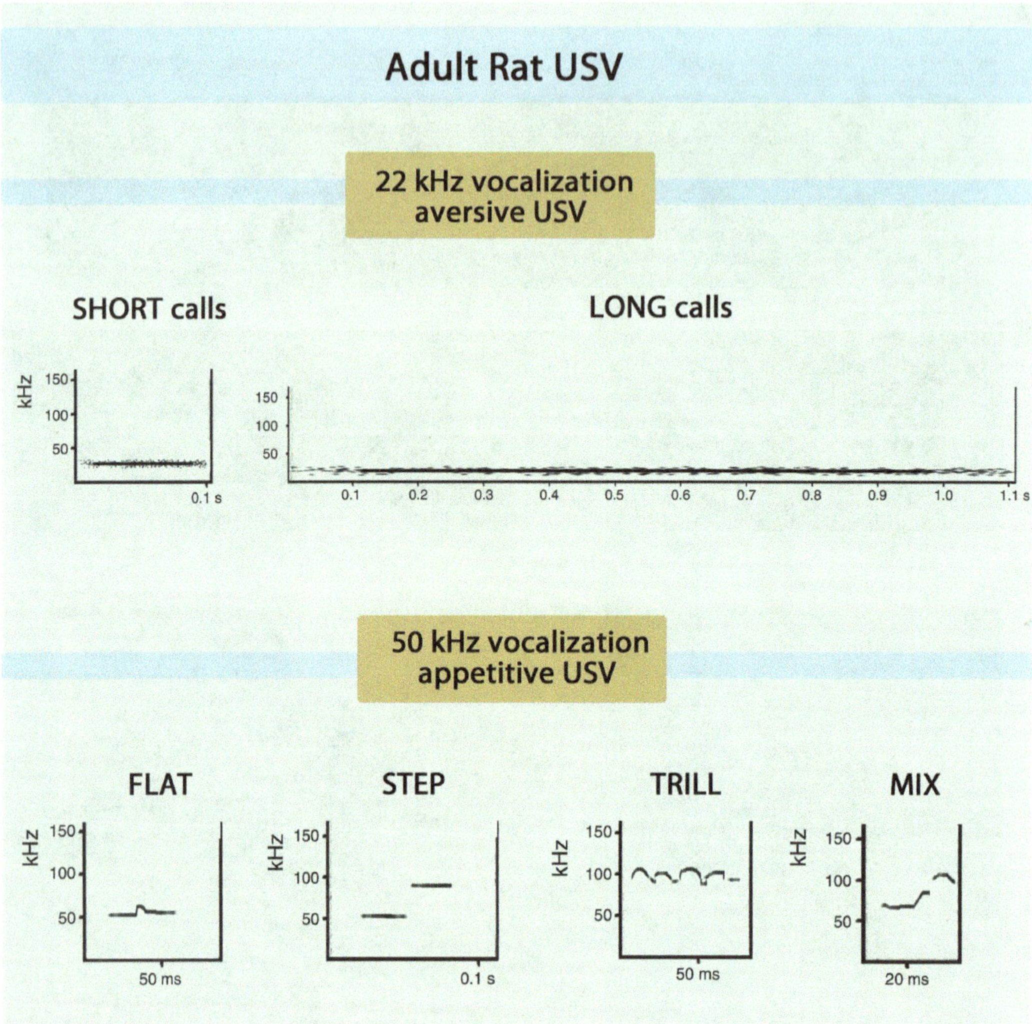

Fig. 5 Example of sonograms of 22-kHz USVs (upper panel) and of flat and frequency-modulated (FM) 50-kHz USVs (lower panel)

must be 5-kHz larger than the main call), and mixed calls (calls that did not fit into in the others categories) [51,52–53]. They can also be further categorized into several other categories [54].

2.4 Notes

- Some animals might not vocalize at all since there is substantial interindividual variability in ultrasonic vocalization [55]. To account for this factor, baseline readouts of spontaneous USV emission might be used to determine whether and how much a given animal spontaneously emits USV [51]. Subsequently, animals should be assigned to the different experimental groups, so that each group has similar levels of baseline USV rates.

- While high frequency sounds such as USV do not tend to cross physical barriers, some thought must be put out on how near different subjects should be positioned to each other during USV readouts. Acoustic soundproofing between test cages, sufficient distance from testing areas so that the microphone of one subject does not detect the USV from another, or recording subjects in sequence rather than simultaneously are recommended.
- USV are natural responses to social stimuli. With this in mind, testing cages should be thoroughly cleaned between subjects so that no smell from the previous subject elicits USV from the next one.
- While it is yet undetermined whether circadian rhythm influences USV emission, it is recommended to keep experiments in one period of the day for standardization and randomization of groups for time of testing.
- It is possible to synchronize USV emissions with behavior using a special configuration as done by Laplagne and Costa [56], that locked locomotor activity and USV emission.

3 Concluding Remarks

It is well established that acute pain is related to increased emission of 22-kHz calls by rats and that analgesic treatment is capable of reducing aversive USV emission. However, current knowledge in the field indicates that emission of 22-kHz calls is not considered a direct pain measure, but signals a negative affective state related to pain, serving as alarm calls to warn the conspecific about the danger or reflecting aversiveness, dysphoria, or frustration. Emission of 22-kHz calls may also be detected during chronic pain conditions, depending on the context. For instance, noxious stimulation is rats subjected to chronic pain models induced aversive USV emission. In spite of an initial controversy about the utility of USV measures in chronic pain models, there is cumulating evidence that they can represent a useful tool in the investigation of the emotional pain component. Studies have consistently shown that chronic pain is associated with reduction of 50-kHz appetitive calls. The full implication of this finding remains to be clarified, but it has been suggested that the chronic pain states cause impairment in social interaction, a finding that corroborates observations from several preclinical and clinical studies. Findings with USV measures should be interpreted with caution, as call emissions by rats may have more than one function. However, when USV emission is registered and analyzed following established protocols and together with other pain measures, it may add valuable and complimentary data to pain research. In addition to reflecting the emotional state of rats, aversive and appetitive USV emission reflect distinct patterns of brain activation. For instance, 50-kHz calls, to both social and nonsocial contexts, are associated with an augmentation in the mesolimbic dopaminergic activity. Interestingly, chronic pain results in a hypodopaminergic tone that impairs motivated behavior and, probably, emission of appetitive calls. Thus, USV analysis allows not only to capture the emotional component of pain, but also may contribute to elucidate the neuronal mechanisms related to it. Finally, recent data indicate sex differences in USV emission in chronic pain models, suggesting that female rats exhibit higher durations of evoked aversive USV. This aspect certainly deserves to be further explored and reinforces the utility of USV measure in pain research.

Acknowledgments

This study was financed in part by the Coordination for the Improvement of Higher Education Personnel—Brazil (CAPES, PROBRAL GRANT 88881.198683/2018-01) and by the Deutsche Forschungsgemeinschaft (DFG, German Research Foundation, Project-ID 422744262—TRR 289).

CPS and VBPL are recipients of PhD fellowships from Coordenação de Aperfeiçoamento de Pessoal de Nível Superior (CAPES, Brasil, Finance Code 001).

JGC and RA are recipients of researcher fellowships from The Brazilian National Council for Scientific and Technological Development (CNPq, grants 303781/2020-3 and 312201/2019-2, respectively).

Conflict of Interest The authors have no conflicts of interest to declare that are relevant to the content of this chapter.

References

1. Brudzynski SM (2013) Ethotransmission: communication of emotional states through ultrasonic vocalization in rats. Curr Opin Neurobiol 23(3):310–317
2. Anderson JW (1954) The production of ultrasonic sounds by laboratory rats and other mammals. Science 119(3101):808–809
3. MacDonald T, Brudzynski SM (2018) Chapter 2 - Ultrasonic vocalizations, their recording, and bioacoustic analysis. Brudzynski SM (eds), Handbook of Behavioral Neuroscience, vol 25. Elsevier, pp 7–19. https://doi.org/10.1016/B978-0-12-809600-0.00002-0. ISSN 1569-7339, ISBN 9780128096000.
4. Brudzynski SM. Biological Functions of Rat Ultrasonic Vocalizations, Arousal Mechanisms, and Call Initiation. Brain Sci. 2021 May 9;11(5):605. doi: 10.3390/brainsci11050605. PMID: 34065107; PMCID: PMC8150717.
5. Jourdan D, Ardid D, Eschalier A (2002) Analysis of ultrasonic vocalisation does not allow chronic pain to be evaluated in rats. Pain 95(1):165–173
6. Miczek KA, Tornatzky W, Vivian J (1991) Ethology and neuropharmacology: rodent ultrasounds. In: Animal models in psychopharmacology. Springer, Birkhäuser, Basel. pp 409–427. https://doi.org/10.1007/978-3-0348-6419-0_41
7. Brudzynski SM (2015) Pharmacology of ultrasonic vocalizations in adult rats: significance, call classification and neural substrate. Curr Neuropharmacol 13(2):180–192
8. Wöhr M, Schwarting RK (2013) Affective communication in rodents: ultrasonic vocalizations

as a tool for research on emotion and motivation. Cell Tissue Res Oct;354(1):81–97. https://doi.org/10.1007/s00441-013-1607-9. Epub 2013 Apr 11. PMID: 23576070
9. Han JS, Bird GC, Li W, Jones J, Neugebauer V (2005) Computerized analysis of audible and ultrasonic vocalizations of rats as a standardized measure of pain-related behavior. J Neurosci Methods 141(2):261–269
10. Levine JD, Clark R, Devor M, Helms C, Moskowitz MA, Basbaum AI (1984) Intraneuronal substance P contributes to the severity of experimental arthritis. Science 226(4674):547–549
11. Tonoue T, Ashida Y, Makino H, Hata H (1986) Inhibition of shock-elicited ultrasonic vocalization by opioid peptides in the rat: a psychotropic effect. Psychoneuroendocrinology 11(2):177–184
12. van der Poel AM, Noach EJ, Miczek KA (1989) Temporal patterning of ultrasonic distress calls in the adult rat: effects of morphine and benzodiazepines. Psychopharmacology (Berl) 97(2):147–148
13. Ardid D, Jourdan D, Eschalier A, Arabia C, Le Bars D (1993) Vocalization elicited by activation of A delta-and C-fibres in the rat. Neuroreport 5(2):105–108
14. Abbott FV, Franklin KB, Westbrook RF (1995) The formalin test: scoring properties of the first and second phases of the pain response in rats. Pain 60(1):91–102
15. Tjølsen A, Berge OG, Hunskaar S, Rosland JH, Hole K (1992) The formalin test: an evaluation of the method. Pain 51(1):5–17
16. Dallel R, Raboisson P, Clavelou P, Saade M, Woda A (1995) Evidence for a peripheral origin of the tonic nociceptive response to subcutaneous formalin. Pain 61(1):11–16
17. Oliveira AR, Barros HM (2006) Ultrasonic rat vocalizations during the formalin test: a measure of the affective dimension of pain?. Anesth Analg 102(3):832–839
18. Lim DW, Kim JG, Han D, Kim YT (2014) Analgesic effect of harpagophytum procumbens on postoperative and neuropathic pain in rats. Molecules 19(1):1060–1068
19. Jourdan D, Ardid D, Chapuy E, Le Bars D, Eschalier A (1998) Effect of analgesics on audible and ultrasonic pain-induced vocalization in the rat. Life Sci 63(20):1761–1768
20. Martino G, Perkins M (2008) Tactile-induced ultrasonic vocalization in the rat: a novel assay to assess anti-migraine therapies in vivo. Cephalalgia 28(7):723–733
21. Bobade V, Bodhankar SL, Aswar U, Vishwaraman M, Thakurdesai P (2015) Prophylactic effects of asiaticoside-based standardized extract of Centella asiatica (L.) Urban leaves on experimental migraine: Involvement of 5HT1A/1B receptors. Chin J Nat Med 13(4):274–282
22. Vivian JA, Miczek KA (1999) Interactions between social stress and morphine in the periaqueductal gray: effects on affective vocal and reflexive pain responses in rats. Psychopharmacology (Berl) 146(2):153–161
23. Araya EI, Baggio DF, Koren LO, Andreatini R, Schwarting RKW, Zamponi GW et al. (2020) Acute orofacial pain leads to prolonged changes in behavioral and affective pain components. Pain 161(12):2830–2840
24. Barroso AR, Araya EI, de Souza CP, Andreatini R, Chichorro JG (2019) Characterization of rat ultrasonic vocalization in the orofacial formalin test: influence of the social context. Eur Neuropsychopharmacol 29(11):1213–1226
25. Vivian JA, Miczek KA (1993) Morphine attenuates ultrasonic vocalization during agonistic encounters in adult male rats. Psychopharmacology (Berl) 111(3):367–375
26. Bardin L, Gregoire S, Aliaga M, Malfetes N, Vitton O, Ladure P et al. (2010) Comparison of milnacipran, duloxetine and pregabalin in the formalin pain test and in a model of stress-induced ultrasonic vocalizations in rats. Neurosci Res 66(2):135–140
27. Porreca F, Navratilova E (2017) Reward, motivation, and emotion of pain and its relief. Pain 158(Suppl 1):S43–S49
28. Bushnell MC, Ceko M, Low LA (2013) Cognitive and emotional control of pain and its disruption in chronic pain. Nat Rev Neurosci 14(7):502–511
29. Naito H, Okumura T, Inoue M, Suzuki Y (2006) Ultrasonic vocalization response elicited in adjuvant-induced arthritic rats as a useful method for evaluating analgesic drugs. Exp Anim 55(2):125–129
30. Naito H, Inoue M (2008) Quantitative analysis of the ultrasonic vocalization responses elicited in adjuvant-induced arthritic rats for screening analgesic drugs. Exp Anim 57(4):419–422
31. Calvino B, Besson JM, Boehrer A, Depaulis A (1996) Ultrasonic vocalization (22-28 kHz) in a model of chronic pain, the arthritic rat: effects of analgesic drugs. Neuroreport 7(2):581–584
32. Kim H, Thompson J, Ji G, Ganapathy V, Neugebauer V (2017) Monomethyl fumarate inhibits pain behaviors and amygdala activity in a rat arthritis model. Pain 158(12):2376–2385
33. Han JS, Neugebauer V (2005) mGluR1 and mGluR5 antagonists in the amygdala inhibit different components of audible and ultrasonic vocalizations in a model of arthritic pain. Pain 113(1-2):211–222
34. Grégoire S, Neugebauer V (2013) 5-HT2CR blockade in the amygdala conveys analgesic efficacy to SSRIs in a rat model of arthritis pain. Mol Pain 9:41
35. Mazzitelli M, Neugebauer V (2019) Amygdala group II mGluRs mediate the inhibitory effects of systemic group II mGluR activation on

behavior and spinal neurons in a rat model of arthritis pain. Neuropharmacology 158:107706
36. Burgdorf JS, Ghoreishi-Haack N, Cearley CN, Kroes RA, Moskal JR (2019) Rat ultrasonic vocalizations as a measure of the emotional component of chronic pain. Neuroreport 30(13):863–866
37. Araya EI, Carvalho EC, Andreatini R, Zamponi GW, Chichorro JG (2022) Trigeminal neuropathic pain causes changes in affective processing of pain in rats. Mol Pain Jan Dec;18:17448069211057750. https://doi.org/10.1177/17448069211057750. PMID: 35042377; PMCID: PMC8777332
38. Barker DJ, Simmons SJ, West MO (2015) Ultrasonic vocalizations as a measure of affect in preclinical models of drug abuse: a review of current findings. Curr Neuropharmacol 13(2):193–210
39. Barros HM, Miczek KA (1996) Withdrawal from oral cocaine in rate: ultrasonic vocalizations and tactile startle. Psychopharmacology (Berl) 125(4):379–384
40. Covington 3rdHE, Miczek KA (2003) Vocalizations during withdrawal from opiates and cocaine: possible expressions of affective distress. Eur J Pharmacol 467(1–3):1–13
41. Brudzynski SM, Bihari F, Ociepa D, Fu XW (1993) Analysis of 22 kHz ultrasonic vocalization in laboratory rats: long and short calls. Physiol Behav 54(2):215–221
42. Presto P, Ji G, Junell R, Griffin Z, Neugebauer V (2021) Fear extinction-based inter-individual and sex differences in pain-related vocalizations and anxiety-like behaviors but not nocifensive reflexes. Brain Sci Oct 11;11(10):1339. https://doi.org/10.3390/brainsci11101339. PMID: 34679403; PMCID: PMC8533751
43. Araya EI, Carvalho EC, Andreatini R, Zamponi GW, Chichorro JG (2022) Trigeminal neuropathic pain causes changes in affective processing of pain in rats. Mol Pain 18: 17448069211057750
44. Gambeta E, Batista MA, Maschio GP, Turnes JM, Araya EI, Chichorro JG (2018) Anxiety- but not depressive-like behaviors are related to facial hyperalgesia in a model of trigeminal neuropathic pain in rats. Physiol Behav 191:131–137
45. Araya EI, Baggio DF, de Oliveira Koren L, Schwarting RKW, Chichorro JG (2022) Trigeminal neuropathic pain reduces 50-kHz ultrasonic vocalizations in rats, which are restored by analgesic drugs. Eur J Pharmacol 922:174905
46. Lim DW, Kim JG, Kim YT (2016) Analgesic effect of Indian gooseberry (Emblica officinalis fruit) extracts on postoperative and neuropathic pain in rats. Nutrients Nov 26;8(12):760. https://doi.org/10.3390/nu8120760. PMID: 27898027; PMCID: PMC5188415
47. Lenell C, Broadfoot CK, Schaen-Heacock NE, Ciucci MR (2021) Biological and acoustic sex differences in rat ultrasonic vocalization. Brain Sci Apr 4;11(4): 459. https://doi.org/10.3390/brainsci11040459. PMID: 33916537; PMCID: PMC8067311.
48. Coffey KR, Marx RG, Neumaier JF (2019) DeepSqueak: a deep learning-based system for detection and analysis of ultrasonic vocalizations. Neuropsychopharmacology 44(5):859–868
49. Brudzynski SM, Chiu EM (1995) Behavioural responses of laboratory rats to playback of 22 kHz ultrasonic calls. Physiol Behav 57(6):1039–1044
50. Brudzynski SM (2001) Pharmacological and behavioral characteristics of 22 kHz alarm calls in rats. Neurosci Biobehav Rev 25(7-8):611–617
51. Pereira M, Andreatini R, Schwarting RK, Brenes JC (2014) Amphetamine-induced appetitive 50-kHz calls in rats: a marker of affect in mania?. Psychopharmacology (Berl) 231(13):2567–2577
52. Wendler E, de Souza CP, Vecchia DD, Kanazawa LKS, de Almeida Soares Hocayen P, Wöhr M et al. (2016) Evaluation of 50-kHz ultrasonic vocalizations in animal models of mania: ketamine and lisdexamfetamine-induced hyperlocomotion in rats. Eur Neuropsychopharmacol 26(12):1900–1908
53. Wendler E, de Souza CP, Dornellas APS, Santos LE, Ferreira ST, Galduróz JCF et al. (2019) Mania-like elevated mood in rats: enhanced 50-kHz ultrasonic vocalizations after sleep deprivation. Prog Neuropsychopharmacol Biol Psychiatry 88:142–150
54. Wright JM, Gourdon JC, Clarke PB (2010) Identification of multiple call categories within the rich repertoire of adult rat 50-kHz ultrasonic vocalizations: effects of amphetamine and social context. Psychopharmacology (Berl) 211(1):1–13
55. Taracha E, Hamed A, Krząścik P, Lehner M, Skórzewska A, Płaźnik A et al. (2012) Inter-individual diversity and intra-individual stability of amphetamine-induced sensitization of frequency-modulated 50-kHz vocalization in Sprague-Dawley rats. Psychopharmacology (Berl) 222(4):619–632
56. Laplagne DA, Elías Costa M (2016) Rats synchronize locomotion with ultrasonic vocalizations at the subsecond time scale. Front Behav Neurosci 10:184

Check for updates

Chapter 12

Identifying the Placebo and Nocebo Responders: Two Methods for Eliciting Placebo Analgesia or Nocebo Hyperalgesia in Rats with a Neuropathic Injury

Damien C. Boorman and Kevin A. Keay

Abstract

The ongoing shortfalls of analgesic treatments for managing chronic pain conditions have pushed research toward nonpharmacological approaches. Identifying the underlying neurobiology of placebo analgesia and nocebo hyperalgesia would allow for new therapies that exploit and inhibit these phenomena, respectively. Human evidence to date provides clear indications that there are populations of responders and nonresponders for both placebo and nocebo effects, with most studies focused on psychosocial explanations. It is also possible to elicit placebo analgesia and nocebo hyperalgesia in rats with chronic pain. Several recent studies have found that as in people, placebo analgesia and nocebo hyperalgesia only develop in subpopulations of rodents. These animal models will be essential for elucidating the cellular and subcellular mechanisms of these phenomena as well as for assisting in the preclinical pipeline for novel treatments. This chapter details two methodological approaches that have successfully elicited placebo analgesia or nocebo hyperalgesia in subpopulations of rats and highlights the advantages and limitations of these approaches. The first approach uses pharmacological conditioning, which pairs the experience of drug-induced analgesia (morphine) with contextual cues to elicit context-specific placebo analgesia. The second uses response conditioning, which pairs stimulus intensity with contextual cues to elicit either context-specific placebo analgesia or nocebo hyperalgesia.

Keywords Pharmacological conditioning, Response conditioning, Chronic constriction injury, Cold plate, Hind paw withdrawal, Estimation statistics

1 Introduction

Placebo analgesia is the phenomenon in which a pharmacologically and physiologically inert treatment is able to elicit a reduction in one's pain experience [1]. Identifying ways to enhance placebo analgesia would have widespread therapeutic ramifications, especially in the case of chronic pain disorders, in which the management of pain is the primary goal. Systematic research into placebo analgesia began in the 1940s and 1950s, which quickly found large individual variability in the strength of placebo responses, with some individuals reporting

Isaura Tavares and Hugo Leite-Almeida (eds.), *Translational Research Methods for Neuropathic Pain*, Neuromethods, vol. 235, https://doi.org/10.1007/978-1-0716-5312-8_12,

immense pain relief from a placebo intervention and others reporting little or no effect. In an early meta-analysis of these studies, Henry K. Beecher calculated that the average percentage of "patients satisfactorily relieved by placebo" was 35.2% ± 2.2% (SEM) across a range of different conditions (e.g., postoperative pain, headache, seasickness, anxiety, and the common cold) [2]. This individual variability has remained one of the most consistent findings in more contemporary studies as well, most of which report similar proportions of "placebo responders" and "nonresponders" in their study populations. These studies include 39% of patients with postoperative pain [3], 27% of participants in an experimental model of ischemic arm pain [4], 35% and 55% of participants using acute thermal pain [5, 6], and 32% of knee osteoarthritis patients [7]; see [8, 9] for reviews.

This individual variability in placebo responsiveness appears to be a characteristic feature of rodent models of placebo analgesia as well. In rats with neuropathic injuries, Zeng et al. [10] reported strong placebo analgesic responses in 36% (9/25) of their rats (males), which is remarkably similar to the proportions of placebo responders we found in our recent study [11], in which 36% of male rats (5/14) and 25% of female rats (6/24) demonstrated significant placebo analgesic responses. These findings are also supported by the observations from Nolan et al. [12] and McNabb et al. [13], who noted large within-group variance and skewness in placebo treatment groups, likely representing subpopulations of strong placebo responders. Identification of these subpopulations thus requires large sample sizes in the placebo group and may explain why some other studies have failed to report significant placebo effects [14, 15]. Since both humans and rodents clearly stratify into placebo responder and nonresponder phenotypes, one of the most common approaches for investigating the neural substrates of placebo analgesia is to compare functional activity and sometimes brain structure between these phenotypes. Neuroimaging studies using this design have so far identified a range of brain regions that appear to be involved in the production of placebo analgesia, including the medial prefrontal cortex, thalamus, anterior cingulate cortex, and brainstem nuclei.

To date, the vast majority of placebo analgesia research has been performed in humans using acute pain stimulation in healthy participants. However, since the development of chronic pain is associated with a collection of maladaptive changes in the circuitry and systems that regulate pain, there is a strong possibility that these changes may impact one's ability to generate placebo analgesic responses. Promisingly, some recent research indicates that people with chronic pain are able to experience placebo analgesia (to acute pain) to the same magnitude as healthy controls [16, 17], and that placebo interventions can reduce chronic pain and associated disability in a clinically meaningful way [18, 19]. However, further research is needed to identify the neurobiological mechanisms that underlie placebo analgesia in both healthy controls and people with chronic pain, which would allow for the development of novel treatments that harness or exploit these mechanisms. Such treatments would likely be safer, more effective, and elicit fewer side effects than current pharmacotherapeutics (e.g., opioids, gabapentin).

Animal models of placebo analgesia have at least three clear advantages over human studies. First, they allow for tractable experiments that can precisely interrogate the neurobiological mechanisms that underpin placebo analgesia at the cellular and subcellular levels. These experiments can utilize powerful techniques unavailable to human research, including electrophysiology, lesioning, tract tracing, optogenetics, chemogenetics, and calcium imaging. Second, many of the factors that have been shown to influence the strength of placebo responses in people can be controlled or do not apply to rodent models. These include the history of the subject (e.g., prior drug exposure), the environments they are tested in, their personality traits, and the expectations created by verbal suggestions or social observations. Finally, as has been case for preclinical drug trials, animal models will be critical for the development of novel therapeutic treatments that harness or exploit the neurobiology of placebo analgesia.

However, the obvious challenge faced by those interested in developing and using preclinical models of placebo analgesia is summarized by the following question: how can we elicit and recognize the development of placebo analgesia in animals that (1) are not susceptible to verbal suggestions and (2) are unable to directly report their pain experience? With respect to (1), we know that both Pavlovian and operant conditioning procedures are able to elicit placebo analgesia in both humans and rodents. This includes *pharmacological conditioning*, in which neutral contextual cues are paired with the experience of drug-induced analgesia and subsequently used to measure pain sensitivity in the presence of these cues alone [10, 11, 20–22]; and *response conditioning*, in which contextual cues are paired with a low-intensity stimulus and then used to measure pain sensitivity to a moderate- to high-intensity stimulus in the presence of these contextual cues [23, 24].

With respect to (2), as with any preclinical model that measures "pain" and "analgesia," we must rely on changes in behaviors during pain assays to serve as a proxy, and there is good evidence to suggest that certain pain assays are better than others for eliciting and detecting placebo analgesia (see [11, 13, 15, 24]). In the context of neuropathic pain, using noxious thermal stimuli appears to be more susceptible to the development and/or manifestation of placebo analgesia compared to noxious mechanical stimuli.

Our lab has now successfully combined the sciatic nerve chronic constriction injury (CCI) model of neuropathic pain with pharmacological conditioning procedures in male and female rats to reveal subpopulations of placebo responders and nonresponders [11]. We have also combined the CCI model with response conditioning procedures in male rats to reveal both group-level placebo analgesic effects and nocebo hyperalgesic effects [25]. This chapter details each of these experimental designs, highlighting the strengths and weaknesses of each approach.

2 Materials and Procedures

The following materials and procedures are common to both pharmacological conditioning and response conditioning protocols.

2.1 Animals, Housing, and Testing Conditions

We use sexually mature, outbred male and female Sprague–Dawley rats sourced from the Animal Resource Centre in Perth, Australia. Rats arrive in the laboratory at 6 weeks old, weighing 180–200 g for males and 160–180 g for females. Upon arrival, rats are inspected, and their general health, appearance, and activity levels are recorded. Rats are rehoused into standard, individually ventilated cages, three to four rats per cage, retaining the same cage mates they were transported with to reduce stress induced by changes in their social environment. Rats are given ad libitum access to water and food (standard laboratory chow: 19.6% protein, 4.4% fat, 4.6% crude fiber, 7.7% acid detergent fiber, 15.5% neutral detergent fiber, 59.3% total carbohydrate; 14.0 MJ/kg digestible energy). The housing room is kept on a reversed 12 h:12 h dark/light cycle (lights off at 0700 h) to align behavioral testing with the period when the rats are most active, and the animal facility is temperature (21 ± 1°C) and humidity (40%–70%) controlled.

Rats are given 7 days to acclimate to these new laboratory conditions, during which they are handled at least once per day for 5 min each by the experimenter who will be performing the behavioral testing. All husbandry, behavioral testing, and conditioning procedures should be performed by the same experimenter to control for experimenter effects on the rats' sensory thresholds [26]. All husbandry, handling during the acclimatization period, and general animal monitoring are conducted during the dark phase under dim (~10 lux), red light conditions, using wavelengths undetectable by Sprague–Dawley rats [27]. To achieve this, we cover standard desk lamps with LEE™ filters (medium red), which only allow for wavelengths >600 nm.

2.2 Chronic Constriction Injury of the Sciatic Nerve

This protocol is based on the procedures first described by Bennett and Xie [28].

Materials

2× curved-tip forceps; 2× straight-tip forceps; 2× small hemostats; 1× pair of curved blunt-tip scissors; Michel clips (~4/animal) and

Michel clip applicator; isoflurane and vaporizer; 100% oxygen tank; small animal anesthesia nosecone; anesthetic induction box; electric clippers; 3× underpads; catgut ligatures (Chromic 5-0; Johnson & Johnson); rectal thermometer; 2× heat blanket; 70% ethanol; bead sterilizer; topical antibiotic (e.g., Tricin®, Jurox); Vaseline/petroleum jelly; cotton tips; povidone iodine; gauze; sterile saline in syringe for washing out incision; quinine; lidocaine in syringe (if required); electrical tape.

Procedure

Spray all surfaces and equipment with 70% ethanol and sterilize surgical utensils (bead sterilizer) prior to use. Anesthetize the rat using an induction box with 5% isoflurane in 100% oxygen (flow rate: 1–2 L/min). Once unconscious, transfer the rat to nosecone and maintain anesthesia with 2%–3% isoflurane in 100% oxygen (flow rate: 1–2 L/min). The animal has reached the surgical plane of anesthesia only once eyeblink and pedal reflexes have been abolished. Throughout the procedure, closely observe the animal's heart rate, respiratory rate, temperature (via rectal thermometer), and the color of its mucous membranes and paws. Shave the hind limb region using electric clippers and discard any shaved fur. Wipe the shaved region with povidone iodine. Tape the foot down to position the limb appropriately (the knee should be flexed at ~90°).

Make a 2 cm incision just inferior to the femur, along its axis, using the curved blunt-tip scissors. Along the same axis as the incision, bluntly dissect through the biceps femoris to reveal the sciatic nerve. Isolate the nerve from the surrounding connective tissue proximal to its trifurcation, and thread four sutures (each ~5–6 cm long) underneath the nerve using the curved-tip forceps. Tie off each suture using a square knot, tight enough to slightly depress the nerve (to ~70% of its original diameter). Tightening the ligatures often (but not always) elicits brief muscle twitching. Each ligature should be spaced ~1 mm apart. Cut away suture tails (see Fig. 1).

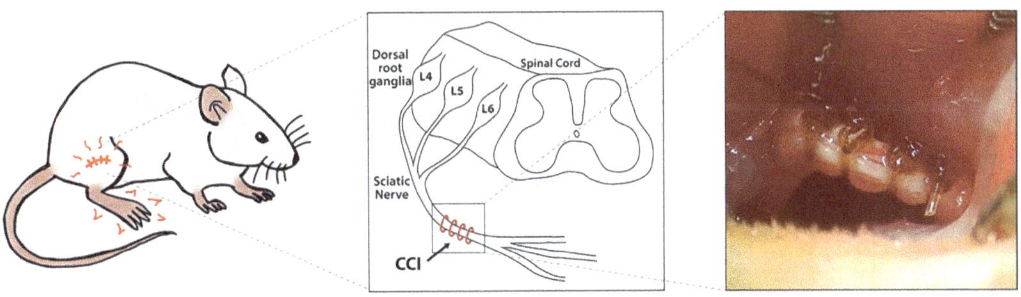

Fig. 1 Sciatic nerve chronic constriction injury (CCI) is preclinical model of peripheral neuropathic pain that induces a compression neuropathy, first developed by Bennett and Xie [28]. Four ligatures are tied around the sciatic nerve, proximal to its trifurcation, which compress the nerve to ~70% of its original diameter but do not occlude the epineural blood flow.

Gently massage the surrounding muscles to allow the nerve to return to its natural position and wash out the incision with sterile saline. Dry with gauze and use Michel clips to close the incision (~2 clips/cm of incision).

Apply topical lidocaine (if required), topical antibiotic powder (e.g., Tricin®), petroleum jelly, and quinine powder (to prevent licking of the wound). Remove the rat from anesthesia and place it into a recovery cage, which should have no food, water, or bedding (to prevent aspiration). Monitor the rat until ambulatory and alert (~20–30 min) before returning to its home cage. Postsurgery, rats should be monitored daily for signs of wound dehiscence and autotomy (i.e., biting of the nails/digits of the injured limb). Michel clips will often fall off after ~7 days as the incision heals.

Notes

- Each surgery takes 10–15 min.
- Hemostats can be used to provide tension on the skin and muscles around the incision to assist during ligature application.
- Sham surgeries should replicate this procedure exactly except without the application of the ligatures. This includes the duration of anesthesia exposure and isolation of the sciatic nerve.
- There is some evidence that the degree of nerve constriction can influence the degree of thermal hypersensitivity that develops (see Boorman and Keay [11], Supplementary Fig. 2). As such, we recommend that the same surgeon perform all surgeries for a given experiment.
- When possible, we house rats that have received the same procedure (CCI or sham surgery) in the same cage. That is, CCI rats are not housed with sham surgery rats. This is to prevent any possible social contagion of pain effects [29].

2.3 Thermal Sensitivity (Hot and Cold Plate) Testing

Given that placebo analgesia is a complex, integrated response known to involve both spinal and supraspinal structures, we chose to use the hot plate and cold plate tests, which engage supraspinal structures [30–32], rather than other pain assays that rely primarily on spinally mediated reflexes (e.g., tail flick [33]). Indeed, McNabb et al. [13] propose that their use of the punctate mechanical paw withdrawal test, which measures the evoked, reflexive, and sensory/discriminatory aspects of pain, was one factor as to their failure to detect placebo analgesic responses in their animal model. It seems likely that such tests might not be sufficiently sensitive to the modulation of the supraspinal cognitive processes that presumably underlie placebo analgesia.

Materials

Hot/cold plate analgesiometer (e.g., Ugo Basile® product code 35150, or equivalent), including transparent Perspex containment cylinder and foot pedal; 70% ethanol and soft paper towels/tissues for cleaning; lightly scented disinfectant wipes (optional); digital camera with infrared capabilities (e.g., Sony PXW-X70) for behavioral recordings and camera tripod; small mirror; mini whiteboard and marker for test identification.

Testing Temperature

Previous studies as well as pilot data from our lab revealed that 4 and 46°C are the minimum and maximum temperatures that do not reliably elicit hind paw withdrawal behavior in uninjured rats [34–36], and thus represent the allodynia–hyperalgesia boundary. Additionally, we have previously demonstrated that 4°C elicits stronger and more reliable hind paw withdrawal behaviors post-CCI than any temperature up to 48°C. As such, we recommend the use of cold plate testing in the context of the CCI model.

Procedure

We use a Ugo Basile hot/cold plate analgesiometer for our thermal sensitivity testing. This device has a central metal plate (22 cm diameter), which can be heated or cooled to a precise temperature between −5 and 65°C (constant temperature function). A clear Perspex cylinder sits on top of and around the plate to contain the rats during testing. The entire procedure is recorded using a digital camera (infrared mode if testing in dim red lighting). The camera should be positioned ~40 cm away from the device, angled slightly above the horizontal plane of the metal plate to give the best view for quantifying hind paw withdrawal behavior. We also position a small mirror (20 cm [w] x 27 cm [h]) alongside the device, angled so that all sides of the rat are visible to the camera. The test identification details (see notes below) are written on a small whiteboard, which is also placed in the camera's frame. Prior to any thermal sensitivity testing, each rat is given 10–15 min of apparatus habituation, in which the device is running and set to 25°C.

To begin each thermal sensitivity test, once the plate has reached the desired temperature, place the rat into the Perspex containment cylinder and begin the timer (using the foot pedal) as the rat's hind paws first touch the plate. Place the lid on the containment cylinder. Once the designated time has passed, press the foot pedal again to stop the timer and remove the rat. Between each test, the plate and Perspex tube should be wiped down with a soft tissue doused in 70% ethanol to remove any traces of feces and urine from the previous rat.

Notes

- The device should be positioned away from walls and enclosed spaces as it can overheat (more common during cold plate testing), during which the metal plate may not remain stable at the designated temperature.
- We use a custom-ordered 40 cm high containment cylinder, as we found that a small proportion of rats (~10%–15%) would repeatedly jump into the lid during testing. The higher cylinder prevents this behavior.

- We use a coding system during thermal sensitivity testing to allow for blinded scoring from video playback (i.e., animal ID and other testing details are not written on the whiteboard).
- The device takes approximately 1–2 min to go from room temperature to reach stable hot temperatures (45–50°C) and 5–7 min to reach stable cold temperatures (4°C).

2.4 Behavioral Scoring

The behavioral repertoire and sequences of rats during the hot plate and cold plate tests have been described in detail previously [37–39]. The more commonly observed behaviors in rats during these tests are walking, rearing, forepaw licking, grooming, hind paw withdrawal, hind paw licking, jumping, immobility, and freezing.

The majority of preclinical studies that employ the hot plate or cold plate tests generally use hind paw withdrawal latency as their measure of pain, i.e., the time taken from when the animal lands on the plate to retract or lick one of its hind paws, after which it is removed from the testing apparatus. However, as noted by Espejo and Mir [38, 40], this procedure might be problematic since the animal could learn with repeated testing that these behaviors prompt the end of the test and henceforth its removal from a presumably aversive situation. Indeed, Gunn et al. (2011) [41] demonstrated that naïve animals showed progressively reduced hind paw withdrawal latencies with repeated tests at the same temperature. As such, our thermal sensitivity tests are of a fixed duration (e.g., 60 s [11]). As Espejo and Mir conclude: "...every parameter (frequency, duration and latency for the first occurrence) and not just latency needs to be evaluated during the hot plate situation" [38].

Nonetheless, there have been a number of studies that have used either the hind paw withdrawal latency or hind paw lick latency on the hot plate test to successfully elicit placebo analgesia from mice and rats [20, 21, 42–48]. However, none of these studies used this test and this behavioral measure in the context of an injury model. Indeed, the only two studies to do so was Yin et al. (2020) [14], which failed to detect placebo effects, and ours [11], in which we found that the "number of paw withdrawals" over the test period was a far more robust measure than hind paw withdrawal latency for detecting placebo analgesia.

Materials

We use a simple behavioral scoring apparatus or software (e.g., ODLog, JWatcher, Noldus Observer XT), an analog TTL input box connected to a PowerLab 8/35 data acquisition device (AD Instruments), and LabChart 7 software. Separate buttons/keyboard keys should be designated for each behavior being scored.

Procedure

During behavioral scoring, we simultaneously record two behaviors—hind paw withdrawal behavior and rearing behavior. For each of these behaviors, we calculate the latency to the first instance, number of times it was performed over the test period, and its cumulative duration over the test period. Hind paw withdrawal behavior is used as the primary measure of pain, whereas rearing behavior is used to gauge general activity levels and exploratory activity, each also indicating the degree of alertness. Our operational definitions of each of these behaviors are:

- **Hind paw withdrawal:** a lifting of one of the hind paws away from the surface of the plate otherwise unnecessary for body repositioning or locomotion, which may include licking of the paw but not jumping.
- **Rearing:** elevation of the head and extension of the vertebral column accompanied by the removal of both forepaws from the plate, which may rest on the Perspex containment cylinder.

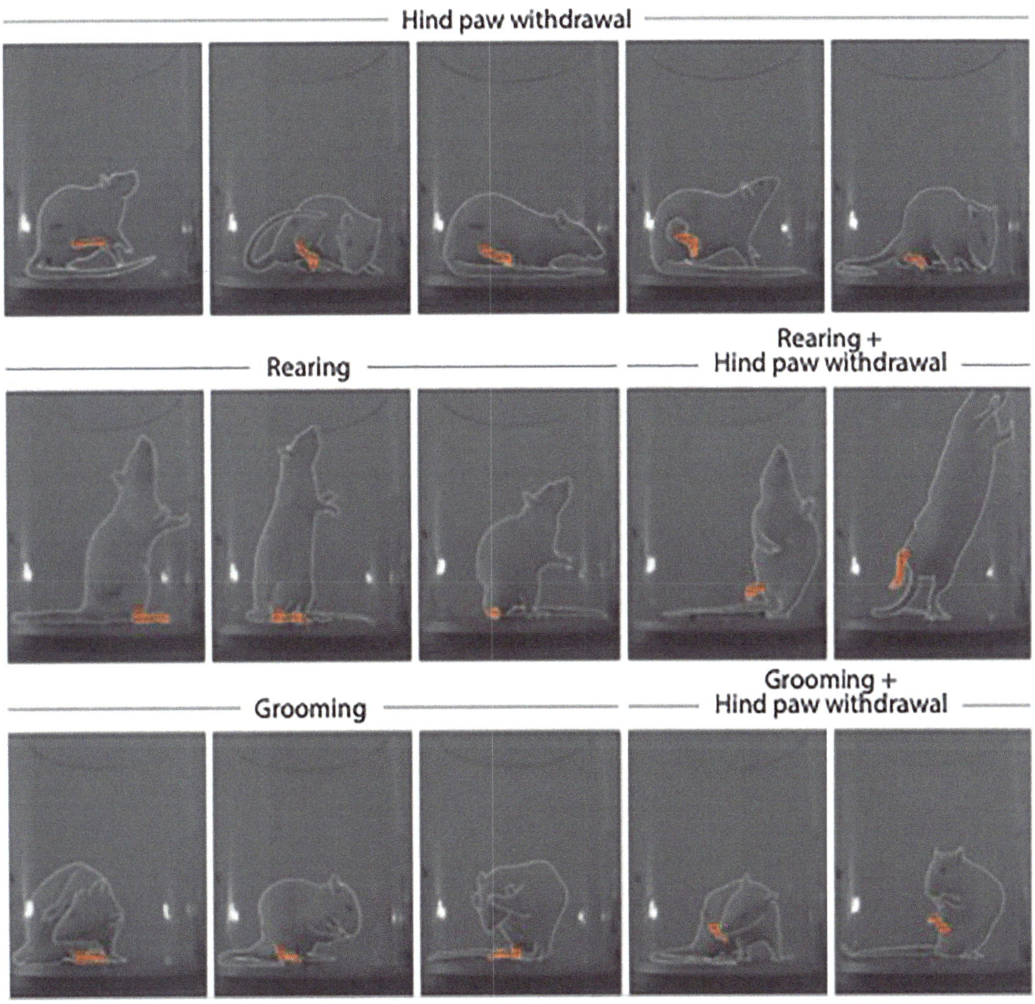

Fig. 2 Photographic examples of hind paw withdrawal behaviors, rearing behaviors, grooming behaviors, and combinations of these behaviors during cold plate testing. The outlines of each rat have been traced, and the injured hind paw has been highlighted in red

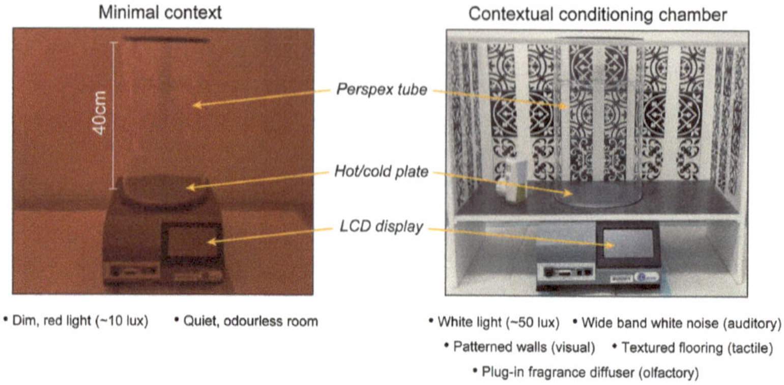

Fig. 3 Photographs of the hot/cold plate analgesiometer in a "minimal context" setting (left), in which thermal sensitivity testing takes place under dim, red lighting within a quiet, odourless room; or placed within our custom-built conditioning chamber (right), in which thermal sensitivity testing takes place amongst a collection of multisensory stimuli (visual, olfactory, auditory and tactile)

Video playback of the hot/cold plate test should be run at normal playback speed. Scoring should begin only once the rat's two hind paws first touch the plate. The chosen scoring setup should be able to precisely record the time (relative to test start) and duration of each instance one of these behaviors is performed. It is important to note that these behaviors are not mutually exclusive, and rats will often perform a paw withdrawal while rearing. Hind paw withdrawals can also be performed during a grooming sequence. Figure 2 illustrates several examples of hind paw withdrawals, rearing, grooming, and combinations of these behaviors.

2.5 Contextual Conditioning Chamber

Work by Swanton (2020) [49], which analyzed all studies investigating placebo analgesia in rodents up to and including 2020, found that conditioning cue type and cue combination significantly influenced the size of the placebo effects. Importantly, studies that used a specific contextual conditioning chamber produced the strongest effects, whereas studies that used two or more diffuse cues produced weak or no placebo effects. We use a custom-built conditioning chamber (dimensions 60 cm [w] x 45 cm [h] x 30 cm [d]), which directly houses the hot/cold plate analgesiometer (see Fig. 3). A collection of multisensory "neutral" cues was installed in the chamber or presented alongside it. These were:

- **Visual**—a fractal-like visual pattern was placed on three walls of the chamber (the fourth wall was transparent to allow filming).
- **Olfactory**—a plug-in fragrance diffuser (Ambi Pur™) filled the testing room with a vanilla-like scent, and the chamber was wiped with a lemon-scented nonalcoholic disinfecting wipe between tests.

- **Tactile**—textured polyvinyl was attached to the chamber floor.
- **Auditory**—wide-band, white noise was played just outside of the conditioning chamber (~80 dB inside the chamber).

3 Methods and Experimental Designs

The protocols described below are adapted from the methods outlined in [11, 50, 51].

3.1 Pharmacological Conditioning

Materials

Analgesic drugs (e.g., morphine, gabapentin, fentanyl) and dilution vehicle (e.g., saline); 1 mL syringes; 25-gauge needles; contextual conditioning chamber; thermal sensitivity testing equipment (see Sect. 2.3); behavioral scoring equipment (see Sect. 2.4).

Procedure

Figure 4 illustrates the experimental timeline for a pharmacological conditioning design, similar to that used in both [11, 50]. In order to elicit placebo analgesia in rats with CCI, the pharmacological conditioning procedure pairs a range of neutral contextual cues with the experience of morphine-induced analgesia and subsequently tests the rats' thermal sensitivity in the presence of these contextual cues without morphine administration. This treatment regimen constitutes the placebo treatment group, which receives 4 days of morphine + neutral cues, immediately followed by a day of saline + neutral cues.

The essential control group that determines the size of any placebo effects is a natural history control group, which receives saline for all 5 days. Additionally, two other control groups may be added to determine the degree of drug tolerance developed: (i) a morphine control group, which receives morphine for all 5 days and (ii) a drug control group, which receives saline during pharmacological conditioning and morphine on Test Day. For simplicity, these groups

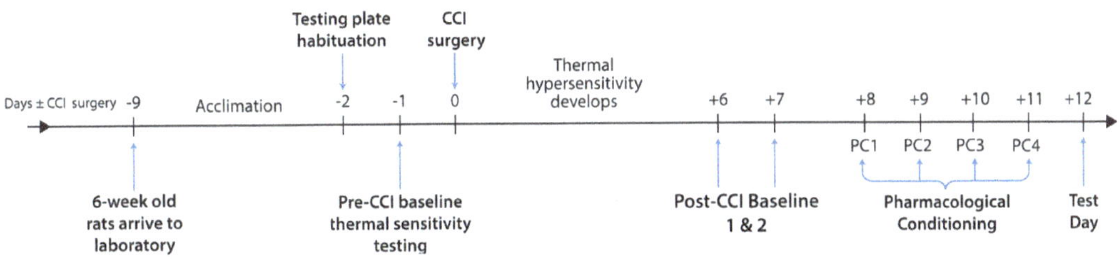

Fig. 4 Example experimental timeline for pharmacological conditioning procedures similar to that used in [11, 50]

are referred to as "treatment during pharmacological conditioning" and "treatment on Test Day," respectively, as indicated in Table 1.

After acclimation to laboratory conditions, rats undergo hot/cold plate apparatus habituation, pre-CCI baseline thermal sensitivity testing, and CCI surgery on three consecutive days. Rats are then given 5–6 days with no intervention or testing to allow for the development of thermal hypersensitivity. On days +6 and +7 post-CCI, rats' thermal sensitivity is retested to confirm the development of thermal allodynia and hyperalgesia. As is well established in the literature, CCI produces large individual variability in both thermal and mechanical allodynia, and a proportion of rats will fail to develop any significant thermal hypersensitivity after CCI (~15% for Sprague–Dawley rats) [52]. Since it is essential that the pharmacological conditioning procedure induces analgesiato allow for the expression of placebo analgesia, it is critical that these nonallodynic rats be excluded from analysis. Based on our pilot data, we exclude rats that do not demonstrate at least five hind paw withdrawals totaling 2 s during the 60 s thermal sensitivity testing procedure on at least one of the post-CCI baseline days. Figure 5 illustrates this individual variability in hind paw withdrawal behaviors to noxious cold stimuli (5°C) seen 6–7 days after CCI in both male and female rats, as well as rats that were excluded based on this criterion. All pre- and post-CCI baseline testing is conducted in the housing room in a "minimal context" setting during the first 2–4 h of the dark phase (i.e., the rat's normal waking cycle).

Following the post-CCI baseline tests, rats undergo pharmacological conditioning (4 days) and then Test Day over 5 consecutive days. Pharmacological conditioning takes place in a different room from the housing room, under medium-level white lighting to provide the visual cues. The conditioning procedure begins when a rat is injected with the chosen analgesic drug. We chose intraperitoneal injections (rather than subcutaneous or intravenous) for their ease of administration and relatively quick absorption rate. Immediately

Table 1

Drug treatments during pharmacological conditioning and Test Day for the placebo treatment and the three control groups

	Pharmacological conditioning				Test Day
Days postsurgery	**+8**	**+9**	**+10**	**+11**	**+12**
Mor/Sal (placebo condition)	Mor	Mor	Mor	Mor	Sal
Sal/Sal (natural history control)	Sal	Sal	Sal	Sal	Sal
Mor/Mor (drug control)	Mor	Mor	Mor	Mor	Mor
Sal/Mor (drug control)	Sal	Sal	Sal	Sal	Mor

Sal Saline, *Mor* Morphine

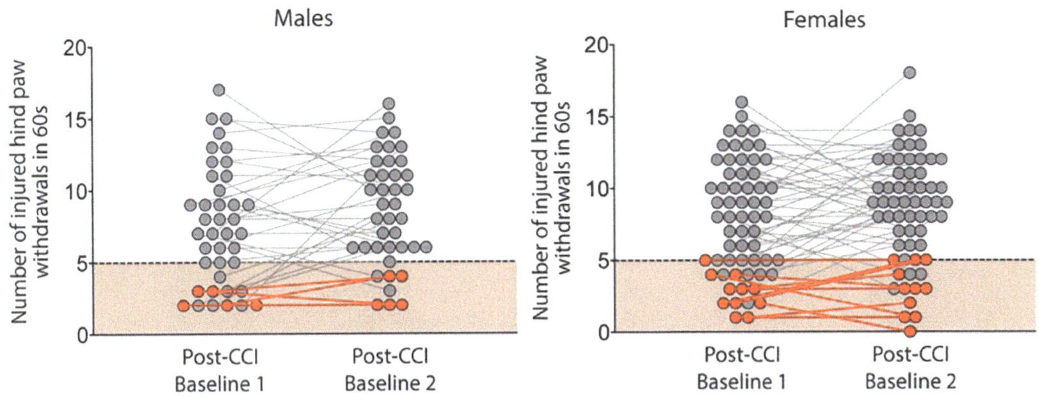

Fig. 5 Individual variability in hind paw withdrawal behavior on the cold plate +6 (Post-CCI baseline 1) and +7 (Post-CCI baseline 2) days after CCI surgery. Graphs show the number of injured hind paw withdrawals performed by individual rats on the cold plate (5°C) in 60 s. Red circles and red lines indicate the rats that were excluded as they did not perform "at least 5 hind paw withdrawals totalling 2 seconds during the 60s thermal sensitivity testing procedure on at least one of the post-CCI baseline days." (Data from Boorman and Keay [11])

after injection, the rat is transferred into the conditioning chamber in the testing room. The rat should be allowed to freely explore the conditioning chamber for the duration that the analgesic drug takes to reach its maximum potential effect (MPE). For i.p morphine, this equates to 30–45 min. At this time, thermal sensitivity testing should be performed (add in the Perspex cylinder; see Sect. 2.3), after which the rat should be returned to its home cage.

Classifying the Placebo Responders

Rats in the placebo treatment group (i.e., morphine/saline) display large individual variability in the degree of hind paw withdrawal behavior during thermal sensitivity testing on Test Day. This variability is significantly larger than the individual variability displayed by the natural history control group (i.e., saline/saline), and thus reflects subpopulations of placebo responders and nonresponders. In order to classify rats into these phenotypes, we use the following criterion: if a rat demonstrates fewer than half of the average of its two post-CCI baseline tests, the rat is classified as a placebo responder; otherwise, it is classified as a nonresponder. This criterion can be described by the following equation:

$$\text{Number}_{(\text{Test Day})>} < \frac{1}{2} \times \frac{\left(\text{Number}_{[\text{Post-CCI1}]} + \text{Number}_{[\text{Post-CCI2}]}\right)}{2}$$

This criterion, which requires greater than a 50% reduction in hind paw withdrawals compared to the rat's own post-CCI baseline,

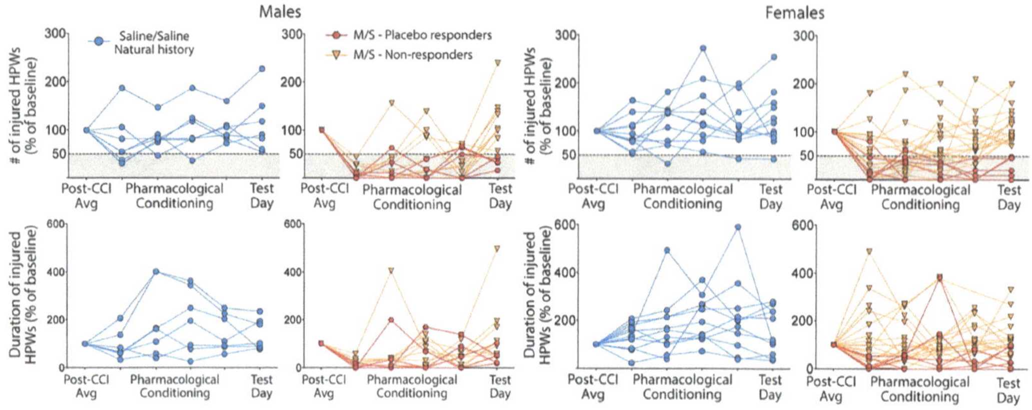

Fig. 6 Individual behavioral profiles of rats in the natural history control group (blue) and the placebo treatment group (orange and red) during the pharmacological conditioning procedure and on Test Day. The number of hind paw withdrawals (top row) and the duration of hind paw withdrawals (bottom row) are shown as a percentage of the rat's own post-CCI baseline average. Rats in the placebo treatment group were either classified as placebo responders (red) or nonresponders (orange) based on the criterion described above. Using this criterion, 5/14 male rats and 6/24 female rats that were given the placebo treatment regimen were classified as placebo responders. In comparison, 0/8 of the male rats and 1/11 of the female rats in the natural history control group would be considered "placebo responders" using this same criterion, indicating that the reduced hind paw withdrawal behavior on Test Day shown by some of the rats in the placebo treatment group is not simply variability in thermal sensitivity across time (Data from Boorman and Keay (2021))

is at the conservative end of the criteria most often used in humans to define a clinically meaningful pain reduction [53–56]. Figure 6 shows individual hind paw withdrawal behavior (as a percentage of their post-CCI baseline average) during pharmacological conditioning and Test Day for the natural history control group and the placebo treatment group for both males and females.

Notes

- Drug injections should take place in a separate room from the testing room so that conditioned stress-induced analgesia cannot develop.
- Conditioning with an escalating morphine dose counteracts the development of tolerance to the analgesic effects of morphine and appears to abolish the placebo responses in all rats [50].
- Rats should be assigned randomly to each experimental group. We use a computer-generated, simple randomization procedure within Microsoft Excel.

3.2 Response Conditioning

Materials

Contextual conditioning chamber; thermal sensitivity testing equipment (see Sect. 2.3); behavioral scoring equipment (see Sect. 2.4).

Procedure

Figure 7 illustrates the experimental timeline for the response conditioning design, which was used in Boorman and Keay (2023) [51]. Response conditioning procedures differ from pharmacological conditioning in that analgesic drugs are not required to "condition" a placebo response. Instead, this procedure relies on rats learning an association between contextual cues and a specific stimulus intensity. For instance, placebo analgesia can be elicited by repeatedly presenting a non-noxious thermal stimulus (via the hot/cold plate) with the contextual conditioning chamber and subsequently testing the rat's thermal sensitivity to a moderate or high noxious stimulus in the presence of the same contextual cues.

There are two main advantages of using response conditioning over pharmacological conditioning. First, no drugs are necessary, which removes one potential source of individual variability in hind paw withdrawal behavior performed during conditioning and on Test Day. That is, individual differences in pharmacokinetic and pharmacodynamic responses to the drug. Second, this procedure can readily elicit both placebo analgesia and nocebo hyperalgesia with simple changes to the conditioning stimulus intensity. However, there is also a notable disadvantage when using response conditioning rather than pharmacological conditioning—we cannot perform post-CCI baseline testing to characterize each rat's individual thermal sensitivity profile. This is because any exposure to the hot/cold plate would interfere with the response conditioning procedure itself. As such, identifying the placebo responder subpopulation is more challenging, since we are unable to construct a "placebo responder" criterion that is relative to each rat's thermal sensitivity profile, as we did for the pharmacological conditioning design.

The experimental timeline for response conditioning procedures is similar to that used for pharmacological conditioning. After the 7-day acclimation period, rats are habituated to the hot/cold plate apparatus for 10 min each. This habituation takes place in the testing room in a minimal context setting. Additionally, to prevent exposure of the rats to the metal hot/cold plate (which is the unconditioned stimulus), the plate is covered with 3 mm medium-density fiberboard during the habituation. The following day, rats undergo pre-CCI

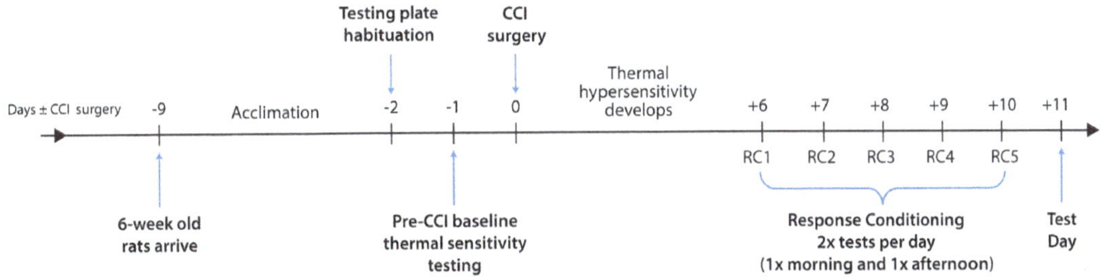

Fig. 7 Experimental timeline for response conditioning procedure, which was used in Boorman and Keay [51]

thermal sensitivity testing, in which the hot/cold plate should be set to the same temperature as will be experienced during the response conditioning procedure (see Table 2). This test should also be conducted in the testing room under the minimal context setting (see Sect. 2.5 for details). CCI surgery is performed the next day, followed by a 5-day no-intervention period to allow for the development of thermal hypersensitivity. Rats then undergo the response conditioning procedure, which repeatedly pairs the same thermal stimulus ten times over five consecutive days (five morning and five afternoon tests) within the contextual conditioning chamber. For the placebo group, the conditioning temperature is set to a non-noxious thermoneutral temperature (30°C), whereas for the nocebo group, the conditioning temperature is set to a noxious cold temperature (4°C). On the following day (Test Day), placebo and nocebo groups are tested on a mildly noxious cold stimulus (20°C). The hind paw withdrawal behaviors of these groups are then compared to a mildly noxious cold stimulus (20°C) natural history control group to determine the size of any placebo and nocebo effects. Two additional natural history control groups should also be included to further determine the relative strength of the conditioning effects of the placebo and nocebo groups. These are: (i) a thermoneutral (30°C) control and (ii) a noxious cold (4°C) control, which receives the same stimulus intensity on Test Day as during response conditioning. These groups indicate the maximal possible effect of conditioning. Therefore, differences between the placebo group and the thermoneutral control indicate the relative magnitude of placebo analgesia, whereas differences between the nocebo group and the noxious cold control

Table 2
Thermal sensitivity testing temperatures during the response conditioning protocol for the placebo group, nocebo group, and the three natural history control groups

	Pre-CCI	Response conditioning					Test Day
Days postsurgery	**−1**	**+6**	**+7**	**+8**	**+9**	**+10**	**+11**
Placebo condition	30°C	30°C × 2	30°C × 2	30°C × 2	30°C × 2	30°C × 2	20°C
Nocebo condition	4°C	4°C × 2	4°C × 2	4°C × 2	4°C × 2	4°C × 2	20°C
Mild cold natural history control	20°C	20°C × 2	20°C × 2	20°C × 2	20°C × 2	20°C × 2	20°C
Noxious cold natural history control	4°C	4°C×2	4°C × 2	4°C × 2	4°C × 2	4°C × 2	4°C
Thermoneutral natural history control	30°C	30°C×2	30°C× 2	30°C×2	30°C×2	30°C×2	30°C

indicate the relative magnitude of nocebo hyperalgesia exhibited. The testing regimen for all experimental groups is shown in Table 2.

For each thermal sensitivity testing session, rats are tested for 150 s each, with all rats from the same cage tested in the same order in succession (~2 min between tests). The hot/cold plate apparatus should be thoroughly cleaned with 70% ethanol and lightly scented disinfecting wipes between each test.

Analysis

Since we are unable to set a similar "placebo responder" criterion as we can for the pharmacological conditioning design, we instead calculate group-level placebo and nocebo effects. To do this, we use *estimation statistics*, which provides both an estimate of the effect size and an uncertainty interval (most often expressed as the 95% confidence interval). These statistics can be easily calculated using the web application built by Hung Nguyen (estimationstats.com), which uses the Python code developed by Ho et al. [57]. Effect sizes (Cohen's *d*) and their 95% confidence intervals are calculated using bias-corrected and accelerated bootstrap resampling (5000 bootstrap samples per test). When reporting these results, effect sizes between 0.2 and 0.5 are considered "small," effect sizes between 0.5 and 0.8 are considered "moderate," and effect sizes above 0.8 are considered "large." However, the uncertainty estimates should also contribute to the interpretation of the statistical results when reporting and making conclusions about the data.

To estimate the direction and size of any placebo and nocebo effects, the Test Day hind paw withdrawal behavior of the placebo and nocebo groups is compared to the mild cold (20°C) natural history control group. A placebo effect is defined as a reduction in hind paw withdrawal behavior of these groups compared to this natural history control group, whereas a nocebo effect is defined as an increase in hind paw withdrawal behavior compared to this natural history control group. Additionally, the "combined conditioning effect" can be calculated, which is the difference in hind paw withdrawal behavior between the placebo and nocebo groups. Figure 8 shows the hind paw withdrawal behaviors of rats in the placebo group (red), the nocebo group (purple), and the mild cold natural history control group (blue) during the response conditioning procedure and on Test Day (data from [25]). The response conditioning procedures described in this chapter produced moderate placebo effects and large nocebo effects. By comparing the hind paw withdrawal behavior of individual rats on Test Day to their behaviors during the final days of response conditioning, it is possible to identify rats as likely placebo/nocebo responders and likely non-responder phenotypes (panels C, D).

Notes

- We performed a pilot study with 24 rats to determine both the optimal plate temperatures and the testing frequency. The data

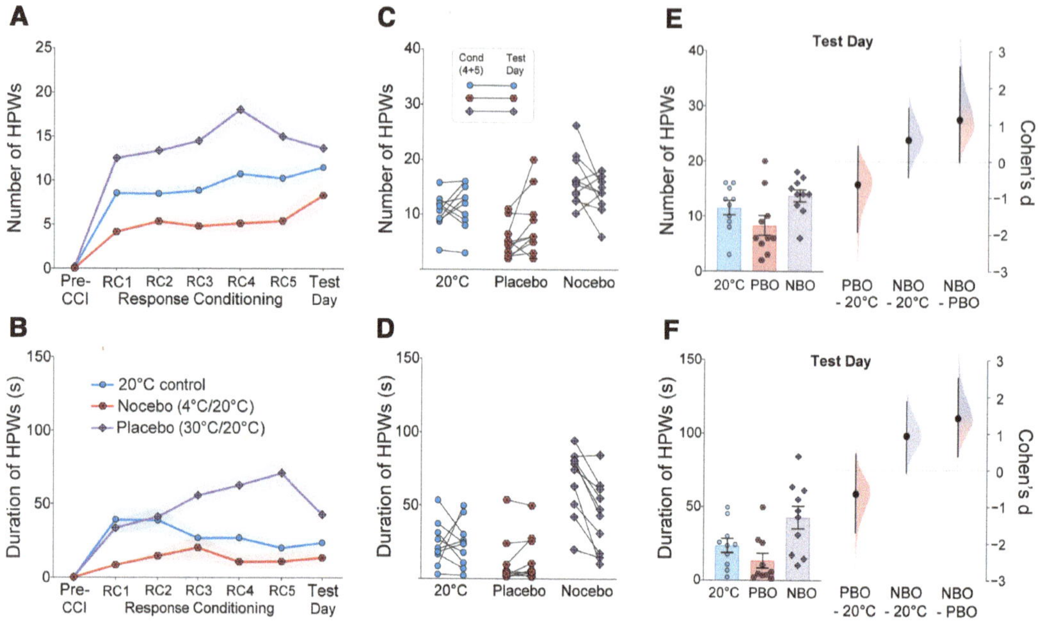

Fig. 8 Hind paw withdrawal behavior of rats that underwent placebo conditioning (PBO, red strikethrough hexagons, $n = 10$) and nocebo conditioning (NBO, purple strikethrough diamonds, $n = 10$). (A) The number of hind paw withdrawals, (B) the cumulative duration of hind paw withdrawals prior to injury (pre-CCI), during conditioning and on Test Day. Shaded regions indicate ±SEM. (C, D) Individual data showing changes in hind paw withdrawal behavior from the average of the final 2 days of conditioning (Cond 4 + 5, left side) to the Test Day. (E, F) Comparisons of hind paw withdrawal behavior on Test Day between the groups. The effect sizes (Cohen's *d*) are shown on the Cummings estimation plots, with mean differences between groups plotted as a bootstrap sampling distribution. Dots indicate mean differences, and the 95% confidence intervals are indicated by the ends of the vertical error bars

from this study confirmed that 4°C elicited frequent injured paw withdrawal responses with long durations in nearly all rats post-CCI, 20°C elicited more moderate paw withdrawal behaviors, and 30°C was the least noxious temperature tested. There does not appear to be a plate temperature that does not elicit hind paw withdrawals in at least some CCI rats. Additionally, we found that the conditioning twice per day and using 150 s thermal sensitivity tests elicited the strongest conditioned responses.

- Statistical analysis comparing the degree of hind paw withdrawals between morning and afternoon tests found negligible "time-of-day" effects, and thus we averaged the data from morning and afternoon thermal sensitivity testing sessions for each day [25].
- The same analysis used to determine the size of any placebo and nocebo effects based on hind paw withdrawal behavior can be used for rearing behavior.

4 Summary

This chapter details two methodological approaches that have been recently used to successfully elicit placebo analgesia or nocebo hyperalgesia in rats with a neuropathic injury (i.e., sciatic nerve chronic constriction injury). The first methodological approach, *pharmacological conditioning*, repeatedly pairs the experience of drug-induced analgesia with a contextual conditioning chamber and subsequently tests the rat's thermal sensitivity in the same contextual conditioning chamber without the drug. One of the significant advantages of this study design is that we are more easily able to identify the placebo responders from the non-responders, since we can compare the rat's behavior on Test Day to its own postinjury baseline. We presented data from our study [11], which identified subpopulations of both male (36%) and female (25%) placebo responders. The second methodological approach, *response conditioning*, pairs a particular thermal stimulus intensity (low intensity for placebo; high intensity for nocebo) with the contextual conditioning chamber and subsequently tests the rat's thermal sensitivity within the contextual conditioning chamber to a moderate stimulus intensity. One advantage of this approach is that it can be used to elicit either placebo analgesia or nocebo hyperalgesia. However, it can also be more challenging to identify the responder vs. non-responder phenotypes. Importantly, these experimental designs can be combined with a range of powerful preclinical techniques, including electrophysiology, lesioning, tract tracing, optogenetics, chemogenetics, and calcium imaging to investigate the neural substrates of placebo analgesia and nocebo hyperalgesia in the context of chronic neuropathic pain.

Conflict of Interest

The authors have no conflicts of interest to declare that are relevant to the content of this chapter.

References

1. Hoffman GA, Harrington A, Fields HL (2005) Pain and the placebo: what we have learned. Perspect Biol Med 48(2):248–265
2. Beecher HK (1955) The powerful placebo. J Am Med Assoc 159(17):1602–1606
3. Levine JD, Gordon NC, Bornstein JC, Fields HL (1979) Role of pain in placebo analgesia. Proc Natl Acad Sci USA 76(7):3528–3531
4. Benedetti F (1996) The opposite effects of the opiate antagonist naloxone and the cholecystokinin antagonist proglumide on placebo analgesia. Pain 64(3):535–543
5. Petrovic P, Kalso E, Petersson KM, Ingvar M (2002) Placebo and opioid analgesia – imaging a shared neuronal network. Science 295(5560):1737–1740
6. Crawford LS, Mills EP, Hanson T, Macey PM, Glarin R, Macefield VG, Keay KA, Henderson LA (2021) Brainstem mechanisms of pain modulation: a within-subjects 7T fMRI study of placebo analgesic and nocebo hyperalgesic responses. J Neurosci 41(47):9794–9806
7. Tetreault P, Mansour A, Vachon-Presseau E, Schnitzer TJ, Apkarian AV, Baliki MN (2016)

Brain connectivity predicts placebo response across chronic pain clinical trials. PLoS Biol 14(10):e1002570
8. Kaptchuk TJ, Kelley JM, Deykin A, Wayne PM, Lasagna LC, Epstein IO, Kirsch I, Wechsler ME (2008) Do "placebo responders" exist?. Contemp Clin Trials 29(4):587–595
9. Koban L, Ruzic L, Wager TD (2013) Chapter 10 - brain predictors of individual differences in placebo responding. In: Colloca L, Flaten MA, Meissner K (eds) Placebo and pain. Academic Press, San Diego, pp 89–102
10. Zeng Y, Hu D, Yang W, Hayashinaka E, Wada Y, Watanabe Y, Zeng Q, Cui Y (2018) A voxel-based analysis of neurobiological mechanisms in placebo analgesia in rats. Neuroimage 178:602–612
11. Boorman DC, Keay KA (2021) Morphine-conditioned placebo analgesia in female and male rats with chronic neuropathic pain: c-Fos expression in the rostral ventromedial medulla. Neuroscience 457:51–73
12. Nolan TA, Price DD, Caudle RM, Murphy NP, Neubert JK (2012) Placebo-induced analgesia in an operant pain model in rats. Pain 153(10):2009–2016
13. McNabb CT, White MM, Harris AL, Fuchs PN (2014) The elusive rat model of conditioned placebo analgesia. Pain 155(10):2022–2032
14. Yin XS, Yang JY, Cao S, Wang Y (2020) Failure of placebo analgesia model in rats with inflammatory pain. Neurosci Bull 36(2):121–133
15. Akintola T, Tricou C, Raver C, Castro A, Colloca L, Keller A (2019) In search of a rodent model of placebo analgesia in chronic orofacial neuropathic pain. Neurobiol Pain 6:100033
16. Colloca L, Akintola T, Haycock NR, Blasini M, Thomas S, Phillips J, Corsi N, Schenk LA, Wang Y (2020) Prior therapeutic experiences, not expectation ratings, predict placebo effects: an experimental study in chronic pain and healthy participants. Psychother Psychosom 89(6):371–378
17. Wang Y, Chan E, Dorsey SG, Campbell CM, Colloca L (2022) Who are the placebo responders? A cross-sectional cohort study for psychological determinants. Pain 163(6):1078–1090
18. Vase L, Skyt I, Hall KT (2016) Placebo, nocebo, and neuropathic pain. Pain 157(Suppl 1):S98–S105
19. Kaptchuk TJ, Hemond CC, Miller FG (2020) Placebos in chronic pain: evidence, theory, ethics, and use in clinical practice. BMJ 370:m1668
20. Bryant CD, Roberts KW, Culbertson CS, Le A, Evans CJ, Fanselow MS (2009) Pavlovian conditioning of multiple opioid-like responses in mice. Drug Alcohol Depend 103(1–2):74–83
21. Zhang RR, Zhang WC, Wang JY, Guo JY (2013) The opioid placebo analgesia is mediated exclusively through mu-opioid receptor in rat. Int J Neuropsychopharmacol 16(4):849–856
22. Guo JY, Wang JY, Luo F (2010) Dissection of placebo analgesia in mice: the conditions for activation of opioid and non-opioid systems. J Psychopharmacol 24(10):1561–1567
23. Lee IS, Lee B, Park HJ, Olausson H, Enck P, Chae Y (2015) A new animal model of placebo analgesia: involvement of the dopaminergic system in reward learning. Sci Rep 5: 17140
24. Xu L, Wan Y, Ma L, Zheng J, Han B, Liu FY, Yi M, Wan Y (2018) A context-based analgesia model in rats: involvement of prefrontal cortex. Neurosci Bull 34(6):1047–1057
25. Boorman DC, Keay KA (2023) Learning pain in context: response-conditioned placebo analgesia and nocebo hyperalgesia in male rats with chronic neuropathic pain. Physiol Behav 263:114116
26. Sorge RE, Martin LJ, Isbester KA, Sotocinal SG, Rosen S, Tuttle AH, Wieskopf JS, Acland EL, Dokova A, Kadoura B, Leger P, Mapplebeck JC, McPhail M, Delaney A, Wigerblad G, Schumann AP, Quinn T, Frasnelli J, Svensson CI, Sternberg WF, Mogil JS (2014) Olfactory exposure to males, including men, causes stress and related analgesia in rodents. Nat Methods 11(6):629–632
27. Neitz J, Jacobs GH (1986) Reexamination of spectral mechanisms in the rat (Rattus norvegicus). J Comp Psychol 100(1):21–29
28. Bennett GJ, Xie YK (1988) A peripheral mononeuropathy in rat that produces disorders of pain sensation like those seen in man. Pain 33(1):87–107
29. Baptista-de-Souza D, Rodrigues Tavares LR, Canto-de-Souza L, Nunes-de-Souza RL, Canto-de-Souza A (2022) Behavioral, hormonal, and neural alterations induced by social contagion for pain in mice. Neuropharmacology 203:108878
30. Jensen TS, Yaksh TL (1986) Comparison of antinociceptive action of morphine in the periaqueductal gray, medial and paramedial medulla in rat. Brain Res 363(1):99–113
31. Gardmark M, Hoglund AU, Hammarlund-Udenaes M (1998) Aspects on tail-flick, hot-plate and electrical stimulation tests for morphine antinociception. Pharmacol Toxicol 83(6):252–258
32. Craft RM, Bernal SA (2001) Sex differences in opioid antinociception: kappa and 'mixed action' agonists. Drug Alcohol Depend 63(3):215–228
33. Le Bars D, Gozariu M, Cadden SW (2001) Animal models of nociception. Pharmacol Rev 53(4):597–652
34. Desmeules JA, Kayser V, Weil-Fuggaza J, Bertrand A, Guilbaud G (1995) Influence of the sympathetic nervous system in the development

of abnormal pain-related behaviours in a rat model of neuropathic pain. Neuroscience 67(4):941–951
35. Yeomans DC, Pirec V, Proudfit HK (1996) Nociceptive responses to high and low rates of noxious cutaneous heating are mediated by different nociceptors in the rat: behavioral evidence. Pain 68(1):133–140
36. Tanimoto-Mori S, Nakazato-Imasato E, Toide K, Kita Y (2008) Pharmacologic investigation of the mechanism underlying cold allodynia using a new cold plate procedure in rats with chronic constriction injuries. Behav Pharmacol 19(1):85–90
37. Espejo EF, Stinus L, Cador M, Mir D (1994) Effects of morphine and naloxone on behaviour in the hot plate test: an ethopharmacological study in the rat. Psychopharmacol (Berl) 113(3–4):500–510
38. Espejo EF, Mir D (1993) Structure of the rat's behaviour in the hot plate test. Behav Brain Res 56(2):171–176
39. Jasmin L, Kohan L, Franssen M, Janni G, Goff JR (1998) The cold plate as a test of nociceptive behaviors: description and application to the study of chronic neuropathic and inflammatory pain models. Pain 75(2-3):367–382
40. Espejo EF, Mir D (1994) Differential effects of weekly and daily exposure to the hot plate on the rat's behavior. Physiol Behav 55(6):1157–1162
41. Gunn A, Bobeck EN, Weber C, Morgan MM (2011) The influence of non-nociceptive factors on hot-plate latency in rats. J Pain 12(2):222–227
42. Bardo MT, Valone JM (1994) Morphine-conditioned analgesia using a taste cue: dissociation of taste aversion and analgesia. Psychopharmacol (Berl) 114(2):269–274
43. Bevins RA, Valone JM, Bradley MC, Bardo MT (1995) Morphine taste conditioning and analgesia: assessing conditioned and novelty-induced analgesia. Exp Clin Psychopharmacol 3(1):9–14
44. Krank MD (1987) Conditioned hyperalgesia depends on the pain sensitivity measure. Behav Neurosci 101(6):854–857
45. Miller JS, Kelly KS, Neisewander JL, McCoy DF, Bardo MT (1990) Conditioning of morphine-induced taste aversion and analgesia. Psychopharmacol (Berl) 101(4):472–480
46. Randall CK, Kraemer PJ, Valone JM, Bardo MT (1993) Odor conditioning with morphine: Conditioned preference, aversion, and analgesia. Psychobiology 21(3):215–220
47. Valone JM, Randall CK, Kraemer PJ, Bardo MT (1998) Olfactory cues and morphine-induced conditioned analgesia in rats. Pharmacol Biochem Behav 60(1):115–118
48. Guo JY, Yuan XY, Sui F, Zhang WC, Wang JY, Luo F, Luo J (2011) Placebo analgesia affects the behavioral despair tests and hormonal secretions in mice. Psychopharmacol (Berl) 217(1):83–90
49. Swanton R (2020) Learned analgesic cue association and placebo analgesia in rats: an empirical assessment of an animal model and meta-analysis. University of Sydney.
50. Boorman DC, Keay KA (2021) Escalating morphine dosage fails to elicit conditioned analgesia in a preclinical chronic neuropathic pain model. Behav Pharmacol 32(6):479–486
51. Boorman DC, Keay KA (2023) Learning pain in context: Response-conditioned placebo analgesia and nocebo hyperalgesia in male rats with chronic neuropathic pain. Physiol Behav 263:114116
52. De Felice M, Sanoja R, Wang R, Vera-Portocarrero L, Oyarzo J, King T, Ossipov MH, Vanderah TW, Lai J, Dussor GO, Fields HL, Price TJ, Porreca F (2011) Engagement of descending inhibition from the rostral ventromedial medulla protects against chronic neuropathic pain. Pain 152(12):2701–2709
53. Farrar JT, Portenoy RK, Berlin JA, Kinman JL, Strom BL (2000) Defining the clinically important difference in pain outcome measures. Pain 88(3):287–294
54. Rowbotham MC (2001) What is a "clinically meaningful" reduction in pain?. Pain 94(2):131–132
55. Cepeda MS, Africano JM, Polo R, Alcala R, Carr DB (2003) What decline in pain intensity is meaningful to patients with acute pain?. Pain 105(1-2):151–157
56. Olsen MF, Bjerre E, Hansen MD, Tendal B, Hilden J, Hrobjartsson A (2018) Minimum clinically important differences in chronic pain vary considerably by baseline pain and methodological factors: systematic review of empirical studies. J Clin Epidemiol 101:87–106 e2
57. Ho J, Tumkaya T, Aryal S, Choi H, Claridge-Chang A (2019) Moving beyond P values: data analysis with estimation graphics. Nat Methods 16(7):565–566

Check for updates

Chapter 13

Burrowing Behavior as a Pain-Behavioral Readout in Rodent Pain Models

Andy Kuo, Mohammad Zafar Imam, Arjun Muralidharan and Maree T. Smith

Abstract

In the past three decades, there has been a huge collective global research effort aimed at unravelling the complex neurobiology of various chronic pain conditions and a concomitant large investment in novel analgesics discovery. However, promising preclinical efficacy data generated in rodent pain models has typically not translated to analgesic efficacy in proof-of-concept human clinical trials. One aspect of preclinical pain research that has attracted criticism is that stimuli-evoked pain behavioral readouts, such as mechanical allodynia, mechanical hyperalgesia, noxious heat hyperalgesia, or non-noxious cold allodynia, are often used in rodent studies. To address this issue, assessment of nonevoked and ethologically-relevant rodent behavior, such as burrowing, has been proposed as a surrogate measure of spontaneous pain in rodents. In this paradigm, rodents spontaneously empty a tube filled with substrates, such as food pellets or gravel, and the amount of substrate displaced from the tube over certain period of time is assessed as a measure of burrowing behavior. In the following sections of this chapter, we review the literature on burrowing behavior as a behavioral readout of spontaneous pain in rodent pain models, and we provide a step-by-step protocol for evaluating burrowing behavior in rats.

Keywords Rats, Burrowing, Pain, Behavioral, Rodent, Spontaneous, Surrogate, Allodynia, Hyperalgesia

1 Introduction

Chronic pain is a leading cause of disability and a major socioeconomic burden affecting more than 30% of people worldwide [1], with prevalence increasing with age [2, 3]. Arthritis, nerve damage, and cancer-related pain are the top three leading causes of chronic pain [4].

The economic costs of chronic pain that afflicts approximately one in three Americans are large, estimated at US $560 to US $635 billion per year in medical costs and lost productivity alone in 2010 [1]. In Australia, the average cost per year for one of the 15.4% of Australians aged 15 years or older with chronic pain was estimated at AU $22,588 per person or AU $42,979 per person when

Isaura Tavares and Hugo Leite-Almeida (eds.), *Translational Research Methods for Neuropathic Pain*, Neuromethods, vol. 235, https://doi.org/10.1007/978-1-0716-5312-8_13,

nonfinancial costs were also considered [1, 5]. In the European Union, the indirect and direct costs associated with chronic pain are estimated at 2%–3% of GDP [6]. This is equivalent to 441 billion Euros based on the 2016 GDP [6].

Currently, effective pharmacological treatments for the relief of chronic pain that significantly improve the quality of life (QOL) of patients are limited [7, 8]. This is due to either inadequate symptom relief and/or dose-limiting side-effects [9]. Chronic pain conditions that are poorly alleviated by clinically available analgesic/adjuvant agents include osteoarthritis [10], painful diabetic neuropathy [11], and postherpetic neuralgia [12]. Hence, there is a large unmet and urgent medical need for better analgesics.

Drug discovery efforts aimed at generating highly effective and well-tolerated novel pain relief medications has stalled in the past several decades. This is largely underpinned by the high attrition rate due to lack of efficacy in clinical trials [13–15]. One of the main concerns is that the current preclinical pain models rely heavily on stimulus-evoked reflex pain-like responses [13, 16, 17]. These include responses evoked in rodent pain models by (i) von Frey filaments (mechanical allodynia), (ii) the Hargreaves apparatus (thermal hyperalgesia), (iii) the Randall–Selitto apparatus (mechanical hyperalgesia), and (iv) the cold plate (thermal hyperalgesia/allodynia). Although these methods are widely used for their validity in animals, the translatability of these methods is debated, as they measure spinal reflexes rather than integrating complex pain-like behaviors, where cerebral processing of the nociceptive signal is required [18–20].

In both animals and humans, pain can be divided into three major components, spontaneous pain, constant pain, and evoked pain [13, 16, 17, 21]. Pain is a subjective phenomenon that is inferred based upon behavioral responses in rodents, whereas in humans, pain severity ratings are self-reported, encompassing intensity of the nociceptive stimulus and its resultant affective/emotional response [14, 22]. Hence, the term antinociception is used in animals for describing pain relief, whereas in humans, the term, analgesia, is used [23]. Additionally, evoked responses are usually the primary readouts in animal pain models, whereas in humans, randomized controlled trials normally focus on readouts relating to spontaneous pain, as well as other comorbidities that impact patient's QOL [7, 13]. This mismatch between the goal in the clinical trials to restore function *c.f.* largely evoked experimental readouts in preclinical studies has potentially contributed to the limited success in novel pain therapeutics development [24]. Therefore, nonstimulus evoked tests have received growing interest [20]. Example of nonevoked based behavioral assessments, such as burrowing, the grimace scale, conditioned place preference, weight bearing, and gait analysis, have been proposed and developed as a readouts to improve translatability of candidate analgesics into the clinic [16]. Burrowing is of a particular interest as it is an ethologically relevant rodent behavior. Rodents are noted for their burrowing ability, which is innate, natural, and highly conserved [13, 25, 26]. Laboratory mice, similar to wild rodents, have high motivation and a behavioral need to burrow [27]. The assessment of a rodent's ability to clear material from an enclosed space was initially conducted [26] to explore behavioral deficits associated with central nervous system neurodegeneration [28]. Deacon et al. formally standardized the burrowing protocol in an endeavor to ensure simplicity, sensitivity, and robustness [25]. The term "burrowing" coined by Deacon et al., attempted to measure hoarding in mice. They observed that mice were "digging the pellets out of a jar with coordinated hind-limb and fore-limb movements and kicking them outside the container, and not carrying pellets carefully to a selected spot as in true hoarding." Hence, they termed this behavior "burrowing" [25]. For burrowing, rats prefer earthy substrates in contrast to mice where food pellets may be used [25]. Pea shingle and sand being very dense are useful in minimizing ceiling effects for burrowing [25].

In contrast to the stimulus-evoked behavioral assessments that are largely mediated by spinal reflexes, burrowing behavior is under supraspinal control; for instance, lesions to the hippocampus and the medial prefrontal cortex were shown to reduce burrowing activity [25, 29, 30].

The original burrowing protocols were conducted for 2 h (commencing 3 h before the dark cycle) and after that overnight [25]. Many rodents are extremely vigorous burrowers, so they would completely empty virtually any size burrow overnight, and so for this reason a 2 h burrowing period is more sensitive [25]. These observations underpin why most burrowing tests in rodents in the published literature have been conducted over a period of ≤2 h. Intersubject variability in burrowing is considerable [25], as attested to by the relatively large number of animals required in a study [31]. This is clearly illustrated in a study where burrowing activity was assessed in two commonly utilized rat pain models, namely the Freund's Complete Adjuvant (FCA)-induced rat model of unilateral chronic inflammatory pain in one hindpaw and in the chronic constriction injury (CCI) rat model of peripheral neuropathic pain [31], relative to evoked pain-like behaviors in the ipsilateral (injured side) hindpaws. In this example [31], a 1 h burrowing period was utilized, which resulted in relatively large interanimal variability in the burrowing outcomes.

In this chapter, we provide a detailed step-by-step burrowing protocol used in our laboratory. We also discuss burrowing data previously published by our laboratory for both FCA- and CCI-rat models in male rats [31]. The primary objective was to assess the validity and robustness of burrowing as an ethologically-relevant, non-evoked behavioral readout that could be used as a surrogate measure of spontaneous pain in rodent pain models. Specifically, we compared the utility of burrowing behavior relative to evoked mechanical hyperalgesia in the ipsilateral hindpaws of FCA-rats and evoked mechanical allodynia in the ipsilateral hindpaws of CCI-rats [31].

2 Materials

2.1 Animals

The "burrowing" study [31] was conducted in rat models of chronic inflammatory pain (FCA-rats) and peripheral neuropathic pain (CCI-rats), in accordance with the guidelines set out in the Australian Code of Practice for the Care and Use of Animals for Scientific Purposes, eighth edition [32]. Animal ethics approval was obtained from the University of Queensland Animal Ethics Committee prior to initiation of the study. Male Sprague–Dawley (SD) rats were purchased from the Animal Resources Centre (Perth, WA, Australia).

Male SD rats weighing 180–200 g (6–7 weeks of age) upon arrival were housed in a purpose-built Physical Containment Level 2 (PC2) animal holding facility. Animals were housed in groups of two or three in individually ventilated cages (Bio-Zone Global, Thornehill, UK) in a temperature-controlled facility (23°C ± 3°C) equipped with a 12 h/12 h light/dark cycle. Animals were maintained in cages with recycled paper bedding material (FibreCycle Pty Ltd., Yatala, Australia). Kimwipes (Kimberly-Clark Professional, Milsons Point, Australia), red rodent hutches (Able Scientific, Welshpool, Australia), and rat chewsticks (Able Scientific, Welshpool, Australia) were available in the cages as environmental enrichments. Rats had access to food (Specialty Feeds, Glen Forrest, Western Australia, Australia) and tap water ad libitum. Rats were acclimatized for at least 4 days before commencing any experiments.

3 Methods

3.1 Induction of FCA-Induced Inflammatory Pain

Inflammatory pain was induced in rats by unilateral intraplantar (i.pl.) injection of FCA, as previously described [33–35]. Briefly, whilst anesthetized with 3% isoflurane delivered in oxygen, SD rats (200–250 g) received an i.pl. injection of 150 μL of FCA (Sigma Aldrich, MO, USA) into their left hindpaws. The corresponding groups of sham-rats received 150 μL i.pl. injections of sterile saline. Following i.pl. injection, rats were returned to their home cages and were monitored thrice weekly for general health and body weight changes until study completion at day 14 post-FCA or saline injection.

3.2 Induction of CCI-Induced Peripheral Neuropathy Pain

Chronic constriction injury (CCI) of the sciatic nerve was used to induce peripheral neuropathy in rats, according to a published method [36]. Briefly, male SD rats (200–250 g) were anesthetized with 3% isoflurane delivered in oxygen. After shaving the left thigh, the skin was cleaned using 70% ethanol followed by betadine. Next, a small incision was made through the biceps femoris to expose the sciatic nerve. Subsequently, four loose ligatures (~1 mm apart) were tied proximal to the trifurcation of the sciatic nerve using silk sutures (4/0). The muscle and skin were closed, and the animals were monitored closely during surgical recovery. For rats that underwent sham surgery, the sciatic nerve was exposed but not ligated. Following surgery, the rats were returned to their home cages and were monitored for general health and body weight changes twice weekly until study completion at day 14 post-CCI or -sham surgery.

3.3 Behavioral Assessments

All behavioral experiments described herein, except burrowing, were conducted between 09:00 and 14:00 h. Burrowing experiments were conducted for 1 h between 15:30 and 17:00 h towards the end of the 12 h light phase [12 h light (06:00–18:00 h)/dark cycle (18:00–06:00 h)]. Burrowing sessions took place at the same time each day.

All behavioral assessments were conducted before model induction (i.pl. FCA injection or CCI of the sciatic nerve), and then at scheduled postmodel induction timepoints (Fig. 1A, B). Animals had 2 days of social facilitation (SF) and 3 days of individual burrowing training (IT) prior to any other experimentation. Baseline burrowing behavior, mechanical hyperalgesia in the bilateral hindpaws, and hindpaw volumes were assessed in rats prior to i.pl. FCA or saline (sham group) injection on day 0, and thereafter, thrice weekly until study completion on day 13/14 postmodel induction (Fig. 1A). For CCI- or sham-surgery rats, baseline burrowing behavior and von Frey paw withdrawal thresholds were assessed in the bilateral hindpaws prior to surgery on day −1 and day 0, respectively, and then at twice-weekly intervals until study completion on day 13/14 postmodel induction (Fig. 1B). For convenience, the pre-CCI burrowing baseline was determined on day −1, in contrast to FCA-rats where this assessment was performed on day 0.

3.4 Assessment of Hindpaw Volumes (FCA Model)

For rats administered an i.pl. injection of either FCA or saline, ipsilateral (injected side) and contralateral (noninjected side) hindpaw volumes (PV) were measured using a plethysmometer (Ugo Basile, Italy). Measurements were undertaken prior to i.pl. FCA or saline injection on day −1, and thereafter thrice-weekly until study completion on day 13/14.

3.5 Assessment of Mechanical Hyperalgesia (FCA Model)

Baseline paw pressure thresholds (PPTs) for each of the ipsilateral and contralateral hindpaws of FCA- and sham-rats were measured using the Randall–Selitto apparatus (Ugo Basile, Italy), as previously described [37]. Briefly, a noxious mechanical stimulus of increasing force was applied to the medial portion of the hindpaw until a withdrawal response was elicited. The maximum force applied was 250 g to prevent tissue damage. Baseline PPT values for each of the ipsilateral and contralateral hindpaws were the mean of three readings for the

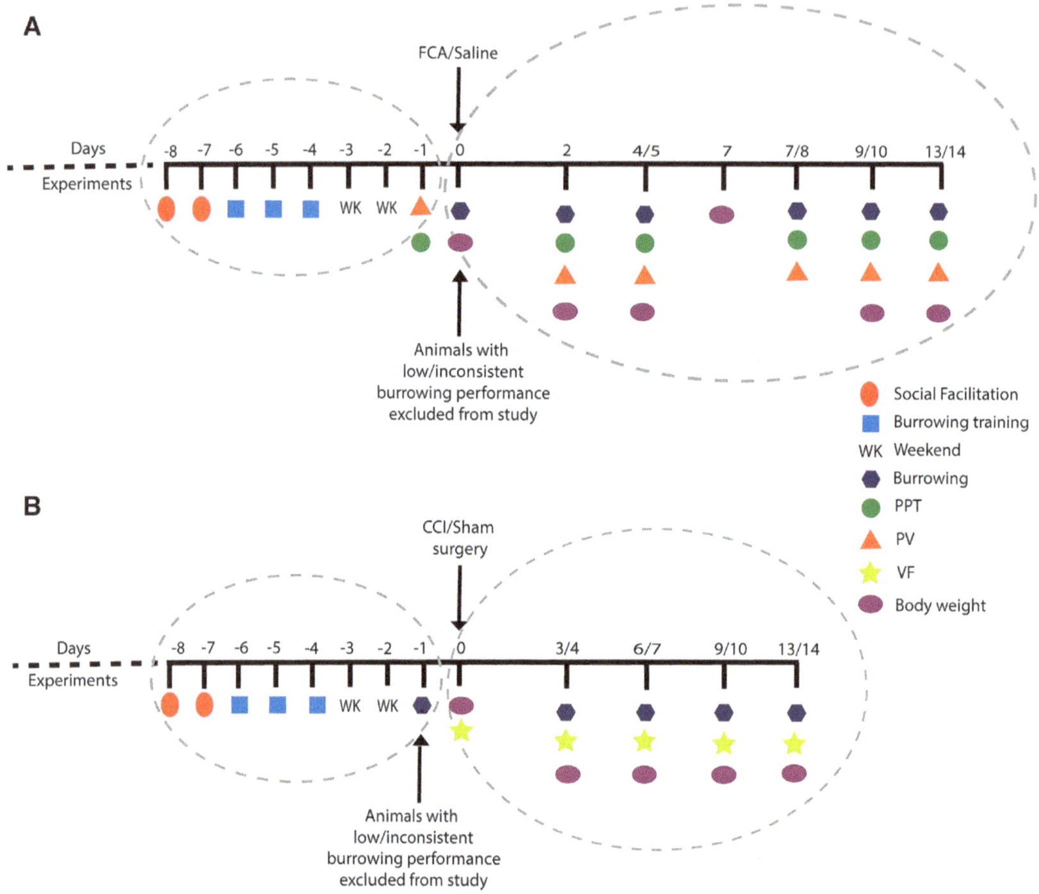

Fig. 1 Schematic timeline for (**A**) FCA and (**B**) CCI experimental procedures. *FCA* Freund's complete adjuvant, *CCI* chronic constriction injury, *PPT* paw pressure threshold, *PV* paw volume *VF* von Frey paw withdrawal threshold

corresponding hindpaw, with a 5-min interval between consecutive measurements. The baseline PPTs were determined in both hindpaws prior to i.pl. FCA or saline injection on day −1, and thereafter, up to thrice weekly until study completion at day 13/14.

3.6 Assessment of Mechanical Allodynia (CCI Model)

Baseline paw withdrawal thresholds (PWTs) for each of the ipsilateral and contralateral hindpaws of CCI- or sham-rats were measured using calibrated von Frey filaments, as previously described [38]. Baseline PWTs were determined in both hindpaws prior to CCI- or sham-surgery on day 0 and twice weekly thereafter until study completion at day 13/14 postsurgery.

3.7 Assessment of Burrowing Behavior (FCA and CCI Models)

Burrowing behavior is known to be altered by factors such as motivation [39], anxiety, and distress [40]. Burrowing performance was measured in the absence of an experimenter in the room so as to avoid any effects of predator–prey relationships and subjectivity bias

[13]. The gravel (smooth, 3–5 mm in diameter) used in the study was washed with chlorhexidine and tap water, then thoroughly rinsed with tap water and dried (40°C) prior to initiation of experiments for each rat Cohort (Fig. 2). The gravel was not washed between different experimental days of the same Cohort. On completion of studies for each Cohort, the gravel was washed and dried, and then used for studies in another Cohort in a similar fashion.

In detail, the burrowing apparatus comprised a clean transparent Plexiglas rat cage [49.8 (L) × 38 (W) × 21.5 (H) cm] (Bio-Zone Global, Thornehill, UK) containing a hollow black colored plastic (polyvinyl chloride) burrowing tube [32 (L) × 10 (D) cm] sealed at one end and open at the other end that was custom-prepared by the University of Queensland, Faculty of Science Workshop (Brisbane, QLD, Australia). The burrowing tube was filled with 2 kg of gravel, and the open entrance was raised 6 cm above the floor in the cage (Fig. 3). The burrowing tube was placed diagonally in the burrowing cage with the sealed end touching the back wall of the cage.

Our burrowing method [31] was based upon a previously published method [25, 41], with slight modifications. All animals were habituated and trained in the burrowing apparatus for 5 days prior to measurement of their baseline burrowing behavior. The training phase comprised social facilitation (SF; Days −8 and −7) and individual training (IT; Days −6, −5, and −4). All experiments were conducted toward the end of the light cycle (15:30–17:00 h). On the first day of social facilitation (Day −8), rats were placed in pairs into the testing apparatus with a burrowing tube filled with 2 kg of gravel for a period of 1 h. After 1 h, the amount of gravel displaced from the tube was weighed. At the end of experimentation, all rats were returned to

Fig. 2 Burrowing substrate, gravel (3–5 mm diameter) used in our burrowing assay

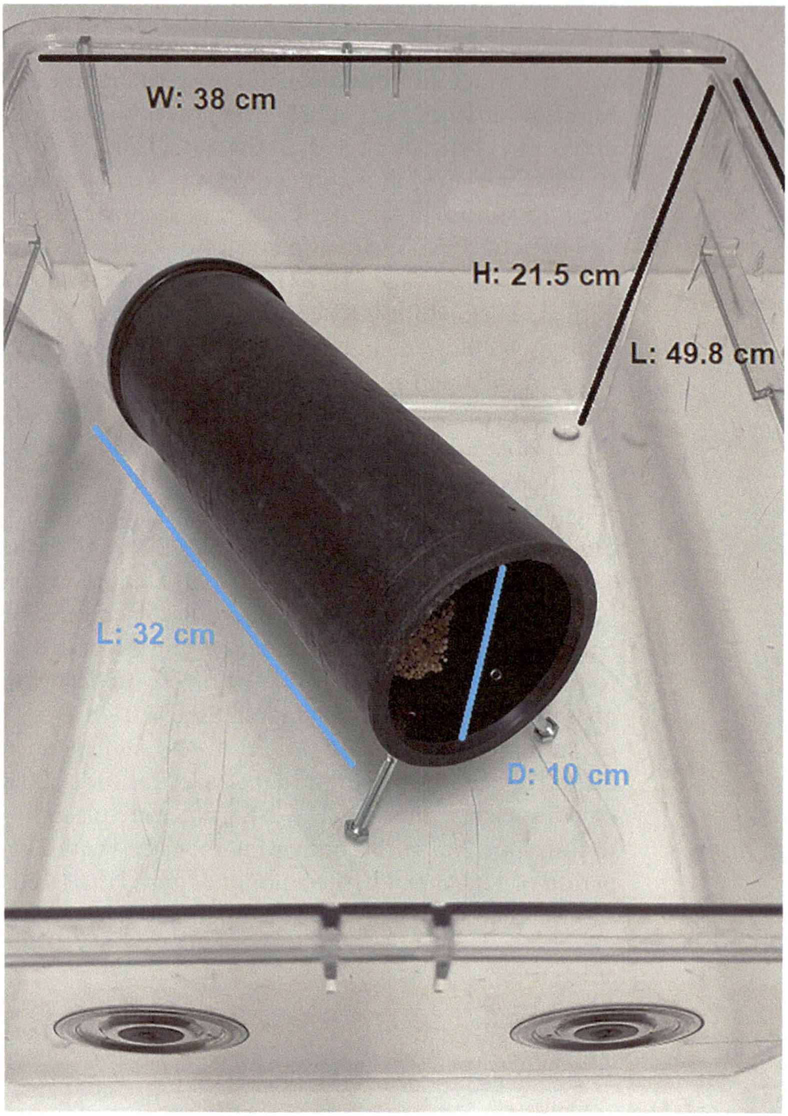

Fig. 3 Burrowing apparatus prepared for the burrowing assay

their home cages. The same procedure was repeated on the second day of social facilitation (Day −7). If any rat pairs did not burrow at all on Day −8, then one rat from this pair was swapped with a rat from a burrowing pair (i.e., a rat that burrowed on Day −8) for social facilitation on Day −7. Burrowing has been shown to be enhanced with practice or when other animals are present [24, 26, 42], and this is the rationale for the 2 days of SF training.

On each day of individual training (Days −6, −5, and −4), rats were placed individually in a test cage containing a burrowing tube filled with 2 kg of gravel. The amount of gravel displaced from the tube after 1 h was weighed. At the end of 3 days of IT, rats that burrowed ≤450 g of gravel were excluded from further participation in the study. Baseline burrowing data were collected from the remaining rats that burrowed >450 g on at least 2 out of 3 days of IT. These parameters were selected based upon our in-house pilot study, wherein we assessed the impact of time of testing [a.m. (8:00–9:00 h and 10:30–11:30 h) *c.f.* p.m. (15:30–16:30 h)] and lighting conditions of testing room (lights-on *c.f.* lights-off) on burrowing behavior. Our data showed no discernible differences in the burrowing outcomes under these various conditions (data not shown).

The baseline burrowing behavior was assessed prior unilateral i.pl. FCA/saline injections on day 0 or unilateral CCI/sham surgery on day −1, and then up to twice-weekly thereafter until study completion at day 13/14 post model induction.

The chronological order of each step in the burrowing protocol is summarized below.

Burrowing Protocol

Day ~−12 (Animal Arrival Day)

- At least 4 days acclimatization in the animal holding facility with animals housed in groups of two to three.

Day −8 and −7 (Paired Social Facilitation; SF)

- On Day −8, rats are placed in pairs in the testing apparatus for a period of 1 h with a burrowing tube filled with 2 kg of gravel. After 1 h, the amount of the displaced gravel is weighed. The light intensity of the burrowing room is ~235 LUX.
- The same procedure is repeated on the second day of social facilitation (Day −7). If any rat pair does not burrow on Day −8, then one rat of this pair is swapped with a rat from a burrowing pair for the testing on Day −7.

Day −6, −5, and −4 (Individual Training; IT)

- Rats are placed individually in the test apparatus with a burrowing tube filled with 2 kg of gravel, and the amount of gravel displaced at the end of a 1 h period is weighed.
- At the end of 3 days of individual training, only rats that burrow >450 g on at least 2 out of 3 days of individual training are included in the study.

Day −3 and −2 (Weekend)

- Rats are group housed in individually ventilated cages undisturbed during the weekend.

Day −1

FCA Cohorts:

- Baseline paw volume (PV) and Randall–Selitto paw pressure thresholds (PPTs) are determined.

CCI Cohorts:

- Pre-CCI/sham burrowing baselines are determined.

Day 0

- Baseline clinical observations (including animals' body weights) determined.

FCA Cohorts:

- Pre-FCA/saline burrowing baselines are determined.
- Animals are anaesthetized, and a unilateral intraplantar injection of FCA (150 μL) or saline (150 μL) is given into one hindpaw.

CCI Cohorts:

- Baseline von Frey paw withdrawal thresholds (PWTs) are determined.
- Animals are anaesthetized, and a unilateral CCI of the sciatic nerve or sham surgery is performed.

Day 1–14
CCI Cohorts:

- Twice-weekly clinical observations (including body weights), PWTs and burrowing behavior are assessed for the CCI- and sham-group of rats.

FCA Cohorts:

- Thrice-weekly clinical observations (including body weights), PPTs, PVs, and burrowing behavior are assessed for the FCA- or sham-groups of rats.

Day 14

- Animals are humanely euthanized post-study with an overdose of pentobarbitone (325 mg/kg; intraperitoneal) or CO_2 asphyxiation.

3.8 Statistical Analyses

Statistical analyses are performed using repeated measures two-way analysis of variance (ANOVA) followed by the Bonferroni test for repeated measures, to assess between-group differences in pain hypersensitivity behaviors (Randall–Selitto mechanical hyperalgesia and von Frey mechanical allodynia in FCA- and CCI-rats, respectively), hindpaw volume (FCA-rats only), burrowing behavior, and body weight data [31]. Pearson's correlation analysis was used to determine the correlation between burrowing behavior and mechanical allodynia (in individual CCI-rats) or mechanical hyperalgesia (in individual FCA-rats) [31]. The F-values together with their associated degrees of freedom (df of treatment, time, interaction, and residual) are reported as F (df of treatment, time, and interaction/residual). Statistical analyses were performed using GraphPad Prism™ v6.04 (GraphPad software Inc., San Diego, CA, USA), and the statistical significance criterion was $P \leq 0.05$.

Table 1
Details of cohort sizes and number of exclusions for each rat pain model described in our previously published study [31]

Cohort	Number of rats used	Number of rats excluded from the studya
Rat model of chronic inflammatory pain (FCA)		
1	10	3
2	10	5
3	10	3
Total	30	11/30 (~37%)
Rat model of neuropathic pain (CCI)		
1	16	3
2	10	4
3	10	0
4	10	2
Total	46	9/46 (~20%)

[a]Number of animals excluded because they did not meet the burrowing criterion (≥450 g)

3.9 Animal Exclusion

After the acclimatization period, SD rats were subjected to 2 days of SF and then 3 days of IT, as mentioned above. The number of rats excluded due to poor burrowing performance in each cohort of our study are shown in Table 1. Overall, between 20%–37% of rats used in our study were classified as "poor burrowers" [31]. This result is suggestive of considerable interanimal variability in the motivation for naïve SD rats to burrow. This is in contrast to work by others where no animals were excluded [43], <10% of animals were excluded [14] or 10% of mice ($n=4$) and rats ($n=7$) with the lowest burrowing scores at baseline were excluded from the experiment [20]. In other work, Wordaski and colleagues [44] recommended not to exclude "poor" burrowers.

3.10 General Animal Welfare

General animal health was assessed by monitoring temporal changes in the body weights of rats. We found that there were no significant differences ($P>0.05$) in the body weights of FCA- or CCI-rats *c.f.* their corresponding saline-injected or sham-group rats (Fig. 4A, B) [31].

3.11 Development of Hindpaw Edema (FCA Model)

The paw volume data measured using a plethysmometer showed significant development of ipsilateral hindpaw edema in FCA rats between days 2 and 14 post i.pl. FCA injection [$F(3, 5, 15/170)=582.8$, 63.3, 64.4; $P\leq 0.05$], relative to the corresponding data for the bilateral hindpaws of rats administered unilateral i.pl. injections of saline or the contralateral hindpaws of FCA-rats for the 14-day study period (Fig. 5A) [31].

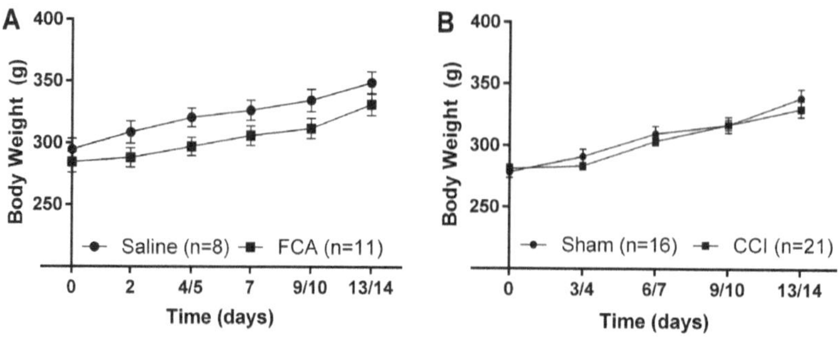

Fig. 4 Cumulative mean (±SEM) body weight versus time curves for (**A**) Cohorts 1–3 rats administered a unilateral i.pl. injection of FCA (n = 11) or saline (n = 8), and (**B**) Cohort 1–4 rats that underwent CCI (n = 21) or sham (n = 16) surgery. For FCA- and CCl-rats, there were no significant (P>0.05) differences in the mean (±SEM) body weights throughout the experimental period when compared with that of their respective sham-control rats [31]

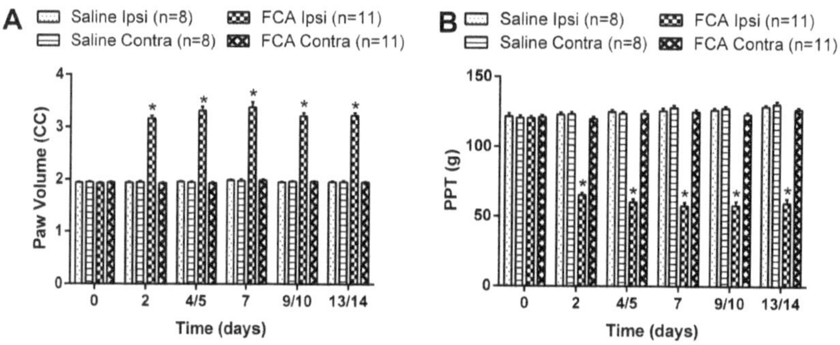

Fig. 5 The cumulative (**A**) mean (±SEM) PV and (**B**) mean (±SEM) PPT versus time curves for rats administered a unilateral i.pl. injection of FCA (n = 11), relative to that of rats that received a unilateral i.pl. injection of saline (n = 8). Between days 2 and 14 post-i.pl. injection, there was significant (P≤0.05) temporal development of edema and mechanical hyperalgesia in the ipsilateral hindpaws of FCA-rats *c.f.* the corresponding values for the ipsilateral hindpaws of sham-rats [31]. *Ipsi* ipsilateral hindpaw, *Contra* Contralateral hindpaw. [*P ≤0.05 (Two-way ANOVA, post-hoc: Bonferroni) relative to saline-injected sham-rats]

3.12 Development of Mechanical Hyperalgesia (FCA Model)

In the same Cohorts of FCA-, or saline-injected rats, the development of mechanical hyperalgesia in the ipsilateral hindpaws was assessed using the Randall–Selitto apparatus. The mean (±SEM) paw pressure threshold (PPT) data (Fig. 5B) for FCA-rats showed that mechanical hyperalgesia was fully developed in the ipsilateral hindpaws from day 2 until study completion on day 14 [$F(3, 5, 15/170) = 586.6, 31.6, 64.7$; $P \le 0.05$] [31]. In contrast, mechanical hyperalgesia did not develop in the ipsilateral hindpaws of sham-rats or the contralateral hindpaws of FCA-rats [31].

3.13 Development of Mechanical Allodynia (CCI Model)

The development of mechanical allodynia in the rat hindpaws was measured using calibrated von Frey filaments. There was a significant [$F(3, 4, 12/280) = 165.6, 55.2, 24.7$; $P \le 0.05$] temporal development of mechanical allodynia in the ipsilateral hindpaws of CCI-rats (Fig. 6) between days 3 and 14 postsurgery. In contrast, mechanical

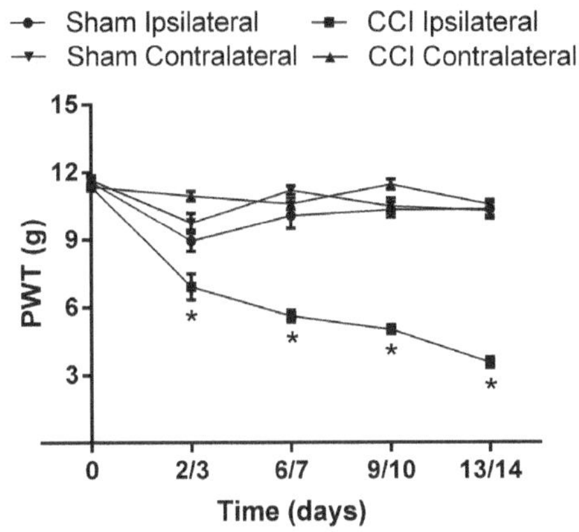

Fig. 6 The mean (±SEM) PWT versus time curves for CCI- ($n=21$) and sham ($n=16$)-control rats over the 13/14-day study period. For CCI-rats, there was a significant ($P > 0.05$) temporal reduction in the mean (± SEM) PWTs of the ipsilateral hindpaw between days 2 and 14 post-CCI surgery, relative to that of the mean (± SEM) PWT values of the ipsilateral hindpaw of sham-control rats or the contralateral hindpaw of CCI-rats [31]. [$^*P \leq 0.05$ (Two-way ANOVA, post-hoc: Bonferroni) relative to sham-group rats]

allodynia did not develop in the contralateral hindpaws of CCI-rats or the bilateral hindpaws of sham-operated rats [31].

3.14 Development of Burrowing Deficits (FCA and CCI Models)

In our work [31], there was a significant reduction [F(1, 8, 8/136) = 9.15, 6.03, 4.36; $P \leq 0.05$] in the weight of gravel burrowed by FCA-rats between days 2 and 10 (Fig. 7A). In contrast, there were no significant differences in the weight of gravel burrowed by saline-injected rats over the 14-day duration of the experiment. Similarly, there was a significant reduction in the weight of gravel burrowed by CCI-rats from day 3 to day 14, but this was not the case for the group of sham-operated rats (Fig. 7B).

However, there was significant between-cohort variability in the burrowing behavior of both FCA- and CCI-rats relative to their respective sham-control groups (Figs. 8A–C and 9A–D, respectively).

3.15 Significant Between-Cohort Variability in Burrowing Behavior (FCA and CCI Models)

Comparison of the burrowing behavior of FCA-rats between the different cohorts showed significant [F(2, 5, 10/40) = 3.2, 9.3, 2.2; $P \leq 0.05$] between-cohort variability in burrowing behavior on days 2 and 4/5 post-i.pl. FCA injection (Fig. 10A) [31]. Likewise, for CCI-rats, there was significant [$F_{(3, 4, 12/68}$ = 2.4, 7.3, 2.3; $P \leq 0.05$] between cohort variability in burrowing behavior on day 14 post-CCI surgery (Fig. 10B) [31]. By contrast, there were no between-cohort differences in the burrowing behavior of saline-injected rats [$F_{(2, 5, 10/25}$ = 2.3, 1.4, 0.6; Fig. 10C] or sham surgery animals [$F_{(3, 4, 12/48}$ = 1.4, 0.8, 1.2; Fig. 10D] [31].

Although different cohorts had different experimenters, the between-experimenter differences are mitigated as testers were not present in the room during the burrowing sessions. Additionally, the burrowing readouts are quantitative and so minimize subjective interpretation.

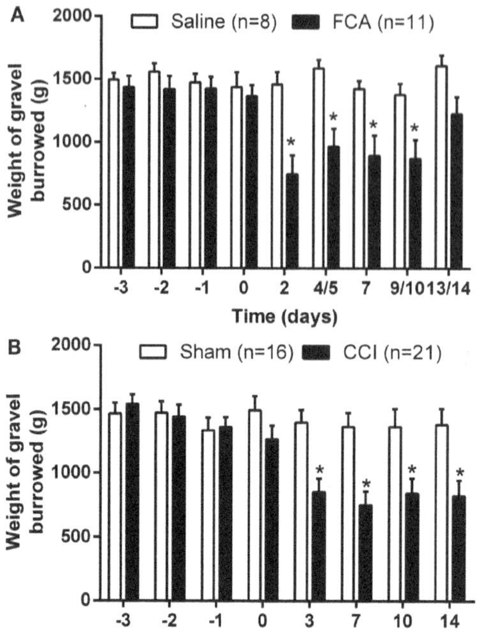

Fig. 7 Cumulative mean (±SEM) weight of gravel burrowed versus time curves for (**A**) rats administered a unilateral i.pl. injection of FCA ($n = 11$) or saline ($n = 8$), and (**B**) rats that underwent CCI ($n = 21$) or sham ($n = 16$) surgery [31][*$P \leq 0.05$ (Two-way ANOVA, post-hoc: Bonferroni) relative to that observed in the corresponding groups of sham-control rats]

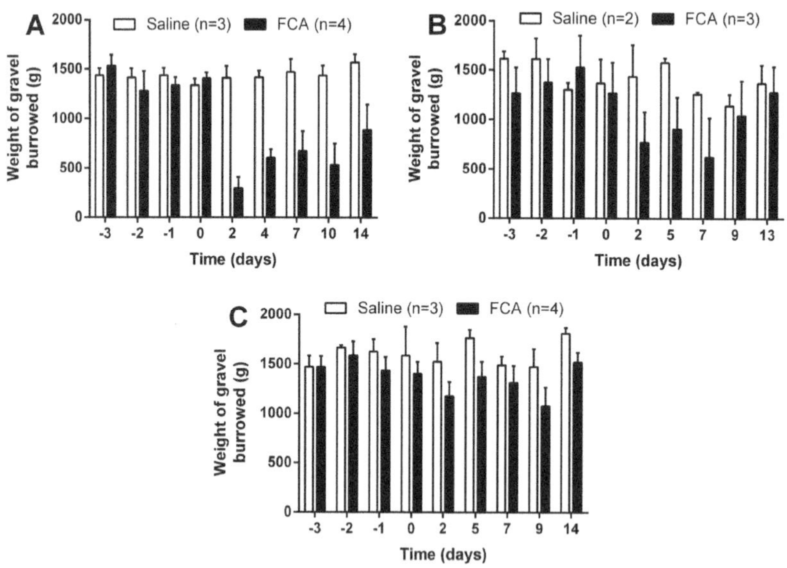

Fig. 8 The mean (±SEM) weight of gravel burrowed versus time curves for FCA (n = 3–4 per cohort) and saline-rats (n = 2–3 per cohort) in Cohorts (**A**) one [F(1, 8, 8/40) = 13.4, 7.01, 7.9], (**B**) two [F(1, 8, 8/24) = 0.9, 1.5, 1.1], and (**C**) three [F(1, 8, 8/40) = 5.7, 1.7, 0.5], show considerable between-cohort variability in burrowing behavior [31]

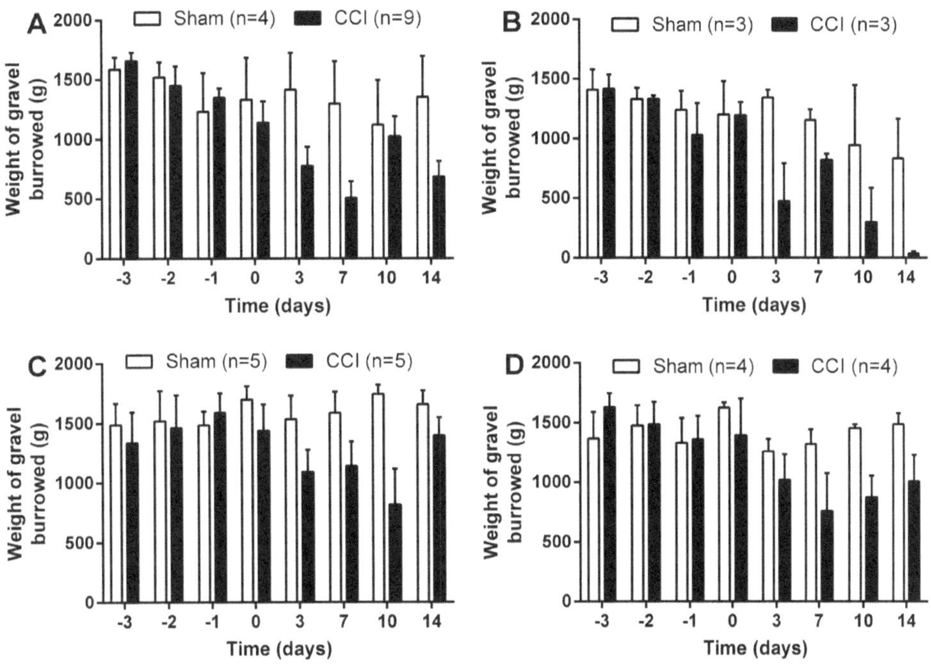

Fig. 9 The mean (± SEM) weight of gravel burrowed versus time curves for CCI- ($n = 3$–9 per cohort) and sham-rats ($n = 3$–5 per cohort) in Cohorts (**A**) one [F(1, 7, 7/77) = 1.5, 6.8, 3.7], (**B**) two [F(1, 7, 7/28) = 5.7, 5.2, 1.5], (**C**) three [F(1, 7, 7/56) = 2.2, 1.2, 2.4], and (**D**) four [F(1, 7, 7/42) = 1.9, 2.6, 1.8], show considerable between-cohort variability in burrowing behavior [31]

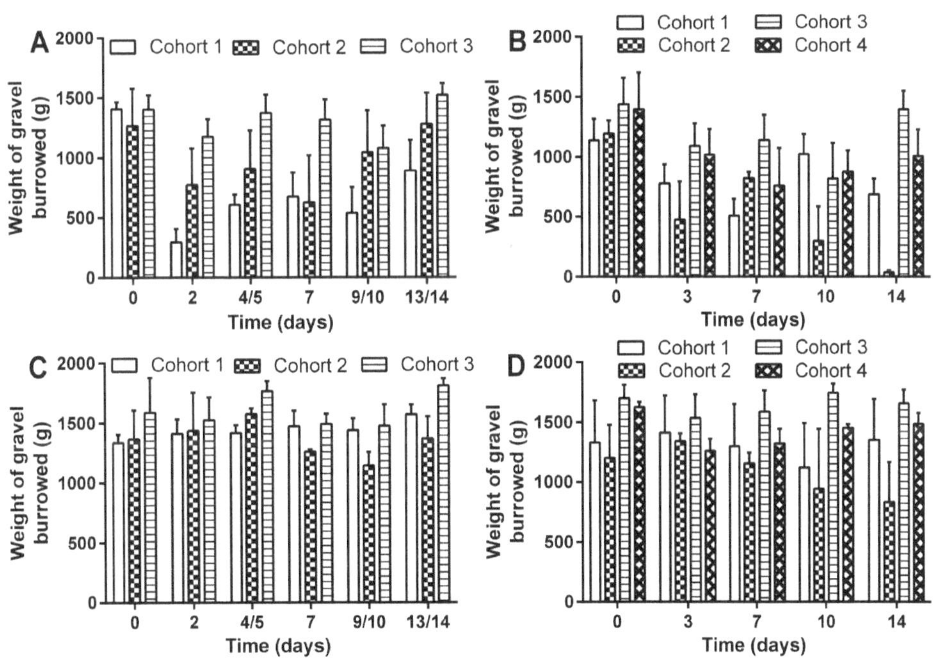

Fig. 10 The mean (±SEM) weight of gravel burrowed versus time curves for various cohorts of (**A**) FCA- ($n =$ 3–4 per cohort), (**B**) CCI- ($n = 3$–9 per cohort), (**C**) saline- ($n = 2$–3 per cohort), and (**D**) sham-surgery group ($n = 3$–5 per cohort) rats. There were significant between cohort variability in the burrowing behavior of FCA- (day 2 and 4/5) and CCI-rats (day 14). In contrast, no between-cohort variability was observed for saline-injected rats and sham-surgery group rats [31]

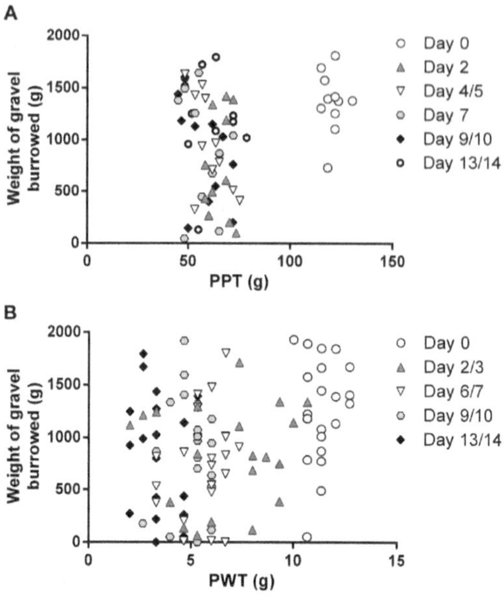

Fig. 11 Significant correlation between burrowing performance and mechanical hyperalgesia was observed only on (**A**) day 4/5 in FCA-rats, in contrast to the lack of significant correlation between burrowing behavior and the extent of mechanical allodynia in the (**B**) ipsilateral hindpaws of CCI-rats [31]

Further data analyses showed that there was a significant correlation ($P \leq 0.05$) between burrowing performance and the extent of development of mechanical hyperalgesia in the ipsilateral hindpaws on day 4/5 post-i.pl. FCA injection in FCA-rats (Fig. 11A) [31]. However, for CCI-rats, burrowing performance was not correlated significantly ($P > 0.05$) with von Frey PWT values in the ipsilateral hindpaws throughout the experimental period (Fig. 11B) [31]. Our results are aligned with those of Shepherd et al. [28], who reported that the burrowing test might not be universally suitable for all types of rodent pain models. The authors proposed that impaired locomotor activity due neuropathic pain has limited impact on burrowing activity in mice. In other work [43], it was concluded that impairment of spontaneous ethological behaviors including burrowing reflects the occurrence of maximal evoked mechanical and cold hypersensitivity. In comparison, the evoked behavioral measurements did distinguish subtle differences that commenced in the early phase of the development of pain behavior. This lack of correlation could also suggest that suppressed burrowing behavior may reflect a pain component not directly linked to sensory gain [44].

3.16 Burrowing Notes

Key factors with potential to affect burrowing performance are summarized in Table 2 and briefly discussed in the following paragraphs.

Laboratory Conditions

According to Deacon [25], the laboratory environment used for testing should be suitably quiet. Also, animals should be undisturbed and experimenters should avoid testing on cage-cleaning days, as the

Table 2
Summary of recently published or keystone papers involving burrowing (BW) as a pain-like behavioral readout

Species and strain	Pain model	Burrowing (BW) protocol	Conclusions	Reference
Male and female SD Rats (≥ 6 weeks of age; male: 183–293 g, female: 144–289 g) Note: rats were house singly with a 14:10 h light/dark cycle.	Dextran sulfate sodium (DSS) induced colitis • Acute colitis (4D DSS in drinking water) • Chronic (4D DSS, 7D water, 3D DSS • Control (14D water)	• BW cage: 58.4 cm×41.3 cm × 31.4 cm. • BW tube: 32 cm × 10 cm diameter, raised 6 cm from open end, filled with 2.5 kg of gravel (2–5 mm, Premium Aquarium gravel). • 2 days of habituation in pairs in cage with an empty BW tube for 30 min. After 30 min, BW tube filled with 2.5 kg gravel was introduced for 1 h. • If a pair of rats did not burrow sufficiently (<100 g) on the first day of habituation (D-4), a new pair was created including a BW rat (identified on D-4) and the protocol repeated. • Baseline assessments were made over the next 3 days (D-2 to 0) with rats placed individually in the BW box with the gravel-filled BW tube for 60 min daily. The amounts of gravel displaced over these 3 days were averaged to produce a BW score for each rat. • Exclusion criteria: BW score <100 g of gravel.	• BW↓ in acute but not chronic phase of colitis model. Authors stated that "the absence of changes in BW behavior from baseline during the chronic phase may reflect a lack of study power (due to wide 95% CI), …… "the effect of chronic pain on BW behavior is currently unknown". • Other pain-like behavioral readouts included Disease Activity Index (DAI), Rat Grimace Scale (RGS), and a Composite Behavior Score. Of these, it was concluded that RGS could be applied to a wider range of pain types and disease chronicity in rats. • In acute colitis, DAI↑ cc RGS↑, whereas BW↓. The changes in RGS scores coincided with model severity as assessed with the DAI, suggesting these parameters can be used as a proxy measure of pain.	[21]

(continued)

Table 2 (continued)

Species and strain	Pain model	Burrowing (BW) protocol	Conclusions	Reference
Male and female C57BL/6J and FVB/NJ (8–14 weeks) mice	• Spared nerve injury (SNI) neuropathic pain model • Freund's Complete adjuvant (FCA) inflamed hindpaw model • Systemic nitroglycerin induced migraine (SN) model—10 mg/kg, i.p., every 2 D for 5 doses.	• BW cage: cage length 25 cm but full dimensions not specified • Red or transparent BW tube. Size: 12.5 cm × 4.4 cm (inner diameter). • Acclimatized group of 5 mice with (empty tube) for 1 h, on day –1, and 30 min on day 0. • Each mouse was tested (15 min) individually with acrylic tube filled with 50 g corncob bedding placed in one corner.	• "…male and female mice of multiple backgrounds displaced similar amounts of bedding after 15 min of access to a bedding-filled tube, with relatively little interindividual variability". • This much shorter (15 min) BW protocol likely avoids inducing chronic stress in mice due to social isolation, a potential confounding factor in repeated measurement-based study designs. • SNI impaired BW and this was correlated with increased mechanical allodynia, that was attenuated by gabapentin administration. • SNI induced neuropathic pain depressed BW activity in both male and female mice, but only temporarily (days 5–8 post-SNI) when BW was performed once daily. When BW activity was no longer performed once daily and changed to once per 5 days, the BW deficit was maintained (days 5–25 post-SNI). • No BW deficits in FCA and SN models, in contrast to work by others [31, 41, 44]. Authors concluded that BW may not be universally suitable as a behavioral measure of pain for all rodent pain models. • "The color of the BW tube affected the variability of BW performance when tested in a lit room, suggesting that light aversion is at least a partial driver of BW behaviors." • Authors suggested that impairment in locomotor activity due to neuropathic pain has limited impact on BW activity in mice. • Although SNI did depress BW behavior in male and female mice, it did so only temporarily in contrast to mechanical allodynia that is well-known to persist for several weeks post-SNI surgery.	[28]

(continued)

Table 2
(continued)

Species and strain	Pain model	Burrowing (BW) protocol	Conclusions	Reference
Male SD rats (225–250 g arrival). 9 D acclimatization.	Infraorbital nerve ligation (INL) model of neuropathic pain	• 32 cm × 10 cm hollow plastic tube sealed with a medium-density fiberboard plug at one end. Open end raised 65 mm above the floor. • Cage size: 56 cm × 34 cm × 19 cm. • Tube filled with 1 kg food pellets. Wood chip bedding on the floor. • Habituation/training and testing were conducted in a darkened room (light provided by a 60 W red light bulb suspended 1 m above the observation area), 45 dB background noise. • Three habituation/training sessions were performed before preoperative testing. BW behavior was assessed on preoperative day −1 and on day 5 post-surgery. • Rats were placed individually in the test cages and allowed to burrow for 4 h.	• INL rats had significantly reduced BW *c.f.* sham-rats (recorded on day 5 post-surgery). • A significant correlation was found between isolated facial grooming and BW. • At day 5, INL rats developed hyporesponsiveness to mechanical stimulation over the ipsilateral infraorbital nerve. The extent to which intact vibrissal sensitivity and subsequent thigmotactic scanning is important to the BW behavior, is unclear.	[54]

(continued)

Table 2
(continued)

Species and strain	Pain model	Burrowing (BW) protocol	Conclusions	Reference
Male SD rats (180–200 g arrival). At least 3 D of acclimatization.	• Chronic constriction injury (CCI) rat model of neuropathic pain • Freund's complete adjuvant (FCA) – by i.pl. injection to one hindpaw as model of chronic inflammatory pain	• The gravel used was washed and dried for each cohort of rats. • 32 cm × 10 cm hollow tube sealed at one end, filled with 2 kg of smooth gravel (3–5 mm diameter) with open entrance raised ~6 cm above the floor in the cage. • Cage size: 49.8 cm × 38 cm× 21.5 cm. • Rats were habituated and trained to burrow for 5 days before measurement of baseline BW. • Training phase comprised social facilitation (Days −5 and −4) and individual training (Days −3, −2, and −1). All experiments were conducted toward the end of the light phase (15:30 h). • BW time for training/testing = 1 h. • Exclusion criteria: Rats that burrowed ≤ 450 g of gravel after 3 days of individual training were excluded.	• Sensitivity of BW was comparable with noxious mechanical stimulus testing in the same rats. However, intercohort variability in both CCI and FCA rats were much greater for BW compared with mechanical allodynia and mechanical hyperalgesia in CCI- and FCA-rat models, respectively. • There was a significant correlation between BW and mechanical hyperalgesia in the ipsilateral hindpaws of FCA-rats. However, there was no significant correlation between BW and mechanical allodynia in the ipsilateral hindpaws of CCI-rats.	[31]
Male SD rats (175–199 g). At least 4 D acclimatization.	• Complete Freund's adjuvant (CFA)—unilateral intrarticular (i.a.) injection (arthritis)	• Rats were tested during the light phase, 7:00 a.m. to noon. • Animals were acclimatized to the testing room for at least 1 h before the start of experiments. • Hollow plastic tube, 32 cm × 10 cm, filled with 2.5 kg pea shingle gravel (2–4 mm diameter), tube raised 6 cm from floor at the open end. • Rats received 4 training sessions. D 1–2 in pairs, D3–4 individually. • BW time = 1 h. • Exclusion criteria: BW <500 g of gravel.	• Rats were assessed for changes in weight-bearing (WB), open-field (OF), or BW 3 days after i.a. injection. • Rats with CFA-induced knee arthritis burrowed significantly less gravel *c.f.* naïve rats. • BW is as sensitive as WB and more sensitive than the OF test. • Single bolus doses of naproxen and ibuprofen (NSAIDs) and pregabalin (anticonvulsant) significantly improved BW performance in rat model of knee arthritis.	[13]

(continued)

Table 2 (continued)

Species and strain	Pain model	Burrowing (BW) protocol	Conclusions	Reference
Male SD rats (175–200 g).	• Paclitaxel (2 mg/kg; intraperitoneally on D 0, 2, 4, and 6)) induced peripheral neuropathy	• A phased "sunset" and "sunrise" lighting period over ~30 min preceded complete dark/light periods. By 7 a.m., the room was fully illuminated. • Test cage contained sawdust bedding with food/water available ad libitum. Test cages were kept within the same room where rats were housed. • The rats were habituated in pairs for 24 h before first training session for a total of two training sessions. • Plastic tube, 32 cm × 10 cm, sealed at one end with open end raised ~6 cm above floor. • BW tube filled with 2500 g of 5 mm pea shingle. • Animals were trained in pairs for 2 h, from 4 to 6 p.m., after which they were returned to their original home cage with sawdust and environmental enrichment materials, without the BW tube. • If any pairs of rats did not burrow in the first training session, the pairings were changed so at least one rat did show BW activity. • Individual training occurred at 24 h after second paired training session. Rats were observed to record latency to burrow. • After 2 h, the amount of gravel left in the tube was determined. • All animals burrowed more than 500 g at baseline.	• BW activity was assessed before paclitaxel treatment and then on D7 and 28. • Latency to start BW was increased on D7 compared with baseline (prepaclitaxel), which further increased at D28 but not significantly (due to large interindividual variability). • Amount of pea shingle burrowed significantly reduced on D28 but not on D7 when compared to the baseline. • Paclitaxel treatment reduced BW and wheel-running activity. Time courses of evoked and on-going pain-like behaviors were similar, with delayed onset. • Several variables were identified that had a significant impact on sensitivity, including the number of training sessions provided and the animal weight at time of initiation of paclitaxel-induced peripheral nerve injury. • Wheel running is more sensitive when assessed in animals overnight (7:10 p.m.–6:20 a.m.) compared with during the dusk phase (6–7 p.m.). • It was concluded that BW can identify ongoing pain-like behavior, but it may be confounded by factors that produce variability in evoked-pain testing.	[43]

(continued)

Table 2
(continued)

Species and strain	Pain model	Burrowing (BW) protocol	Conclusions	Reference
• Male BALB/c (21–25 g upon arrival) • Male SD rats (100–124 g upon arrival)	Cancer-induced bone pain: • Mouse: intratibial (i.t.) 4T1–Luc2 cells • Rat: i.t. NRMT-Luc2 cells	**MICE** • BW tube: 20 cm × 7.2 cm, sealed at one end. The open end raised 3 cm above the floor. • Transparent plastic cage: 12.5 cm × 26.6 cm × 18.5 cm without bedding. • 5 days training for BW. D1 habituation in pairs in cage with empty BW tube. D2-3, training in pairs with tube filled with 500 g of sand (0–3 mm diameter particles) D4–5, individual testing with BW tube filled with 500 g sand. • On test days, mice were placed individually in a cage with a BW tube filled with 500 g sand. • All sessions were conducted for 2 h; cages and BW tubes assigned to each mouse remained unchanged throughout the entire experiment. • Ten percent of mice ($n = 4$) with the lowest BW scores at baseline were excluded from the experiment. • All BW sessions were conducted between 08:00 a.m. and 12:00 p.m. **RATS** • BW tube: 32 cm × 10 cm, sealed at one end. The open end raised 6 cm above floor.	• BW for mice and rats assessed before bone cancer model induction (D0), D5, and D9; before cancer induction (D0), D8, D17, and D20, respectively. • For both mouse and rat models of bone cancer pain, there was impaired BW performance in parallel with reduced limb use and an alteration in weight bearing (the latter not tested in rats). However, in rats, development of a pain-like phenotype in the BW test was more subtle than that in mice.	[20]

(continued)

Table 2
(continued)

Species and strain	Pain model	Burrowing (BW) protocol	Conclusions	Reference
		• Standard Makrolon Type IV cage: 59.5 cm x 38 cm x 20 cm. • 4 days training for BW. • SF in pairs for first 2 D. D1 for 60 min, D2 for 30min with same pair using empty BW tube, then followed by 1h training using BW tube filled with 2500g gravel (2-4 mm diameter). • When a pair of individual rats showed poor BW performance performance and 1 extra SF session was conducted. • D3-4 (IT): rats acclimatised individually in empty cage for 30 min then with BW tube filled with 2500g grave for 1 h. • D5: BW baseline determined as per IT Day. • Ten percent of rats (n=7) with the lowest BW scores at baseline were excluded from the experiment. • All BW sessions were conducted between 08:00 AM and 12:00 PM.		

(continued)

Table 2
(continued)

Species and strain	Pain model	Burrowing (BW) protocol	Conclusions	Reference
• Male Wistar Han rats (180–200 g) for CCI. • Zucker diabetic fatty (ZDF; fa/fa) rats (7 weeks) for Type II diabetes. [55] • Zucker lean (ZL; fa/+) control rats • Wistar rats (290–380 g). [55]	• Chronic constriction injury (CCI) • Zucker diabetic fatty rat (model of type 2 diabetes) • Streptozotocin (65 mg/kg; i.p.) model of Type 1 diabetes	**CCI and ZDF rat models of peripheral neuropathic pain (BW performed by Boehringer Ingelheim):** • Steel tubes: 32 cm × 10 cm filled with 2.5 kg of quartz sand. Open end raised 6 cm above floor. • Plexiglas cage: 60 cm × 34 cm × 20 cm. • Social facilitation (SF): rats paired in cages with a sand filled tube for 2 h on 2 consecutive days. • Swapped rat if a pair burrowed <1.5 kg on second SF day with one from a pair that burrowed <1.5 kg. • Individual training (IT): rats were placed • individually in a cage with tube filled with 2.5 kg sand for 30 min per day and the average amount burrowed over 3 days was calculated to attain a baseline value. • Before model induction animals were randomly assigned to each induction group. Animals were regrouped to ensure that each group had a comparable baseline BW value. **CCI and STZ-induced rat models of peripheral neuropathic pain (BW performed by Grunenthal):** • Plastic tube: 32 cm × 10 cm filled with 2.5 kg pea shingle gravel (diameter 2–4 mm). Open end raised 6 cm above floor.	Drugs tested: morphine (0.1–6 mg/kg), tramadol (10–100 mg/kg), gabapentin (25–60 mg/kg), pregabalin (3.16–61.6 mg/kg). **Boehringer Ingelheim:** • BW tested on days 8, 9, 16, 17, and 20 post-CCI surgery. In CCI-rats, BW activity significantly reduced *c.f.* sham-rats on D8, 17, and 20. • ZDF rats developed hyperglycemia spontaneously at 8 weeks that plateaued at 13 weeks of age. • Mechanical sensitivity (assessed using a Randall Selitto device and von Frey filaments), was not different between ZDF rats and ZL rats aged 7 week. In contrast, development of significant mechanical hypersensitivity was observed in ZDF rats compared to ZL rats at 15 weeks of age. • ZDF rats (7 weeks) burrowed slightly less sand than ZL rats. The BW deficit was more pronounced at 15 weeks of age. • No significant correlations between BW deficit and either mechanical allodynia or mechanical hyperalgesia at similar time points. Similar to BW, locomotor activity in ZDF rats also reduced *c.f.* ZL rats at 7 and 15 week. • In ZDF rats (≥15 week), single bolus doses of morphine (1, 3, and 6 mg/kg; s.c.), pregabalin (10 and 30 mg/kg; PO), gabapentin (30 and 45 mg/kg; i.p.) or tramadol (10, 30 and 10 mg/kg; PO) recovered the BW activity. Instead of rescuing the BW deficit, tramadol at 30 or 100 mg/kg significantly reduced BW activity. **Grunenthal:** • BW tested on 4, 7, 11, 15, 19, 22, and 25 D post-CCI surgery. In CCI rats BW activity significantly reduced *c.f.* sham rats at each timepoint assessed.	[14]

(continued)

Table 2
(continued)

Species and strain	Pain model	Burrowing (BW) protocol	Conclusions	Reference
		• All rats received 4 D training (2× SF + 2× IT) and 1 D BW test. • Training: rats were in cages with empty tubes for 1h (D1). • On D2–4, animals spent 30 min in empty cage, after which the gravelfilled tube was introduced for 60-min. Rats were placed into cages in pairs for training sessions D1 and 2 (SF) and on their own for IT on D3 and 4. • BW baseline was determined on D5. Rats that burrowed.	• In CCI rats, the BW activity significantly recovered with morphine at 0.1 mg/kg (i.p.) but not at higher single bolus doses (0.316 & 1 mg/kg). By contrast, single bolus doses of pregabalin (3.16-31.6 mg/kg; i.p.) and gabapentin (25 & 50 mg/kg; presumably i.p.) dose-dependently recovered BW activity. • Gravel was selected instead of sand as it provided more consistent BW deficits in CCI rats. • STZ rats (3 wk) had markedly reduced baseline BW. Single bolus doses of pregabalin (10 & 30 mg/kg; i.p.) had no effect on BW performance.	
• C57BL/6 mice • Lister Hood rats	• Presumably naïve	**For mice BW:** • 20 cm × 6.8 cm diameter gray plastic tube filled with 200 g of food pellets. Open end raised 3 cm above floor. BW tube placed against the longer wall of a clean cage with thin layer of bedding. The closed end of the tube (wooden; mdf) was against the back wall of the cage. The cage dimensions were not described. • Water was provided and food hopper of cage was full as usual. **For rats BW:** • 32 cm × 10 cm diameter plastic tube filled with 2.5 kg of pea shingle. Open end raised 6 cm above floor. Place BW tube against the longer wall of a clean cage with thin layer of bedding. The closed end of tube (wooden; mdf) was against the back wall of the cage. The cage dimensions were not described. • Water provided and the food hopper of the cage was full as usual.	• BW activity in mice and rats was tested in two phases; individually for 2 h starting 3 h before the dark cycle. Then, the BW material was replaced prior to the overnight phase until next morning (no specific duration, as the animals will have burrowed all they wanted earlier in the night). After completing the two individual phases of BW activity, animals returned to their group cages. • C57BL/6 mice normally burrowed about 60 g over 2 h and 150 g overnight. • Hooded Lister rats burrowed about 1500 g after 2 h and 2400 g overnight.	[25]

animals will have increased activity re-exploring their clean cages, and subsequently will be less active with respect to burrowing [25].

Experimental Designs and Housing Conditions

We recommend that model establishment be optimized for each individual laboratory geo-location and experimental settings before undertaking the main study. This is because there may be between-species, between-strain [25], and between-animal vendor differences in terms of pain behavioral responses, including burrowing [45], and social/burrowing and/or behavioral/motivation [39]. Encouragingly, however, a multicenter study has reported a lack of differences in the burrowing behavior of SD and Wistar rats [44]. Also, Shepherd and colleagues [28] reported relatively little interindividual variability in burrowing in mice (Table 2). By contrast, Leung and colleagues [21] reported large inter-animal variability during the chronic but not the acute phase in a rat colitis model. It is unclear whether these observations are due to differences in burrowing duration (15 min *c.f.* 60 min), burrowing exclusion criteria (no exclusion *c.f.* rats excluded when <100 g was burrowed), burrowing substrate composition (50 g corncob *c.f.* 2.5 kg gravel), pain model/pain phase (systemic nitroglycerin induced migraine model *c.f.*chronic dextran sulfate sodium-induced colitis model), and/or between-species differences. In other work in rodent models of cancer-induced bone pain [20], the development of a pain-like phenotype in the burrowing test in rats was more subtle compared with that in mice.

The color of the burrowing tube is also a critical aspect of the burrowing test [28]. However, this was only observed when the burrowing assay was undertaken in a well-lit room but not during lights off, suggesting that light intensity is a modulator of burrowing behavior.

Shepherd and colleagues [28] reported that spared nerve injury (SNI) induced neuropathic pain reduced burrowing activity in both male and female mice, but only temporarily (days 5–8 post-SNI) when burrowing was performed once daily. When the burrowing interval was increased from once daily to once per 5 days, the burrowing deficit was maintained from days 5 to 25 post-SNI. These results suggested that the frequency of testing likely impacted the motivation to burrow such that it was inversely proportional to the frequency with which the burrowing test was performed. Thus, reducing the frequency with which the burrowing test is performed appears to be a strategy for potentially reducing high interanimal variability.

Shepherd et al. [28] also reported that the burrowing behavior was only impaired temporarily in contrast to mechanical allodynia, which has been known to persist for several months in the peripheral nerve-injured male and female mice [46].

Burrowing tests are typically conducted during normal business hours for durations of ≤2 h but not overnight or during the night-phase (Table 2). As rodents are nocturnal animals that are most active during the night phase [47], this factor may also contribute to inter-animal variability in the weight of substrate burrowed.

Most burrowing studies do not provide food or water during the burrowing period, as this could distract the animals from burrowing [44]. However, there are some geolocation/experimental designs that require access to food and/or water. For example, in the United Kingdom, local regulations permit a maximum of 2 h for animals to be without access to water [44].

Laboratory rodents are typically cage-housed under standard temperatures in the range of 20°C–26°C, which may contribute to cold stress [48]. A thermal preference study showed that mice prefer temperatures nearer to 30°C [49]. Burrows are advantageous for rodents in a laboratory setting [50] as retreating into a burrow allows rats to withdraw from perceived threats, such as unfamiliar humans or loud noises [51]. Burrows also provide shelter from light, which is aversive to rats and it leads to retinal atrophy and blindness at levels commonly used in laboratories. Burrows also enable rodents to regulate ambient temperature, which in most laboratories is likely to be below the thermoneutral zone for rats and mice [52, 53]. Although burrowing is not exactly a burrow, as defined by Deacon et al. [25], the afore-mentioned burrowing behavior is relevant. Typically, in the burrowing test environment, there is no bedding or environmental enrichment materials available, except a darkened burrowing tube with burrowing material placed inside the tube. Rodents prefer shelter, with minimal light and noise and a warm environment [48]. Burrowing and/or/then hiding in the burrowing tube is commonly observed after the burrowing test when the experimenter enters the room to weigh the burrowed material (authors unpublished observations).

In other work by Rutten and colleagues [14], Zucker Lean rats had increased burrowing performance at 15-weeks of age compared with 7-weeks of age, and so it is important that control groups are age-matched.

Burrowing Materials

Burrowing substrate composition is another factor to consider as CCI-rats burrowed more consistently when gravel was used as the burrowing substrate compared with sand [14]. In the original work by Deacon [25], rats were found to prefer earthy substrates rather than food pellets in contrast to mice. In our own work [31], we used gravel that was washed and dried for each new cohort of rats but not within the same cohort. This dimension of the burrowing substrate has not been addressed in most publications on burrowing as an ethologically-relevant pain behavior and so this is an area for improvement. To address this issue and minimize the potential confounding effects of differing olfactory cues, Shepherd and colleagues [28] transferred ~5 g of bedding from the home cage to the testing cage for their mouse burrowing test. Typically, however, bedding material is not added to the testing cage to minimize the potential for disruption of burrowing behavior in a short-term experiment (e.g., ≤2 h). More work is needed on defining the balance between minimizing differences in olfactory cues relative to minimizing the disruption of burrowing activity due to the presence of bedding material in the burrowing testing cages.

Burrowing Exclusion Criteria

Regarding burrowing exclusion criteria, for rats, it is typically defined as burrowing ≤500 g of substrate (Table 2). Based on pilot work in our laboratory, we defined the exclusion criterion as rats that burrowed <450 g of substrate [31]. However, in other work, the exclusion criterion was 1500 g, as strong burrowers displaced nearly all of the 2500 g of quartz sand in 1 h [44]. These same authors [44], also recommended that "poor" burrowers not be excluded.

4 Conclusion

In our hands [31], there was a temporal decrease in burrowing behavior for both FCA- and CCI-rats, but there was considerable interanimal variability. By contrast, mechanical hyperalgesia and mechanical allodynia in the ipsilateral hindpaws of FCA- and CCI-rats, respectively, exhibited much less interanimal variability. Our data collectively show that burrowing behavior is altered in rodent models of chronic inflammatory pain and peripheral neuropathic pain. However, larger group sizes are needed, compared with evoked pain behavioral measures, to ensure studies are adequately powered due to the considerable interanimal variability. Additionally, the burrowing test may not be universally applicable to all rodent pain models as there was no significant correlation between burrowing deficit in CCI-rats and the corresponding von Frey paw withdrawal thresholds.

Acknowledgments

A.K., M.Z.I., and the Centre for Integrated Preclinical Drug Development (CIPDD) in the School of Biomedical Sciences (SBMS) at the University of Queensland (UQ) are supported by Therapeutic Innovation Australia (TIA). TIA is supported by the Australian Government through the National Collaborative Research Infrastructure Strategy (NCRIS) programme. The authors also acknowledge The Queensland Government Smart State Research Facilities Programme for supporting the CIPDD/SBMS infrastructure at UQ. The authors also thank Ms. Vy Truong for drawing the schematic timeline.

Conflict of Interest The authors have no conflicts of interest to declare that are relevant to the content of this chapter.

References

1. Cohen SP, Vase L, Hooten WM (2021) Chronic pain: an update on burden, best practices, and new advances. Lancet 397(10289):2082–2097
2. Stompor M, Grodzicki T, Stompor T, Wordliczek J, Dubiel M, Kurowska I (2019) Prevalence of chronic pain, particularly with neuropathic component, and its effect on overall functioning of elderly patients. Med Sci Monit 25:2695–2701
3. Schofield P (2007) Pain in older adults: epidemiology, impact and barriers to management. Rev Pain 1(1):12–14
4. Galea S (2017) Chronic pain and the health of populations: Boston University School of Public Health. https://www.bu.edu/sph/news/articles/2017/chronic-pain-and-the-health-of-populations/
5. Streatfeild J (2019) The cost of pain in Australia. https://www2.deloitte.com/content/dam/Deloitte/au/Documents/Economics/deloitte-au-economics-cost-pain-australia-040419.pdf
6. Platform S (2017) Impact of pain on society costs the EU up to 441 billion euros annually: societal impact of pain (SIP). https://www.sip-platform.eu/press-area/article/impact-of-pain-on-society-costs-the-eu-up-to-441-billion-euros-annually
7. Rice ASC, Cimino-Brown D, Eisenach JC, Kontinen VK, Lacroix-Fralish ML, Machin I et al. (2008) Animal models and the prediction of efficacy in clinical trials of analgesic drugs: a critical appraisal and call for uniform reporting standards. Pain 139(2):243–247
8. Hadi MA, McHugh GA, Closs SJ (2019) Impact of chronic pain on patients' quality of life: a comparative mixed-methods study. J Patient Exp 6(2):133–141
9. Shenoy PA, Kuo A, Vetter I, Smith MT (2016) The walker 256 breast cancer cell- induced bone pain model in rats. Front Pharmacol 7(286):286
10. Laires PA, Lains J, Miranda LC, Cernadas R, Rajagopalan S, Taylor SD et al. (2017) Inadequate pain relief among patients with primary knee osteoarthritis. Rev Bras Reumatol Engl Ed 57(3):229–237
11. Juster-Switlyk K, Smith AG (2016) Updates in diabetic peripheral neuropathy. F1000Research, 5, F1000 Faculty Rev-738.
12. Gudin J, Fudin J, Wang E, Haylon T, Patel K, Goss TF (2019) Treatment patterns and medication use in patients with postherpetic neuralgia. J Manag Care Spec Pharm 25(12):1387–1396
13. Rutten K, Schiene K, Robens A, Leipelt A, Pasqualon T, Read SJ, et al. (2014) Burrowing as a non-reflex behavioural readout for analgesic action in a rat model of sub-chronic knee joint inflammation. Eur J Pain 18(2):204–212.
14. Rutten K, Gould SA, Bryden L, Doods H, Christoph T, Pekcec A (2018) Standard analgesics reverse burrowing deficits in a rat CCI model of neuropathic pain, but not in models of type 1 and type 2 diabetes-induced neuropathic pain. Behav Brain Res 350:129–138
15. Hijma HJ, Groeneveld GJ (2021) Analgesic drug development: proof-of-mechanism and proof-of-concept in early phase clinical studies. Med Drug Disco 10:100083
16. Deuis JR, Dvorakova LS, Vetter I (2017) Methods used to evaluate pain behaviors in rodents. Front Mol Neurosci 10:284
17. Cunha AM, Pereira-Mendes J, Almeida A, Guimaraes MR, Leite-Almeida H (2020) Chronic pain impact on rodents' behavioral repertoire. Neurosci Biobehav Rev 119:101–127
18. Mogil JS (2009) Animal models of pain: progress and challenges. Nat Rev Neurosci 10(4):283–294
19. Woolf CJ (1984) Long term alterations in the excitability of the flexion reflex produced by peripheral tissue injury in the chronic decerebrate rat. Pain 18(4):325–343
20. Sliepen SHJ, Diaz-Delcastillo M, Korioth J, Olsen RB, Appel CK, Christoph T et al. (2019) Cancer-induced bone pain impairs burrowing behaviour in mouse and rat. In Vivo 33(4):1125–1132
21. Leung VSY, Benoit-Biancamano MO, Pang DSJ (2019) Performance of behavioral assays: the Rat Grimace Scale, burrowing activity and a composite behavior score to identify visceral pain in an acute and chronic colitis model. Pain Rep 4(2):e718
22. Jirkof P (2014) Burrowing and nest building behavior as indicators of well-being in mice. J Neurosci Methods 234:139–146
23. Filippov V, Nozdrachev D (2010) Modern concepts on the mechanisms of encoding visceral nociceptive stimuli human. Physiology 36(1):107–117
24. Tappe-Theodor A, King T, Morgan MM (2019) Pros and cons of clinically relevant methods to

assess pain in rodents. Neurosci Biobehav Rev 100:335–343
25. Deacon RM (2006) Burrowing in rodents: a sensitive method for detecting behavioral dysfunction. Nat Protoc 1(1):118–121
26. Deacon RMJ, Raley JM, Perry VH, Rawlins JNP (2001) Burrowing into prion disease. Neuroreport 12(9):2053–2057
27. Sherwin CM, Haug E, Terkelsen N, Vadgama M (2004) Studies on the motivation for burrowing by laboratory mice. Appl Anim Behav Sci 88(3):343–358
28. Shepherd AJ, Cloud ME, Cao YQ, Mohapatra DP (2018) Deficits in burrowing behaviors are associated with mouse models of neuropathic but not inflammatory pain or migraine. Front Behav Neurosci 12:124
29. Deacon RM, Croucher A, Rawlins JN (2002) Hippocampal cytotoxic lesion effects on species-typical behaviours in mice. Behav Brain Res 132(2):203–213
30. Deacon RM, Penny C, Rawlins JN (2003) Effects of medial prefrontal cortex cytotoxic lesions in mice. Behav Brain Res 139(1–2):139–155
31. Muralidharan A, Kuo A, Jacob M, Lourdesamy JS, Carvalho LM, Nicholson JR et al. (2016) Comparison of burrowing and stimuli-evoked pain behaviors as end-points in rat models of inflammatory pain and peripheral neuropathic pain. Front Behav Neurosci 10:88
32. NHMR C (2013) Australian code for the care and use of animals for scientific purposes, 8th edn. Australian Government
33. Edwards SR, Mather LE, Smith MT (2007) Studies with ketamine and alfentanil following Freund's complete adjuvant-induced inflammation in rats. Clin Exp Pharmacol Physiol 34(5–6):414–420
34. Smith MT, Wyse BD, Edwards SR, El-Tamimy M, Gaetano G, Gavin P (2015) Topical application of a novel oxycodone gel formulation (tocopheryl phosphate mixture) in a rat model of peripheral inflammatory pain produces localized pain relief without significant systemic exposure. J Pharm Sci 104(7):2388–2396
35. Stein C, Millan MJ, Herz A (1988) Unilateral inflammation of the hindpaw in rats as a model of prolonged noxious stimulation: alterations in behavior and nociceptive thresholds. Pharmacol Biochem Behav 31(2):445–451
36. Bennett GJ, Xie YK (1988) A peripheral mononeuropathy in rat that produces disorders of pain sensation like those seen in man. Pain 33(1):87–107
37. Randall LO, Selitto JJ (1957) A method for measurement of analgesic activity on inflamed tissue. Arch Int Pharmacodyn Ther 111(4):409–419
38. Muralidharan A, Wyse BD, Smith MT (2013) Optimization and characterization of a rat model of prostate cancer-induced bone pain using behavioral, pharmacological, radiological, histological and immunohistochemical methods. Pharmacol Biochem Behav 106:33–46
39. Netser S, Meyer A, Magalnik H, Zylbertal A, de la Zerda SH, Briller M et al. (2020) Distinct dynamics of social motivation drive differential social behavior in laboratory rat and mouse strains. Nat Commun 11(1):5908
40. Jirkof P, Cesarovic N, Rettich A, Nicholls F, Seifert B, Arras M (2010) Burrowing behavior as an indicator of post-laparotomy pain in mice. Front Behav Neurosci 4:165
41. Andrews N, Legg E, Lisak D, Issop Y, Richardson D, Harper S et al. (2012) Spontaneous burrowing behaviour in the rat is reduced by peripheral nerve injury or inflammation associated pain. Eur J Pain 16(4):485–495
42. McLinden KA, Kranjac D, Deodati LE, Kahn M, Chumley MJ, Boehm GW (2012) Age exacerbates sickness behavior following exposure to a viral mimetic. Physiol Behav 105(5):1219–1225
43. Griffiths LA, Duggett NA, Pitcher AL, Flatters SJL (2018) Evoked and ongoing pain-like behaviours in a rat model of paclitaxel-induced peripheral neuropathy. Pain Res Manag 2018:8217613
44. Wodarski R, Delaney A, Ultenius C, Morland R, Andrews N, Baastrup C et al. (2016) Cross-centre replication of suppressed burrowing behaviour as an ethologically relevant pain outcome measure in the rat: a prospective multicentre study. Pain 157(10):2350–2365
45. Kuo A, Smith MT (2018) In vivo profiling of four centrally administered opioids for antinociception, constipation and respiratory depression: between-colony differences in Sprague Dawley rats. Clin Exp Pharmacol Physiol 45(10):1056–1066
46. Muralidharan A, Sotocinal SG, Austin JS, Mogil JS (2020) The influence of aging and duration of nerve injury on the antiallodynic efficacy of analgesics in laboratory mice. Pain Rep 5(3):e824
47. Stryjek R, Modlinska K, Turlejski K, Pisula W (2013) Circadian rhythm of outside-nest activity in wild (WWCPS), albino and pigmented laboratory rats. PLOS One 8(6):e66055
48. Hankenson FC, Marx JO, Gordon CJ, David JM (2018) Effects of rodent thermoregulation on animal models in the research environment. Comp Med 68(6):425–438
49. Gaskill BN, Gordon CJ, Pajor EA, Lucas JR, Davis JK, Garner JP (2012) Heat or insulation: behavioral titration of mouse preference

for warmth or access to a nest. PLOS One 7(3):e32799
50. Patterson-Kane EG (2003) Shelter enrichment for rats. Contemp Top Lab Anim Sci 42(2):46–48
51. Blanchard RJ, Blanchard DC (1990) Antipredator defense as models of animal fear and anxiety. Fear and defence. Ettore Majorana international life sciences series, vol. 8. Harwood Academic Publishers, Amsterdam, Netherlands, 89–108
52. Poole S, Stephenson JD (1977) Body temperature regulation and thermoneutrality in rats. Q J Exp Physiol Cogn Med Sci 62(2):143–149
53. Makowska IJ, Weary DM (2016) The importance of burrowing, climbing and standing upright for laboratory rats. R Soc Open Sci 3(6):160136
54. Deseure K, Hans G (2018) Orofacial neuropathic pain reduces spontaneous burrowing behavior in rats. Physiol Behav 191:91–94
55. Schuelert N, Gorodetskaya N, Just S, Doods H, Corradini L (2015) Electrophysiological characterization of spinal neurons in different models of diabetes type 1- and type 2-induced neuropathy in rats. Neuroscience 291:146–154

Check for updates

Chapter 14

Racing Against Time: Understanding the Temporal Dynamics of Anxiety and Depression in Chronic Pain

Vandana Shree Vedartham Srinivasan, Susana Lima, Victor Mathis and Ipek Yalcin

Abstract

Major depressive disorder (MDD) and generalized anxiety disorder (GAD) are some of the leading causes of disability worldwide, contributing to the global burden of diseases. Although chronic stress is considered as the major risk factor for such conditions, chronic pain is an important determinant for MDD and GAD, with a prevalence rate ranging from 30% to 80% in chronic pain patients, highlighting the important interplay between these pathologies. This chapter describes commonly used animal models and methods to study the emotional consequences in rodent models by taking into consideration the time factor. We also discuss the limitations of the current models along with future directions to better understand this comorbidity and improve the translatability of preclinical studies.

Keywords Chronic pain, Mood disorders, Behavior, Comorbidity, Rodents

1 Introduction

The International Association for the Study of Pain (IASP) defines pain as an unpleasant sensory and emotional experience associated with, or resembling that associated with, actual or potential tissue damage [1, 2]. In its nonpathological definition, the experience of pain is subjective and a beneficial alert signal for the organism to avoid potential tissue damage(s). This crucial survival function has multiple dimensions (Fig. 1). First, the sensory-discriminative dimension encodes the spatiotemporal information as well as the intensity of the painful stimulus. This sensory aspect requires high-threshold sensory receptors (nociceptors) of the peripheral somatosensory nervous system that are capable of transducing and encoding noxious stimuli. Hence, nociception is defined as the neural process of encoding noxious stimuli (IASP). A second aspect of pain is its affective dimension that represents the feeling of unpleasantness of

S. Lima and V. Vedartham Srinivasan are equally contributed to this work.

Isaura Tavares and Hugo Leite-Almeida (eds.), *Translational Research Methods for Neuropathic Pain*, Neuromethods, vol. 235, https://doi.org/10.1007/978-1-0716-5312-8_14,

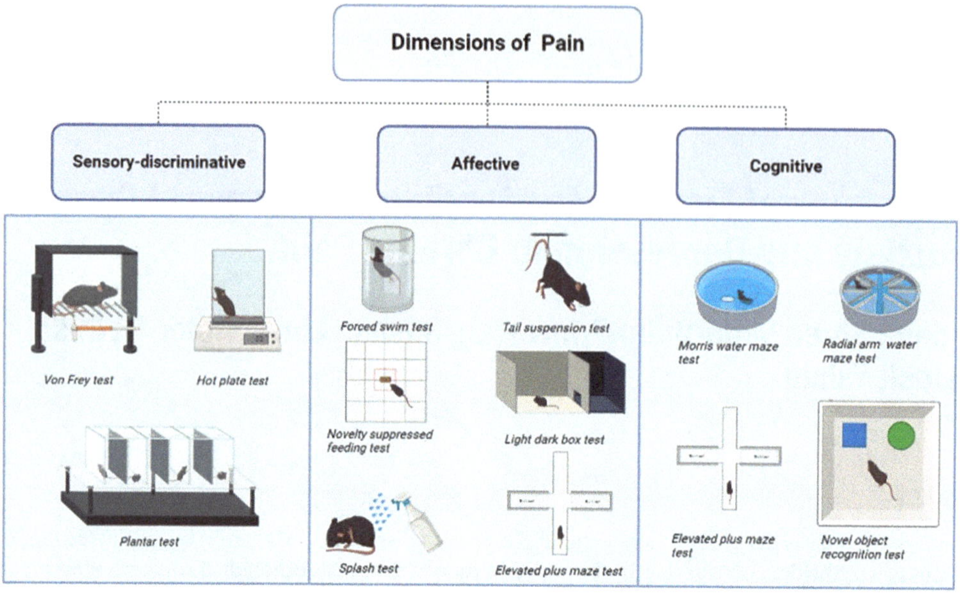

Fig. 1 How to evaluate different dimensions of pain in preclinical studies? This schematic figure illustrates the classic tests used to evaluate sensory-discriminative, affective, and cognitive dimensions of pain in preclinical studies

painful stimuli. Finally, the cognitive dimension of pain represents the perception, thoughts, and motivation toward pain. All these dimensions give an individual all the information to process painful stimuli and adapt its behavior toward protection (Fig. 1).

In addition to the aforementioned dimensions, pain can be subcategorized into acute and chronic pain. Acute pain drives the motivation toward self-protection and disappears when the underlying cause has been treated. On the other hand, chronic pain that affects around 20% of the population worldwide [3] is defined as a pain that persists more than 3 months, and even after the normal period of healing. Among others, neuropathic [4] and inflammatory pain, which affect 7%–10% and 10% [5] of the population, respectively, are common causes of chronic pain. An important distinction between acute and chronic pain is the possible development of a myriad of other symptoms in individuals affected by chronic pain. Indeed, because of the persistent feeling of pain, chronic pain patients can start to develop different comorbidities, such as anxiety, depression, or irregularity of sleep cycle, that affect the patient's everyday life. Among these comorbidities, major depressive disorder (MDD) and anxiety appear to be the most recurrent, with 32.4%–56.8% [6] of chronic pain patients displaying depressive symptoms and 6%–35% of patients developing anxiety disorders [7]. Such prevalence of psychiatric disorders confirms the impact of chronic pain on brain functions; however, the precise mechanisms leading to psychiatric consequences need to be studied further.

In that purpose, multiple animal models have been developed, bringing new knowledge about pain processing as well as the development of psychiatric symptoms accompanying chronic pain. This book chapter aims at describing and discussing the animal models and behavioral tests that are commonly used to study chronic pain and associated mood disorders. A final section will be dedicated to some perspectives about the new tools in development for the study of this comorbidity.

2 Animal Models of Chronic Pain Used to Study the Affective Consequences

Even though it is widely accepted that depression and painful symptoms are common comorbidities and that their combination is more disabling than either condition alone, the interaction between the two is not fully understood. Important advances have been achieved in the

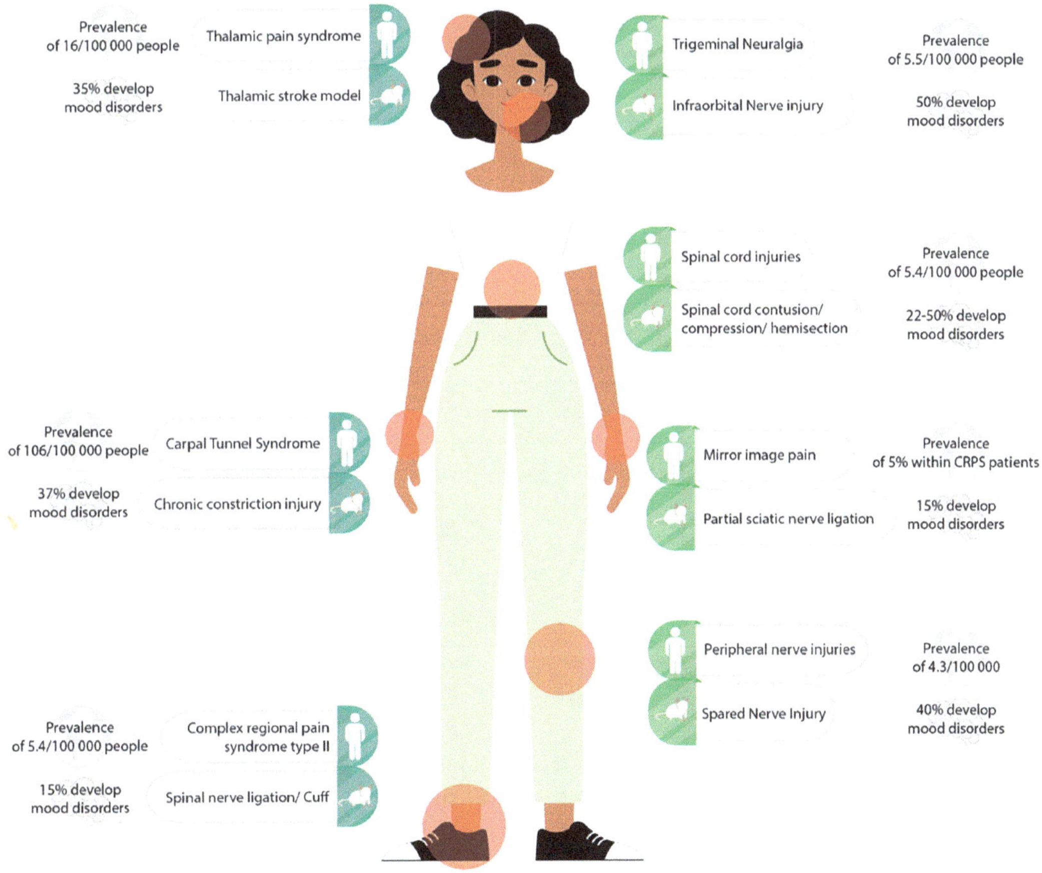

Fig. 2 Neuropathic pain in humans and preclinical animal models. Representation of numerous neuropathic pain syndromes in humans, their prevalence, comorbidity with mood disorders, and the most common preclinical animal model used for their study

past decade, thanks to the development of animal models allowing the study of anxiety- and depressive-like consequences of chronic

Table 1
Description of animal models of inflammatory, neuropathic, cancer, and fibromyalgia pain

Type chronic pain	Animal model	Brief description	Reference protocol
Inflammatory pain	Inflammatory agents - CFA, uric acid, carrageenan	Injection of inflammatory agents into the knee, the tibiotarsal or temporomandibular joint	[10, 12, 13]
Peripheral neuropathic pain	Chronic constriction injury	Four ligatures are tied loosely around the sciatic nerve at 1 mm intervals	[22]
	Partial sciatic nerve ligation	Ligation of 1/3 to 1/2 of the sciatic nerve with silk suture	[23]
	Spared nerve injury	Ligation of peroneal and tibial nerves	[24]
	Spinal nerve ligation	Ligation of the lumbar segmental spinal nerve L5 or L5/L6	[25]
	Sciatic nerve cuffing	Polyethylene cuff placed on the sciatic nerve	[51]
	Trigeminal neuralgia	Ligation of the buccal branch of the facial nerve	[27]
Central neuropathic pain	Thalamic pain syndrome	Injection of a vasoconstrictor or thrombogenic substance into the thalamus	[65]
	Spinal cord injury	Spinal cord contusion, compression, excitatory neurotoxin administration, photochemically induced ischemia, spinal cord hemisection, or spinal cord crushing	[69–73]
Fibromyalgia	Chronic musculoskeletal pain	Repeated muscle insults with acid or amine depletion by reserpine	[85, 88, 87]
	Stress-induced fibromyalgia	Intermittent cold stress, repeated cold stress, unpredictable sound stress, or subchronic swim stress	[89, 85, 91, 90, 93]
Cancer pain	Peripheral neuropathy by cancer	Transplantation of tumor cell lines or chronic injection of oxaliplatin	[101, 103]

pain. All these studies point out the fact that time is an important factor to model anxiodepressive consequences of chronic pain in animals. Besides the time factor, the chosen animal models and tests, but also protocols used, strongly affect the behavioral outputs and can explain some contradictory results published in the field. In the next sections, after a brief introduction about each pain condition, we will explore the different chronic pain models in rodents used to study the comorbidity between chronic pain and mood disorders (Fig. 2). Since an overview of anxiety and depressive comorbidities in rodent models of chronic pain was described in Kremer et al., based on literature until 2019 [8], here, we present a collection of recent studies from 2019 to 2023 (Table 1).

3 Inflammatory Pain

Inflammatory pain is commonly caused by a pathogenic infection or an injury. It is characterized by increased sensitivity of the injured tissue as well as the adjacent site. Immune response triggered at the site of injury further activates the primary nociceptors, thereby inducing hypersensitivity. When the inflammatory reactions become persistent, they cause peripheral and central sensitization, which leads to the chronification of pain over time. Among the inflammatory pain conditions, arthritis (joint inflammation), which is a rheumatic disease, is strongly associated with depression, with the prevalence reaches up to 66% [9].

4 Animal Models of Inflammatory Pain

Different animal models are used to study the anxiodepressive consequences of chronic inflammatory pain. Monoarthritic models consist of the injection of inflammatory agents such as CFA, uric acid, or carrageenan into the knee [10, 11], the tibiotarsal [12], or temporomandibular [13] joint and have been extensively used to study the anxiodepressive consequences of inflammatory pain. Another model is the use of obesity animal model that also displays inflammatory pain [14, 15]. Among these pain models, the most frequently used is the CFA-induced paw inflammation model, which leads to long-lasting (>3 weeks) mechanical and thermal hyperalgesia in both female and male rodents [16] and has been described to display anxiodepressive-like behavior early on after pain induction [17]. In the CFA model, the development of anxiety- and depressive-like behavior is mostly reported between 1 and 4 weeks after induction [8]. Carrageenan injection, known to induce inflammation when injected into the hind paw, joints, or muscles in rodents, is also a common model of inflammatory pain, even though its effect on mechanical

Table 2
Description of frequently used tests to evaluate the anxiety- and depressive-like behaviors in rodents

Behavioral test	Dimension	Description	Evaluated parameter	Reference protocol
Open field test		The apparatus consists of an enclosed open arena with or without light. The animal is allowed to freely explore the apparatus.	Mobility/immobility time, time spent in different zones (i.e., center, corners) of the apparatus	[177]
Light-dark box test		The apparatus consists of two connected compartments. One compartment is lit (>1000 lux) while the other remains dark (<5 lux). The animal is allowed to freely explore the apparatus.	Time spent and number of visits in both chambers, latency to enter the lit chamber	[123, 124]
Elevated plus maze test	Anxiety	The apparatus is an elevated plus maze with two open and two closed arms, with walls on the periphery. The animal is allowed to freely explore the apparatus.	Time spent and number of visits in different arms, latency to enter an open arm	[131, 178]
Hole-board test		The apparatus consists of a closed arena with holes on the floor. The animal is allowed to freely explore the apparatus.	Number of head dips/nose pokes	[134]
Burrowing test		After acclimatization to the presence of a tube in their cage, a tube filled with bedding material is placed in the mouse cage for a specific time period (usually 15 min).	Amount of bedding burrowed from the tube	[137]
Marble burying test		Marbles are placed inside the mouse home cage.	Number of marbles buried over a specific time period	[141, 142]

Behavioral test	Dimension	Description	Evaluated parameter	Reference protocol
Novelty-suppressed feeding test	Anxiety/Depression	After a food restriction, mice are placed in a novel open field with a food pellet placed in the center. The animal is allowed to freely explore the apparatus and eat the pellet.	Latency to touch and eat the food, number of touches	[161, 179]
Splash test		A sucrose solution (10%–20%) is sprayed on the back of the test animal.	Latency to start grooming, time spent, and number of grooming episodes	[144]
Nesting test		Nesting material is placed in the mouse cage.	Evaluation of the nest quality after different time periods (2, 4, 24 h)	[146]
Sucrose preference test		Mice have free access to water and sucrose solution (1%–2%) over 24 h.	Amount of sucrose solution consumed over time	[149]
Urine sniffing test	Depression	Two cotton tips are inserted in the test animal cage. One is filled with female urine (age- and strain-matched mice), while the other is filled with water. The animal is allowed to freely explore for a specific time period.	Amount of time spent sniffing both tips	[154]
Forced swim test		The animal is placed in a beaker filled with water (~22/24°C).	Total mobility/immobility time	[155, 180]
Tail suspension test		The test animal is suspended upside down by taping the tail up.	Total immobility time	[159]

and thermal thresholds lasts no longer than 72 h [18]. Inflammatory agents such as capsaicin, formalin, mustard oil, and zymosan can also induce hyperalgesia and mechanical allodynia; nonetheless, their anxiodepressive consequences have not been studied, which might be due to their acute effects (Table 2).

5 Neuropathic Pain

Neuropathic pain (NP), defined as a pain caused by a lesion or disease affecting the somatosensory system, is a broad category of pain syndromes that covers an extensive variety of peripheral and central disorders. The cause of neuropathic pain is substantially diverse and linked to a disease or lesion. According to epidemiological studies, neuropathic pain affects about 7%–8% of the general population and accounts for 20%–25% of people who suffer from chronic pain [19].

The symptomatology of neuropathic pain can have up to five different components, namely superficial burning pain, deep pain, paroxysmal pain, evoked pain, and dysesthesia/paresthesia [20]. Based on the patients' sensory profile, consisting of combinations of particular symptoms, different patient subgroups can be identified [19]. As a result, due to the complexity of the diseases and the patient variances, it should be emphasized that therapeutic approaches are challenging and should be tailored to each patient.

In addition, the majority of people who suffer from neuropathic pain also have other symptoms such as mood changes, sleep disturbances, and cognitive impairment, which negatively affect the patient's quality of life. Notably, the prevalence of anxiety and major depressive disorders is almost three times higher in neuropathic pain patients than in the normal population (Fig. 2) [21]. In addition, longer duration of pain and baseline pain severity appear to increase the likelihood of developing an emotional disorder in patient experiencing persistent pain. Therefore, for the treatment and management of chronic pain, it is crucial to recognize the comorbidity between mental health issues and neuropathic pain (Table 3).

Taking this information into consideration, the next sections will discuss the existing animal models that have reported the diverse manifestations of neuropathic pain and mood disorders in three segments: (1) peripheral neuropathic pain, (2) central neuropathic pain, and (3) central and peripheral neuropathic syndromes (Fig. 3).

Table 3
Time-dependent development of anxiety- and depressive-like behavior in different animal models of chronic pain

Pain model	Species	Behavioral tests	Results	Reference
Animal models of inflammatory pain				
CFA	Sprague–Dawley rats	EPM, LDB	ALB at 10 days PI	[125]
Osteoarthritis	Sprague–Dawley and Wistar Kyoto rats	OF, EPM	ALB	[181]
Osteoarthritis	Rat	FST	DLB at 28 days PI	[182]
Carrageenan	Wistar albino rats	OF, EPM, LDB	ALB at 3 and 7 days PI	[126]
Animal models of peripheral neuropathic pain				
SNI	Sprague–Dawley rats	TST, FST, OF, SPT	ALB and DLB at 21 days PO	[151]
PSNL	Genetically modified mice	EPM, LDB, FST, NOR	ALB at 11 and 23 days PO	[42]
PSNL	Genetically modified mice	NOR, FST	DLB at 19 days PO; decreased cognitive performance at 12 days PO	[43]
PSNL	C57BL/6J	OF, EPM, TST	ALB at 35 days PO	[41]
PSNL	Transgenic mice	OF, NSF, FST, SI	Decreased SI at 6 and 8 weeks PO; DLB and ALB at 8 weeks PO	[121]
PSNL	Swiss mice	FST, ST	DLB at 4 weeks PO	[40]
PSNL	Transgenic mice	EPM, LDB, OF, FST, TST, CPP	ALB at 30 days PO	[127]
CCI	Wistar rats	MB, TST, FST	ALB and DLB at 21 days PO	[156]
CCI	Sprague–Dawley rats	EPM, FST	ALB at 5 weeks PO; no depression	[33]
CCI	Transgenic Long Evans rats	EZM, OF	ALB at 5–6 weeks PO	[31]
CCI	Wistar Han rats	OF, EPM, SPT, FST	No ALB or DLB	[30]
CCI	Wistar albino rats	EPM, OF	ALB	[183]
CCI	Wistar albino rats	FST	DLB at 20 days PO	[32]

(continued)

Table 3
(continued)

Pain model	Species	Behavioral tests	Results	Reference
CCI	Sprague–Dawley rats	SPT, FST	DLB at week 4 (FST); week 8 (SPT)	[34]
CCI	Sprague-Dawley rats	FST, SPT	DLB at 3 weeks PO	[79]
cuff	C57BL/6J mice	CPP, NSF, MB, splash, FST	ALB and DLB at 4–8 weeks PO	[53]
SNL	Sprague–Dawley rats	OF	ALB at 20 and 30 days PO	[50]
SNL	Sprague– Dawley rats	CPA, SPT, FST	Increased CPA score at 3 weeks PO; DBL at 4 and 5 weeks PO	[47]
SNL	Sprague– Dawley rats	MWM, FST, SPT	Decreased cognitive performance at 20 days PO; DLB at 20 days PO	[46]
SNL	Transgenic Wistar rats	EPM, OF, vocalizations	ALB at 1 and 4 weeks PO; vocalizations evoked by noxious stimuli increased at 1 and 4 weeks PO	[49]
SNL	ICR mice	FST	DLB at 14 days PO	[48]
trigeminal neuralgia	Sprague–Dawley rats	LDB, EPM	ALB ar 8–9 weeks PO	[59]
trigeminal neuralgia	C57BL/6J mice	OF, EPM, FST, TST, SPT	ALB and DLB at 14 days after PO	[60]
trigeminal neuralgia	Wistar rats	Vocalizations	ALB at 15 days PO	[56]
trigeminal neuralgia	Wistar rats	EPM, LDB	ALB at 14 days PO	[61]
trigeminal neuralgia	C57BL/6	EPM, OF	ALB at 14 and 21 days PO	[58]
trigeminal neuralgia	C57BL/6	TST, FST, OF, SPT, SI	DLB at 4 weeks PO	[62]
trigeminal neuralgia	C57BL/6	OF, EPM, FST	ALB at 14–19 days PO	[57]

Table 3 (continued)

Pain model	Species	Behavioral tests	Results	Reference
Animal models of central neuropathic pain				
Thalamic hemorrhage	CD1 mice	TST, FST, ST	DLB at 21 days PI	[66]
Cancer-related chronic pain				
Oxaliplatin	C57BL/6J and BALB/cJ mice	LDB, SPT, burrowing, nesting	ALB at 4 weeks PI; DLB at 1 day PI	[103]

ALB anxiety-like behavior, *DLB* depressive-like behavior, *CPA* conditioned place aversion, *CPP* conditioned place preference, *EPM* elevated plus maze, *EZM* elevated zero maze, *FST* forced swim test, *LDB* light-dark box, *MB* marble burying, *MWM* Morris water maze, *NSF* novelty-suppressed feeding, *OF* open field, *PI* postinduction, *PO* postoperative, *SI* social interaction, *SPT* sucrose preference test, *ST* splash test, *TST* tail suspension test

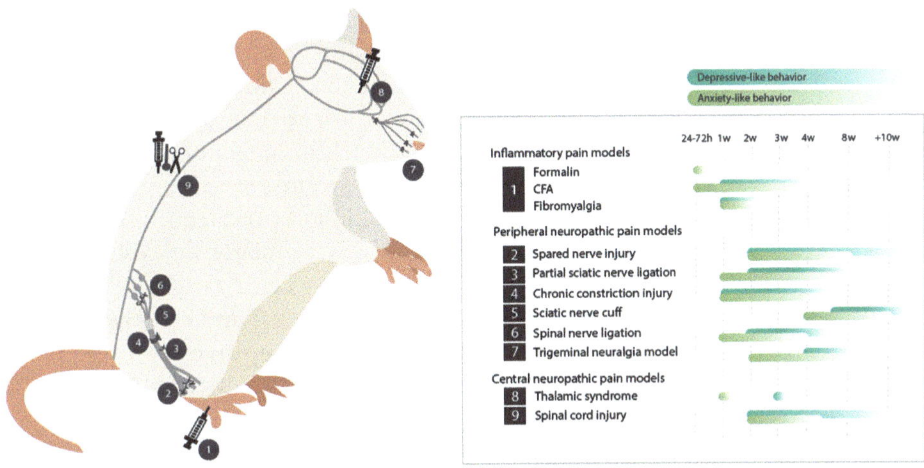

Fig. 3 Animal models of chronic pain and the time course of the development of depressive and anxiety-like behavior. Schematic representation of the sites of injury for the most common animal models of chronic pain (left). Schematic representation of the time course of the depressive- and anxiety-like symptoms observed in different animal models (right)

6 Animal Models for Studying Peripheral Neuropathic Pain–Induced Anxiodepressive-Like Behavior

Over the years, the development of chronic pain models expanded and originated a variety of models for peripheral neuropathy such as chronic constriction injury (CCI) [22], partial sciatic ligation (PSNL) [23], spared nerve injury (SNI) [24], spinal nerve ligation (SNL) [25], sciatic nerve cuffing [26], and trigeminal neuralgia [27], which will be further discussed in this section (Fig. 3). All these models appear to have a good etiological validity since they are directly based on observations made on patients affected by neuropathic pain.

The chronic constriction injury (CCI) was the first animal model based on a partial injury to a peripheral nerve that is analogous to the clinical condition of chronic nerve compression [22]. In patients, chronic nerve compression injury can cause a variety of complications, including neuropathic pain, numbness, skeletal muscle atrophy, and motor disability. Carpal tunnel syndrome (CST), with a prevalence of 5% and incidence of 1–2 per 1000 person-years, is the most common mononeuropathy [28]. Patients with CTS, for example, present significantly higher anxiety and depression when compared to healthy subjects [29]. In animal models of CCI, the development of anxiety-like behavior is mostly reported between 1 and 4 weeks after induction, while depressive-like behavior is observed between 1 and 6 weeks. Besides emotional dysregulation, cognitive impairments such as working memory, spatial and social recognition, and fear memory have been reported [30–34].

Seltzer et al. [23] later developed a rodent partial sciatic nerve ligation (PSNL) injury model, described as an accurate model for "mirror image" pain syndromes [35, 23]. The PSNL model is described to exhibit a wide range of symptoms, including allodynia to touch, hyperalgesia, mirror image phenomena, and sympathetic dependency, which characterize causalgia in humans [35, 23]. For instance, complex regional pain syndrome, a disorder part of the mirror image pain syndromes, exhibits a prevalence of mood disorders estimated to be around 15% [36]. Similarly, in PSNL models, anxiety-like behaviors are observed between 1 and 8 weeks post-surgery and depressive-like behaviors between 2 and 8 weeks postinduction [37–39]. Additionally, decreased cognitive performance during memory tasks is observed as early as 12 days after surgery [40–43].

Ten years later, similarly to the PSNL model, Decosterd and Woolf [24] developed another model of injury, the spared nerve injury (SNI) model. In this model, the tibial and common peroneal nerves, two of the three terminal branches of the sciatic nerve, are damaged, while the sural nerve is left intact. In humans, peripheral nerve injuries (PNI) are relatively common pathologies in clinical practice and occur in approximately 3% of all trauma patients. Within these patients, about 40% report the development of depressive disorders and/or anxiety persisting over the course of the disease [44]. In rodents, anxiety-like behavior following SNI has been found in both rats and mice from week 2 to week 8 post-surgery, and depressive-like behavior was reported between 2 and 11 weeks post-surgery. Furthermore, in this model, declarative memory is impaired from 1 to 6 weeks after surgery [45].

Kim and Chung presented a new model of nerve ligation, the spinal nerve ligation (SNL) or Chung model in 1992 [25]. The L5 or both L5 and L6 segmental spinal nerves distal to the DRG are unilaterally tight-ligated during SNL surgery, leaving the L4 and L3 components of the sciatic nerve unharmed [25]. Spinal nerve ligation is mostly used to mimic the symptoms of human patients suffering from complex regional pain syndrome type II, or causalgia, developed after an injury to the peripheral nerves. As mentioned before, this disorder also has the capacity to negatively impact the physical, mental, and social health of the patients. In SNL models, anxiety-like behavior has been observed between 1 and 4 weeks post-surgery, while depressive-like behavior is reported after 2 weeks until 5 weeks. Recognition memory is also impaired at 14 days post-surgery [46–50].

Since several groups reported that ligation models induce variability that can decrease the reproducibility of data, Mosconi and Kruger developed the cuff model [51]. In this model, a short polyethylene tubing cuff is placed unilaterally around the major branch of the sciatic nerve, causing ipsilateral heat thermal hyperalgesia 3 weeks post-surgery and ipsilateral mechanical allodynia for at least 2 months [52]. Mice exposed to this model develop anxiety- and depressive-like behavior in a time-dependent manner. Indeed, animals first develop anxiety-related behavior from 4 weeks after induction of the neuropathy until 9 weeks, followed by depression-related behaviors from 6 weeks that persist up to 16 weeks [53, 54].

Finally, over the years, some models have also been developed with the specific intent of further modeling specific pathologies, such as trigeminal neuralgia (TN). Trigeminal neuralgia, a neuropathic pain state triggered by vascular compression of the trigeminal nerve, is usually initiated by an atypical loop of artery or vein. TN is estimated to cause mild-to-severe depression in 35.7% patients and anxiety disorder in over 50% of the patients [55]. In rodents, damage to the infraorbital nerve is the most common way to model this pathology [27]. Using this model, it has been shown that the anxiety-like behavior can be observed between 2 and 9 weeks post-surgery, while depressive-like behavior is developed later around 4–8 weeks [56–62].

7 Animal Models for Studying Central Neuropathic Pain-Induced Anxiodepressive-Like Behavior

Besides the peripheral nervous system, neuropathic pain can also result from conditions that affect the brainstem, spinal cord, or brain. Indeed, vascular, infectious, or traumatic damage to the central nervous system can cause central neuropathic pain [63] and induce comorbid anxiety, depression, or sleep disorders. Epidemiological studies showed that around 67% of patients are estimated to exhibit some degree of anxiety, depression, or sleep disorders, comorbid with central neuropathic pain [64].

Central post-stroke pain or thalamic pain syndrome, a type of central pain that typically follows a thalamic stroke, is characterized by attacks of allodynia, dysesthesia, and spontaneous pain. Approximately 7.3% of people experience central post-stroke pain, and 1 year after the stroke, quality of life is estimated to decrease by 40% [65]. Inducing a stroke in the ventral posterolateral (VPL) nucleus of the thalamus by injecting a vasoconstrictor or thrombogenic substance into the thalamus, or by performing a surgery that obstructs the blood flow to the thalamus, are regularly used thalamic stroke animal model. Following stroke induction, animals commonly exhibit symptoms such as spontaneous pain, hyperalgesia, and allodynia. Animal models of this painful syndrome produce anxiety-like behavior 1 week following induction, while depressive-like behavior was found at week 3 [66].

Spinal cord injury (SCI), affecting 20–40 people per million annually, may occur acutely (traumatic) or chronically (spinal tumors), and usually results in symptoms of pain, disability, loss of function, and neurologic dysfunction [67]. In addition, it is estimated that between 22% and 50% of patients develop depression after diagnosis [68, 67]. Numerous spinal cord injuries were modeled using contusion, spinal cord compression, excitatory neurotoxin administration, photochemically induced ischemia, spinal cord hemisection, or spinal cord crushing in rodents [69–73]. Among them, the spinal cord contusion is the most extensively used model. In addition to motor dysfunction, this model usually provokes sensory alterations including neuropathic pain, such as tactile allodynia and thermal hyperalgesia [74, 75]. The impact of the injury tends to vary. Therefore, making a consistent phenotype can be a challenge [74]. Anxiety-like behavior following spinal cord contusion has been reported between 2 and 6 weeks post-surgery and depressive-like behavior between 2 and 10 weeks, while declarative memory impairments are observed at 6 weeks post-surgery [76–80].

8 Animal Models for Studying Fibromyalgia-Induced Anxiodepressive-Like Behaviors

Fibromyalgia is a chronic musculoskeletal pain characterized by chronic widespread tenderness and stiffness [81, 82]. MDD and anxiety are frequently observed in fibromyalgia patients, with the prevalence rates ranging between 20%–80% and 13%–64% [83]. While the main cause of fibromyalgia is still unknown, it is suggested that infections, malfunctioning of HPA axis, stressful early-life events, and central sensitization [84] can trigger this pain condition. Due to the lack of precise etiology, the validity of animal models remains imperfect and generally based on symptoms and on the response to treatment such as antidepressant drugs. Accordingly, the most common models for studying anxiodepressive behaviors are the ones mimicking the muscle fatigue using repeated muscle insults such as injections of acids [85–87] or amine depletion by reserpine [88]. Besides, stress-induced fibromyalgia models have been developed, such as the intermittent (ICS) [89, 90] or repeated cold stress (RCS) [91], the unpredictable sound stress [85], or the subchronic swim stress [92, 93] models. All of these models induce chronic widespread-like pain and have been used for studying the comorbidity of fibromyalgia and mood disorders. In general, animal models of fibromyalgia have been observed to develop anxiety- and depressive-like behavior 2 weeks after the fibromyalgia induction [94]. However, although an increased prevalence of fibromyalgia in women has been observed [95], most of the animal studies used male individuals; hence, future animal studies should take into consideration sex differences.

9 Animal Models for Studying Cancer-Related Chronic Pain-Induced Anxiodepressive-Like Behavior

Chronic pain can be triggered when the neurological system is directly harmed by a primary tumor/metastases (direct invasion into peripheral or central nervous system), or by cancer treatments such as chemotherapy. According to various studies, neuropathic pain affects between 24% and 60% of all cancer patients, and it has been linked to worsening physical, mental, and social functioning [96–99]. In cancer patients, there is also a significant correlation between the existence and intensity of pain and anxiodepressive symptoms. While cancer patients with no pain had a prevalence of anxiety and major depression of approximately 9.2% and 17.7%, respectively, patients with severe pain displayed about 32.8% and 36.7%, respectively [100]. To better understand peripheral neuropathy induced by cancer, the disease is usually modeled in rodents by transplanting tumor cell lines [101]. Animal models of cancer pain, specifically mammary tumors, have been observed to produce anxiety-like behavior between 1 and 3 weeks after tumor palpation [102]. To understand chemotherapy-related pain, oxaliplatin model is used. Here, chronic injection of oxaliplatin induces peripheral neuropathic pain and anxiety-like behaviors 1 week post-injection in C57BL/6 females [103].

10 Methods to Evaluate Emotional Component in Chronic Pain

Besides the sensory-discriminative dimension of pain stimuli, numerous tests have been developed in rodents to assess the affective dimension [104]. Since ongoing pain creates a motivational drive to seek relief, preference for a context associated with relief of pain is commonly used as a measure of pain averseness [105, 104]. Hence, two tests are commonly used in preclinical studies to evaluate averseness:

conditional place aversion/preference or place escape/avoidance paradigm.

10.1 Conditioned Place Preference, Aversion, and Avoidance

Conditioned place preference (CPP) and conditioned place aversion (CPA) are behavioral paradigms rooted in classical Pavlovian conditioning principles. In these paradigms, animals learn to associate a specific environment or location with a particular experience [106].

CPA is a behavioral phenomenon in which an animal develops an aversion to a specific location or environment as a result of its association with an unpleasant or aversive stimulus. An electric shock, a noxious medication, or any other unpleasant experience could serve as the aversive stimulus [107]. The CPA paradigm typically includes a conditioning phase and a testing phase. During the conditioning phase, the animal is exposed to two unique settings, with and without the unpleasant signal. During the second testing phase, the animal is free to explore both surroundings without being subjected to the aversive stimulus. The level of aversion is determined by comparing the amount of time the animal spends in the aversion-paired environment to the unpaired environment [107].

CPP, the inverse of CPA, is a behavioral phenomenon in which an animal develops a preference for a specific location or environment linked with a rewarding or pleasurable stimulus. It serves as a useful paradigm in research for studying the rewarding effects of various stimuli, including drugs, natural rewards such as food, or social interactions. The CPP paradigm, like CPA, has a conditioning phase and a testing phase. In the conditioning phase, one environment is paired with a rewarding stimulus, such as medication administration (i.e., analgesics in pain field), while the other environment remains neutral. The rewarding signal is associated with the specific surroundings by the animal [107].

Finally, the place escape avoidance paradigm is a variation of the conditioned place aversion paradigm. Similar to CPA, the animal is exposed to two compartments or environments. However, instead of associating one compartment with an aversive stimulus, both compartments are linked to aversive sensations.

In the PEAP, a chamber with one dark and one bright region is typically used. When the animal is in the dark side of the chamber, the injured paw (subject to nerve injury) is stimulated with a von Frey monofilament, and when the animal is in the light side, the unaffected paw is stimulated. This paradigm evaluates the animal's learning and decision-making ability when it comes to selecting the less unpleasant setting. The latency to transfer compartments or the number of transitions made is commonly used to assess the efficiency of escape or avoidance behavior [108].

Inflammatory models, such as the injection of CFA [109], and nerve injury models, such as CCI [110], have demonstrated a notable switch in preference. Specifically, animals experiencing these conditions show a preference for the illuminated compartment as opposed to the dark compartment associated with a noxious stimulus. Furthermore, in animals with nerve injuries, a significant preference for the compartment linked to analgesia has been observed between 2 and 6 weeks after the peripheral nerve injury [111].

11 Methods to Study Anxiodepressive Consequences of Pain

Some of the core behavioral and physiological characteristics of mood disorders, known as endophenotypes [112], can be observed in preclinical models, including anhedonia, appetite and sleep disturbances, behavioral despair, and anxiety-like behaviors. In preclinical settings, a wide range of behavioral tests are used to assess such characteristics.

12 Tests to Evaluate Anxiety-Like Behaviors

Most of the behavioral paradigms used to evaluate anxiety in rodents are based on the approach–avoidance conflict between exploration of novel environments and avoidance of unsafe, brightly lit, open spaces [113, 114]. The following paragraphs will highlight the most important aspects of each test.

12.1 Open Field Test

The open field (OF) test is based on a conflict between rodents' desire to explore novel environments and their fear of unsafe areas [115]. The most common open field apparatus is a square enclosure with surrounding walls. Typically, the amount of time spent mobile, the time spent in different zones of the apparatus (i.e., corners, center, thigmotaxis), and the time spent rearing are evaluated. Time spent in the center versus the time spent in thigmotaxis (referring to the corridor along the walls) is classically used as indicators of anxiety-like behavior [116–118]. For instance, in both acute [119] and chronic [120] inflammatory pain models, a reduction of the time spent in the center of the open field and the open arms of the EPM can be observed at 24–72 h post formalin and 29 days post CFA injection, respectively. Similarly, mice that received an intramuscular injection of acid-saline (fibromyalgia model) display decreased time spent in the center zone in the OF test from day 14 to 17 post insult [94]. In neuropathic pain models, rodents with spinal nerve injuries, such as CCI, also show a reduced time spent in the center of the OF as well as the number of entries in the central area [30]. Rats that received spinal nerve ligation or partial sciatic nerve ligation also showed anxiety-like behavior from 1 to 4 weeks post operation [41, 121, 49, 50]. This temporal course is also observed in rodents subjected to a model of trigeminal neuralgia, which present an anxiety-like phenotype between 2 and 4 weeks after the pain induction [57, 58,

60]. Finally, post-stroke mice also display a robust anxious phenotype in the open field test, 1 week post-surgery [122].

12.2 Light-Dark Box Test

Another test aiming at assessing anxiety-like behavior, based on the rodents' natural avoidance of lit areas, is the light-dark box (LDB) test developed by Crawley and Goodwin [123]. The setup consists of two connected chambers, with one chamber being brightly lit (1000–1500 lux) and the other remaining in the dark. The test solely consists of evaluating the exploration of the full apparatus for a specific amount of time, usually 5 min. The common variables are the latency to enter the light side, the overall amount of time spent, and the number of entries in each chamber. In comparison to control animals, animals with anxiety-like behavior exhibit a reduction in their time spent and number of entries in the lit chamber; an increased latency to enter the lit side is sometimes also observed [124].

The use of the light-dark box test in acute inflammatory pain models, such as CFA and carrageenan, a strong reduction in the time spent in the lit chamber was observed between day 3 and 10 postinduction [125, 126]. Similarly, rodents with spinal nerve injuries, such as CCI and PSNL, have reported decrease in time spent in the lit compartment between weeks 2 and 4 postinduction [127, 42, 128]. Moreover, when using the sciatic nerve cuffing model, authors observed a long-lasting reduction in the time spent in the lit compartment, from 4 to 9 weeks post-surgery [54], which correlates with the mechanical allodynia observed in this model [111].

12.3 Elevated Plus Maze Test

The elevated plus maze (EPM) is based on the conflict between the rodents' innate exploratory behavior and the fear generated by open and elevated environments, where rodents are the most vulnerable to predators [129]. The elevated plus maze consists of two open arms, two closed arms (with walls), and a center platform that is elevated 40–80 cm above the ground. Classically, for the test, the animal is placed on the central platform and then it is left free to explore the apparatus for 5 min. The amount of time spent and the number of entries into closed and open arms are assessed. It is also possible to evaluate additional behavioral traits, including head dipping and stretching (when animals stay in the closed arms by stretch their body into open arms). Compared to control animals, animals exhibiting anxiety-like behavior spend less time exploring the open arms [130, 131]. In acute inflammatory pain models, a reduction in the time spent exploring the open arms can be observed 24–72 h post formalin injection [119]. Similarly, in chronic inflammation models, the same reduction is reported 4 weeks after the induction [120]. A reduction in the number of entries and time spent in the open arms was also observed in the PSNL and SNI models from 5 to 47 days and 19 to 24 days, respectively [132]. In the stroke model, mice also displayed robust alterations in the EPM at 1 week post-surgery [133].

Finally, decreased frequency to enter the open arms in the EPM test was also reported from 14 to 17 days post insult with acid-saline in a fibromyalgia model [94].

12.4 Hole-Board Test

Although less commonly used, the hole-board test can also give an insight into anxiety-like behaviors. Based on the innate exploration behavior, the test consists of an enclosed arena with holes in the floor into which an animal can poke its head, referred to as head-dipping [134]. The duration and frequency of head-dipping or nose-pokes are thought to provide indicators of neophilia (or purposeful exploration) that are separate from the animal's overall locomotor activity. A lack of neophilia can subsequently reflect the manifestation of anxiety-like behavior in the rodent [134]. At 4–6 post-operative weeks, a decline in nose-poke activity has been observed in a spared nerve injury model [135]. Additionally, this impact is already visible 3 days after the induction in a fibromyalgia model [136].

12.5 Burrowing/Burying Behaviors

Finally, burrowing behavior in rodents has also been used to study animal well-being and has been tested as a potential surrogate parameter for pain. Burrowing behavior is measured by the amount of bedding removed from an object (usually a tube) over a 2-h testing period or overnight [137]. While it has been shown that several neuropathic pain models (SNI, PSNL, CFA) reduce the burrowing behavior at various time points such as 10, 15, and 20 days PO [138–140], Freund's adjuvant (a model of inflammatory pain) administration did not induce any burrowing deficit [137]. Although easily implementable in laboratories, this test also reflects the animals' well-being and so might be impacted by animal facility conditions (stress, other strains, or sexes housed in the same room) and thus not always precisely reflects pain.

When provided a suitable substrate, such as deep bedding, mice will naturally begin to dig. One behavioral test based on this innate behavior is the marble burying test, first described by Broekkamp et al. [141]. A pattern of marbles is spread on the surface of a cage filled with bedding, and the number of marbles that are at least two-thirds buried after a predetermined amount of time is counted. Anxiolytic substances such as benzodiazepines have been shown to lessen marble burying [142]. Using this test, it has been shown that neuropathic animal models such as PSNL, SNI, SNL, and cuff decrease the number of buried marbles 10 days post-PSNL and SNL [138, 8], between 4 and 7 weeks post cuff [54], as well as up to 1 year in the SNI [143]. In a cancer-related pain model, 1–3 weeks after tumor palpation, rodents with mammary tumors also presented anxiety-like behaviors in the marble burying test [102].

12.6 Depressive-Like Behaviors

Several tests have been developed to evaluate depressive-like behaviors in rodents.

12.7 Splash Test

In rodents, grooming is an ethologically rich behavior that can be observed and analyzed in wild and laboratory animals and is extremely sensitive to adversities [144]. Artificial grooming can be elicited using the splash test, in mice. Common patterns of grooming include elliptical stroke, unilateral stroke, bilateral stroke, and body and tail licking. For this test, mice are placed in their home cages, and a sucrose solution (10%-20%) is sprayed onto their back. Then, grooming activity measures such as the total grooming time, the latency to initiate the first grooming episode, and the number of grooming episodes (head and body) are evaluated [144]. When compared to control animals, animals exhibiting depressive-like behaviors groom less frequently with longer latency to initiate grooming [145]. Chronic neuropathic pain, using the sciatic nerve cuffing model, induces a reduced grooming time between week 6 and 15 postinduction [54]. Similarly, the development of a depressive-like state through reduced grooming in the splash test was also observed at week 4 post partial sciatic nerve ligation [40].

12.8 Nesting Test/ Nest Building Test

Rodents exhibit a wide range of unconditioned innate behaviors that are relevant to their habitat. Among them, we can cite nesting and burrowing behavior. As it helps to conserve heat and is used for refuge and reproduction, nest construction is crucial for rodents. Therefore, a decreased concern for nest construction may be viewed as depressive-like behavior [146]. For the nesting test, animals are separately housed in their standard home cages with a cotton square in the center and left free to build the nest for a certain amount of time. The duration of this test can vary according to different protocols from 120 min to overnight. A poor nest score is thought to indicate depressive-like behavior. This test in rodents is usually performed in their home cage environment, which excludes the stress variable caused in novel environments. After the test period, the nests are evaluated based on the appearance and build and scored from 0 to 4 [146]. Poor nesting behavior has been shown in mice models of early-life stress [147] and irritable bowel syndrome. In the mouse oxaliplatin model, deficits in nesting were observed starting at day 1 post-injection [103]. In the rat CCI model, nesting deficits were observed at 14 days post-surgery [148].

12.9 Sucrose Preference Test

A crucial sign of depression in humans is anhedonia, which is defined as the inability to experience pleasure from rewarding or enjoyable activities. Sucrose preference test is one of the most frequently used tests to assess rodent anhedonia. For this test, the animal is given the choice to consume either water or a sucrose solution (1%–2%). A ratio of the preference between these two options is then calculated by tracking the total amount of water or sucrose solution consumed over time (2–24 h). A decrease in the sucrose preference ratio between experimental and control groups is a sign of anhedonia

[149]. The consumption of sucrose traditionally has been measured by weighing the sucrose bottle and the water bottle before and after the experimental time period. To overcome the potential biases in measuring the consumption, automated systems such as Intellicage and Phenoworld have been developed, which include lickometers to count the number of licks. Contrary to the nonautomated tests, these latest systems also allow to keep animals in groups with minimal interference by the human experimenter [150]. In the SNI model in rats, a reduction in the sucrose preference has been observed 3 weeks post-surgery [151], while the CCI model combined with the unpredictable chronic mild stress (UCMS) model caused a reduction in the sucrose preference from week 1 to week 5 post-surgery [79].

Anhedonia has also been reported in the CFA and carrageenan models; nevertheless, the time component of the depressive comorbidities in these models are underreported [120, 152, 153]. Besides inflammatory and neuropathic pain models, a reduced sucrose preference was also observed from day 14 to day 17 in an animal model of fibromyalgia [94].

12.10 Urine Sniffing Test

The female urine-sniffing test (FUST) is also used to evaluate reward-seeking behavior or anhedonia in male rodents. In this test, the animals have choice between two cotton tips dipped in water or urine of the opposite sex. Time spent sniffing both cotton tips is analyzed to check the preference. It was shown that this preference was accompanied with biochemical changes, such as extracellular dopamine levels increase in the nucleus accumbens following urine exposure, which might be associated with pleasurable and hedonic behaviors. The FUST was also discovered to be sensitive to genetic, behavioral, and pharmacological therapies for mania and depression [154]. Unpublished data from Yalcin lab have also showed an impact of sciatic nerve cuffing in the FUST in male mice.

12.11 Forced Swim Test

The forced swim test (FST) has also been extensively used for assessing antidepressant-like behavior in rodents. It is founded on the premise that when exposed to an unescapable situation, here being placed in a cylinder filled with water, rodents intensely struggle to escape and show an increased mobility. Contrary to this, the immobility is thought to reflect despair and depressive-like behavior in rodents. Practically, the test protocol can vary from 5 to 15 min during which mobility (swimming and climbing on the edge) and immobility time are evaluated [155]. In the CCI model, increased immobility time is usually observed at 2–3 weeks post-surgery [8, 79, 32, 156, 34]. On the other hand, the sciatic nerve cuffing model presents alterations in the FST between 8 and 9 weeks post-surgery [54]. Additionally, in rodents with mammary tumors, depressive-like behaviors were also observed in the FST [102].

An important limitation of this test is that, although it has been validated for depression models, the immobility per se can also be interpreted as a passive coping strategy to the inescapable stress rather than despair [157, 158]. Another major limitation is the compatibility of this test with different strains since some strains show little to no immobility, making the test difficult to interpret.

12.12 Tail Suspension Test

Similar to the FST, the tail suspension test is based on the fact that animals subjected to an acute inescapable stressful situation (here being suspended by their tail) will progressively display despair and so stop struggling to escape the situation. Practically, the animals are taped by their tail upside down on a surface around 50 cm above the ground. The test lasts usually between 5 and 10 min, and the immobility time is recorded [159]. Peripheral nerve injury models such as SNI [160] show depressive-like behavior through increased immobility period in the tail suspension test 3 weeks after injury. In the PSNL model, despair-like behavior was observed in the TST from day 20 until day 62 [132].

13 Novelty-Suppressed Feeding Test

Finally, the novelty-suppressed feeding (NSF) test, based on the conflict between hunger and exploration of an unfamiliar open environment, can be used to assess both anxiety- and depressive-like behaviors. For this test, after a 24 h food deprivation, animals are placed in an unfamiliar open field with a single food pellet positioned in the center. The latency to touch and then eat the food pellet is measured; animals displaying a longer latency to touch/eat being considered as less motivated compared to control animals [161]. Anxiolytic or antidepressant drug treatments shorten the latency to first contact or eat the pellet [162]. The SNI model has been shown to induce anxiety-like behavior, with an increase in latency to feed in the NSF test at day 23 post surgery [163]. In another peripheral nerve injury model, CCI, increase in latency to eat has been observed consistently at around 3 weeks post-surgery as well [30, 128, 164]. In the sciatic nerve cuffing model, alterations in the NSF test were present between 4 and 9 weeks post cuff implantation in male mice [54].

13.1 Limitations of Animal Models and Tests to Study Anxiodepressive-Like Consequences of Chronic Pain

As seen from the studies presented in this chapter, the past 20 years have yielded a growing number of studies that tried to model the comorbidity of chronic pain and mood disorders in rodents. While most published studies demonstrated the possibility to model this comorbidity in animals by highlighting the critical role of the time factor, contradictory results have also been reported. Besides the time factor, these discrepancies might be due to the chosen animal models of pain and test protocols used, which reinforce the idea that behavioral profiling based on performing several tests evaluating anxiety and depression should be prioritized rather than relying on a single parameter in animal models. Accordingly, choosing the appropriate tests and the animal models are the milestones in studying the emotional consequences of pain. One of the limitation of aforementioned studies is that the majority of data have been obtained using trauma models of neuropathic pain, even though recent studies on anxio-depressive consequences of pain showed the potential involvement of distinct mechanisms with different neuropathic pain etiologies, i.e., diabetic vs. trauma [165]. There is thus an unmet need to study this comorbidity challenge by taking into consideration different etiologies such as metabolic diseases, viral infection, or neurotoxicity in neuropathic pain. This disequilibrium and lack of information are also observed in other chronic pain conditions such as fibromyalgia.

Another critical point concerns the possible bias of sexual dimorphism. Indeed, even though sex differences in the genetic, neurochemical, and neuroimmune mediation of pain are well known, the majority of studies cited in this chapter used only male rodents, since in the past, females in general were not used based on the premise that estrous cycle could affect the behavior. However, current studies comparing both male and female mice completely refute this foresight [166], which led to a recent explosion of studies in the pain field.

Depression is well characterized in the DSM-V, with different levels of severity. Unfortunately, it is a challenge to characterize it similarly in rodents. Tests on rodents mainly focus on two specific aspects of depressive symptoms: anhedonia and despair-like behavior. It is thus challenging to mimic the overall human phenotype and strengthen the reproducibility in rodent studies. The order in which different behavioral tests are conducted in mice can also influence the results of subsequent tests, especially when recording depressive-like phenotypes. Therefore, it is common practice to conduct behavioral tests in the order of least to most stressful, but it is important to remember that the more these tests are repeated, the more interactions exist that could induce the bias. One of the possibilities is to launch different batches with their own control groups.

Another overlooked issue is that tests commonly used to assess anxiety- and depressive-like behaviors have originally been designed with male animals. Recent studies point out potential natural sex differences that play an important role in understanding the various facets of pain pathology, including differences in the epigenome and different immune responses to pain [167–169]. To understand these differences better, traditional behavioral paradigms should be re-evaluated and refined to develop sex-specific tests.

13.2 New Tools for Future Directions

One fundamental aspect of pain-associated pathologies/comorbidities is their progressive development, at a distance from the initial cause of pain. Thus, it appears critical to follow the development and maintenance of pain and its associated comorbidities at different stages in a long-term fashion.

The recent development of new imaging tools has revolutionized brain research and now opens a new era of research for the study of pain and most specifically on pain-induced brain alterations in rodents. In that purpose, magnetic resonance imaging (MRI) in small animals has been developed and improved [170]. This allows longitudinal studies of brain alterations observed after the initial pain stimulus. The two most commonly used brain imaging methods in rodents are magnetic resonance imaging (MRI) and positron emission tomography (PET). MRI provides better spatial and temporal resolution than PET, but PET allows for the imaging of neurotransmitters and nonneuronal cells, such as astrocytes, in addition to functional imaging [171]. Although human MRI is by far an important source of information, rodent MRI studies overcome many of the limitations encounter with humans. Indeed, the homogeneity of individuals allows ruling out potential effect due to patient's history. Thus, neuroscience research starting from rodents (back-translation research) has the potential to provide more detailed analyses about how the brain integrates and propagates noxious stimuli. For example, most human studies examining the long-term effects of chronic pain on brain structure and function are cross-sectional in nature, and thus cannot provide much information about cause versus effect [171]. Rodent MRI studies also allow the study of pain pharmacology. For instance, it has been shown that morphine, one of the most used analgesics, modulates noxious-evoked changes in signal intensity in discrete brain regions such as the cingulate cortex, PAG, amygdala, and thalamus [172]. However, one important limitation of MRI is the need to keep the individuals still. In small animals, this can only be obtained by using anesthetics. Hence, despite the development of new anesthetics, MRI studies interrogate brain states that might be altered or at least different compared to vigil states.

Recently, the development of virally transfected biosensors and optical tools [173, 174] also brought a valuable new neuroscientific toolbox. These biosensors allow the recording of neuronal calcium signals or specific neurotransmitters, such as dopamine, serotonin, GABA, or glutamate, on freely moving animals. Hence, among others, calcium imaging–based fiber photometry and miniscopes specifically assess the cerebral activity of specific areas and/or specific neuronal subtypes while awake animals are subjected to noxious stimuli, giving new powerful insights about how brain areas integrate nociceptive information. Furthermore, longitudinal studies are also feasible to describe how specific brain areas and/or networks progressively change their activity in response to pain. Using these technologies, it has been demonstrated that lateral hypothalamic parvalbumin-positive glutamatergic neurons respond to acute thermal stimuli and a persistent inflammatory irritant [175]. Another recent study described two discrete neuronal circuits mediating allodynia in mice [176]; one circuit, involving projections from the posterior thalamic nucleus to primary somatosensory cortex glutamatergic neurons, mediates mechanical allodynia, whereas another one, involving the pathway from the parafascicular thalamic nucleus to anterior cingulate cortex, is involved in allodynia associated with a depression-like state. These studies bring new information about how one brain area is altered when pain becomes chronic and participates in the development of other comorbidities such as depression or anxiety. Finally, these technical advancements will also help deciphering the role of specific neurotransmission systems in pain processing and pain-related comorbidities.

Acknowledgments

This work was supported by the Centre National de la Recherche Scientifique (UPR3212), the University of Strasbourg (UPR3212), the French National Research Agency (ANR-18-CE37-0004, IY), the Fondation pour la recherche médicale (FRM, ARF202120013920, VM), and the European Union's Horizon 2020 Research and Innovation Programme under the Marie Sklodowska-Curie grant agreement No. 955684 (SL, VS). The authors would like to acknowledge that images were adapted from BioRender.com (2023).

Conflict of Interest The authors have no conflicts of interest to declare that are relevant to the content of this chapter.

References

1. Cohen M, Quintner J, Rysewyk S (2018) Reconsidering the international association for the study of pain definition of pain. Pain Rep 3:634. https://doi.org/10.1097/PR9.0000000000000634
2. Raja SN, Carr DB, Cohen M, Finnerup NB, Flor H, Gibson S (2020) The revised international association for the study of pain definition of pain: concepts, challenges, and compromises. Pain 161:1976–1982. https://doi.org/10.1097/j.pain.0000000000001939
3. Goldberg DS, McGee SJ (2011) Pain as a global public health priority. BMC Public Health 11: 770. https://doi.org/10.1186/1471-2458-11-770
4. van Hecke O, Austin SK, Khan RA, Smith BH, Torrance N (2014) Neuropathic pain in the general population: a systematic review of epidemiological studies. Pain 155:654–662. https://doi.org/10.1016/j.pain.2013.11.013
5. Mork H, Motte J, Fisse AL, Grüter T, Brünger J, Stoykova Z, Bulut Y, Athanasopoulos D, Sturm D, Tegenthoff M, Gold R, Enax-Krumova E, Pitarokoili K (2022) Prevalence and determinants of pain in chronic inflammatory demyelinating polyneuropathy: Results from the German INHIBIT registry. Eur J Neurol 29:2109–2120. https://doi.org/10.1111/ene.15341
6. Meda RT, Nuguru SP, Rachakonda S, Sripathi S, Khan MI, Patel N (2022) Chronic pain-induced depression: a review of prevalence and management. Cureus 14:e28416. https://doi.org/10.7759/cureus.28416
7. Kim S, Lee J, Boone D (2022) Protective and risk factors at the intersection of chronic pain, depression, anxiety, and somatic amplification: a latent profile approach. J Pain Res 15:1107–1121. https://doi.org/10.2147/JPR.S340382

8. Kremer M, Becker L, Barrot M, Yalcin I (2020) How to study anxiety and depression in rodent models of chronic pain. Eur J Neurosci 53:236–270. https://doi.org/10.1111/ejn.14686
9. Fiest KM, Hitchon CA, Bernstein CN, Peschken CA, Walker JR, Graff LA, Zarychanski R, Abou-Setta A, Patten SB, Sareen J, Bolton J, Marrie RA, CIHR Team "Defining the Burden and Managing the Effects of Psychiatric Comorbidity in Chronic Immunoinflammatory Disease" (2017) Systematic review and meta-analysis of interventions for depression and anxiety in persons with rheumatoid arthritis. J Clin Rheumatol 23:425–434. https://doi.org/10.1097/RHU.0000000000000489
10. López-Muñoz FJ, Salazar LA, Castañeda-Hernández G, Villarreal JE (1993) A new model to assess analgesic activity: Pain-induced functional impairment in the rat (PIFIR). Drug Dev Res 28:169–175. https://doi.org/10.1002/ddr.430280207
11. Radhakrishnan R, Moore SA, Sluka KA (2003) Unilateral carrageenan injection into muscle or joint induces chronic bilateral hyperalgesia in rats. Pain 104:567–577. https://doi.org/10.1016/S0304-3959(03)00114-3
12. Butler SH, Godefroy F, Besson J-M, Weil-Fugazza J (1992) A limited arthritic model for chronic pain studies in the rat. Pain 48:73–81. https://doi.org/10.1016/0304-3959(92)90133-V
13. Harper RP, Kerins CA, McIntosh JE, Spears R, Bellinger LL (2001) Modulation of the inflammatory response in the rat TMJ with increasing doses of complete Freund's adjuvant. Osteoarthritis Cartilage 9:619–624. https://doi.org/10.1053/joca.2001.0461
14. Georgiev T, Angelov AK (2019) Modifiable risk factors in knee osteoarthritis: treatment implications. Rheumatol Int 39:1145–1157. https://doi.org/10.1007/s00296-019-04290-z
15. Griffin TM, Fermor B, Huebner JL, Kraus VB, Rodriguiz RM, Wetsel WC, Cao L, Setton LA, Guilak F (2010) Diet-induced obesity differentially regulates behavioral, biomechanical, and molecular risk factors for osteoarthritis in mice. Arthritis Res Ther 12:R130. https://doi.org/10.1186/ar3068
16. Rusznák K, Horváth ÁI, Pohli-Tóth K, Futácsi A, Kemény Á, Kiss G, Helyes Z, Czéh B (2022) Experimental arthritis inhibits adult hippocampal neurogenesis in mice. Cells 11:791. https://doi.org/10.3390/cells11050791
17. Hsu Y-C, Ma K-H, Guo S-L, Lin B-F, Tsai C-S, Yeh C-C (2021) The occurrence of pain-induced depression is different between rat models of inflammatory and neuropathic pain. J Clin Med 10:4016. https://doi.org/10.3390/jcm10174016
18. Mert T, Ocal I, Cinar E, Yalcin MS, Gunay I (2014) Pain-relieving effects of pulsed magnetic fields in a rat model of carrageenan-induced hindpaw inflammation. Int J Radiat Biol 90:95–103. https://doi.org/10.3109/09553002.2013.835501
19. Bouhassira D (2019) Neuropathic pain: definition, assessment and epidemiology. Rev Neurol 175:16–25. https://doi.org/10.1016/j.neurol.2018.09.016
20. Bouhassira D, Attal N, Fermanian J, Alchaar H, Gautron M, Masquelier E, Rostaing S, Lanteri-Minet M, Collin E, Grisart J, Boureau F (2004) Development and validation of the neuropathic pain symptom inventory. Pain 108:248–257. https://doi.org/10.1016/j.pain.2003.12.024
21. Bair MJ, Robinson RL, Katon W, Kroenke K (2003) Depression and pain comorbidity: a literature review. Arch Intern Med 163:2433–2445. https://doi.org/10.1001/archinte.163.20.2433
22. Bennett G, Xie Y (1988) A peripheral mononeuropathy in rat that produces disorders of pain sensation like those seen in man. Pain 33:87–107. https://doi.org/10.1016/0304-3959(88)90209-6
23. Seltzer Z, Dubner R, Shir Y (1990) A novel behavioral model of neuropathic pain disorders produced in rats by partial sciatic nerve injury. Pain 43:205–218. https://doi.org/10.1016/0304-3959(90)91074-s
24. Decosterd I, Woolf C (2000) Spared nerve injury: an animal model of persistent peripheral neuropathic pain. Pain 87:149–158. https://doi.org/10.1016/s0304-3959(00)00276-1
25. Ho Kim S, Mo Chung J (1992) An experimental model for peripheral neuropathy produced by segmental spinal nerve ligation in the rat. Pain 50:355–363. https://doi.org/10.1016/0304-3959(92)90041-9
26. Benbouzid M, Gavériaux-Ruff C, Yalcin I, Waltisperger E, Tessier L, Muller A (2008) Delta-opioid receptors are critical for tricyclic antidepressant treatment of neuropathic allodynia. Biol Psychiatry 63:633–636. https://doi.org/10.1016/j.biopsych.2007.06.016
27. Luo D-S, Zhang T, Zuo C-X, Zuo Z-F, Li H, Wu S-X, Wang W, Li Y-Q (2012) An animal model for trigeminal neuralgia by compression of the trigeminal nerve root. Pain Physician 15:187–196
28. Hanewinckel R, Ikram M, Doorn P (2016) Peripheral neuropathies. Neuroepidemiology

138:263–282. doi: https://doi.org/10.1016/b978-0-12-802973-2.00015-x
29. Alsharif A, Al Habbal A, Daaboul Y, Al Hawat L, Al Habbal O, Kakaje A (2022) Is psychological distress associated with carpal tunnel syndrome symptoms and nerve conduction study findings? A case–control study from Syria. Brain Behav 12(2):e2493. https://doi.org/10.1002/brb3.2493
30. Fonseca-Rodrigues D, Amorim D, Almeida A, Pinto-Ribeiro F (2021) Emotional and cognitive impairments in the peripheral nerve chronic constriction injury model (CCI) of neuropathic pain: a systematic review. Behav Brain Res 399:113008. https://doi.org/10.1016/j.bbr.2020.113008
31. Llorca-Torralba M, Suárez-Pereira I, Bravo L, Camarena-Delgado C, Garcia-Partida JA, Mico JA, Berrocoso E (2019) Chemogenetic silencing of the locus coeruleus-basolateral amygdala pathway abolishes pain-induced anxiety and enhanced aversive learning in rats. Biol Psychiatry 85:1021–1035. https://doi.org/10.1016/j.biopsych.2019.02.018
32. Malvestio RB, Medeiros P, Negrini-Ferrari SE, Oliveira-Silva M, Medeiros AC, Padovan CM, Luongo L, Maione S, Coimbra NC, de Freitas RL (2021) Cannabidiol in the prelimbic cortex modulates the comorbid condition between the chronic neuropathic pain and depression-like behaviour in rats: The role of medial prefrontal cortex 5-HT1A and CB1 receptors. Brain Res Bull 174:323–338. https://doi.org/10.1016/j.brainresbull.2021.06.017
33. Murasawa H, Kobayashi H, Saeki K, Kitano Y (2020) Anxiolytic effects of the novel α2δ ligand mirogabalin in a rat model of chronic constriction injury, an experimental model of neuropathic pain. Psychopharmacology (Berl) 237:189–197. https://doi.org/10.1007/s00213-019-05356-3
34. Sun Y-M, Shen Y, Huang H, Liu Q, Chen C, Ma L-H, Wan J, Sun Y-Y, Zhou C-H, Wu Y-Q (2021) Downregulated SIRT1 in the CeA is involved in chronic pain-depression comorbidity. Brain Res Bull 174:339–348. https://doi.org/10.1016/j.brainresbull.2021.07.002
35. Dowdall T, Robinson I, Meert T (2005) Comparison of five different rat models of peripheral nerve injury. Pharmacol Biochem Behav 80:93–108. https://doi.org/10.1016/j.pbb.2004.10.016
36. Brinkers M, Rumpelt P, Lux A, Kretzschmar M, Pfau G (2018) Psychiatric disorders in complex regional pain syndrome (CRPS): the role of the consultation-liaison psychiatrist. Pain Res Manag 2018:2894360. https://doi.org/10.1155/2018/2894360
37. Gai BM, Bortolatto CF, Brüning CA, Zborowski VA, Stein AL, Zeni G, Nogueira CW (2014) Depression-related behavior and mechanical allodynia are blocked by 3-(4-fluorophenylselenyl)-2,5-diphenylselenophene in a mouse model of neuropathic pain induced by partial sciatic nerve ligation. Neuropharmacology 79:580–589. https://doi.org/10.1016/j.neuropharm.2014.01.020
38. Narita M, Kaneko C, Miyoshi K, Nagumo Y, Kuzumaki N, Nakajima M, Nanjo K, Matsuzawa K, Yamazaki M, Suzuki T (2006) Chronic pain induces anxiety with concomitant changes in opioidergic function in the amygdala. Neuropsychopharmacology 31:739–750. https://doi.org/10.1038/sj.npp.1300858
39. da Silva MD, Guginski G, Sato KL, Sanada LS, Sluka KA, Santos ARS (2020) Persistent pain induces mood problems and memory loss by the involvement of cytokines, growth factors, and supraspinal glial cells. BBI - Health 7:100118. https://doi.org/10.1016/j.bbih.2020.100118
40. Birmann PT, Sousa FSS, Domingues M, Brüning CA, Vieira BM, Lenardão EJ, Savegnago L (2019) 3-(4-Chlorophenylselanyl)-1-methyl-1H-indole promotes recovery of neuropathic pain and depressive-like behavior induced by partial constriction of the sciatic nerve in mice. J Trace Elem Med Biol 54:126–133. https://doi.org/10.1016/j.jtemb.2019.04.014
41. Chen Q, Ma K, Wang B, Chen Y (2022) The possibility of alleviating chronic neuropathic pain and related behaviors by the direct suppression of the parabrachial nucleus. J Clin Neurosci 95:180–187. https://doi.org/10.1016/j.jocn.2021.11.019
42. Martínez-Navarro M, Cabañero D, Wawrzczak-Bargiela A, Robe A, Gavériaux-Ruff C, Kieffer BL, Przewlocki R, Baños JE, Maldonado R (2020) Mu and delta opioid receptors play opposite nociceptive and behavioural roles on nerve-injured mice. Br J Pharmacol 177:1187–1205. https://doi.org/10.1111/bph.14911
43. Ramírez-López A, Pastor A, de la Torre R, La Porta C, Ozaita A, Cabañero D, Maldonado R (2021) Role of the endocannabinoid system in a mouse model of Fragile X undergoing neuropathic pain. Eur J Pain 25:1316–1328. https://doi.org/10.1002/ejp.1753
44. Heinzel JC, Dadun LF, Prahm C, Winter N, Bressler M, Lauer H, Ritter J, Daigeler A, Kolbenschlag J (2021) Beyond the knife—reviewing the interplay of psychosocial factors and peripheral nerve lesions. J Pers

Med 11:1200. https://doi.org/10.3390/jpm11111200

45. Guida F, Gregorio D, Palazzo E, Ricciardi F, Boccella S, Belardo C (2020) Behavioral, biochemical and electrophysiological changes in spared nerve injury model of neuropathic pain. Int J Mol Sci 21:3396. https://doi.org/10.3390/ijms21093396
46. Chen C, Liu A, Lu Q, Luo L, Li J, Ke J, Liu Y, Feng X (2022) HDAC6 inhibitor ACY-1215 improves neuropathic pain and its comorbidities in rats of peripheral nerve injury by regulating neuroinflammation. Chem Biol Interact 353:109803. https://doi.org/10.1016/j.cbi.2022.109803
47. Jiang H, Liu J-P, Xi K, Liu L-Y, Kong L-Y, Cai J, Cai S-Q, Han X-Y, Song J-G, Yang X-M, Wan Y, Xing G-G (2021) Contribution of AMPA receptor-mediated LTD in LA/BLA-CeA Pathway to comorbid aversive and depressive symptoms in neuropathic pain. J Neurosci 41:7278–7299. https://doi.org/10.1523/JNEUROSCI.2678-20.2021
48. Jing P-B, Chen X-H, Lu H-J, Gao Y-J, Wu X-B (2022) Enhanced function of NR2C/2D-containing NMDA receptor in the nucleus accumbens contributes to peripheral nerve injury-induced neuropathic pain and depression in mice. Mol Pain 18: 17448069211053255. https://doi.org/10.1177/17448069211053255
49. Mazzitelli M, Yakhnitsa V, Neugebauer B, Neugebauer V (2022) Optogenetic manipulations of CeA-CRF neurons modulate pain- and anxiety-like behaviors in neuropathic pain and control rats. Neuropharmacology 210:109031. https://doi.org/10.1016/j.neuropharm.2022.109031
50. Shen C-L, Wang R, Ji G, Elmassry MM, Zabet-Moghaddam M, Vellers H, Hamood AN, Gong X, Mirzaei P, Sang S, Neugebauer V (2022) Dietary supplementation of gingerols- and shogaols-enriched ginger root extract attenuate pain-associated behaviors while modulating gut microbiota and metabolites in rats with spinal nerve ligation. J Nutr Biochem 100:108904. https://doi.org/10.1016/j.jnutbio.2021.108904
51. Mosconi T, Kruger L (1996) Fixed-diameter polyethylene cuffs applied to the rat sciatic nerve induce a painful neuropathy: ultrastructural morphometric analysis of axonal alterations. Pain 64:37–57. https://doi.org/10.1016/0304-3959(95)00077-1
52. Benbouzid M, Pallage V, Rajalu M, Waltisperger E, Doridot S, Poisbeau P (2008) Sciatic nerve cuffing in mice: a model of sustained neuropathic pain. Eur J Pain 12:591–599. https://doi.org/10.1016/j.ejpain.2007.10.002
53. Kamoun N, Gazzo G, Goumon Y, Andry V, Yalcin I, Poisbeau P (2021) Long-lasting analgesic and neuroprotective action of the non-benzodiazepine anxiolytic etifoxine in a mouse model of neuropathic pain. Neuropharmacology 182:108407. https://doi.org/10.1016/j.neuropharm.2020.108407
54. Yalcin I, Bohren Y, Waltisperger E, Sage-Ciocca D, Yin J, Freund-Mercier M, Barrot M (2011) A time-dependent history of mood disorders in a murine model of neuropathic pain. Biol Psychiatry 70:946–953. https://doi.org/10.1016/j.biopsych.2011.07.017
55. Zakrzewska JM, Wu J, Mon-Williams M, Phillips N, Pavitt SH (2017) Evaluating the impact of trigeminal neuralgia. Pain 158:1166–1174. https://doi.org/10.1097/j.pain.0000000000000853
56. Araya EI, Baggio DF, Koren L, de O, Schwarting RKW, Chichorro JG (2022) Trigeminal neuropathic pain reduces 50-kHz ultrasonic vocalizations in rats, which are restored by analgesic drugs. Eur J Pharmacol 922:174905. https://doi.org/10.1016/j.ejphar.2022.174905
57. Chen W-J, Niu J-Q, Chen Y-T, Deng W-J, Xu Y-Y, Liu J, Luo W-F, Liu T (2021) Unilateral facial injection of botulinum neurotoxin a attenuates bilateral trigeminal neuropathic pain and anxiety-like behaviors through inhibition of TLR2-mediated neuroinflammation in mice. J Headache Pain 22: 38. https://doi.org/10.1186/s10194-021-01254-2
58. Cui W-Q, Zhang W-W, Chen T, Li Q, Xu F, Mao-Ying Q-L, Mi W-L, Wang Y-Q, Chu Y-X (2020) Tacr3 in the lateral habenula differentially regulates orofacial allodynia and anxiety-like behaviors in a mouse model of trigeminal neuralgia. Acta Neuropathol Commun 8:44. https://doi.org/10.1186/s40478-020-00922-9
59. McIlwrath SL, Montera MA, Gott KM, Yang Y, Wilson CM, Selwyn R, Westlund KN (2020) Manganese-enhanced MRI reveals changes within brain anxiety and aversion circuitry in rats with chronic neuropathic pain- and anxiety-like behaviors. Neuroimage 223:117343. https://doi.org/10.1016/j.neuroimage.2020.117343
60. Sheng H-Y, Lv S-S, Cai Y-Q, Shi W, Lin W, Liu T-T, Lv N, Cao H, Zhang L, Zhang Y-Q (2020) Activation of ventrolateral orbital cortex improves mouse neuropathic pain-induced anxiodepression. JCI Insight 5(e133625):133625. https://doi.org/10.1172/jci.insight.133625

61. Turnes J, de M, Araya EI, Barroso AR, Baggio DF, Koren L, de O, Zanoveli JM, Chichorro JG (2022) Blockade of kappa opioid receptors reduces mechanical hyperalgesia and anxiety-like behavior in a rat model of trigeminal neuropathic pain. Behav Brain Res 417:113595. https://doi.org/10.1016/j.bbr.2021.113595
62. Zhang L, Wang J, Niu C, Zhang Y, Zhu T, Huang D, Ma J, Sun H, Gamper N, Du X, Zhang H (2021) Activation of parabrachial nucleus - ventral tegmental area pathway underlies the comorbid depression in chronic neuropathic pain in mice. Cell Rep 37:109936. https://doi.org/10.1016/j.celrep.2021.109936
63. Watson JC, Sandroni P (2016) Central neuropathic pain syndromes. Mayo Clin Proc 91:372–385. https://doi.org/10.1016/j.mayocp.2016.01.017
64. Failde I, Dueñas M, Ribera MV, Gálvez R, Mico JA, Salazar A, de Sola H, Pérez C (2018) Prevalence of central and peripheral neuropathic pain in patients attending pain clinics in Spain: factors related to intensity of pain and quality of life. J Pain Res 11:1835–1847. https://doi.org/10.2147/JPR.S159729
65. Şahin-Onat Ş, Ünsal-Delialioğlu S, Kulaklı F, Özel S (2016) The effects of central post-stroke pain on quality of life and depression in patients with stroke. J Phys Ther Sci 28:96–101. https://doi.org/10.1589/jpts.28.96
66. Infantino R, Schiano C, Luongo L, Paino S, Mansueto G, Boccella S, Guida F, Ricciardi F, Iannotta M, Belardo C, Marabese I, Pieretti G, Serra N, Napoli C, Maione S (2022) MED1/BDNF/TrkB pathway is involved in thalamic hemorrhage-induced pain and depression by regulating microglia. Neurobiol Dis 164:105611. https://doi.org/10.1016/j.nbd.2022.105611
67. Lim S-W, Shiue Y-L, Ho C-H, Yu S-C, Kao P-H, Wang J-J, Kuo J-R (2017) Anxiety and depression in patients with traumatic spinal cord injury: a nationwide population-based cohort study. PLoS One 12:e0169623. https://doi.org/10.1371/journal.pone.0169623
68. Khazaeipour Z, Taheri-Otaghsara S-M, Naghdi M (2015) Depression following spinal cord injury: its relationship to demographic and socioeconomic indicators. Top Spinal Cord Inj Rehabil 21:149–155. https://doi.org/10.1310/sci2102-149
69. Kanno H, Ozawa H, Sekiguchi A, Itoi E (2009) Spinal cord injury induces upregulation of Beclin 1 and promotes autophagic cell death. Neurobiol Dis 33:143–148. https://doi.org/10.1016/j.nbd.2008.09.009
70. Marques S, Garcez V, Del Bel E, Martinez A (2009) A simple, inexpensive and easily reproducible model of spinal cord injury in mice: Morphological and functional assessment. J Neurosci Methods 177:183–193. https://doi.org/10.1016/j.jneumeth.2008.10.015
71. Starkey M, Davies M, Yip P, Carter L, Wong D, McMahon S, Bradbury E (2009) Expression of the regeneration-associated protein SPRR1A in primary sensory neurons and spinal cord of the adult mouse following peripheral and central injury. J Comp Neurol 513:51–68. https://doi.org/10.1002/cne.21944
72. Ung R, Landry E, Rouleau P, Lapointe N, Rouillard C, Guertin P (2008) Role of spinal 5-HT_2receptor subtypes in quipazine-induced hindlimb movements after a low-thoracic spinal cord transection. Eur J Neurosci 28:2231–2242. https://doi.org/10.1111/j.1460-9568.2008.06508.x
73. Yu W, Liu T, Fehlings T, Fehlings M (2009) Involvement of mitochondrial signaling pathways in the mechanism of Fas-mediated apoptosis after spinal cord injury. Eur J Neurosci 29:114–131. https://doi.org/10.1111/j.1460-9568.2008.06555.x
74. Nakae A, Nakai K, Yano K, Hosokawa K, Shibata M, Mashimo T (2011) The animal model of spinal cord injury as an experimental pain model. J Biomed Biotechnol 2011:939023. https://doi.org/10.1155/2011/939023
75. Tanabe M, Ono K, Honda M, Ono H (2009) Gabapentin and pregabalin ameliorate mechanical hypersensitivity after spinal cord injury in mice. Eur J Pharmacol 609:65–68. https://doi.org/10.1016/j.ejphar.2009.03.020
76. Brakel K, Aceves AR, Aceves M, Hierholzer A, Nguyen Q-N, Hook MA (2019) Depression-like behavior corresponds with cardiac changes in a rodent model of spinal cord injury. Exp Neurol 320:112969. https://doi.org/10.1016/j.expneurol.2019.112969
77. do Espírito Santo CC, da Silva Fiorin F, Ilha J, Duarte MMMF, Duarte T, Santos ARS (2019) Spinal cord injury by clip-compression induces anxiety and depression-like behaviours in female rats: the role of the inflammatory response. Brain Behav Immun 78:91–104. https://doi.org/10.1016/j.bbi.2019.01.012
78. Fukutoku T, Kumagai G, Fujita T, Sasaki A, Wada K, Liu X, Tanaka T, Kudo H, Asari T, Nikaido Y, Ueno S, Ishibashi Y (2020) Sex-Related Differences in Anxiety and Functional Recovery after Spinal Cord Injury in Mice. J Neurotrauma 37:2235–2243. https://doi.org/10.1089/neu.2019.6929

79. Li Y, Ritzel RM, Khan N, Cao T, He J, Lei Z, Matyas JJ, Sabirzhanov B, Liu S, Li H, Stoica BA, Loane DJ, Faden AI, Wu J (2020) Delayed microglial depletion after spinal cord injury reduces chronic inflammation and neurodegeneration in the brain and improves neurological recovery in male mice. Theranostics 10:11376–11403. https://doi.org/10.7150/thno.49199
80. Luedtke K, Bouchard SM, Woller SA, Funk MK, Aceves M, Hook MA (2014) Assessment of depression in a rodent model of spinal cord injury. J Neurotrauma 31:1107–1121. https://doi.org/10.1089/neu.2013.3204
81. Clauw DJ (2014) Fibromyalgia: a clinical review. JAMA 311:1547–1555. https://doi.org/10.1001/jama.2014.3266
82. Goldenberg DL (1987) Fibromyalgia syndrome. An emerging but controversial condition. JAMA 257:2782–2787. https://doi.org/10.1001/jama.257.20.2782
83. Munipalli B, Allman ME, Chauhan M, Niazi SK, Rivera F, Abril A, Wang B, Wieczorek MA, Hodge DO, Knight D, Perlman A, Abu Dabrh AM, Dudenkov D, Bruce BK (2022) Depression: a modifiable risk factor for poor outcomes in fibromyalgia. J Prim Care Community Health 13:21501319221120738. https://doi.org/10.1177/21501319221120738
84. Begum N, Taylor JR, Brown C, Rajan J, Keevil B, Pye E, Rainey T, Jones A (2022) Morning and evening salivary cortisol levels in patients with chronic widespread pain and those at high risk. Eur J Pain 26:197–206. https://doi.org/10.1002/ejp.1854
85. Khasar SG, Dina OA, Green PG, Levine JD (2009) Sound stress-induced long-term enhancement of mechanical hyperalgesia in rats is maintained by sympathoadrenal catecholamines. J Pain 10:1073–1077. https://doi.org/10.1016/j.jpain.2009.04.005
86. Onuora S (2020) Stress triggers fibromyalgia-like pain in mice. Nat Rev Rheumatol 16:602. https://doi.org/10.1038/s41584-020-00514-w
87. Sluka KA, Rasmussen LA (2010) Fatiguing exercise enhances hyperalgesia to muscle inflammation. Pain 148:188–197. https://doi.org/10.1016/j.pain.2009.07.001
88. Nagakura Y, Oe T, Aoki T, Matsuoka N (2009) Biogenic amine depletion causes chronic muscular pain and tactile allodynia accompanied by depression: a putative animal model of fibromyalgia. Pain 146:26–33. https://doi.org/10.1016/j.pain.2009.05.024
89. Araki K, Nishiyori M, Ueda H (2012) Intermittent cold stress-induced experimental fibromyalgia model in mice - pharmacology and neurobiology. Arthritis Res Ther 14:P6. https://doi.org/10.1186/ar3607
90. Nishiyori M, Ueda H (2008) Prolonged gabapentin analgesia in an experimental mouse model of fibromyalgia. Mol Pain 4: 52. https://doi.org/10.1186/1744-8069-4-52
91. Nasu T, Kubo A, Queme LF, Mizumura K (2019) A single administration of Neurotropin reduced the elongated immobility time in the forced swimming test of rats exposed to repeated cold stress. Behav Pharmacol 30:547–554. https://doi.org/10.1097/FBP.0000000000000488
92. Nazeri P, Shirani Rad AH, ValadAbadi SA, Mirakhori M, Hadidi Masoule E (2018) Effect of sowing dates and late season water deficit stress on quantitative and qualitative traits of canola cultivars. Outlook Agric 47:291–297. https://doi.org/10.1177/0030727018793658
93. Quintero L, Moreno M, Avila C, Arcaya J, Maixner W, Suarez-Roca H (2000) Long-lasting delayed hyperalgesia after subchronic swim stress. Pharmacol Biochem Behav 67:449–458. https://doi.org/10.1016/s0091-3057(00)00374-9
94. Xu Y, Zhang L, Shao T, Ruan L, Wang L, Sun J, Li J, Zhu X, O'Donnell JM, Pan J (2013) Ferulic acid increases pain threshold and ameliorates depression-like behaviors in reserpine-treated mice: behavioral and neurobiological analyses. Metab Brain Dis 28:571–583. https://doi.org/10.1007/s11011-013-9404-4
95. Wolfe F, Walitt B, Perrot S, Rasker JJ, Häuser W (2018) Fibromyalgia diagnosis and biased assessment: sex, prevalence and bias. PLoS One 13:e0203755. https://doi.org/10.1371/journal.pone.0203755
96. Bennett MI, Rayment C, Hjermstad M, Aass N, Caraceni A, Kaasa S (2012) Prevalence and aetiology of neuropathic pain in cancer patients: a systematic review. Pain 153:359–365. https://doi.org/10.1016/j.pain.2011.10.028
97. van den Beuken-van Everdingen MHJ, de Rijke JM, Kessels AG, Schouten HC, van Kleef M, Patijn J (2007) Prevalence of pain in patients with cancer: a systematic review of the past 40 years. Ann Oncol 18:1437–1449. https://doi.org/10.1093/annonc/mdm056
98. Edwards HL, Mulvey MR, Bennett MI (2019) Cancer-related neuropathic pain. Cancers (Basel) 11:373. https://doi.org/10.3390/cancers11030373
99. Rayment C, Hjermstad MJ, Aass N, Kaasa S, Caraceni A, Strasser F, Heitzer E, Fainsinger R, Bennett MI, European Palliative Care Research Collaborative (EPCRC) (2013)

Neuropathic cancer pain: prevalence, severity, analgesics and impact from the European palliative care research collaborative-computerised symptom assessment study. Palliat Med 27:714–721. https://doi.org/10.1177/0269216312464408

100. Arora A, Saini SK, Nautiyal V, Verma S, Gupta M, Kalra B, Ahmad M (2019) Cancer pain, anxiety, and depression in admitted patients at a tertiary care hospital: a prospective observational study. Indian J Palliat Care 25:562–566. https://doi.org/10.4103/IJPC.IJPC_4_19
101. Pineda-Farias JB, Saloman JL, Scheff NN (2020) Animal models of cancer-related pain: current perspectives in translation. Front Pharmacol 11:610894. https://doi.org/10.3389/fphar.2020.610894
102. Pyter LM, Pineros V, Galang JA, McClintock MK, Prendergast BJ (2009) Peripheral tumors induce depressive-like behaviors and cytokine production and alter hypothalamic-pituitary-adrenal axis regulation. Proc Natl Acad Sci U S A 106:9069–9074. https://doi.org/10.1073/pnas.0811949106
103. Warncke UO, Toma W, Meade JA, Park AJ, Thompson DC, Caillaud M, Bigbee JW, Bryant CD, Damaj MI (2021) Impact of dose, sex, and strain on oxaliplatin-induced peripheral neuropathy in mice. Front Pain Res (Lausanne) 2:683168. https://doi.org/10.3389/fpain.2021.683168
104. Porreca F, Navratilova E (2017) Reward, motivation, and emotion of pain and its relief. Pain 158:43–49. https://doi.org/10.1097/j.pain.0000000000000798
105. Baliki M, Apkarian A (2015) Nociception, pain, negative moods, and behavior selection. Neuron 87:474–491. https://doi.org/10.1016/j.neuron.2015.06.005
106. Bardo MT, Rowlett JK, Harris MJ (1995) Conditioned place preference using opiate and stimulant drugs: a meta-analysis. Neurosci Biobehav Rev 19:39–51. https://doi.org/10.1016/0149-7634(94)00021-r
107. Cunningham CL, Gremel CM, Groblewski PA (2006) Drug-induced conditioned place preference and aversion in mice. Nat Protoc 1:1662–1670. https://doi.org/10.1038/nprot.2006.279
108. Fuchs PN, McNabb CT (2012) The place escape/avoidance paradigm: a novel method to assess nociceptive processing. J Integr Neurosci 11:61–72. https://doi.org/10.1142/S0219635212500045
109. Refsgaard LK, Hoffmann-Petersen J, Sahlholt M, Pickering DS, Andreasen JT (2016) Modelling affective pain in mice: effects of inflammatory hypersensitivity on place escape/avoidance behaviour, anxiety and hedonic state. J Neurosci Methods 262:85–92. https://doi.org/10.1016/j.jneumeth.2016.01.019
110. Slattery D, Cryan J (2012) Using the rat forced swim test to assess antidepressant-like activity in rodents. Nat Protoc 7:1009–1014. https://doi.org/10.1038/nprot.2012.044
111. Sellmeijer J, Mathis V, Hugel S, Li X-H, Song Q, Chen Q-Y (2018) Hyperactivity of anterior cingulate cortex areas 24a/24b drives chronic pain-induced anxiodepressive-like consequences. J Neurosci 38:3102–3115. https://doi.org/10.1523/JNEUROSCI.3195-17.2018
112. Hasler G, Drevets WC, Manji HK, Charney DS (2004) Discovering endophenotypes for major depression. Neuropsychopharmacology 29:1765–1781. https://doi.org/10.1038/sj.npp.1300506
113. Arrant A, Schramm-Sapyta N, Kuhn C (2013) Use of the light/dark test for anxiety in adult and adolescent male rats. Behav Brain Res 256:119–127. https://doi.org/10.1016/j.bbr.2013.05.035
114. Crawley J (1985) Exploratory behavior models of anxiety in mice. Neuroscience &Amp. Biobehav Rev 9:37–44. https://doi.org/10.1016/0149-7634(85)90030-2
115. Walsh RN, Cummins RA (1976) The open-field test: a critical review. Psychol Bull 83:482–504
116. Gould TD (2009) Mood and anxiety related phenotypes in mice: characterization using behavioral tests. Humana Press, Totowa, NJ
117. Treit D, Fundytus M (1988) Thigmotaxis as a test for anxiolytic activity in rats. Pharmacol Biochem Behav 31:959–962. https://doi.org/10.1016/0091-3057(88)90413-3
118. Walz N, Mühlberger A, Pauli P (2016) A human open field test reveals thigmotaxis related to agoraphobic fear. Biol Psychiatry 80:390–397. https://doi.org/10.1016/j.biopsych.2015.12.016
119. Luo C, Zhang Y-L, Luo W, Zhou FH, Li C-Q, Xu J-M, Dai R-P (2015) Differential effects of general anesthetics on anxiety-like behavior in formalin-induced pain: involvement of ERK activation in the anterior cingulate cortex. Psychopharmacology (Berl) 232:4433–4444. https://doi.org/10.1007/s00213-015-4071-2
120. Parent AJ, Beaudet N, Beaudry H, Bergeron J, Bérubé P, Drolet G, Sarret P, Gendron L (2012) Increased anxiety-like behaviors in rats experiencing chronic inflammatory pain. Behav Brain Res 229:160–167. https://doi.org/10.1016/j.bbr.2012.01.001
121. Hisaoka-Nakashima K, Tomimura Y, Yoshii T, Ohata K, Takada N, Zhang F (2019) High-mobility group box 1-mediated microglial

activation induces anxiodepressive-like behaviors in mice with neuropathic pain. Prog Neuropsychopharmacol Biol Psychiatry 92:347–362. https://doi.org/10.1016/j.pnpbp.2019.02.005

122. Gritsch S, Bali KK, Kuner R, Vardeh D (2016) Functional characterization of a mouse model for central post-stroke pain. Mol Pain 12: 1744806916629049. https://doi.org/10.1177/1744806916629049

123. Crawley J, Goodwin FK (1980) Preliminary report of a simple animal behavior model for the anxiolytic effects of benzodiazepines. Pharmacol Biochem Behav 13:167–170. https://doi.org/10.1016/0091-3057(80)90067-2

124. Hascoët M, Bourin M, Nic Dhonnchadha BA (2001) The mouse light-dark paradigm: a review. Prog Neuropsychopharmacol Biol Psychiatry 25:141–166. https://doi.org/10.1016/s0278-5846(00)00151-2

125. Ducret E, Jacquot F, Descheemaeker A, Dallel R, Artola A (2022) Chronic facial inflammatory pain-induced anxiety is associated with bilateral deactivation of the rostral anterior cingulate cortex. Brain Res Bull 184:88–98. https://doi.org/10.1016/j.brainresbull.2022.03.012

126. Šutulović N, Grubač Ž, Šuvakov S, Jerotić D, Puškaš N, Macut D, Rašić-Marković A, Simić T, Stanojlović O, Hrnčić D (2021) Experimental chronic prostatitis/chronic pelvic pain syndrome increases anxiety-like behavior: the role of brain oxidative stress, serum corticosterone, and hippocampal parvalbumin-positive interneurons. Oxid Med Cell Longev 2021:6687493. https://doi.org/10.1155/2021/6687493

127. Du Y, Xu C-L, Yu J, Liu K, Lin S-D, Hu T-T, Qu F-H, Guo F, Lou G-D, Nishibori M, Hu W-W, Chen Z, Zhang S-H (2022) HMGB1 in the mPFC governs comorbid anxiety in neuropathic pain. J Headache Pain 23: 102. https://doi.org/10.1186/s10194-022-01475-z

128. Thomas A, Burant A, Bui N, Graham D, Yuva-Paylor LA, Paylor R (2009) Marble burying reflects a repetitive and perseverative behavior more than novelty-induced anxiety. Psychopharmacology (Berl) 204:361–373. https://doi.org/10.1007/s00213-009-1466-y

129. Kulkarni SK, Sharma AC (1991) Elevated plus-maze: a novel psychobehavioral tool to measure anxiety in rodents. Methods Find Exp Clin Pharmacol 13:573–577

130. Belovicova K, Bogi E, Csatlosova K, Dubovicky M (2017) Animal tests for anxiety-like and depression-like behavior in rats. Interdiscip Toxicol 10:40–43. https://doi.org/10.1515/intox-2017-0006

131. Carola V, D'Olimpio F, Brunamonti E, Mangia F, Renzi P (2002) Evaluation of the elevated plus-maze and open-field tests for the assessment of anxiety-related behaviour in inbred mice. Behav Brain Res 134:49–57. https://doi.org/10.1016/s0166-4328(01)00452-1

132. González-Sepúlveda M, Pozo O, Marcos J, Valverde O (2015) Chronic pain causes a persistent anxiety state leading to increased ethanol intake in CD1 mice. J Psychopharmacol 30:188–203. https://doi.org/10.1177/0269881115622238

133. Vahid-Ansari F, Lagace DC, Albert PR (2016) Persistent post-stroke depression in mice following unilateral medial prefrontal cortical stroke. Transl Psychiatry 6: e863. https://doi.org/10.1038/tp.2016.124

134. Brown GR, Nemes C (2008) The exploratory behaviour of rats in the hole-board apparatus: is head-dipping a valid measure of neophilia. Behav Processes 78:442–448. https://doi.org/10.1016/j.beproc.2008.02.019

135. Sieberg CB, Taras C, Gomaa A, Nickerson C, Wong C, Ward C, Baskozos G, Bennett DLH, Ramirez JD, Themistocleous AC, Rice ASC, Shillo PR, Tesfaye S, Edwards RR, Andrews NA, Berde C, Costigan M (2018) Neuropathic pain drives anxiety behavior in mice, results consistent with anxiety levels in diabetic neuropathy patients. Pain Rep 3:e651. https://doi.org/10.1097/PR9.0000000000000651

136. Montserrat-de la Paz S, García-Giménez MD, Ángel-Martín M, Fernández-Arche A (2015) Validation and additional support for an experimental animal model of fibromyalgia. Mod Rheumatol 25:116–122. https://doi.org/10.3109/14397595.2014.924194

137. Shepherd AJ, Cloud ME, Cao Y-Q, Mohapatra DP (2018) Deficits in burrowing behaviors are associated with mouse models of neuropathic but not inflammatory pain or migraine. Front Behav Neurosci 12:124. https://doi.org/10.3389/fnbeh.2018.00124

138. Andrews N, Legg E, Lisak D, Issop Y, Richardson D, Harper S, Pheby T, Huang W, Burgess G, Machin I, Rice ASC (2012) Spontaneous burrowing behaviour in the rat is reduced by peripheral nerve injury or inflammation associated pain. Eur J Pain 16:485–495. https://doi.org/10.1016/j.ejpain.2011.07.012

139. Lau W, Dykstra C, Thevarkunnel S, Silenieks LB, de Lannoy IaM, Lee DKH, Higgins GA (2013) A back translation of pregabalin and carbamazepine against evoked and non-evoked endpoints in the rat spared nerve injury model of neuropathic pain. Neuropharmacology

73:204–215. https://doi.org/10.1016/j.neuropharm.2013.05.023
140. Tappe-Theodor A, Kuner R (2014) Studying ongoing and spontaneous pain in rodents–challenges and opportunities. Eur J Neurosci 39:1881–1890. https://doi.org/10.1111/ejn.12643
141. Broekkamp CL, Rijk HW, Joly-Gelouin D, Lloyd KL (1986) Major tranquillizers can be distinguished from minor tranquillizers on the basis of effects on marble burying and swim-induced grooming in mice. Eur J Pharmacol 126:223–229. https://doi.org/10.1016/0014-2999(86)90051-8
142. Njung'e K, Handley SL (1991) Evaluation of marble-burying behavior as a model of anxiety. Pharmacol Biochem Behav 38:63–67. https://doi.org/10.1016/0091-3057(91)90590-x
143. D'Aniello A, Luongo L, Romano R, Iannotta M, Marabese I, Boccella S (2017) d -Aspartic acid ameliorates painful and neuropsychiatric changes and reduces β-amyloid Aβ 1-42 peptide in a long lasting model of neuropathic pain. Neurosci Lett 651:151–158. https://doi.org/10.1016/j.neulet.2017.04.041
144. Smolinsky AN, Bergner CL, LaPorte JL, Kalueff AV (2009) Analysis of grooming behavior and its utility in studying animal stress, anxiety, and depression. In: Gould TD (Ed.)Eds. Mood and anxiety related phenotypes in mice: characterization using behavioral tests. Humana Press, Totowa, NJ, 21–36
145. Isingrini E, Camus V, Le Guisquet A-M, Pingaud M, Devers S, Belzung C (2010) Association between repeated unpredictable chronic mild stress (UCMS) procedures with a high fat diet: a model of fluoxetine resistance in mice. PLoS One 5:e10404. https://doi.org/10.1371/journal.pone.0010404
146. Deacon R (2006) Assessing nest building in mice. Nat Protoc 1:1117–1119. https://doi.org/10.1038/nprot.2006.170
147. Veenit V, Zhang X, Ambrosini A, Sousa V, Svenningsson P (2021) The effect of early life stress on emotional behaviors in GPR37KO mice. Int J Mol Sci 23:410. https://doi.org/10.3390/ijms23010410
148. Garrone B, di Matteo A, Amato A, Pistillo L, Durando L, Milanese C, Di Giorgio FP, Tongiani S (2021) Synergistic interaction between trazodone and gabapentin in rodent models of neuropathic pain. PLoS One 16:e0244649. https://doi.org/10.1371/journal.pone.0244649
149. Liu M, Yin C, Zhu L, Zhu X, Xu C, Luo C (2018) Sucrose preference test for measurement of stress-induced anhedonia in mice. Nat Protoc 13:1686–1698. https://doi.org/10.1038/s41596-018-0011-z
150. Serchov T, Schwarz I, Theiss A, Sun L, Holz A, Döbrössy MD, Schwarz MK, Normann C, Biber K, van Calker D (2020) Enhanced adenosine A1 receptor and Homer1a expression in hippocampus modulates the resilience to stress-induced depression-like behavior. Neuropharmacology 162:107834. https://doi.org/10.1016/j.neuropharm.2019.107834
151. Hu X, Liu Y, Wu J, Liu Y, Liu W, Chen J, Yang F (2020) Inhibition of P2X7R in the amygdala ameliorates symptoms of neuropathic pain after spared nerve injury in rats. Brain Behav Immun 88:507–514. https://doi.org/10.1016/j.bbi.2020.04.030
152. Sotomayor-Sobrino M, Ochoa-Aguilar A, Coffeen U, Velasco-Gonzàlez R, Pérez-Sánchez G, Capula-León G, Luna-Mendoza I, Pavon L, Gómez-Acevedo C (2020) Carrageenan-induced non-painful inflammation reduces voluntary sucrose consumption in a two-bottle choice paradigm: a new model of inflammation induced anhedonia. J Behav Brain Sci 10:153–164. https://doi.org/10.4236/jbbs.2020.103009
153. Su C, D'amour J, Lee M, Lin H-Y, Manders T, Xu D, Eberle SE, Goffer Y, Zou AH, Rahman M, Ziff E, Froemke RC, Huang D, Wang J (2015) Persistent pain alters AMPA receptor subunit levels in the nucleus accumbens. Mol Brain 8: 46. https://doi.org/10.1186/s13041-015-0140-z
154. Malkesman O, Scattoni ML, Paredes D, Tragon T, Pearson B, Shaltiel G, Chen G, Crawley JN, Manji HK (2010) The female urine sniffing test: a novel approach for assessing reward-seeking behavior in rodents. Biol Psychiatry 67:864–871. https://doi.org/10.1016/j.biopsych.2009.10.018
155. Yankelevitch-Yahav R, Franko M, Huly A, Doron R (2015) The forced swim test as a model of depressive-like behavior. J Vis Exp 2(97):52587. https://doi.org/10.3791/52587
156. Medeiros P, Dos Santos IR, Júnior IM, Palazzo E, da Silva JA, Machado HR, Ferreira SH, Maione S, Coimbra NC, de Freitas RL (2021) An adapted chronic constriction injury of the sciatic nerve produces sensory, affective, and cognitive impairments: a peripheral mononeuropathy model for the study of comorbid neuropsychiatric disorders associated with neuropathic pain in rats. Pain Med 22:338–351. https://doi.org/10.1093/pm/pnaa206
157. Commons KG, Cholanians AB, Babb JA, Ehlinger DG (2017) The rodent forced swim test measures stress-coping strategy, not depression-like behavior. ACS Chem Neurosci

8:955–960. https://doi.org/10.1021/acschemneuro.7b00042
158. de Kloet ER, Molendijk ML (2016) Coping with the forced swim stressor: towards understanding an adaptive mechanism. Neural Plast 2016:6503162. https://doi.org/10.1155/2016/6503162
159. Cryan J, Mombereau C, Vassout A (2005) The tail suspension test as a model for assessing antidepressant activity: review of pharmacological and genetic studies in mice. Neuroscience &Amp. Biobehav Rev 29:571–625. https://doi.org/10.1016/j.neubiorev.2005.03.009
160. Suzuki H, Otsuka T, Hitora-Imamura N, Ishimura K, Fukuda H, Fujiwara K, Shuto S, Deyama S, Minami M (2021) Resolvin E1 attenuates chronic pain-induced depression-like behavior in mice: possible involvement of chemerin receptor chemR23. Biol Pharm Bull 44:1548–1550. https://doi.org/10.1248/bpb.b21-00461
161. Stedenfeld KA, Clinton SM, Kerman IA, Akil H, Watson SJ, Sved AF (2011) Novelty-seeking behavior predicts vulnerability in a rodent model of depression. Physiol Behav 103:210–216. https://doi.org/10.1016/j.physbeh.2011.02.001
162. Yuan Y, Zhen L, Li Z, Xu W, Leng H, Xu W, Zheng V, Luria V, Pan J, Tao Y, Zhang H, Cao S, Xu Y (2020) trans-Resveratrol ameliorates anxiety-like behaviors and neuropathic pain in mouse model of post-traumatic stress disorder. J Psychopharmacol 34:726–736. https://doi.org/10.1177/0269881120914221
163. De Gregorio D, McLaughlin RJ, Posa L, Ochoa-Sanchez R, Enns J, Lopez-Canul M, Aboud M, Maione S, Comai S, Gobbi G (2019) Cannabidiol modulates serotonergic transmission and reverses both allodynia and anxiety-like behavior in a model of neuropathic pain. Pain 160:136–150. https://doi.org/10.1097/j.pain.0000000000001386
164. Yasuda S, Yoshida M, Yamagata H, Iwanaga Y, Suenaga H, Ishikawa K, Nakano M, Okuyama S, Furukawa Y, Furukawa S, Ishikawa T (2014) Imipramine ameliorates pain-related negative emotion via induction of brain-derived neurotrophic factor. Cell Mol Neurobiol 34:1199–1208. https://doi.org/10.1007/s10571-014-0097-y
165. Alba-Delgado C, Cebada-Aleu A, Mico JA, Berrocoso E (2016) Comorbid anxiety-like behavior and locus coeruleus impairment in diabetic peripheral neuropathy: A comparative study with the chronic constriction injury model. Prog Neuropsychopharmacol Biol Psychiatry 71:45–56. https://doi.org/10.1016/j.pnpbp.2016.06.007
166. Levy DR, Hunter N, Lin S, Robinson EM, Gillis W, Conlin EB, Anyoha R, Shansky RM, Datta SR (2023) Mouse spontaneous behavior reflects individual variation rather than estrous state. Curr Biol 33(1358-1364):e4. https://doi.org/10.1016/j.cub.2023.02.035
167. Fiore NT, Debs SR, Hayes JP, Duffy SS, Moalem-Taylor G (2023) Pain-resolving immune mechanisms in neuropathic pain. Nat Rev Neurol 19:199–220. https://doi.org/10.1038/s41582-023-00777-3
168. Lopes DM, Malek N, Edye M, Jager SB, McMurray S, McMahon SB, Denk F (2017) Sex differences in peripheral not central immune responses to pain-inducing injury. Sci Rep 7: 16460. https://doi.org/10.1038/s41598-017-16664-z
169. Sorge RE, Mapplebeck JCS, Rosen S, Beggs S, Taves S, Alexander JK, Martin LJ, Austin J-S, Sotocinal SG, Chen D, Yang M, Shi XQ, Huang H, Pillon NJ, Bilan PJ, Tu Y, Klip A, Ji R-R, Zhang J, Salter MW, Mogil JS (2015) Different immune cells mediate mechanical pain hypersensitivity in male and female mice. Nat Neurosci 18:1081–1083. https://doi.org/10.1038/nn.4053
170. Van der Linden A, Van Camp N, Ramos-Cabrer P, Hoehn M (2007) Current status of functional MRI on small animals: application to physiology, pathophysiology, and cognition. NMR Biomed 20:522–545. https://doi.org/10.1002/nbm.1131
171. Thompson SJ, Bushnell MC (2012) Rodent functional and anatomical imaging of pain. Neurosci Lett 520:131–139. https://doi.org/10.1016/j.neulet.2012.03.015
172. Shah YB, Haynes L, Prior MJW, Marsden CA, Morris PG, Chapman V (2005) Functional magnetic resonance imaging studies of opioid receptor-mediated modulation of noxious-evoked BOLD contrast in rats. Psychopharmacology (Berl) 180:761–773. https://doi.org/10.1007/s00213-005-2214-6
173. Chen S-H, Haam J, Walker M, Scappini E, Naughton J, Martin NP (2019) Recombinant viral vectors as neuroscience tools. Curr Protoc Neurosci 87:e67. https://doi.org/10.1002/cpns.67
174. Chen T-W, Wardill TJ, Sun Y, Pulver SR, Renninger SL, Baohan A, Schreiter ER, Kerr RA, Orger MB, Jayaraman V, Looger LL, Svoboda K, Kim DS (2013) Ultrasensitive fluorescent proteins for imaging neuronal activity. Nature 499:295–300. https://doi.org/10.1038/nature12354
175. Siemian JN, Arenivar MA, Sarsfield S, Borja CB, Erbaugh LJ, Eagle AL, Robison AJ, Leinninger G, Aponte Y (2021) An excitatory

lateral hypothalamic circuit orchestrating pain behaviors in mice. eLife 10:e66446. https://doi.org/10.7554/eLife.66446

176. Zhu X, Tang H-D, Dong W-Y, Kang F, Liu A, Mao Y, Xie W, Zhang X, Cao P, Zhou W, Wang H, Farzinpour Z, Tao W, Song X, Zhang Y, Xue T, Jin Y, Li J, Zhang Z (2021) Distinct thalamocortical circuits underlie allodynia induced by tissue injury and by depression-like states. Nat Neurosci 24:542–553. https://doi.org/10.1038/s41593-021-00811-x

177. Hall CS (1934) Emotional behavior in the rat. I. Defecation and urination as measures of individual differences in emotionality. J Comp Psychol 18(3):385–403. https://doi.org/10.1037/h0071444

178. Pellow S, Chopin P, File SE, Briley M (1985) Validation of open : closed arm entries in an elevated plus-maze as a measure of anxiety in the rat. J Neurosci Methods 14(3):149–167. https://doi.org/10.1016/0165-0270(85)90031-7

179. Soubrié P, Kulkarni S, Simon P, Boissier JR (1975) Effects of antianxiety drugs on the food dntake in trained and untrained rats and mice. Psychopharmacologia 45(2):203–210. https://doi.org/10.1007/BF00429062

180. Porsolt RD, Bertin A, Jalfre M (1978) "Behavioural despair" in rats and mice: Strain differences and the effects of imipramine. Eur J Pharmacol 51(3):291–294. https://doi.org/10.1016/0014-2999(78)90414-4

181. Burston JJ, Valdes AM, Woodhams SG, Mapp PI, Stocks J, Watson DJG, Gowler PRW, Xu L, Sagar DR, Fernandes G, Frowd N, Marshall L, Zhang W, Doherty M, Walsh DA, Chapman V (2019) The impact of anxiety on chronic musculoskeletal pain and the role of astrocyte activation. Pain 160(3):658–669. https://doi.org/10.1097/j.pain.0000000000001445

182. Kędziora M, Boccella S, Marabese I, Mlost J, Infantino R, Maione S, Starowicz K (2023) Inhibition of anandamide breakdown reduces pain and restores LTP and monoamine levels in the rat hippocampus via the CB1 receptor following osteoarthritis. Neuropharmacology 222:109304. https://doi.org/10.1016/j.neuropharm.2022.109304

183. Saffarpour S, Janzadeh A, Rahimi B, Ramezani F, Nasirinezhad F (2021) Chronic nanocurcumin treatment ameliorates pain-related behavior improves spatial memory and reduces hippocampal levels of IL-1β and TNFα in the chronic constriction injury model of neuropathic pain. Psychopharmacology 238(3):877–886. https://doi.org/10.1007/s00213-020-05739-x

Chapter 15

Executive Functions in Chronic Pain Models

Ana Margarida Cunha, Madalena Esteves, Sara Bernardo and Hugo Leite-Almeida

Abstract

Assessment of pain and its impact in chronic pain models is challenging. Evoked responses to noxious and innocuous stimuli have been typically used. However, in the past two decades, it became evident that manifestations of anxiety- and depressive-like behaviors as well as cognitive dysfunction can broaden the characterization of pain impact in preclinical research. Executive functions comprise a set of cognitive control processes including working memory, inhibitory control, and cognitive flexibility. Despite the well-known structural and functional impact of chronic pain on frontal cortical and subcortical structures that subserve executive functions, these remain poorly studied in the context of preclinical chronic pain studies. In this chapter, we describe two behavioral paradigms—variable delay-to-signal (motor impulsivity and delay intolerance) and habit-based to goal-directed decision-making—and their respective protocols. Our goal is to provide a step-by-step guide to assist the implementation of the paradigms and interpretation of their results and limitations.

Keywords Chronic pain, Executive functions, Instrumental learning, Decision-making impulsivity, Behavioral flexibility

1 Introduction: Executive Function in Chronic Pain Conditions

Executive functions (EFs) encompass several mental processes underlying decision-making. In other words, EFs serve our ability to adapt and adjust to changing conditions and to plan future actions. It is generally accepted that inhibition (behavioral and cognitive/selective control), working memory, and cognitive flexibility/set-shifting constitute the three core EFs [1]. It should, however, be noted that different perspectives on what constitutes core and ancillary EFs might differ across authors—see, for instance [2–4].

EFs evolve throughout life, accompanying the maturation and/or senescence of the underlying neural networks. For instance, infants rapidly evolve from behaviors that are mainly guided by habits and external cues to more flexible and self-regulated behaviors [5]; there

Isaura Tavares and Hugo Leite-Almeida (eds.), *Translational Research Methods for Neuropathic Pain*, Neuromethods, vol. 235, https://doi.org/10.1007/978-1-0716-5312-8_15,

is evidence that childhood self-control predicts a number of health-related parameters, wealth, and criminal offenses in the adulthood [6]. Also, EFs can be influenced by the physiological condition of the individual and by factors such as sleep and exercise [7, 8]. Manifestations of impaired EFs can, however, be an indicator of an underlying pathology of neuropsychiatric origin or other causes. These include, among others, attention-deficit/hyperactivity disorder, schizophrenia, autism spectrum disorder, and several neurodegenerative disorders [9–11].

The study of patients with executive dysfunctions has provided important cues on the brain areas/networks serving these processes. The observation of increased impulsivity, irascibility, or planning failures in patients with frontal lesions placed the prefrontal cortex and related networks as a central hub for EF—see for instance [12–14]; the case of Phineas Gage, for instance, is often presented as a prototypical example of such anatomo-functional relation (e.g., [15, 16]). However, it is important to note that not only the prefrontal cortex is functionally heterogenous—see for review [4, 13, 17]—but also that some of its subregions have a differential contribution to many EFs, e.g., working memory/goal-driven attention (dorsolateral prefrontal), response selection/monitoring (ventroateral prefrontal), emotional regulation/updating goal-directed behavior (medial prefrontal), and personality/inhibition (orbitofrontal) [13], and that subcortical areas such as the striatum, nucleus accumbens, or amygdala, as well as the cerebellum, are also known to be involved in many EFs [18].

Similarly, rodent studies have also contributed to our understanding of the neurobiological substrates of EF included in disease models. Particularly regarding chronic pain, studies from our group have been uncovering the impact of chronic (neuropathic) pain on EF, namely working memory [19, 20], cognitive flexibility [20], attentional set-shifting [19, 21], impulsivity [19, 22], and habit-based/goal-directed (unpublished) behaviors—consult for a review of ours and other studies [23, 24].

1.1 Methods to Capture Executive Function Deficits in Experimental Models of Chronic Pain

Neuropsychological tests and questionnaires are among the most used instruments to assess EF in humans; these instruments are necessarily diverse as they cover different functions [12, 25–27]. For the same reason, a number of different behavioral tasks have been developed for rodents. Some of these attempt to emulate human paradigms, see for instance [28] (attentional set-shifting task/Wisconsin Card Sorting Test) and [29] (impulsivity), while others take advantage of species-specific behaviors, such asalternation/Y-maze [30] (working memory). The majority of studies assessing EF in chronic pain focus on working memory [23]. Probably, the simplicity of some of these paradigms might explain this bias. Therefore, in this chapter, we describe two paradigms, the variable delay-to-signal (VDS) and the habit-based/goal-directed transition, covering other EFs.

2 Materials and Experimental Subjects

2.1 Operant Chambers

VDS and habit-based/goal-directed decision shift paradigms are performed in operant chambers (OC): 5-hole (Section: 5-Hole Nose Poke) and lever (Section:Operant Chamber with Retractable Levers) chambers (Fig. 1), respectively. Both protocols require operant conditioning, which is achieved by associating an action, nosepoke (5-hole) and lever pressing, with an outcome, a sugared pellet (or water) being delivered upon the correct action (contingency).

5-Hole Nose Poke

The VDS protocol is performed in five-hole boxes for rats with an approximately square shape (25 cm; TSE, Germany; Fig. 1A). One of the walls is slightly curved and contains five holes (2.5 × 2.5 cm) elevated 2 cm from the grid floor. During the VDS training and test phases, only the central hole is open (holes #1, #2, #4, and #5 are blocked with metallic caps). In the opposite wall, there is a food magazine connected to a pellet dispenser located outside the box. All holes (#1–#5) and food magazine contain a 3 W lamp bulb and an infrared light sensor to detect nosepokes. Different versions are commercially available and can be used with the same objective; adjustments to the protocol might have to be equated for each instrument. Five-hole boxes are placed inside noise-attenuating boxes.

Operant Chamber with Retractable Levers

Goal-directed/habit-based decision-making behavior is performed in an operant chamber (OC) designed for rats with the following dimensions: 30.5 cm (L) × 24.1 cm (W) × 21.0 cm (H) (ENV-467; Med Associates, USA; Fig. 1B). The OC is equipped with two retractable levers and a food magazine in between and is placed inside a noise-attenuating box. Alternative equipment from different companies is

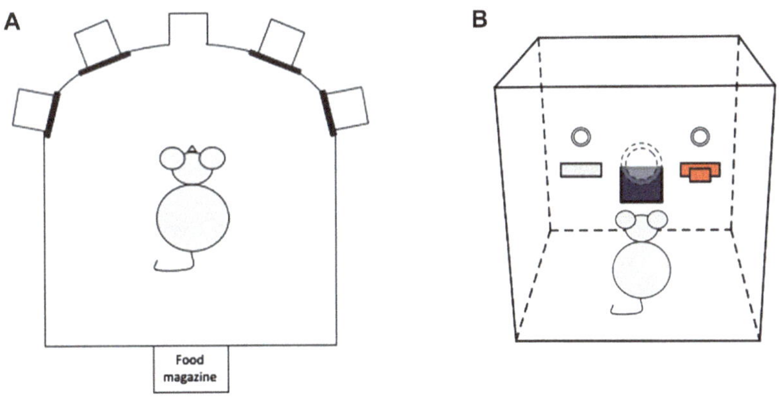

Fig. 1 Operant chambers: 5-hole (**A**) and chamber with retractable levers (**B**). The 5-hole chamber is prepared for the variable delay-to-signal by; just the central aperture (hole #3) is available, and the remaining holes (#1, #2, #4, and #5) are covered with metallic caps (A). The levers operant chamber can be used with either the left or right lever; in specific circumstances, both are simultaneously present. See main text for more information

commercially available and can be used with the same objective; the protocol might have to be adjusted for each instrument.

2.2 Rodents

The protocols presented here for impulsivity and habit-based behaviors were tested in rats [22, 31–33], though studies in mice have been published in the past [34, 35]. Regarding the VDS, behavior in this protocol varies within the population, and biological factors such as age and sex contribute substantially to such heterogeneity [36]; also experimental factors such as testing period might impact behavior. In all cases, pilot studies should be carried out. In our standard experimental conditions, animals are housed in pairs and maintained in a controlled environment—12 h light cycle (lights on at 8 AM), temperature of 21 ± 1°C. and humidity of 50%–60%. Food (4RF21; Mucedola SRL, Italy) and water are provided ad libitum, except during the operant tests (see below).

2.3 Consumables

- Food pellets (4RF21; Mucedola SRL, Italy).
- Sugar pellets (45 mg; Bio+Serv Inc., USA).

3 Methods

In this section, we provide a detailed description of both impulsivity and habit protocols. No mention of the pain model(s) will be made, as these are presented in several other chapters of this book. Some notes are, however, needed. In our studies, we have used the spared nerve injury (SNI) model [37]. Pilot studies are highly advised when other models are to be used to access the implication(s) in the performance of the tests, particularly regarding the manifestation of pain-like phenotypes and motor function. For instance, anhedonia has been described in these chronic pain models and might impact the animals' drive to seek the reward. Also, the use of analgesic drugs, common in this type of study, should (obviously) be well controlled for potential confounders.

3.1 Preparatory Steps

Prior to the beginning of the protocols, several preparatory steps must be carried out.

Randomization: In addition to experimental group randomization (e.g., pain vs. no-pain), it is important to guarantee that aspects related to the behavioral paradigms are also randomized (or at least distributed in a balanced way across groups). This includes, for instance, the testing order, lever side, and reward, among others. Also, if the two tests (impulsivity and habit) are to be used, crossover designs are strongly advised to disperse carryover effects—see, for instance [31].

Food restriction: For both impulsivity and habit protocols, food restriction is essential to increase incentive. This requires careful planning and adherence to ethical guidelines to guarantee both animal welfare and the validity of the collected data. Also, these conditions should be adapted to each experiment.

Habituation: Animals must be habituated to all aspects of the testing conditions. This includes handling by the experimenters involved in the experiments, flavors/odors of the rewards, testing rooms, and equipments, etc. Also, for each protocol, VDS and habit-based/goal-directed decision-making, specific steps must be carried. These are described below in the respective subsections (Sects. 3.1.1 and Section: Acclimatization).

3.2 Impulsivity (Variable Delay-to-Signal)

Acclimatization

Two acclimatization sessions—twice daily (AM/PM)—are performed. In the first two sessions, animals are placed in the operant chamber for 15 min with all lights off. The access to apertures #1–#5 is blocked by metallic caps, and 10–15 sugared pellets are placed in the food magazine. In sessions 3–4, animals spent 30 min in the operant chamber with all lights on, and the cap in the central aperture (hole #3) is removed; sugared pellets are provided there and in the food magazine (approximately 2–3 and 10–15 pellets, respectively).

Training Sessions

Training sessions are initiated by turning on the house light and delivering one sugared pellet in the food magazine. The collection of this pellet triggers the beginning of an intertrial interval (ITI) of 3 s to allow the animal to ingest it. After this, the regular trials start. Each trial consists of a 3-s period with the house light on (delay period) and hole #3 light off. After this period, the hole #3 lights on, which marks the beginning of a 60-s response period. Nosepokes in hole #3 prior to its lighting are punished with a 5-s timeout (TO) period in complete darkness; these are recoded as premature responses. Nosepokes during the response period are rewarded with the delivery of a sugared pellet in the food magazine. As before, the collection of the sugared pellet triggers a 3-s ITI, after which a new trial begins. If the animal ignores or fails to nosepoke hole #3 during the response period, this is recorded as an omission; in this case, a 5-s TO period in complete darkness precedes the beginning of the next trial. Except for the TO periods, the house light is always on. Each training session consists of 100 of these trials—premature, timed, and omissions. Training sessions can be performed once or twice daily; in the latter case, these should be spaced by a minimum interval of 5 h. Ten sessions are sufficient to achieve a stable performance, which, in our conditions, is ~30% average of premature responses in the last four training sessions.

VDS Test

The VDS testing session takes place in a single day and consists of 120 trials, similar to those previously described but with some alterations. First, multiple premature responses are allowed during the delay period; therefore, no TO is initiated. The duration of the delays is also different from the training sessions. Specifically, the first and last 25 trials last 3 s. The middle 70 trials last either 6 or 12 s, randomly attributed by the software. Altogether, the VDS test session is organized in a $3s_i - 6\,s/12\,s - 3s_f$ sequence. The number of premature responses per amount of available time (i.e., delay duration) is calculated according to the formula below and used to compare performances (Fig. 2, upper panel).

$$PR_{min} = \frac{PR_t}{T_t \times t} \times 60 \qquad (1)$$

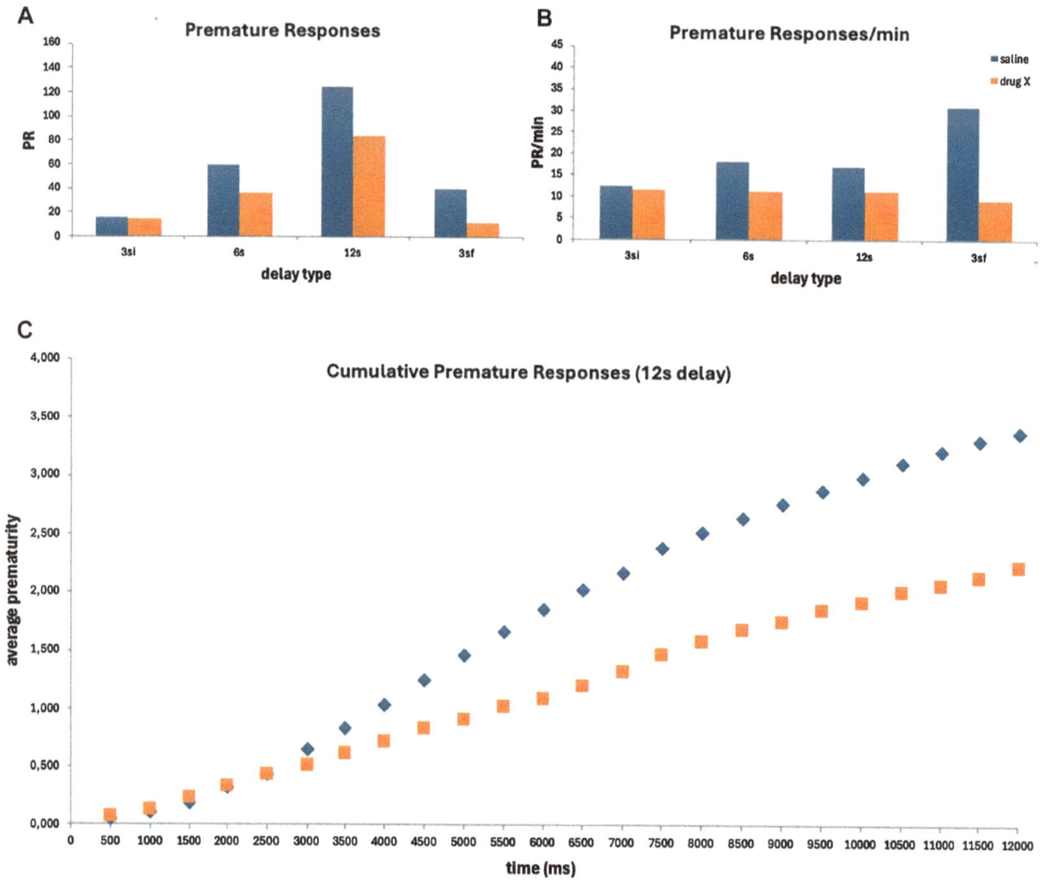

Fig. 2 A hypothetical VDS dataset (based on real data) in which the performance of a saline-injected group (orange) is compared with that of a group injected with an impulsivity-potentiating drug (blue). Total premature responses (PR) in each delay type (**A**) and PR normalized for the available delay time **(B)**. Average pattern of PR (cumulative) along a 12 s delay (**C**). Saline animals present nearly 3× more impulsive responses in $3s_f$ than in $3s_i$, i.e., after the exposure to longer delays. A correlation between this increase and delay intolerance in the classical delay discounting paradigms was demonstrated [38]. This effect changes with the biological factors such as age, sex, etc. [36]) and can be pharmacologically manipulated (e.g., X represents the effect previously observed with methamphetamine [38])

Premature response rate (PR_{min}) is measured as the number of premature responses executed per amount of available delay time. PR_f is the number of premature responses in a given delay type ($3s_i$, 6 s, 12 s, and $3s_f$); T_t is the number of trials for each delay type (for $3s_i$ and $3s_f$ this number is set on 25 trials; for 6 s and 12 s it is randomly attributed); t is theduration in seconds of each delay.

Delay intolerance manifests as an increased rate of premature responses in the 6 s, 12 s, and most notably $3s_f$. Alternatively, the accumulation of premature responses along the delays can provide important information on impulsive behavior, particularly when impulsivity basal values are already high and ceiling effects are a risk (Fig. 2C).

3.3 Flexibility (Habit-Based to Goal-Directed Shift)

Acclimatization

Animals stay in the operant chamber (Fig. 1B) for 10 min on two consecutive days. The respective lever(s), as predefined for each rat (see below), are presented. Also, a limited amount of the reward is made freely available in the food magazine. During this period, equipment should be on, particularly the ventilation fans, to simulate the environmental noises.

Habit Acquisition

Across sessions, animals learn to press a lever for a specific outcome, which will eventually lead to acting in a habitual mode (Fig. 3A). For this, the reward schedule is adjusted as the animal progresses (Sections: One Lever and Two Levers). Animals' ability to transit from habit-based to goal-directed decision-making is tested at sessions 8–9 and 16–17 in early (i.e., prior to a hypothetical habit formation) and late (de)valuation tests (Section: Early and Late (de) Valuation Tests Fig. 3B, C). Alternatively, contingency degradation can be used for the same purpose (Section: Contingency Degradation Fig. 3D, E).

One Lever

Under a continuous reinforcement (CRF; one lever press per reinforcer), rats are trained to press the lever to obtain a specific outcome/ reinforcer (pellets or sucrose); as stated above, the side of the lever and the reinforcer should be balanced within and between experimental groups. After two CRF sessions on consecutive days, random ratio (RR) is introduced and increased in a stepwise fashion from 5 (RR-5; 2 days) to 10 (RR-10; 2 days) and 20 (RR-20; 1 day) average lever presses per reinforcer (Fig. 3A). After the early (de)valuation test (see Section: Early and Late (de) Valuation Tests), six additional sessions on RR-20 are performed (Fig. 3A). In all cases, sessions terminate after 30 reinforcements or when 30 min elapsed. The number of lever presses per minute is used as a behavioral readout.

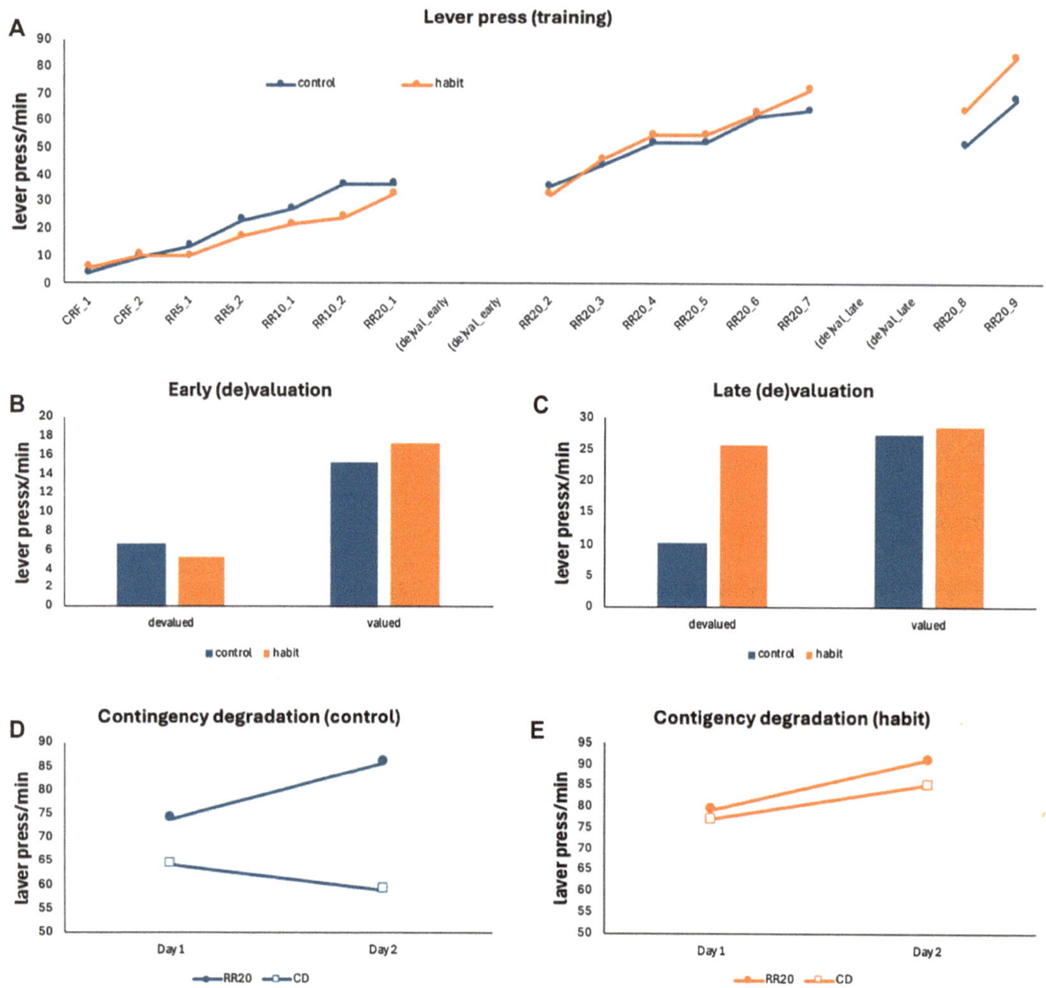

Fig. 3 A hypothetical habit-based/goal-directed lever press decision-making dataset (based on real data using one lever and sugared pellets as reward). Lever-press activity steadily increases across days as the rewarding schedules change from CRF to increasing RR (**A**). Early (**B**) and late (**C**) (de)valuation tests are performed after RR20_1 and RR20_7. Prior to habit formation, animals distinguish valued from devalued conditions (**B**) and decide in a goal-directed manner; however, repetition promotes habit formation, and in some conditions (for instance, chronic stress [35], chronic pain [unpublished data from our group], and its conjunction [39]), the ability to adjust is impaired (**C**). An alternative way to demonstrate this is to degrade the contingency; expectedly, the number of lever presses should reduce substantially if the animal is goal-directed (**D**) but not so much if it is in a habit-based (**F**) decision-making. CRF continuous reinforcement, CD contingency degradation, (de) val devaluation/valuation; RR, random ratio

Two Levers

In a variant of the single-lever protocol, animals can be trained with two levers. Specifically, each animal performs two daily sessions, spaced by a minimum interval of 5 h, with either the right or left lever. Each animal has a specific lever/outcome association that is maintained throughout the protocol (e.g., pellets/right lever and sucrose/left lever or vice versa). This association should be randomized across animals and/or groups; also, for a given animal, the presentation order of the levers should vary in alternate sessions (e.g., morning/right and afternoon/left). The reinforcement schedule for each daily pair of sessions is as described above (Section: One Lever; Fig. 3A). Also as above, sessions terminate after 30 reinforcements or when 30 min elapsed.

Early and Late (de) Valuation Tests

On protocol days 8–9 and 16–17, early and late devaluation/valuation tests are performed to assess contingency acquisition in devalued or valued conditions—see Fig. 3B, C for a hypothetical dataset on (de)valuation. The order—valued/devalued or devalued/valued—is balanced within the groups, i.e., on the first day, half of the controls enter in a valued condition and the remaining in a devalued condition; on the following day, they exchange. During the hour preceding the (de)valuation test, rats are placed individually in a regular cage specifically prepared for this session (each animal should have its own cage for the entire duration of the protocol) and are given free access to one of the two outcomes. Valued condition means that the animal receives the regular reward, whereas the devalued condition means that the animal receives the alternative reward. The rational behind this is that in devalued conditions animals will only maintain the behavior if they fail to transition from an habitual to a goal-directed decision. In all cases, and for control reasons, the amount of sugared pellets or sucrose solution consumed during this hour should be registered. Afterward, animals are then placed in the OC, and the lever on which the animals were trained is presented (the two levers are simultaneously present if trained in the protocol with the two levers; see Section: Two Levers). The number of lever presses is then registered for 5 min; no reward is delivered during this session. In the following day, the procedure is repeated to test the opposite condition. Note that the absence of reward delivery might reduce lever activity in subsequent sessions (extinction), though it can be easily reinstated (see Fig. 3A; cf. RR20_2 and RR20_8).

Contingency Degradation

Contingency degradation essentially refers to the reduction of the association between the behavior (lever pressing) and its result (reward delivery). In other words, in these sessions, the reward is delivered independently of lever press action (ideally close to the rate observed in RR20)—see Fig. 3D, E, for a hypothetical dataset on contingency degradation. In animals trained with two levers, the contingency is degraded in just one lever, while the other should maintain the RR20 schedule. Finally, it is highly recommended that after the late (de)valuation session, animals perform at least two sessions in RR20 in order to restore the average lever-press activity conditions (Fig. 3A; cf. RR20_8).

4 Conclusion

In this chapter, we presented behavioral protocols describing two tasks for executive functions on the realm of inhibitory control and adaptive flexibility, namely the VDS and habit-based/goal-directed tasks, respectively. As stated above, the decision to present these two tasks was mainly due to the fact that they represented core executive functions whose biological underpins remained relatively unexplored in the context of chronic pain. In our laboratory, both tasks proved to be very adequate to be used in the context of our neuropathic chronic pain model with peripheral nerve trauma (SNI). It is yet to be tested its adequacy to other models of neuropathic chronic pain or even to other etiologies.

The VDS is a fast paradigm when compared to other impulsivity paradigms, and it can be repeated (we tested up to three repetitions; unpublished). This is a great advantage because it allows the application of follow-up studies and testing for pharmacological (or other) therapeutics. It should, however, be noticed that trait impulsivity might affect the outcomes [22, 36]. On the other hand, the habit-based/goal-directed decision-making protocol is lengthy (habits take time to form) and more complex in terms of experimental design. A relatively high number of conditions need to be balanced within/between groups to achieve a proper control of the experiment (lever side, reward type, value/devalue order, contingency degradation/habitual schedules), implying the use of large groups. Additionally, it is important to notice that when the two tasks are used in tandem, carryover effects have been demonstrated [31], and crossover designs are greatly advised. One final aspect pertains to anhedonia, which has been largely described in chronic pain models [23, 24] and might ultimately affect the animals' drive to perform the tasks. While we have not observed any substantial differences in the animals' ability to acquire/perform the task, authors should always have this potential confounder in mind in their specific experimental conditions. The same applies to motor performance. Chronic pain is debilitating, and motor deficits have been reported in some conditions (e.g., aged subjects [20]). Particularly in the VDS, in which the animal is required to move inside the OC repeatedly from the decision panel to the food magazine and vice versa as a new trial starts, motor performance can potentially interfere with the outcome. In addition to the readouts described above, response latencies and latencies to feed might provide complementary information regarding motor behavior and should always be collected/analyzed.

The study of executive functions in the context of experimental chronic pain models has relied on the relatively simple and loosely defined behavioral tasks (e.g., alternation protocols) that fail to cover decision-making dimensions such as inhibition or flexibility. The protocols presented here aim to offer an alternative to this scenario.

Acknowledgments

This work has been funded by National funds through the Foundation for Science and Technology (FCT)—projects UIDB/50026/2020 and UIDP/50026/2020.

Conflict of Interest The authors have no conflicts of interest to declare that are relevant to the content of this chapter.

References

1. Diamod A (2013) Executive functions. Annu Rev Psychol 64:135–68
2. Bari A, Robbins TW (2013) Inhibition and impulsivity: behavioral and neural basis of response control. Prog Neurobiol 108:44–79
3. Zink N, Lenartowicz A, Markett S (2021) A new era for executive function research: on the transition from centralized to distributed executive functioning. Neurosci Biobehav Rev 124:235–244
4. Nett KE, LaLumiere RT (2021) Infralimbic cortex functioning across motivated behaviors: can the differences be reconciled?. Neurosci Biobehav Rev 131:704–721
5. Munakata Y, Snyder HR, Chatham CH (2012) Developing cognitive control: three key transitions. Curr Dir Psychol Sci 21(2):71–77
6. Moffitt TE et al (2011) A gradient of childhood self-control predicts health, wealth, and public safety. Proc Natl Acad Sci U S A 108(7):2693–2698
7. Sewell KR et al (2021) Relationships between physical activity, sleep and cognitive function: a narrative review. Neurosci Biobehav Rev 130:369–378
8. Zheng K et al (2021) Concurrent performance of executive function during acute bouts of exercise in adults: a systematic review. Brain Sci 11(10):1364
9. Craig F et al (2016) A review of executive function deficits in autism spectrum disorder and attention-deficit/hyperactivity disorder. Neuropsychiatr Dis Treat 12:1191–1202
10. Pietrzykowski MO et al (2022) The central executive network and executive function in healthy and persons with schizophrenia groups: a meta-analysis of structural and functional MRI. Brain Imaging Behav 16(3):1451–1464
11. Buckner RL (2004) Memory and executive function in aging and AD: multiple factors that cause decline and reserve factors that compensate. Neuron 44(1):195–208
12. Friedman NP, Robbins TW (2022) The role of prefrontal cortex in cognitive control and executive function. Neuropsychopharmacology 47(1):72–89
13. Jones DT, Graff-Radford J (2021) Executive dysfunction and the prefrontal cortex. Continuum (Minneap Minn) 27(6):1586–1601
14. Menon V, D'Esposito M (2022) The role of PFC networks in cognitive control and executive function. Neuropsychopharmacology 47(1):90–103
15. Cato MA et al (2004) Assessing the elusive cognitive deficits associated with ventromedial prefrontal damage: a case of a modern-day Phineas Gage. J Int Neuropsychol Soc 10(3):453–465
16. Cowell RA, Barense MD, Sadil PS (2019) A roadmap for understanding memory: decomposing cognitive processes into operations and representations. eNeuro 6(4):ENEURO.0122-19.2019
17. Zamani A, Carhart-Harris R, Christoff K (2022) Prefrontal contributions to the stability and variability of thought and conscious experience. Neuropsychopharmacology 47(1):329–348
18. Clark SV et al (2021) Cerebellar-subcortical-cortical systems as modulators of cognitive functions. Neuropsychol Rev 31(3):422–446
19. Leite-Almeida H et al (2012) Differential effects of left/right neuropathy on rats' anxiety and cognitive behavior. Pain 153(11):2218–25
20. Leite-Almeida H et al (2009) The impact of age on emotional and cognitive behaviours triggered by experimental neuropathy in rats. Pain 144(1-2):57–65
21. Leite-Almeida H et al (2014) Asymmetric c-fos expression in the ventral orbital cortex is associated with impaired reversal learning in a right-sided neuropathy. Mol Pain 10(1):41
22. Cunha AM et al (2020) High trait impulsivity potentiates the effects of chronic pain on impulsive behavior. Neurobiol Pain 7:100042
23. Cunha AM et al (2020) Chronic pain impact on rodents' behavioral repertoire. Neurosci Biobehav Rev 119:101–127
24. Leite-Almeida H, Pinto-Ribeiro F, Almeida A (2015) Animal models for the study of comorbid pain and psychiatric disorders. In: Finn DP, Leonard BE (eds) Pain in psychiatric disorders.

Mod trends pharmacopsychiatry, vol 30. Karger, Basel, pp 1–21.
25. Chan RC et al (2008) Assessment of executive functions: review of instruments and identification of critical issues. Arch Clin Neuropsychol 23(2):201–216
26. Nyongesa MK et al (2019) Assessing executive function in adolescence: a scoping review of existing measures and their psychometric robustness. Front Psychol 10:311
27. Kusi-Mensah K et al (2021) Assessment tools for executive function and adaptive function following brain pathology among children in developing country contexts: a scoping review of current tools. Neuropsychol Rev 32(3):459–482
28. Brown VJ, Tait DS (2016) Attentional set-shifting across species. Curr Top Behav Neurosci 28:363–395
29. Esteves M et al (2021) Assessing impulsivity in humans and rodents: taking the translational road. Front Behav Neurosci 15:647922
30. Roberts WW, Dember WN, Brodwick M (1962) Alternation and exploration in rats with hippocampal lesions. J Comp Physiol Psychol 55:695–700
31. Cunha AM et al (2020) Unilateral accumbal dopamine depletion affects decision-making in a side-specific manner. Exp Neurol 327:113221
32. Cunha AM et al (2020) Mesocorticolimbic monoamines in a rodent model of chronic neuropathic pain. Neurosci Lett 737:135309
33. Amorim L et al (2022) Shaping social behavior in an enriched environment. Front Behav Neurosci 16:999325.
34. Trueman RC et al (2012) Five choice serial reaction time performance in the HdhQ92 mouse model of Huntington's disease. Brain Res Bull 88(2-3):163–170
35. Dias-Ferreira E et al (2009) Chronic stress causes frontostriatal reorganization and affects decision-making. Science 325(5940): 621–625
36. Soares AR et al (2018) Trait determinants of impulsive behavior: a comprehensive analysis of 188 rats. Sci Rep 8(1):17666
37. Decosterd I, Woolf CJ (2000) Spared nerve injury: an animal model of persistent peripheral neuropathic pain. Pain 87(2):149–58
38. Leite-Almeida H et al (2013) Variable delay-to-signal: a fast paradigm for assessment of aspects of impulsivity in rats. Front Behav Neurosci 7:154
39. Mor D et al (2017) Peripheral nerve injury impairs the ability to maintain behavioural flexibility following acute stress in the rat. Behav Brain Res 328:123–129

Chapter 16

Corticostriatal Connectivity and Its Application to Study Neuropathic Pain-Related Impairments of Cognitive Processes

Helder Cardoso-Cruz, Clara Monteiro and Vasco Galhardo

Abstract

Detailed insight into the functionality of the central nervous system is of pivotal importance for understanding how the brain interacts with disease. The in vivo extracellular activity of the corticostriatal loop has been intensively investigated in rodents to understand their role in learning, memory, and sensory and affective components of pain states. However, it is still unclear how this corticostriatal loop is endogenously engaged in pain-related cognitive perturbations. Here, we describe a method for studying functional corticostriatal loop connectivity dynamics in awake rodents during cognitive demand. For that proposal, we used a well-described rodent model of neuropathic pain, which is based on the sciatic nerve partial lesion. Our goal is to provide the reader with guidelines on how to apply an optogenetic neuromodulation approach to study the impact of corticostriatal connectivity signature on the cognitive demand during a neuropathic pain condition and to highlight the advantages and limitations of this approach.

Keywords Neuropathic pain, Corticostriatal loop, In vivo neural recordings, Optogenetic neuromodulation, Working memory deficits

1 Introduction

While dysfunction of the prelimbic area (PrL) of the medial prefrontal cortex (mPFC) has been implicated in chronic pain, the underlying neural circuits and the contribution of specific cellular populations and microcircuits remain underexplored [1–4]. In this regard, data obtained from preclinical studies revealed that chronic pain states areassociated with functional alterations in PrL pyramidal neurons, including morphological changes, decreased excitatory synaptic inputs, impaired excitatory cholinergic modulation, and reduced intrinsic cellular excitability and discharge activity [4–9]. In the same extension, PFC GABAergic inhibitory interneurons also exhibited changes during peripheral neuropathic pain, including a

Isaura Tavares and Hugo Leite-Almeida (eds.), *Translational Research Methods for Neuropathic Pain*, Neuromethods, vol. 235, https://doi.org/10.1007/978-1-0716-5312-8_16,

decreased excitability of somatostatin-expressing interneurons and an increased excitability of parvalbumin-expressing interneurons [7, 10]. At the circuitry level, information to and from the PrL is distributed to other brain areas and disruption of these networks results in behavioral, emotional, and cognitive deficits [11, 12]. In particular, the PrL is thought to play an important role in supporting spatial working memory (WM) processes. During the cognitive demand, the PrL pyramidal neurons exhibit a sustained elevation of their firing activity, which may reflect the holding of goal-related information or the preparation of forthcoming responses [13]. This persistent firing activity of PrL pyramidal neurons during the delay phase of WM tasks has been considered to be the neural substrate for information maintenance [14]. In addition, the timing and spatial selectivity of these neurons are critically determined by an inhibitory drive enhancement of a local group of PrL GABAergic interneurons [15], which are critically affected by dopamine signaling [16]. These studies support the notion that PrL local microcircuits are important in the regulation of spike timing and information synchronization during cognitive demand.

Nevertheless, the PrL functioning cannot be dissociated from that of its densely interconnected areas, such as the nucleus accumbens core region (NAcC) that is a component of the central reward system [17–19] and important to the motivational value coding necessary to support PFC-dependent executive functions [20]. Both regions are interconnected [21], revealing a strong synchrony mediated by theta oscillations [20]. The PrL exerts a regulatory effect on the NAcC through the direct glutamatergic excitatory input of the neurons projecting to the NAcC [22] and indirectly through the action over other neurons involving a dopamine-dependent control [21, 23]. Perturbations of this corticostriatal loop have been associated with disruptions in cognitive performance [24] and nociceptive information integration [25, 26]. Altered corticostriatal functional connectivity has also been reported in drug addiction [21], age-related studies [27], major depression conditions [28], and patients with neurodegenerative diseases or chronic pain [26, 29, 30]. Despite our knowledge about the negative impact of persistent pain, there are still major gaps in matching cognitive deficits to specific morphological and functional changes in the corticostriatal loop triggered by the pain conditions.

Through the combined use of optogenetics, neuroelectrophysiological recordings, and behavior analyses in this experimental setup, we propose a strategy to manipulate the corticostriatal loop to elucidate its functional role associated with pain-related maladaptive perturbations during cognitive demand. In the following experiment setup, we compared the in vivo corticostriatal connectivity during the performance of a WM task after the induction of a rodent neuropathic pain model, and finally, we explored the spontaneous intraconnectivity of this microcircuit as a general marker of central cognitive alterations induced by persistent neuropathic pain.

2 Materials

2.1 Animal Model and Ethical Statement

For behavioral evaluation, we used Sprague–Dawley adult male rats (weight 250–350 g; Charles River Laboratories, Saint-Germain-Neulles, France). The animals were maintained under standard laboratory conditions on individually ventilated cages (model IVC Smart Flow; Tecniplast, Italy) with an artificial 12-h light/dark cycle (lights on from 07:30 to 19:30 h) and constant controlled temperature of 22 ± 2°C and relative humidity of 50% ± 5%. Before surgical procedures, we housed the animals in collective cages containing environmental enrichment (four per cage; model GR1800 Double decker, Tecniplast, Italy). However, after the surgical interventions (see details below), particular care has to be taken to avoid direct contact with other animals to improve the postoperative recovery, housing the animals individually. To reinforce the motivational level, all animals were food deprived to 90%–95% of their ad libitum feeding body weights with unrestricted access to water during the course of behavioral experiments. In this context, their weights were checked daily and their growth monitored weekly against a standard curve for Sprague–Dawley male rats. We performed the training and recording sessions at approximately the same time each day during the lights-on portion of the cycle (13:00 h). In order to minimize any distress related to manipulation, animals were habituated to handling by the experimenters before starting any experimental procedure. At the ethical level, all experiments illustrated in this experimental setup were performed in accordance with European Union Directive 2010/63/CE guidelines and Research and Ethical Issues of the International Association for the Study of Pain [31], and approved by local ethical committee. Finally, we made all efforts to implement the 3Rs recommendations for animal experimentation, to minimize animal suffering and to reduce the number of animals used.

2.2 In Vivo Optogenetic Neuromodulation of Corticostriatal Loop Neural Activity

The existence of a diversity of mPFC excitatory and inhibitory neurons, together with a complex neurotransmitter network, presents a challenge for the selective control of intra-mPFC activity and output needed to restore cognitive impairments. Recently, optogenetic techniques have opened the possibility of manipulating central microcircuits with great spatial and temporal resolution. Here, we performed combined photostimulation and recording in awake rats using custom-made optrodes and an anatomical approach to promote the selective transfection of the neuronal population of interest. An illustration of the strategy used for viral particles infusion in the PrL and light photostimulation in the NAcC is shown in (Fig. 1a).

The adeno-associated virus (AAV, serotype 5) for halorhodopsin (eNpHR3.0 variant) used in this protocol was obtained from the University of North Carolina Vector Core. Viral titer was 3.0×10^{12} particles/ml for AAV5-hSyn1-eNpHR3.0-enhanced red fluorescent protein (mCherry). The use of a hSyn1 promoter enables transgene expression favoring both excitatory and inhibitory neurons. Virus were housed in a −80°C freezer until the day of infusion (see details below regarding the viral particles infusion in Sect. 3.1). Each custom-made optrode was composed of one optical fiber cannula and two multielectrode arrays (one per recording area), each containing eight filaments of formvar-coated tungsten wire (35 μm diameter;

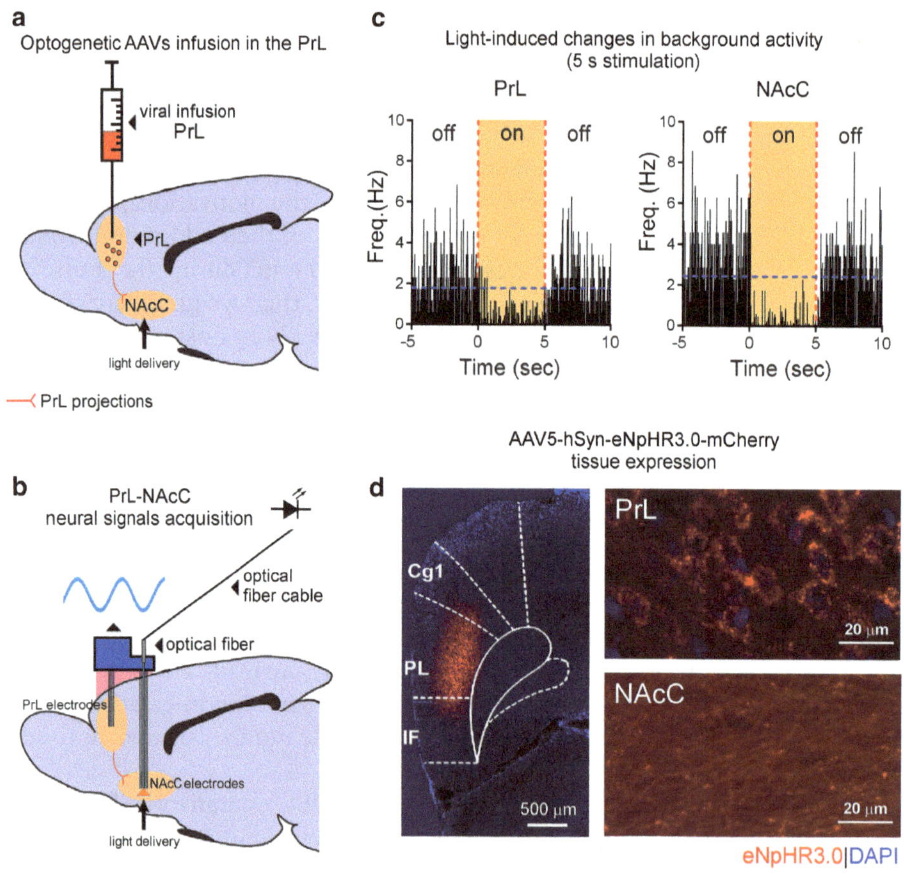

Fig. 1 Infusion of eNpHR3.0-mCherry into the PrL and expression in the corticostriatal loop, neurophysiological recording, and light delivery setup. (**a**) Schematic illustration of the infusion of AAV5-hSyn-eNpHR3.0-mCherry viral particles into the PrL. (**b**) Experimental setup used for optogenetic light delivery in the NAcC and neural activity recording in the PrL and NAcC. (**c**) Histograms illustrating examples of the in vivo background activity of a PrL- and an NAcC- recorded unit. Vertical orange spots indicate the interval of optical light delivery. The blue dotted horizontal lines indicate mean firing activity. Optogenetic selective photoinhibition using an orange LED light (continuous solid pulse at 5–6 mW, 620 nm). (**d**) Coronal illustration of PrL. The red fluorescent protein (mCherry) labeling indicates eNpHR3.0 transfected area in the PrL (left panel). The blue dots represent DAPI DNA-labeling used for nuclear counterstain. Right panels illustrating the magnification of PrL and NAcC areas, respectively

California Fine Wire Company, CA, USA) with impedances varying between 0.7 and 1.2 MΩ, at 1 kHz, and a reference silver wire per array (150 μm diameter; AM Systems, WA, USA) for connection with the implanted screws. Each multielectrode array was assembled in a 4 × 2 architecture, with spacing of 250 μm between filaments and 400 μm between each row [32]. The silver wires were anchored and connected to the implanted screws and covered with conductive epoxy glue (see details below regarding surgical protocol in Sect. 3.1). Each optical fiber cannula was composed of a zirconia ferrule (2.5 mm in diameter and 10.5 mm in length) and an optical fiber (0.62 NA, 200/230 μm in diameter and 10 mm in length). Prior to the optrode structure implantation, the final light output of the LED power source was adjusted to deliver 5–6 mW (irradiance = 158–189.6 mW/mm^2) of intensity at the end of the optical fiber (model PM160; Thorlabs, Munich, Germany). This step is important because there are small physical variations resulting from the polishing of the optical fiber tip. During behavioral probe sessions, the neuromodulation stimulation parameters were controlled with a TTL pulse generator (Prizmatix Pulser 2.0, Israel) connected to an LED driver (Model LD-1 single channel; Plexon Inc., Dallas, TX, USA). Animals expressing inhibitory halorhodopsin (eNpHR3.0 variant) were unilaterally illuminated using an orange LED light ($\lambda = 620$ nm; PlexonBright dual LED commutator; Plexon Inc.) to produce a continuous solid pulse (square wave). This stimulation commutator system has the advantage of avoiding the restraint of movements induced by the optical patch cord during the behavioral performance that requires free navigation.

The light stimulation was delivered during the cognitive demand into the NAcC (see details in Sect. 3.4) in a focal area around the recording multielectrode array (Fig 1b). To ensure that eNpHR3.0 expression at the axon terminals of the PrL is functional, we recorded simultaneously the extracellular neural activity in the PrL and NAcC areas (Fig. 1c). To test the efficiency of viral transfection and light stimulation, we compared the background activity during a 5 s light-on versus light-off (baseline period) (Fig. 1c). These recordings were done 14 days after surgery to enable proper viral transfection in the tissue. Two weeks after AAV5-hSyn-eNpHR3.0-mCherry infusion into the PrL, a robust opsin expression was observed in PrL neurons (with minimal expression in anterior primary cingulate cortex (Cg1) and infralimbic cortex neurons (IF)) and within NAcC, indicating that eNpHR3.0 opsin is expressed in the axon terminals of PrL neurons that project to the NAcC (Fig. 1d).

2.3 Behavioral Procedures: Apparatus, Learning Phase, and Working Memory Task

The apparatus we use for behavioral testing and in vivo electrophysiology consists of mostly off-the-shelf components and a few custom parts and software. To evaluate the impact of neuropathic pain on behavior, we used a classic spatial delayed non-match-to-sample (DNMS) working memory (WM) task [4]. The DNMS testing

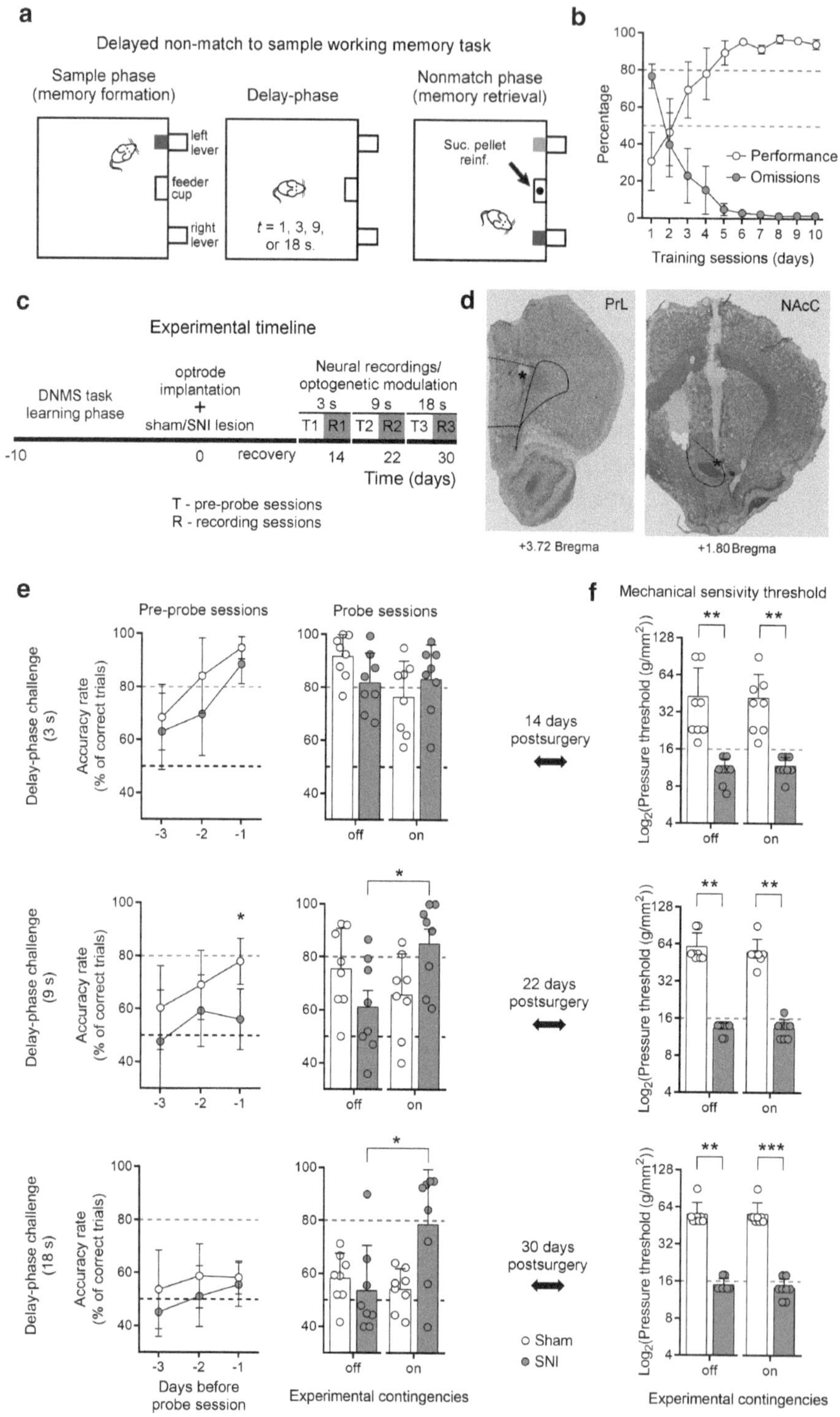

Fig. 2 Delayed non-match-to-sample working memory task, learning phase, experimental timeline, and optogenetic neuromodulation impact on behavioral accuracy rate and pain responses. (**a**) Diagram of delayed non-match-to-sample task used in this study. Four DNMS delay-phase challenges were used: 1 s (learning phase)

protocol consisted of a custom-made square arena (dimensions: 45 cm × 45 cm long and wide, and walls with 40 cm high), containing two retractable levers (Model H23-17RA; Coulbourn Instruments, USA) and a feeder cup placed between them (Model H14-23M; Coulbourn Instruments, USA; Fig. 2a). Each trial comprised a sample phase (memory formation), in which one of the retractable levers were exposed until the animal pressed the lever (with a time limit of 15 s). After the lever press, a short period was initiated (delay phase). After the delay phase, both retractable levers were exposed (non-match phase or memory retrieval) until the animal pressed one (with a time limit of 30 s). A trial was considered correct when the animal pressed the opposite lever pressed in the sample phase. Correct trials were rewarded with one 45 mg sucrose dustless precision pellet (Cat. No. F0023; BioServ, Frenchtown, NJ, USA) delivered in the food cup. Trials in which the animal failed to make a lever response before the time limit imposed were considered omissions. Trials were separated by an interval of 15 s. Control of pellet dispenser was fully automated using the OpenControl software adapted to this particular task [33]. In order to avoid lateralization biases, each testing session contained an equal number of left and right rewarded trials. Behavioral accuracy is expressed as the percentage of correct trials per session.

3 Methods

3.1 Stereotaxic Surgical Procedures: Optogenetic Viral Particles Infusion and Optrode Structure Implantation

Two steps were required to prepare animals for optogenetic neuromodulation and extracellular recording: (1) the optogenetic viral particles infusion in the brain area of interest and (2) the implantation of an optrode for stimulation and neural activity recording. These procedures can be performed during the same surgery. For this purpose, we anesthetized the animals with a mixture of ketamine

Fig. 2 (continued) 3, 9, and 18 s (probe sessions with optogenetic neuromodulation). (**b**) DNMS task learning-phase accuracy level using a delay-phase challenge of 1 s. Only animals that reach at least 80% of correct trials were selected to be candidates to receive the optrode implantation. (**c**) Timeline of the experimental setup. Briefly, each animal was implanted with an optrode in the PrL and NAcC and subjected to a sham or spared nerve injury (SNI) lesion. After postoperative recovery, the animal was subjected to three daily pre-probe training sessions (T) with the respective DNMS delay-period challenge to be applied during the subsequent probe session (R). (**d**) Thionine-labeling microphotographs of a coronal slice of the PrL (left panel) and NAcC (right panel) illustrating the track and final location of an optrode structure. Asterisks indicate the final position. (**e**) The inhibition of contralateral PrL output into the NAcC of animals transfected with AAV5-hSyn1-eNpHR3.0-mCherry vector induced an important recovery of neuropathic pain-related WM impairment in SNI-treated animals at higher DNMS task complexity challenges: pre-probe sessions (left panels) and probe sessions with optogenetic stimulation (right panels) performed during day 14, 22, and 30 after peripheral lesion. (**f**) Level of mechanical sensitivity to von Frey filaments test of eNpHR3.0-expressing rats across different time points after peripheral nerve lesion. Comparisons between experimental groups and photostimulation protocols are based on nonparametric Kruskal–Wallis test followed by Dunn's post hoc test. Values are presented as mean ± standard deviation, $^{*}P < 0.05$, $^{**}P < 0.01$, and $^{***}P < 0.001$

and medetomidine (75 and 0.5 mg/kg in saline, intraperitoneally (IP), respectively). Anesthesia was maintained with small additional injections of ketamine (one-third of the initial dosage). The depth of anesthesia and muscular paralysis were assessed by regularly testing the corneal reflex, hindpaw withdrawal, and tail-pinch reflexes. After the anesthesia induction, the animal also receives a dose of atropine sulfate (0.02 mg/kg, subcutaneous (SC)) and 1 ml of serum (sucrose 2% w/v in saline, SC) every hour during the procedure. After the animals were anesthetized, particular care has to be taken to maintain the body temperature. Core body temperature was maintained in a range between 36 and 38°C through use of a homoeothermic blanket system (Harvard Apparatus). This step is particularly important to prevent hypothermia during long surgical procedures. The anesthetized animal was secured in a stereotaxic frame (Model 900LS; David Kopf Instruments, Tujunga, CA, USA) using ear bars, the mandibles were fixed, and the tongue was folded toward left or right side of the mouth to avoid obstructing the airways during the intervention. The animal's eyes were covered with ophthalmic ointment for protection, and the head was shaved and cleaned using a triple application of alcohol (70% v/v) and betadine. A midline injection of 0.3 ml of 1% lignocaine (B. Braun, Melsungen, Germany) was applied to the scalp for local analgesia.

The surgery was performed under a microscope with light illumination. An incision was made caudally to the animal's eyes and ending between the ears. The connective tissue was blunt-dissected from the skull, and the top of the skull was exposed and cleaned with cotton swabs using hydrogen peroxide so that bregma and lambda could be clearly seen. After the scalp was excised, a hole (4 × 4 mm) was drilled in the skull for viral particles infusion and optrode implantation. Additionally, up to five screws (Cat. No. 19010-10; Fine Science Tools, Heidelberg, Germany) were implanted in the skull for securing probes and for grounding purposes. The skull was covered with mineral oil to avoid dehydration during surgical intervention. The following stereotaxic coordinates relative to the skull bregma point were used to infuse the viral particles into the PrL and to implant the optrode structure in the PrL (3.0–3.7 mm anterior to bregma, 0.7–0.9 mm lateral to midline, depth −3.4 mm) and in the NAcC (1.6–1.8 mm posterior to bregma, 1.8–2.0 mm lateral to midline, depth −7.2 mm) [34] after *dura mater* removal. Prior to optrode implantation, an injection of 1 µl of AAV5-hSyn-eNpHR3.0-mCherry virus was infused in the PrL at a rate of 0.1 µl/min using a 2 µl microsyringe (Model 7002KH – tip type 2; Hamilton, Reno, NV, USA). To promote optimal diffusion of the viral particles in the tissue, the needle was maintained in place for at least 10 min and then slowly withdrawn. After the microsyringe removal, the optrode structure was mounted in the holder of a hydraulic micropositioner (FHC, Bowdoin, ME, USA), oriented rostro-caudally, and subsequently slowly driven (100 μm/min, 10 μm/step) into the PrL and NAcC. After the electrodes reached the correct position, the craniotomy hole was sealed with light-sensitive epoxy glue, and the structure was fixed to the skull using dental acrylic cement. Covering the craniotomy hole with epoxy glue dramatically prevents scar tissue buildup and long-term reossification, resulting in damage to the implanted optrode structure. When dental acrylic cement was almost completely cured, the stereotaxic optrode holder was removed, the animal was sutured, and a topical antibiotic was applied to the wound margins. The optrode structures were implanted unilaterally and counterbalanced between brain hemispheres across animals.

Finally, each animal was subjected to the spared nerve injury (SNI) or to a control sham intervention involving the same extent of skin incision and muscle dissection, leaving the sciatic nerve plexus intact (see details below). After the surgical protocol, an analgesic (ketoprofen, 5 mg/kg) and an antibiotic (enrofloxacin, 5 mg/kg) were administered SC (diluted 1/10 in saline) every 24 h for 5–7 days to improve postoperative recovery. Animals were allowed to recover for 7 days before behavioral testing sessions began. During this period, their overall health status was checked daily. Specifically, careful attention should be paid to the general level of activity of the animal, including grooming behavior, vocalization, respiration, urination, defecation, eating, and drinking. Additionally, the animal should be monitored for any signs of ataxia, and if the animal develops ataxia, the induction of a euthanasia protocol should be considered .

3.2 Model of Neuropathic Pain and Evaluation of Peripheral Mechanical Sensory Responses

The classical SNI rodent model was a robust rodent model of persistent neuropathic pain [35, 36]. The SNI model mimics human neuropathic pain related to peripheral nerve injury and was based upon an invasive but simple surgical procedure. This model was characterized by a reliable and long-lasting neuropathic pain-like behavior (hyperalgesia and allodynia). In this experimental protocol, the evaluation of mechanical sensory threshold responses was performed 1 h after the end of every recording session using von Frey filaments (Somedic, Sweden), with the common Dixon up-down protocol [37] of filaments with logarithmically incremental stiffness (0.064, 0.085, 0.145, 0.32, 0.39, 1.10, 1.70, 3.30, 5.10, 8.30, 17.0, 24 g) as previously described [38]. Von Frey filaments were applied to the lateral one-third of the hindpaw (in the distribution of the sural nerve). After each behavioral probe session, to ascertain the role of PrL excitatory output into the NAcC in peripheral pain responses, we evaluated the mechanical withdrawal threshold using the von Frey test. Animals were individually placed in a plexiglass chamber over a mesh table and acclimated for 30 min. The measurements were performed in independent sessions without and with optogenetic photoinhibition. To this aim, we used a 10 s continuous pulse per trial. Notably, in this protocol, we found that selective photoinhibition of

PrL glutamatergic neuron terminals into the NAcC did not induce any significant effect in peripheral pain responses (Fig. 2f).

3.3 In Vivo Electrophysiological Neural Recordings

The extracellular recording is an important technique that not only allows a real-time feedback about the correlation of neural activity and behavioral responses, but also provides a functional validation complementary control of the optogenetic manipulations in the areas of interest. Here, extracellular single-cell and local field potentials (LFPs) neural activity were obtained from intracranial multielectrode arrays during the DNMS task performance. The 16-channel optrode structure was connected to a wireless head-stage transmitter (W16; Triangle Biosystems, Durham, NC, USA) that sent continuous analog signals to a Multineuron Acquisition Processor system (MAP-16; Plexon Inc.). Neural signals were preamplified (10,000–25,000×) and digitized at 40 kHz. Each optrode channel was manually checked for artifacts (e.g., 50 or 60 Hz noise and harmonics). There are several steps involved in sorting spikes (simple or complex spikes), some of which depend on the particular properties of the cell type. In an extracellular recording, simple spikes typically appear as brief negative and/or positive deflections in the extracellular field potential that last for only a few microseconds, whereas complex spikes begin with a similar deflection in the field potential that can last for several milliseconds. Unfortunately, some spike profiles are difficult to sort using only automatic methods. Here, we used voltage–time threshold windows to identify single-unit waveforms, and up to three neuronal action potentials per recording channel were sorted online (SortClient 2.6 software; Plexon Inc.), and validated offline using automatic and manual sorting techniques (Offline Sorter 2.8 software; Plexon Inc.) according to the cumulative criteria described in Ref. [39]. Furthermore, to ensure that the same cell was being recorded in repeated sessions, only cells that exhibited stable spike amplitude and consistent waveforms were included for analysis. Waveforms with an inter-spike interval of <1.2 ms were excluded. Extracellular LFP signals were simultaneously recorded from the same implanted optrode by low-frequency filtering of the raw signals (0.3–200 Hz). LFP signals were preamplified and digitized at a 1 KHz sampling rate. These data were subsequently exported and analyzed with custom routines (MatLab 2013; MathWorks, Natick, MA, USA).

3.4 Experimental Timeline

Before starting the learning phase, each animal goes through a series of different steps beginning with habituation to the testing arena and experimenter. Training began with 2–4 days of testing arena (15 min per session), in which the animal was allowed to freely explore the arena and received a pellet when pressed on any of the exposed retractable levers with no time limit. Next, animals entered the forced-choice training (4–6 days), in which only one retractable lever was exposed at a time with a 15 s timeout. Finally, each animal initiated the learning curve of the DNMS task using a delay-phase challenge of 1 s for 10 days (50 trials per session). Throughout the training phase, animals showed a progressive improvement in their accuracy rate and a reduction in their omissions (Fig. 2b). At the end of this phase, we only selected animals that reached an accuracy rate of at least 80% during the last three training sessions to continue the testing protocols. In this protocol, we used several delay-phase complexity challenges: 1 s (learning curve, before surgery), 3 s (day 14 after surgery), 9 s (day 22 after surgery), and 18 s (day 30 after surgery) (Fig. 2c). Since the behavioral testing takes 2 weeks to be initiated after postsurgery, postsurgery training was required. In this sense, probe sessions were preceded by three training sessions in the respective delay-phase challenge. In respect to pre-probe training sessions, SNI-treated animals revealed a clear lower performance when compared with control animals during DNMS delay-phase lower complexity challenges (3 and 9 s), whereas both experimental groups revealed a chance-level activity during the high-complexity challenge (18 s) (Fig. 2e, left panels).

During the probe sessions (Fig. 2e, right panels), optogenetic photoinhibition of PrL glutamatergic neurons projecting into the NAcC was delivered in the NAcC during delay phase of the DNMS task. The light neuromodulation protocols were tested in separate sessions. The connection between the optical patch cable and the implanted optrode was completely covered with black adhesive tape to prevent light scattering from becoming a cue to the animals during behavioral tests. The experimenters were blind to group treatment. We used an overhead video-tracking system (CinePlex 2; Plexon Inc.) to provide information about the animal navigation and behavioral events, and to synchronize the video recordings with the acquired neural signals. The most significant effects between experimental groups were reached during higher delay-phase complexity challenges (Fig. 2e, left panels). In the case of probe sessions with photostimulation, there were no differences between the experimental groups when exposed to a lower complexity challenge, but these differences were present in higher complexity challenges.

The following procedure was used to confirm that the virus and optrode targeted the desired brain areas. After all electrophysiological and optogenetic experiments were concluded, animals were deeply anesthetized with pentobarbital sodium (150 mg/kg, IP) and transcardially perfused with 0.01 M phosphate buffer (pH = 7.2) in saline solution, followed by 4% paraformaldehyde (v/v). Brains were removed, post-fixed in 4% paraformaldehyde for 4 h, and stored in 30% sucrose solution before freezing. Brains were sectioned at 30 μm to identify recording and optical fiber position, and to verify tissue opsin expression. The optrode structure position was identified using a rat brain atlas [34]. Stained sections were observed using a Zeiss Z1 Apotome microscope. Only animals with correct location of implanted optrode structure in the PrL and NAcC brain regions, as well as correct viral expression in the tissue, should be included for analysis. A representative example of optrode location is given in Fig. 2d. In order to help the identification of recording location and depending on the experimental needs, we can promote the injection of DC current (10–20 μA for 10–20 s) through one microfilament per array. This step should be done before the perfusion protocol and allows marking the area below the electrode tips.

3.5 The Impact of Neuropathic Pain on Corticostriatal Loop Connectivity

In the past, several authors have demonstrated that pain is one of the leading factors that can induce disruptions of spontaneous oscillations in different cortical and subcortical microcircuits [40–44], including the potentiation or inhibition of oscillatory rhythms [45, 46]. Oscillatory interplay across the different brain pathways reflects a global functional state of the system and is used as an alert trigger to find normal functional state perturbations [46]. Interestingly, it has been suggested that the transition from acute to chronic pain in humans was correlated with an initially greater connectivity level between the ventral striatum and the prefrontal cortex [47]. Here, to study the PrL and NAcc interplay in the absence and during the cognitive demand, we used the statistical partial directed coherence (PDC) technique to quantify the dynamics of information flow between these brain areas. The PDC method has been described in detail elsewhere [48–50]. Briefly, PDC is an alternative statistical representation of Granger causality involving multivariate processes to uncover direct influences in the frequency domain. The allied PDC from structure j to i at frequency λ is defined as:

$$\pi_{ij}(\lambda) = \frac{\overline{A}_{ij}(\lambda)}{\sqrt{\sum_{n=1}^{N} \overline{A}_{nj}(\lambda) . \overline{A}_{nj}^{*}(\lambda)}} \tag{1}$$

where,

$$\overline{A}_{ij}(\lambda) = \begin{cases} 1 - \sum_{r=1}^{p} a_{ij}(r)\exp(-j2\pi\lambda r), & \text{if} i = j \\ - \sum_{r=1}^{p} a_{ij}(r)\exp(-j2\pi\lambda r), & \text{otherwise,} \end{cases} \tag{2}$$

a_{ij} are multivariate autoregressive coefficients and p being the model order. Nullity of PDC $\left(\pi_{ij}(\lambda) = 1\right)$ can be interpreted as absence of effective connectivity from the j-th structure to the i-th structure at frequency λ, while higher PDC activity $\left(\pi_{ij}(\lambda) \sim 1\right)$ indicates strong connectivity between both structures. This response can be interpreted as existence of information flow from brain structure j to i.

In the literature, there are also other measures of neural connectivity based on frequency domain. When compared with other measures, PDC technique has the advantage of being able to distinguish direct and indirect pathways linking different structures, providing a frequency specification, intensity interpretation, and partial directionality and relationship. Here, we discover that selective SNI-treated animals reached higher levels of ascending (NAcC-to-PrL) functional connectivity in the *theta* frequency band (between 4 and 9 Hz) (Fig. 3a, bottom panels), which can be attenuated by selective inhibition of PrL glutamatergic neuron terminals into the NAcC. This suggests that the information that reaches the PrL from the NAcC may hypothetically be modified by the neuropathic pain condition, probably by reducing their quality, accuracy, and fine integration by the PrL and lastly the behavioral performance. In fact, this altered balance between excitatory and inhibitory activity has been reported in several disorders in which the corticostriatal loop dysfunction and cognitive impairments are also found [30, 51].

Moreover, to evaluate if the neuropathic pain condition alters the spontaneous corticostriatal connectivity, we recorded the spontaneous corticostriatal LFP activity for a period of 1 h in an open-field (OF) arena in the absence of cognitive demand. The total dimension of the OF arena was 45 cm^2, with opaque walls (45 cm high with 110° of inclination). The virtual central area considered was 22.5 cm^2 (Fig. 3b). This protocol was implemented without optogenetic neuromodulation. Likewise the previous observation, the peripheral nerve lesion induced a clear enhancement of the spontaneous ascending NAcC-to-PrL connectivity mediated by theta oscillations and independent from the motor activity (Fig. 3c, d). Similarly to these observations, a reorganization of the corticostriatal circuit has been indicated to be an accurate signature to predict a possible shift from an acute to a chronic pain condition [29, 52]. Likewise, back pain patients when subjected to thermal stimulation and perform a self-report pain-monitoring task, they exhibit greater functional connectivity between the ventral striatum and the PFC when compared with healthy patients [53]. Altered PFC–striatal connectivity has also been identified in patients subjected to prolonged opioid treatments [54]. In humans, theta oscillations in frontocentral nodes have been used as a marker to understand the significance of subjective pain sensivity [55]. This result is also in agreement with our past findings that showed a significant decrease of PFC–hippocampal axis theta activity in neuropathic pain rats associated with an important decline in behavioral performance [56], in contrast to what we observed in a transgenic rodent model of congenital hypoalgesia [43, 57]. Together, these findings highlight the importance of the corticostriatal loop balance, not only as marker of cognitive impairments, but also as a marker to identify the central impact of pain.

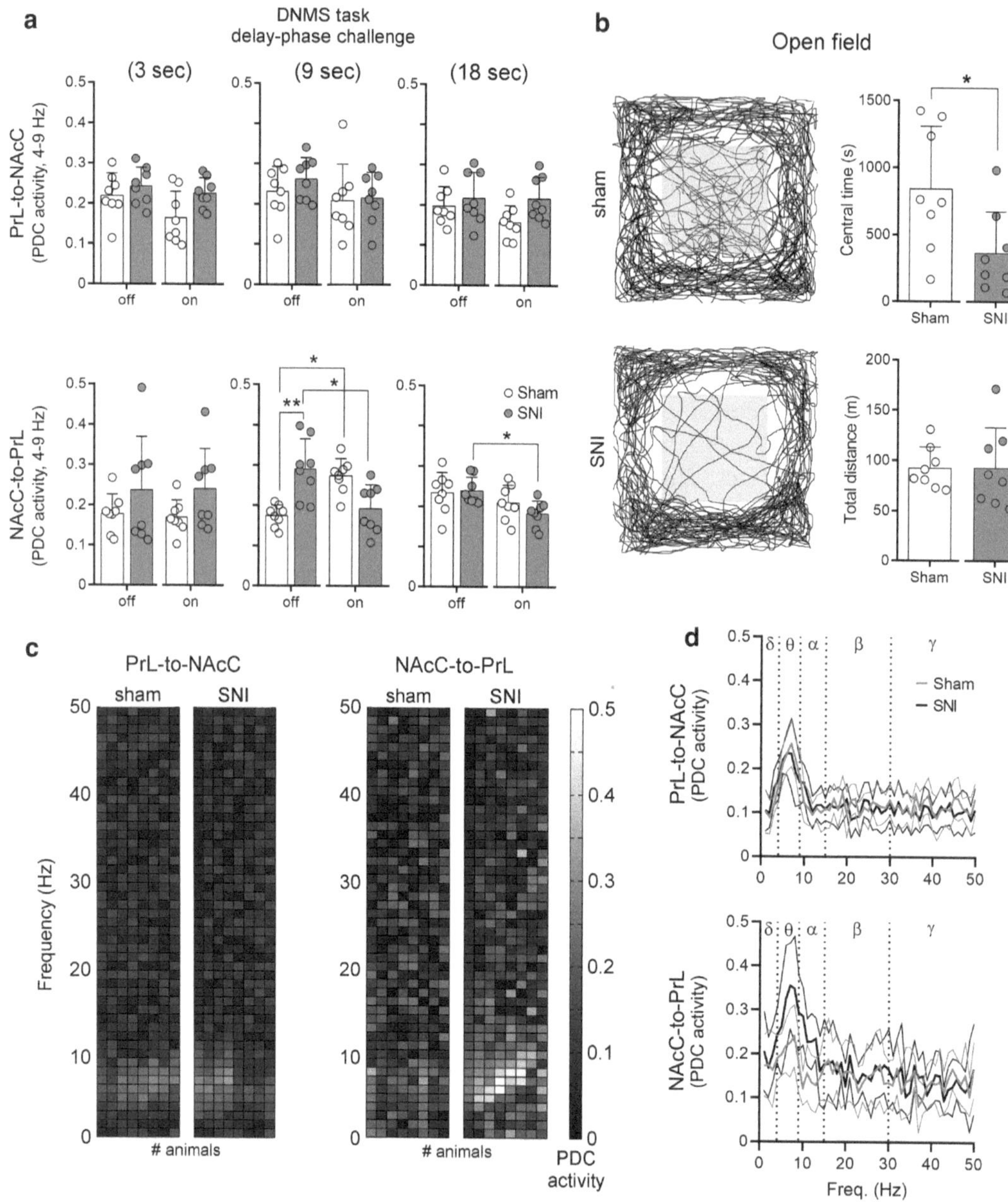

Fig. 3 Neuropathic pain can lead to a higher ascending spontaneous connectivity from NAcC to PrL. (**a**) Optogenetic inhibition of PrL glutamatergic neuron terminals into the NAcC attenuates enhanced theta frequency band functional connectivity in SNI-treated animals during the DNMS task delay-phase. This effect occurs in the ascending direction of the corticostriatal loop, from NAcC to PrL, during higher DNMS task complexity challenges. Top panels represent the theta frequency band PDC activity from the PrL to the NAcC, whereas bottom panels represent the activity from the NAcC to the PrL. Panels were represented by increased delay-phase complexity, from left to right, respectively. Comparisons between experimental groups and photostimulation protocols are based on the nonparametric Kruskal–Wallis test followed by Dunn's post hoc test. (**b**) Left panels illustrate an example of the navigation track of a sham (top) and an SNI-treated animal (bottom). The right top panel represents the total distance performed in the OF during 1 h, whereas the right bottom

4 Conclusions

Nowadays, it is recognized that the identification of neural key markers using preclinical models is essential to develop new and more effective approaches to reverse the central impact of chronic pain, improving also the validity of new translational therapeutic approaches at different dimensions. Electrophysiological and imaging techniques have been utilized to observe neural signals at various scales to understand brain functioning. However, it has been difficult to neuromodulate the activity of specific neurons or microcircuits with the spatial and temporal resolutions relevant to neural coding. To understand how function arises from the interaction between different brain microcircuits, it is necessary to use methods that allow the monitoring of brain activity and the targeted selective manipulation of neural populations. Large-scale recording of neural activity combined with selective optogenetic neuromodulation techniques has emerged as a suitable method for such tasks in awake animals for identifying induced changes in neural activity with coincident changes in behavioral phenotypes [4], or simply confirming that an optogenetic manipulation has the desired effect on neurons within the circuit [56]. Another important extension of current methods is real-time signal processing and closed-loop neuromodulation of microcircuits. In this regard, illumination could be triggered by combining neural activity makers and behavioral parameters. This creates, for instance, the ability to alter timing of neural signatures in freely moving animals during the cognitive demand and pain-related responses. However, extracellular recordings also have several limitations when they are combined with optogenetics like in this experimental setup.

The first limitation is associated with the complex intrinsic difficulty in achieving pain relief protocols without interference in the cognitive component. Moreover, many of the behavioral and functional changes take place relatively early in the course of the disease. This is a particularly interesting timeframe because it likely represents the transition from acute to chronic pain. Yet, brain networks continue to change in the presence of neuropathic lesions, either

Fig. 3 (continued) panel represents the time spent in the central zone (internal gray virtual square). Behavioral activity in the OF arena was compared using an unpaired *t* test ($P < 0.05$). (**c**) Left panels represent the spontaneous PDC activity from PrL to NAcC, and right panels represent the PDC activity from NAcC to PrL. Each vertical column represents the averaged activity of an animal. PDC activity is represented using a resolution of 1 Hz. Data correspond to 1 h of spontaneous LFP signals recorded simultaneously in the PrL and NAcC areas on an open-field arena (OF). OF recordings were performed 30 days after peripheral nerve lesion. (**d**) Spontaneous corticostriatal connectivity trace distributions revealed that SNI-treated animals increased significantly their ascending activity when compared with the control animals. Frequency bands: delta (δ, 1–4 Hz), theta (θ, 4–9 Hz), alpha (α, 9–15 Hz), beta (β, 15–30 Hz), and slow gamma (γ, 30–50 Hz). Values are presented as mean $\pm$ standard deviation. $^*P < 0.05$, and $^{**}P < 0.01$

continuing along their pathological trajectory or due to adaptive responses [58], and that increases the difficulty to identify diagnostic markers to apply more precise therapeutic approaches. In addition, another limitation also present in the described protocol can be associated with a differential gender effect of neuropathic pain. It is important to note that pain also exhibits sex dimorphism, including in PrL-dependent tasks [59]. In fact, recent data revealed that sex-specific differences are apparent in pain-associated effects on local PFC network modulation [10]. However, whether these sex-specific differences in PrL function are relevant for the neuropathic pain phenotype or not, this will constitute a field to be explored in the future. Finally, the most important limitation is the potential translational level. The use of optogenetics in humans would present a completely new paradigm in clinical practice and would be associated with several unprecedented technical considerations. In this regard, to be applied for neuromodulation of neurons in humans, there are some assumptions that should be satisfied. An ideal optogenetic tool would need to be non-immunogenic, highly precise, and should be activated with a light source able to maximize light penetration.

In conclusion, neuroscience has witnessed one of its greatest revolutions with the introduction of optogenetics in the field. However, a significant challenge lies in translating research findings from the laboratory to the clinic, particularly in terms of delivering optogenetic tools to patients (see, for review [60, 61]). A recent successful attempt involved partial vision recovery in a patient diagnosed with retinitis pigmentosa [62]. Based on this progress, we anticipate that therapeutic applications of circuit-directed optogenetic neuromodulation in humans will no longer remain a distant goal but can become a common reality.

Acknowledgments

This research was supported by FEDER funds through the Operational Competitiveness (POCI) Program – COMPETE2020, National Funds through Fundação para a Ciência e Tecnologia – FCT (Project POCI-01-0145-FEDER-028181/2017, Grant PTDC/MED-NEU/28181/2017; and Project POCI-01-0145-FEDER-028498/2017, Grant PTDC/MED-NEU/28498/2017), and FCT Project 2022.05193.PTDC supported by National Funds, and by Norte2020 (project Norte-01-0145-FEDER-000008, PortoNeuroDRIve@i3S).

Conflict of Interest The authors have no conflicts of interest to declare that are relevant to the content of this chapter.

References

1. Moriarty O, McGuire BE, Finn DP (2011) The effect of pain on cognitive function: a review of clinical and preclinical research. Prog Neurobiol 93(3):385–404
2. Baliki MN et al (2014) Functional reorganization of the default mode network across chronic pain conditions. PLOS ONE 9(9):e106133
3. Barroso J, Branco P, Apkarian AV (2021) Brain mechanisms of chronic pain: critical role of translational approach. Transl Res 238:76–89
4. Cardoso-Cruz H et al (2019) Selective optogenetic inhibition of medial prefrontal glutamatergic neurons reverses working memory deficits induced by neuropathic pain. Pain 160(4):805–823
5. Kelly CJ, Martina M (2018) Circuit-selective properties of glutamatergic inputs to the rat prelimbic cortex and their alterations in neuropathic pain. Brain Struct Funct 223(6):2627–2639
6. Metz AE et al (2009) Morphological and functional reorganization of rat medial prefrontal cortex in neuropathic pain. Proc Natl Acad Sci USA 106(7):2423–2428
7. Zhang Z et al (2015) Role of prelimbic GABAergic circuits in sensory and emotional aspects of neuropathic pain. Cell Rep 12(5):752–759
8. Ji G, Neugebauer V (2011) Pain-related deactivation of medial prefrontal cortical neurons involves mGluR1 and GABA(A) receptors. J Neurophysiol 106(5):2642–2652
9. Cheriyan J, Sheets PL (2018) Altered excitability and local connectivity of mPFC-PAG neurons in a mouse model of neuropathic pain. J Neurosci 38(20):4829–4839
10. Jones AF, Sheets PL (2020) Sex-specific disruption of distinct mPFC inhibitory neurons in spared-nerve injury model of neuropathic pain. Cell Rep 31(10):107729
11. Floresco SB (2013) Prefrontal dopamine and behavioral flexibility: shifting from an "inverted-U" toward a family of functions. Front Neurosci 7:62
12. Conway AR, Kane MJ, Engle RW (2003) Working memory capacity and its relation to general intelligence. Trends Cogn Sci 7(12):547–552
13. Baeg EH et al (2003) Dynamics of population code for working memory in the prefrontal cortex. Neuron 40(1):177–188
14. Riley MR, Constantinidis C (2015) Role of prefrontal persistent activity in working memory. Front Syst Neurosci 9:181
15. Rao SG, Williams GV, Goldman-Rakic PS (2000) Destruction and creation of spatial tuning by disinhibition: GABA(A) blockade of prefrontal cortical neurons engaged by working memory. J Neurosci 20(1):485–494
16. Seamans JK et al (2001) Bidirectional dopamine modulation of GABAergic inhibition in prefrontal cortical pyramidal neurons. J Neurosci 21(10):3628–3638
17. Yan H et al (2023) Nucleus accumbens: a systematic review of neural circuitry and clinical studies in healthy and pathological states. J Neurosurg 138(2):337–346
18. Soares-Cunha C et al (2016) Reappraising striatal D1- and D2-neurons in reward and aversion. Neurosci Biobehav Rev 68:370–386
19. Ma L et al (2020) Brain-wide mapping of afferent inputs to accumbens nucleus core subdomains and accumbens nucleus subnuclei. Front Syst Neurosci 14:15
20. Donnelly NA et al (2014) Oscillatory activity in the medial prefrontal cortex and nucleus accumbens correlates with impulsivity and reward outcome. PLOS ONE 9(10):e111300
21. McGlinchey EM et al (2016) Prelimbic to accumbens core pathway is recruited in a dopamine-dependent manner to drive cued reinstatement of cocaine seeking. J Neurosci 36(33):8700–8711
22. Stefanik MT et al (2013) Optogenetic inhibition of cocaine seeking in rats. Addict Biol 18(1):50–53
23. Domingo-Rodriguez L et al (2020) A specific prelimbic-nucleus accumbens pathway controls resilience versus vulnerability to food addiction. Nat Commun 11(1):782
24. Floresco SB, Seamans JK, Phillips AG (1997) Selective roles for hippocampal, prefrontal cortical, and ventral striatal circuits in radial-arm maze tasks with or without a delay. J Neurosci 17(5):1880–1890
25. Zhou H et al (2018) Inhibition of the prefrontal projection to the nucleus accumbens enhances pain sensitivity and affect. Front Cell Neurosci 12:240
26. Lee M et al (2015) Activation of corticostriatal circuitry relieves chronic neuropathic pain. J Neurosci 35(13):5247–5259
27. van den Bos W et al (2012) Striatum-medial prefrontal cortex connectivity predicts developmental changes in reinforcement learning. Cereb Cortex 22(6):1247–1255
28. Ledermann K et al (2016) Relation of dopamine receptor 2 binding to pain perception in female fibromyalgia patients with and without depression-A [^{11}C] raclopride PET-study. Eur Neuropsychopharmacol 26(2):320–30
29. Vachon-Presseau E et al (2016) Corticolimbic anatomical characteristics predetermine risk for chronic pain. Brain 139(Pt 7):1958–1970
30. Carriere N et al (2015) Impaired corticostriatal connectivity in impulse control disorders in Parkinson disease. Neurology 84(21):2116–2123

31. Zimmermann M (1983) Ethical guidelines for investigations of experimental pain in conscious animals. Pain 16(2):109–10
32. Cardoso-Cruz H et al (2011) Dynamics of circadian thalamocortical flow of information during a peripheral neuropathic pain condition. Front Integr Neurosci 5:43.
33. Aguiar P, Mendonça L, Galhardo V (2007) OpenControl: a free opensource software for video tracking and automated control of behavioral mazes. J Neurosci Methods 166(1):66–72
34. Paxinos, G, Watson, C, (1998). The rat brain in stereotaxic coordinates. Academic, San Diego, CA
35. Decosterd I, Woolf CJ (2000) Spared nerve injury: an animal model of persistent peripheral neuropathic pain. Pain 87(2):149–58
36. Bourquin AFet al (2006) Assessment and analysis of mechanical allodynia-like behavior induced by spared nerve injury (SNI) in the mouse. Pain 122(1-2):14.el–14.e14
37. Dixon WJ (1991) Staircase bioassay: the up-and-down method. Neurosci Biobehav Rev 15(1):47–50
38. Chaplan SR et al (1994) Quantitative assessment of tactile allodynia in the rat paw. J Neurosci Methods 53(1):55–63
39. Cardoso-Cruz H, Lima D, Galhardo V (2011) Instability of spatial encoding by CA1 hippocampal place cells after peripheral nerve injury. Eur J Neurosci 33(12):2255–2264
40. Corbetta M, Shulman GL (2002) Control of goal-directed and stimulus-driven attention in the brain. Nat Rev Neurosci 3(3):201–215
41. Cardoso-Cruz H, Lima D, Galhardo V (2013) Impaired spatial memory performance in a rat model of neuropathic pain is associated with reduced hippocampus-prefrontal cortex connectivity. J Neurosci 33(6):2465–2480
42. Cardoso-Cruz H et al (2014) Activation of dopaminergic D2/D3 receptors modulates dorsoventral connectivity in the hippocampus and reverses the impairment of working memory after nerve injury. J Neurosci 34(17):5861–5873
43. Monteiro C et al (2016) Increased fronto-hippocampal connectivity in the Prrxl1 knockout mouse model of congenital hypoalgesia. Pain 157(9):2045–56
44. Ohara S et al (2004) Attention to a painful cutaneous laser stimulus modulates electrocorticographic event-related desynchronization in humans. Clin Neurophysiol 115(7):1641–1652
45. Mouraux A, Guérit JM, Plaghki L (2003) Non-phase locked electroencephalogram (EEG) responses to CO_2 laser skin stimulations may reflect central interactions between A partial partial differential- and C-fibre afferent volleys. Clin Neurophysiol 114(4):710–722
46. Ploner M et al (2006) Pain suppresses spontaneous brain rhythms. Cereb Cortex 16(4):537–540
47. Apkarian AV, Hashmi JA, Baliki MN (2011) Pain and the brain: specificity and plasticity of the brain in clinical chronic pain. Pain 152(3 Suppl):S49–S64
48. Sameshima K, Baccalá LA (1999) Using partial directed coherence to describe neuronal ensemble interactions. J Neurosci Methods 94(1):93–103
49. Baccalá LA, Sameshima K (2001) Partial directed coherence: a new concept in neural structure determination. Biol Cybern 84(6):463–474
50. Baccala LA, Sameshima K(2021) Partial directed coherence: twenty years on some history and an appraisal. Biol Cybern 115(3):195–204
51. Shilyansky C et al (2010) Neurofibromin regulates corticostriatal inhibitory networks during working memory performance. Proc Natl Acad Sci USA 107(29):13141–13146
52. Baliki MN et al (2012) Corticostriatal functional connectivity predicts transition to chronic back pain. Nat Neurosci 15(8):1117–1119
53. Baliki MN et al (2010) Predicting value of pain and analgesia: nucleus accumbens response to noxious stimuli changes in the presence of chronic pain. Neuron 66(1):149–160
54. McConnell PA et al (2020) Impaired frontostriatal functional connectivity among chronic opioid using pain patients is associated with dysregulated affect. Addict Biol 25(2):e12743
55. Taesler P, Rose M (2016) Prestimulus theta oscillations and connectivity modulate pain perception. J Neurosci 36(18):5026–5033
56. Cardoso-Cruz H et al (2019) Bidirectional optogenetic modulation of prefrontal-hippocampal connectivity in pain-related working memory deficits. Sci Rep 9(1):10980
57. Monteiro C, Cardoso-Cruz H, Galhardo V (2019) Animal models of congenital hypoalgesia: untapped potential for assessing pain-related plasticity. Neurosci Lett 702:51–60
58. Ren W et al (2021) Adaptive alterations in the mesoaccumbal network after peripheral nerve injury. Pain 162(3):895–906
59. Shiers S et al (2018) Neuropathic pain creates an enduring prefrontal cortex dysfunction corrected by the type II diabetic drug metformin but not by gabapentin. J Neurosci 38(33):7337–7350
60. Shen Y et al (2020) Challenges for therapeutic applications of opsin-based optogenetic tools in humans. Front Neural Circuits 14:41
61. Towne C, Thompson KR (2016) Overview on research and clinical applications of optogenetics. Curr Protoc Pharmacol 75:11.19.1–11.19.21
62. Sahel JA et al (2021) Partial recovery of visual function in a blind patient after optogenetic therapy. Nat Med 27(7):1223–1229

Check for updates

Chapter 17

Assessing Pain and Related Behaviors in the PhenoWorld

Annamária Liptáková, Magda J. Castelhano-Carlos, Nuno Sousa and Hugo Leite-Almeida

Abstract

Enriched environments can play an important role in well-being. While the mechanisms underlying such interplay are not fully understood, an increasing number of rodent studies have been providing essential cues. Specifically, regarding preclinical chronic pain studies, environmental enrichment has been shown to have a positive impact on pain and pain-related outcomes, which has prompted many studies on the interaction between the environment and pain modulatory systems. Enrichment protocols for rodent studies can vary substantially. In this chapter, we present a brief overview of the main protocols and describe in detail how the PhenoWorld—a customizable enriched environment—can be used in the context of chronic neuropathic pain research focusing on the preparation of the setting, controls in standard caging conditions, and obtention of pain and pain-related outcomes.

Keywords Chronic pain, PhenoWorld, Social behavior, Anxiety, Depression

1 Introduction: Environmental Enrichment in Rodents

Environmental enrichment (EE) refers to a laboratory animal housing model providing novelty and complexity that lead to enhanced physical, cognitive, and sensory stimulation. Enrichment paradigms generally consist of physical and/or social components. Physical enrichment usually involves larger cages or increased floor space, additional bedding material, various objects to allow exercise, and/or various toys of different colors and materials to provide cognitive stimulation. On the other hand, social enrichment refers to housing animals of the same sex together in a group (see Table 1 for examples) [1, 2]. The concept of enriched environment was first described in the field of neuroscience by Donald Hebb in the 1940s, when he observed that rats that were allowed to freely roam in his home as pets had higher problem-solving and learning abilities compared to rats reared in standard laboratory conditions [2, 3].

One reason for a growing interest in EE is that it better mimics rodent's natural, real-life conditions [4]. In contrast, living in

Isaura Tavares and Hugo Leite-Almeida (eds.), *Translational Research Methods for Neuropathic Pain*, Neuromethods, vol. 235, https://doi.org/10.1007/978-1-0716-5312-8_17,

Table 1
Classification of environmental enrichment strategies

Category of enrichment	Examples of enrichment strategies
Physical	• Large space/additional cage floor • Running wheel • Climbing ladder/frame • Staircase • Rope
Social	• Cage mates
Cognitive	Toys of different size/shape/color/material: • Ping-pong balls • Marbles • Tunnels • Tubes • Mazes • Fabric hammock • Swings • Plastic hut • Wooden sticks
Sensory (auditory, visual, olfactory, taste, tactile)	• Slow rhythm music • Treats (i.e., sunflower and pumpkin seeds, peanuts) • Essences (i.e., strawberry, passion fruit, carvone, lavender, cumin, cardamom)

standard laboratory environments can prevent animals from performing their natural behaviors (e.g., exploring, foraging, hiding, climbing), which can decrease their well-being and induce maladaptive stereotyped behaviors. Stereotypies may influence learning and/or conditioning studies, which depend on an animal's adaptability, as well as ethology studies aiming to monitor an animal's natural behavior [2]. Another reason for the interest in EE as a discipline since the 1960s is growing evidence that rodents reared in enriched conditions exhibit numerous plastic responses in the brain, including neurogenesis, increased dendrite branching, synaptic density, cell size, and cortical thickness, which provide robust evidence on enhanced learning and memory [2, 5].

Aforementioned beneficial effects and many others make EE being extensively used in neuroscience as a tool for the prevention and treatment of neurological and psychiatric disorders such as Alzheimer's disease, Parkinson's disease, epilepsy, stroke, amyotrophic lateral sclerosis, traumatic brain injury, Down syndrome, Fragile X syndrome, depression, and anxiety—see for a review [3]. Apart from the studies on the nervous system, follow-up research has shown a significant influence on other diseases such as obesity and cancer [6, 7].

1.1 Enriched Environment Protocols

Enrichment protocols can substantially vary in different aspects—see Table 1 for summary—depending on individual experimenter and lab resources. Variability may concern animals, for instance, species, strains, sex, age, and/or mimicked environment itself, for example, the size of the enriched cages, types of objects used, duration of EE exposure, number of animals housed together, and types of controls used [2]. Variability among protocols can contribute to the inconsistency in research results.

1.2 Impact of EE on Animal's Behavior

Cross-laboratory protocol variability can contribute to the inconsistency in behavioral results. For instance, a vast number of studies have found that EE reduces anxiety assessed by the elevated plus maze test (EPM) by greater time spent in open arms than control animals [8–10], but in contrast, there is also evidence that EE did not promote a significant difference in time spent or entries to open arms [11, 12]. The same contradictory results were observed in the light/dark box test (L/D), where Hellemans et al. [13] showed no effect of EE during adolescence on latency to enter the light compartment of the apparatus, while Sah et al. [14] found that EE animals took less time to enter the aversive light part. Previous data suggest that EE housing decreases stress-induced despair in the forced swim test (FST), with significant reductions in immobility time compared to standard-housed animals [11, 15]. However, Cui et al. [16] did not find an effect of housing on immobility time in FST.

Relatively consistent findings indicate that EE animals have the ability to better adapt, reduce locomotor activity, and explore a novel environment more freely than controls in the open field test (OFT) [2]. Zheng et al. [17] showed that mice reared in EE from birth performed better in both sociability and social memory phases of the social interaction test compared to those reared in standard conditions, suggesting that EE rearing from birth made mice more social in their adulthood. Gabriel et al. found that when EE started upon weaning (postnatal day 21), rats housed only in social enrichment (SE) and a group housed in SE + physical enrichment (PE) showed less interest in social interaction compared to the control group; on the contrary, only the SE group spent more time with the novel stranger in the social memory phase of the social interaction test compared to the control, PE, and SE + PE groups [18, 19].

Most of the studies on memory and learning showed a better performance of EE groups in the Morris water maze (MWM) and novel object recognition (NOR) tests; for instance, Hullinger et al. [20] found that after 1 month of enrichment, both EE and SE rats swam significantly shorter distances in hidden platform training of the MWM test and demonstrated superior performance in the NOR test compared to standard-housed controls. Interestingly, 4 months of enriched exposure led to significantly better performance in the MWM and NOR tests only in EE animals compared to SE and standard controls [20]. A few inconsistencies exist; for instance, Ueda et al. [21] report no housing effect on MWM latencies, suggesting it could be due to the performance of standard-housed controls instead of including isolated controls.

1.3 Impact of EE on Chronic Pain

Distinct EE protocols show advantageous effects on several animal models of chronic pain—see [22] for a review. In the carrageenan (CAR)-induced inflammatory pain model, animals exposed to EE after CAR injection showed a reduction in the recovery period by 1 week compared to standard-housed animals [23, 24]. Moreover, when animals were enriched only socially, a reduction in the recovery period of half of a week was observed [24]. EE provided 7 days prior to induction of inflammatory pain induced by Complete Freund's adjuvant (CFA) was not able to significantly modify mechanical allodynia until day 35 after arthritis induction [25]. A similar result was obtained from a study with 2 weeks of EE pretreatment, where the magnitude of mechanical hyperalgesia was not affected by the housing condition, whereas a significant reduction in thermal hyperalgesia was observed [26]. On the contrary, Zheng et al. [27] found 2 weeks of pretreatment with EE sufficient to significantly attenuate CFA-induced thermal and mechanical hyperalgesia from day 7 until day 17 after CFA injection. Interestingly, providing animals with EE for 2 h daily across seven consecutive days showed to be satisfactory to significantly reduce mechanical allodynia in the CFA pain model [28].

Higher mechanical thresholds were also observed in a spinal cord injury (SCI) animal model in groups exposed to EE 1 week prior to and immediately after the injury, with no effect of EE on thermal hyperalgesia [29, 30].

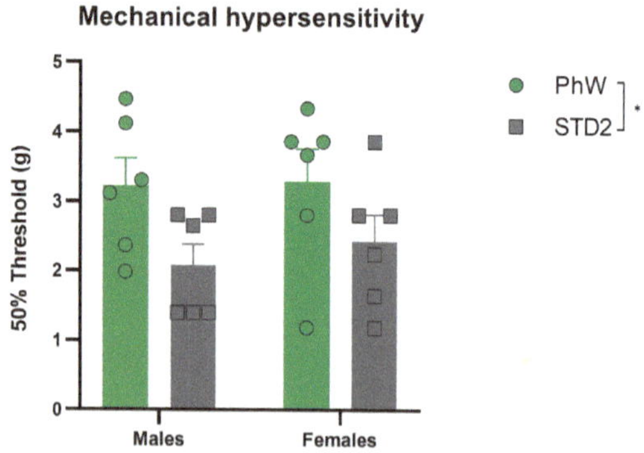

Fig. 1 Preliminary data. Effect of EE—the PhenoWorld (PhW)—on mechanical allodynia 4 weeks after SNI in males and females compared to standard cage with paired animals (STD2). Data are expressed as mean ± S.E.M. $n = 6$/group. Statistical analysis carried out with two-way ANOVA. *$P < 0.05$ male and female PhW vs. male and female STD2

The effect of EE on neuropathic pain was mostly studied in spared nerve injury (SNI) and chronic constriction injury (CCI) animal models. Data on mechanical allodynia in these models are consistent, showing reduced hypersensitivity to mechanical stimuli irrespective of the onset of EE, if EE was present for some time after the surgery [31–37]; see also preliminary data from our group (Fig. 1). On the contrary, animals exposed to EE 6 weeks before the surgery with no continuation after the surgery did not show difference in mechanical thresholds compared to standard-housed animals [35]. Cold hypersensitivity assessed by the acetone test did not vary between different housing conditions in the SNI neuropathic pain model [33, 37].

2 Materials

2.1 Animals and Housing Conditions

Wistar Han male and female rats (2–5 months old), born in the animal facility, were randomly assigned to one of the three environmental groups: standard housing for 2 (STD2) and 6 (STD6) animals, as well as PhW representing a complex enriched housing environment. All animals were maintained under standard laboratory conditions: artificial 12 h light/dark cycle (inverted light cycle—lights off at 8 a.m.), 21 ± 1°C ambient temperature, and 50%–60% relative humidity. Corncob was used as bedding material (Scobis Due; Mucedola S.R.L., Settimo Milanese, Milan, Italy), and sterile cardboard tubes were used as housing refinement. Rats were provided with a standard diet (4RF21; Mucedola S.R.L.) and water ad libitum. Animal's body weight was measured every week along the experiment.

2.2 The PhenoWorld

The PhenoWorld (PhW, TSE Systems GmbH, Bad Homburg vor der Höhe, Germany) is a complex housing and ethological testing paradigm developed in a collaboration between the Life and Health Sciences Research Institute (ICVS), Portugal, and TSE Systems International Group, Germany to screen rodent's behavior in a more naturalistic environment, reducing human interference while providing animals with social and physical enrichment. It consists of a large central cage (1 m^2 and 50 cm height) covered with corncob bedding, which is connected to a box with four running wheels by means of two open access tubes, and to two drinking/feeding boxes. All PhW areas are covered by either perforated Plexiglas or stainless-steel grids (Fig. 2). All animal gates and tubes connecting the central cage to other areas of the PhW can be manually closed. The PhW setup offers a possibility to test anxiety-like and operant behavior by having a light/dark test arena and two boxes with operant walls connected to the central cage. Since we aimed to study pain-like behavior of the animals living in the PhW and compare them with animals living in standard housing, we used only a central cage enriched with one

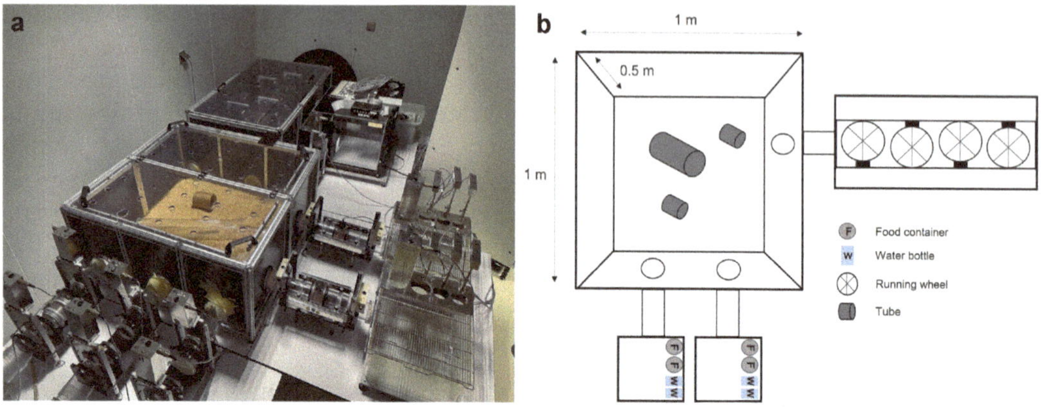

Fig. 2 Photo (**a**) and scheme (**b**) of the PhenoWorld paradigm

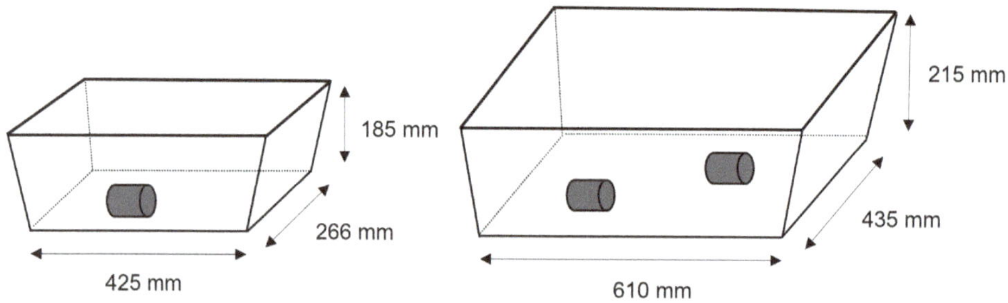

Fig. 3 Scheme representing STD2 (left) and STD6 (right) housing

plastic and two cardboard tubes, connected to the box with running wheels and two drinking/feeding boxes. Animals lived in the PhW in a social group of six rats, with free access to all the connected boxes.

2.3 Standard Housing

For standard housing conditions, we used STD2 and STD6 cages (Fig. 3). The STD2 cage is a standard filter-topped transparent type III cage (425 × 266 × 185 mm; 800 cm^2 floor area; ref. 1291H, Tecniplast, Buguggiate, Italy) with corncob bedding and one cardboard tube. Animals attributed to STD2 environment were pair-housed. The STD6 cage is a standard filter-topped transparent cage (610 × 435 × 215 mm (2065 cm^2 floor area; ref. 2000P, Tecniplast) with corncob bedding and two cardboard tubes. STD6 animals were socially housed in groups of six rats.

3 Methods

3.1 Baseline Measurements

All animals were transported to the room with the PhW paradigm 1 week prior to the the start of the experiment to let the animals acclimatize to the light cycle changed (light on at 8 p.m., light off at

8 a.m.) in the housing room dedicated to this experiment. During this time, animals were handled for three consecutive days for habituation to the experimenter. All animals were housed in the STD2 cages prior to the experiment and during the acclimatization period. One week later, animals were moved to their respective environments; six rats were moved to the PhW, another six rats to the STD6 cage, and the animals assigned to the STD2 cage did not change their environment. One week after exposure to the different housing environments, naïve animals underwent baseline measurements of their anxiety- and depressive-like phenotype in the light/dark test (L/D) and forced swim test (FST) [38].

3.2 Chronic Pain Model

One week after the introduction to the respective housing environments, animals underwent SNI surgery [39] to model neuropathic pain. All animals were anesthetized with an intraperitoneal (i.p.) injection of 75 mg/kg of ketamine (Imalgene®, 100 mg/mL; Merial, Lyon, France) and 1 mg/kg of medetomidine (Dormitor®, 1 mg/mL; Orion Pharma, Espoo, Finland) and were transported to the surgery room in the STD2 cages. SNI model consists of a ligation and a transection of the common peroneal and the tibial branch of the sciatic nerve while sparing the sural branch. Sham-operated animals were used as controls. In this case, surgery involved only nerve exposure without its damage. After surgery, anesthesia was reversed with the administration of 1 mg/kg of atipamezole (Tipafar®, 5 mg/mL; VetPharma Animal Health, Barcelona, Spain) subcutaneously. Animals were left to recover from anesthesia in STD2 cages in the recovery room under the red lamp until being fully awake. Afterward, they were transported to their respective environments.

3.3 Longitudinal Assessment of Pain and Related Behaviors

Mechanical allodynia was assessed weekly using von Frey monofilaments according to the up-and-down paradigm—see [40] for instance. Four weeks after SNI, animals were tested for comorbid, anxiety and depression, using the elevated plus maze test (EPM), FST, and sucrose preference test (SPT) [41]. All behavioral tests were performed in the rooms dedicated to each behavioral paradigm; thus, the PhW animals were transported and housed in STD6 cages while tests were performed. Since animals had an inverted light cycle and behavioral tests were performed during their dark phase of the light/dark cycle, von Frey test and FST had to be carried out with the white lights on.

To analyze behavior and welfare of PhW animals in pain and no-pain conditions, we video-recorded home cage behavior and used a focal sampling analysis approach (Fig. 4). Animals were recorded for 2 h (2–4 p.m.) during the dark phase (with red light on) and for 2 h (2–4 a.m.) during the light phase of the light/dark cycle, once a week, from the baseline week (before SNI) until the fourth week after SNI (pain condition) or sham surgery. Sampling of the videos was made as following: 2 h records were divided into seven videos

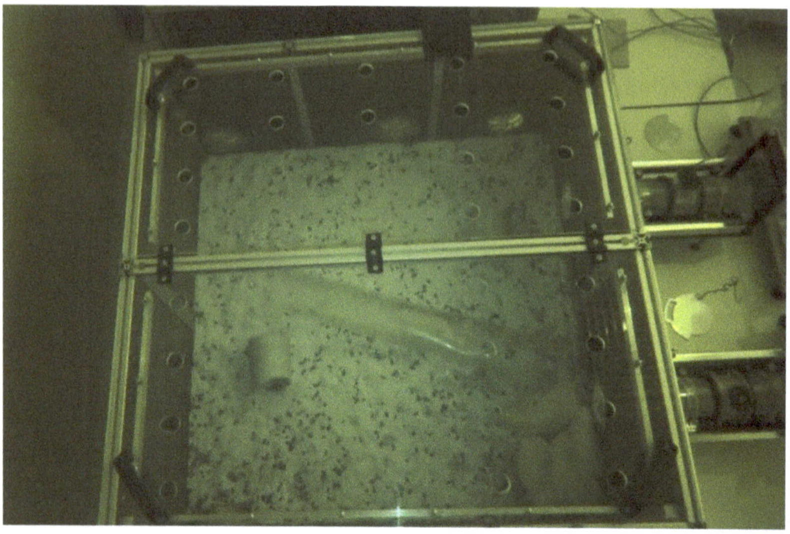

Fig. 4 Home cage behavior recording during dark phase of the light/dark cycle. Animals huddle in the central cage

of 17 min, and from each of these videos, a sample from minute 8 until minute 13 was manually scored. We then counted the time of several patterns of behaviors performed—see Table 2. For individual identification, each animal's tail was pen marked always 2 days before a recording day, so that animals could be videotaped undisturbed on the day of the recording.

3.4 The PhenoWorld Maintenance

1. All the compartments of the PhW used for an experiment were cleaned with tap water and dried with a paper towel. Afterward, 70% alcohol was sprayed to disinfect all the compartments and wiped using a paper towel so as not to let the disinfectant influence animal's behavior.
2. Running wheels were taken out from the paradigm and cleaned with tap water, disinfected with 70% alcohol, and dried using a paper towel. Water bottles were cleaned with tap water and rinse with drinking water.
3. Autoclaved corncob bedding was added to the central cage of the PhW in an amount corresponding to the volume of a full STD2 cage (425 × 266 × 185 mm), so that the bedding covers the whole floor surface of the central cage. Autoclaved plastic tube and two cardboard tubes were placed in the central cage.
4. Every 2 weeks, the soiled bedding was removed and steps 1–3 were repeated.
5. Drinking bottles and food containers were fully filled every two days.
6. Animals stayed in the STD6 cage while PhW was being cleaned.

Table 2
Patterns of behaviors scored in the PhW

Behavior	Description
1. Exploring	Included rearing, sniffing, interaction with objects, walking, jumping, and scanning.
2. Self-grooming	It is a self-maintenance behavior, performed by licking all reachable parts of the body and nibbling its fur.
3. Sleep/rest alone	Scored when a rat slept or rested (not moving) alone.
4. Social behavior	
• Social interaction	Included allogrooming, approaching, sniffing the partner, and interaction with other cage mates without any signs of wrestling and self-defense.
• Huddling	Two or more animals sleeping/resting in proximate contact with each other.
• Fighting (a play fight)	Involved attack and defense of the nape of the neck, wrestling, and pinning.
• Following/chasing	As following/chasing, we scored when both animals were in motion, one chasing another while moving within a tail-length distance.
5. Central cage absence	Scoring time when animals were not present in the central cage provided us with information about place preference between the PhW compartments.

3.5 The Maintenance of Standard Environments

Standard cages (STD2 and STD6) were exchanged weekly, according to the usual animal facility practices, with a replacement of cardboard tubes if needed. Drinking bottles and food containers were being refilled every 2 days.

4 Conclusion

As a principle, all housing conditions should have some enrichment. European guidelines for laboratory animal care and accommodation are stated in the European Directive 2010/63/EU and in the Commission Recommendation 2007/526/EC, where it is indicated that "Animals should be socially housed in stable groups of compatible individuals and provided with sufficient space of adequate complexity to allow expression of a wide range of normal behavior". However, a more spacious area enriched with external objects, running wheels, and a bigger social group reinforces EE effects with notorious effects in the PhW, particularly in animals with ongoing pain. The obvious downsize is that equipment such as the PhW is expensive and require substantial maintenance. Compromises might be achieved providing that a positive effect can be demonstrated when compared with standard conditions. For instance, regarding pain, particularly chronic pain, EE provided by the PhW housing conditions has served not only to study the mechanisms associated with environment-associated pain protection, but also to create conditions where pain phenotypical manifestations become more evident. For instance, control animals in standard caging conditions present depressive-like phenotypes when compared to control animals in enriched conditions.

Acknowledgments

This work was funded by European Union's Horizon 2020 research and innovation programme under the Marie Sklodowska-Curie grant agreement No. 955684 and by national funds through the Foundation for Science and Technology (FCT)— projects UIDB/50026/2020 and UIDP/50026/2020.

Conflict of Interest The authors have no conflicts of interest to declare that are relevant to the content of this chapter.

References

1. Girbovan C, Plamondon H (2013) Environmental enrichment in female rodents: considerations in the effects on behavior and biochemical markers. Behav Brain Res 253:178–190. https://pubmed.ncbi.nlm.nih.gov/23860119/
2. Simpson J, Kelly JP (2011) The impact of environmental enrichment in laboratory rats—behavioural and neurochemical aspects. Behav Brain Res 222(1):246–264. https://pubmed.ncbi.nlm.nih.gov/21504762/
3. Nithianantharajah J, Hannan AJ (2006) Enriched environments, experience-dependent plasticity and disorders of the nervous system. Nat Rev Neurosci 7(9):697–709. https://pubmed.ncbi.nlm.nih.gov/16924259/
4. Castelhano-Carlos M, Costa PS, Russig H, Sousa N (2014) PhenoWorld: a new paradigm to screen rodent behavior. Transl Psychiatry 4(6):e399. https://pubmed.ncbi.nlm.nih.gov/26126181/
5. Han Y, Yuan M, Guo YS, Shen XY, Gao ZK, Bi X (2022) The role of enriched environment in neural development and repair. Front Cell Neurosci 16:890666. https://pubmed.ncbi.nlm.nih.gov/35936498/
6. de León-Guerrero SD, Salazar-León J, Meza-Sosa KF, Valle-Garcia D, Aguilar-León D, Pedraza-Alva G et al (2022) An enriched environment re-establishes metabolic homeostasis by reducing obesity-induced inflammation. Dis Model Mech 15(6):dmm048936. https://pubmed.ncbi.nlm.nih.gov/35112705/
7. Xiao R, Ali S, Caligiuri MA, Cao L (2021) Enhancing effects of environmental enrichment on the functions of natural killer cells in mice. Front Immunol 12:695859. https://pubmed.ncbi.nlm.nih.gov/34394087/
8. Huang H, Wang Q, Guan X, Zhang X, Zhang Y, Cao J et al (2021)Effects of enriched environment on depression and anxiety-like behavior induced by early life stress: a comparison between different periods. Behav Brain Res 411:113389. https://pubmed.ncbi.nlm.nih.gov/34058267/
9. Grippo AJ, Ihm E, Wardwell J, Mcneal N, Scotti MAL, Moenk DA et al (2014) The effects of environmental enrichment on depressive and anxiety-relevant behaviors in socially isolated prairie voles. Psychosom Med 76(4):277–284. https://pubmed.ncbi.nlm.nih.gov/24804886/
10. Muthmainah M, Sari WA, Wiyono N, Ghozali DA, Yudhani RD, Wasita B (2021) Environmental enrichment ameliorates anxiety-like behavior in rats without altering plasma corticosterone level. Open Access Maced J Med Sci 9:1074–1080
11. Brenes JC, Padilla M, Fornaguera J (2009) A detailed analysis of open-field habituation and behavioral and neurochemical antidepressant-like effects in postweaning enriched rats. Behav Brain Res 197(1):125–137. https://pubmed.ncbi.nlm.nih.gov/18786573/
12. Hoffmann LC, Schütte SRM, Koch M, Schwabe K (2009) Effect of "enriched environment" during development on adult

rat behavior and response to the dopamine receptor agonist apomorphine. Neuroscience 158(4):1589–1598. https://pubmed.ncbi.nlm.nih.gov/19110038/
13. Hellemans KGC, Benge LC, Olmstead MC (2004) Adolescent enrichment partially reverses the social isolation syndrome. Dev Brain Res 150(2):103–115. https://pubmed.ncbi.nlm.nih.gov/15158074/
14. Sah A, Rooney S, Kharitonova M, Sartori SB, Wolf SA, Singewald N (2022) Enriched environment attenuates enhanced trait anxiety in association with normalization of aberrant neuro-inflammatory events. Int J Mol Sci 23(21):13052. https://pubmed.ncbi.nlm.nih.gov/36361832/
15. Hattori S, Hashimoto R, Miyakawa T, Yamanaka H, Maeno H, Wada K et al (2007) Enriched environments influence depression-related behavior in adult mice and the survival of newborn cells in their hippocampi. Behav Brain Res 180(1):69–76. https://pubmed.ncbi.nlm.nih.gov/17397941/
16. Cui M, Yang Y, Yang J, Zhang J, Han H, Ma W et al (2006) Enriched environment experience overcomes the memory deficits and depressive-like behavior induced by early life stress. Neurosci Lett 404(1–2):208–212. https://pubmed.ncbi.nlm.nih.gov/16790315/
17. Zheng JJ, Zou R, Huang S, Song TJ, Yu X (2020) Enriched environment rearing from birth reduced anxiety, improved learning and memory, and promoted social interactions in adult male mice. Neuroscience 442:138–150. https://pubmed.ncbi.nlm.nih.gov/32652178/
18. Gabriel P, Mastracchio TA, Bordner K, Jeffrey R (2020) Impact of enriched environment during adolescence on adult social behavior, hippocampal synaptic density and dopamine D2 receptor expression in rats. Physiol Behav 226:113133. https://pubmed.ncbi.nlm.nih.gov/32795458/
19. Eraslan E, Castelhano-Carlos MJ, Amorim L, Soares-Cunha C, Rodrigues AJ, Sousa N (2023) Home-cage behavior is impacted by stress exposure in rats. Front Behav Neurosci 17:1195011. https://www.frontiersin.org/articles/10.3389/fnbeh.2023.1195011/full
20. Hullinger R, O'Riordan K, Burger C (2015) Environmental enrichment improves learning and memory and long-term potentiation in young adult rats through a mechanism requiring mGluR5 signaling and sustained activation of p70s6k. Neurobiol Learn Mem 125:126–134. https://pubmed.ncbi.nlm.nih.gov/26341144/
21. Ueda S, Sakakibara S, Yoshimoto K (2005) Effect of long-lasting serotonin depletion on environmental enrichment-induced neurogenesis in adult rat hippocampus and spatial learning. Neuroscience 135(2):395–402. https://pubmed.ncbi.nlm.nih.gov/16125851/
22. Tai LW, Yeung SC, Cheung CW (2018) Enriched environment and effects on neuropathic pain: experimental findings and mechanisms. Pain Pract 18(8):1068–1082. https://pubmed.ncbi.nlm.nih.gov/29722923/
23. Gabriel AF, Marcus MAE, Honig WMM, Helgers N, Joosten EAJ (2009) Environmental housing affects the duration of mechanical allodynia and the spinal astroglial activation in a rat model of chronic inflammatory pain. Brain Res 1276:83–90. https://pubmed.ncbi.nlm.nih.gov/19406110/
24. Gabriel AF, Paoletti G, Seta DD, Panelli R, Marcus MAE, Farabollini F et al. (2010) Enriched environment and the recovery from inflammatory pain: Social versus physical aspects and their interaction. Behav Brain Res 208(1):90–95. https://pubmed.ncbi.nlm.nih.gov/19914294/
25. Estrázulas M, Freitas RDS, Käfer ET, Dagnino APA, Campos MM (2022) Central and peripheral effects of environmental enrichment in a mouse model of arthritis. Int Immunopharmacol 102:108386. https://pubmed.ncbi.nlm.nih.gov/34824037/
26. Tall JM (2009) Housing supplementation decreases the magnitude of inflammation-induced nociception in rats. Behav Brain Res 197(1):230–233. https://pubmed.ncbi.nlm.nih.gov/18773924/
27. Zheng J, Jiang YY, Xu LC, Ma LY, Liu FY, Cui S et al (2017) Adult hippocampal neurogenesis along the dorsoventral axis contributes differentially to environmental enrichment combined with voluntary exercise in alleviating chronic inflammatory pain in mice. J Neurosci 37(15):4145–4157. https://pubmed.ncbi.nlm.nih.gov/28292830/
28. Jiang S, Zheng C, Wen G, Bu B, Zhao S, Xu X . Down-regulation of NR2B receptors contributes to the analgesic and antianxiety effects of enriched environment mediated by endocannabinoid system in the inflammatory pain mice. Behav Brain Res 435:114062. https://pubmed.ncbi.nlm.nih.gov/35985400/
29. Berrocal Y, Pearse DD, Singh A, Andrade CM, McBroom JS, Puentes R et al (2007) Social and environmental enrichment improves sensory and motor recovery after severe contusive spinal cord injury in the rat. J Neurotrauma 24(11):1761–1772. https://pubmed.ncbi.nlm.nih.gov/18001204/
30. Tai WL, Sun L, Li H, Gu P, Joosten EA, Cheung CW (2021) Additive effects of environmental enrichment and ketamine on neuropathic

pain relief by reducing glutamatergic activation in spinal cord injury in rats. Front Neurosci 15:635187. https://pubmed.ncbi.nlm.nih.gov/33828447/
31. Wang XM, Pan W, Xu N, Zhou ZQ, Zhang GF, Shen JC (2019) Environmental enrichment improves long-term memory impairment and aberrant synaptic plasticity by BDNF/TrkB signaling in nerve-injured mice. Neurosci Lett 694:93–98. https://pubmed.ncbi.nlm.nih.gov/30496785/
32. Wang XM, Zhang GF, Jia M, Xie ZM, Yang JJ, Shen JC et al (2019) Environmental enrichment improves pain sensitivity, depression-like phenotype, and memory deficit in mice with neuropathic pain: role of NPAS4. Psychopharmacology (Berl) 236(7):1999–2014. https://pubmed.ncbi.nlm.nih.gov/30798405/
33. Vachon P, Millecamps M, Low L, Thompsosn SJ, Pailleux F, Beaudry F et al (2013) Alleviation of chronic neuropathic pain by environmental enrichment in mice well after the establishment of chronic pain. Behav Brain Funct 9(1):22. https://pubmed.ncbi.nlm.nih.gov/24025218/ 7
34. Parent-Vachon M, Beaudry F, Carrier D, Di Cristo G, Vachon P (2019) The effects of exercise on pain and reproductive performance in female pregnant mice with neuropathic pain. Biol Res Nurs 21(5):500–509. https://pubmed.ncbi.nlm.nih.gov/31288563/
35. Falkowska M, Ntamati NR, Nevian NE, Nevian T, Acuña MA (2023) Environmental enrichment promotes resilience to neuropathic pain-induced depression and correlates with decreased excitability of the anterior cingulate cortex. Front Behav Neurosci 17:1139205. https://pubmed.ncbi.nlm.nih.gov/37008999/
36. Parent-Vachon M, Vachon P (2018) Environmental enrichment alleviates chronic pain in rats following a spared nerve injury to induce neuropathic pain. A preliminary study. Vet Med (Auckl) 9:69–72. https://pubmed.ncbi.nlm.nih.gov/30349809/
37. Gong X, Chen Y, Chang J, Huang Y, Cai M, Zhang M (2018) Environmental enrichment reduces adolescent anxiety- and depression-like behaviors of rats subjected to infant nerve injury. J Neuroinflammation 15(1):262. https://pubmed.ncbi.nlm.nih.gov/30208926/
38. Guimarães MR, Soares AR, Cunha AM, Esteves M, Borges S, Magalhães R et al (2019) Evidence for lack of direct causality between pain and affective disturbances in a rat peripheral neuropathy model. Genes Brain Behav 18(6):e12542. https://pubmed.ncbi.nlm.nih.gov/30488664/
39. Decosterd I, Allchorne A, Woolf CJ (2002) Progressive tactile hypersensitivity after a peripheral nerve crush: non-noxious mechanical stimulus-induced neuropathic pain. Pain 100(1–2):155–162. https://pubmed.ncbi.nlm.nih.gov/12435468/
40. Guimarães MR, Anjo SI, Cunha AM, Esteves M, Sousa N, Almeida A et al (2021) Chronic pain susceptibility is associated with anhedonic behavior and alterations in the accumbal ubiquitin-proteasome system. Pain 162(6):1722–1731. https://pubmed.ncbi.nlm.nih.gov/33449505/
41. Leite-Almeida H, Almeida-Torres L, Mesquita AR, Pertovaara A, Sousa N, Cerqueira JJ et al (2009) The impact of age on emotional and cognitive behaviours triggered by experimental neuropathy in rats. Pain 144(1–2):57–65. https://pubmed.ncbi.nlm.nih.gov/19398158/

Index

Isaura Tavares and Hugo Leite-Almeida (eds.), *Translational Research Methods for Neuropathic Pain*, Neuromethods, vol. 235, https://doi.org/10.1007/978-1-0716-5312-8,

FSC
www.fsc.org
MIX
Papier | Fördert
gute Waldnutzung
FSC® C083411

Kolibri 360 GmbH
Ferdinand-Jühlke-Straße 7
99095 Erfurt, Deutschland
produktsicherheit@kolibri360.de